Electromechanical Energy Conversion

DAVID R. BROWN
General Electric Company,
formerly University of Texas, Austin

E. P. HAMILTON III
University of Texas, Austin

Electromechanical Energy Conversion

Macmillan Publishing Company
New York

Collier Macmillan Publishers
London

Copyright © 1984, Macmillan Publishing Company,
a division of Macmillan, Inc.

Printed in the United States of America

Macmillan Publishing Company
866 Third Avenue, New York, New York 10022

Collier Macmillan Canada, Inc.

Library of Congress Cataloging in Publication Data

Brown, David
  Electromechanical energy conversion.

  Includes index.
  1. Electromechanical devices. I. Hamilton, E. P.
II. Title.
TK145.B765   1984   621.31′042   83-11300
ISBN 0-02-315590-6

Printing: 1 2 3 4 5 6 7 8   Year: 4 5 6 7 8 9 0 1 2

ISBN   0-02-315590-6

PREFACE

This book is intended primarily for use with a first course in electromechanical energy conversion. Sufficient material is presented for a two-semester sequence. Traditional subjects are covered in Chapters 1 through 5. In instances where programs allow for only a one-semester coverage, these first five chapters will satisfy most requirements. Additional supplementary material from the remaining chapters may then be appropriate depending on the interests of the instructor.

The emphasis of this book is on development of machine characteristics from as fundamental a framework as is practical. This emphasis is chosen in order to provide an appreciation of the subject beyond that which results from an approach based solely on equivalent circuits. The authors have been gratified to find that considerable student interest exists in development of an understanding of fundamental capabilities and limitations from a broader perspective than that presented when other approaches are used. In order to meet this need, this book starts with first principles but recognizes that many students will not have developed a strong feel for those aspects of field theory which are crucial to an understanding of electromechanical energy conversion. The authors feel that definitions and clear explanations in terms of fundamental relationships are critical. The first chapter reflects this viewpoint for what follows in the later chapters. This permits a development that does not later restrict treatment to special cases. Models are developed in such a way that steady-state operation is seen to be simply a special case of device behavior under certain assumed conditions.

It also permits presentation of design considerations where the intent is not to provide a detailed design manual but rather to develop an appreciation for fundamental limitations and capabilities.

Chapter 1 starts with a very deliberate treatment of a simple magnetic circuit. Definitions and fundamental principles are introduced *as required*. Maxwell's equations in integral form are stressed as fundamental to the concept. Consideration is given to field variation in all three dimensions and the role of highly permeable core material is emphasized fully. Examples are presented in a very deliberate step-by-step format. Phasor concepts are introduced where they are appropriate in order to examine steady-state ac behavior. Notation is chosen to emphasize the distinction between a phasor and its magnitude. The section on design considerations emphasizes practical aspects of transformer design without getting bogged down in detailed case-specific design procedures.

Chapter 2 develops the fundamental force and torque relationships. The emphasis is on examples of a practical nature. State-variable formulation of the equations required to describe a system is stressed. The basic action of conventional induction machines, synchronous machines, and dc machines is introduced.

Chapters 3 and 4 cover induction and synchronous machines, respectively. The coupled-coil viewpoint is used to develop mathematical models in each case. Relevant assumptions are explicitly stated. Models are then developed in a rigorous but straightforward manner and are sufficiently general to permit the later analysis of machine transients (Chapter 6) and ac variable-speed drives (Chapter 8). The emphasis is on steady-state characteristics in Chapters 3 and 4, and appropriate models are obtained by assuming sinusoidal time variation in the more general relationships. Examples are presented in a deliberate step-by-step format and are in many cases based on the authors' personal experiences in industry.

Chapter 5 covers the dc machine. Motor and generator characteristics for the basic dc machine types are developed. Emphasis is on modern applications of dc machinery. The dynamics of dc machinery are treated with a state-variable formulation as well as with the more traditional block diagram approach.

Chapter 6 treats the analysis of ac machine transients. Models developed in Chapters 3 and 4 are examined for the general case, that is, without the assumption of steady-state sinusoidal time variation. The classical qdo transformation is introduced. It is presented as a mathematical tool that simplifies the necessary machine equations and provides additional insights on machine behavior. Numerous examples are presented early in the development to aid the student in making the transition from steady-state analysis to a more general analysis. Simplifying assumptions are made to permit analytical solutions (by hand), which are later compared to computer-assisted solutions (Chapter 8). Examples are based on practical applications and stress an intuitive understanding of the phenomena investigated.

Chapter 7 treats a variety of special-purpose electromechanical devices not generally regarded as conventional ac or dc rotating machines. The emphasis is on a general description of each device and its basic operating characteristics. The major thrust of the chapter is to develop an understanding that unusual and sometimes highly complex systems and devices can be developed based on the fundamental framework learned earlier. Topics covered range from servo systems and computer control to magnetically levitated, "super-speed" trains.

Chapter 8 is devoted to advanced applications. The transformer is reexamined with a view toward model refinements required in order to examine various phenomena not generally covered in introductory courses. Examples are chosen to illustrate practical problems encountered. State-variable formulation of the necessary mathematical relationships is stressed throughout, and computer-assisted solutions are presented. The important area of ac machine variable-speed drives is introduced with emphasis on development of analytical methods suitable for computer implementation.

Finally, because of the shifting emphasis in many engineering and science programs, it is no longer possible to assume that all students will have sufficient background in the various analytical methods required. The material presented in the Appendixes provides a useful review of those techniques which are used throughout the text.

A complete solutions manual, with solutions to all of the text problems, is available from the publisher.

We wish to acknowledge the substantial contribution to this work made by our colleagues and students and Debra Hayden for her assistance in typing the manuscript through its many drafts. Additionally, we would like to thank the reviewers of the manuscript during its preparation, including John Grainger, A. G. Potter, Milton R. Johnson, and F. W. Schott. Finally, we would like to thank our families for their continued and patient support throughout this project.

D.R.B.
E.P.H.III

CONTENTS

Electromechanical
Energy Conversion

CHAPTER 1

Fundamental Concepts

1.1

INTRODUCTION

Electromechanical energy conversion involves the interaction of current conducting materials and magnetic fields. Maxwell's equations describe such phenomena and provide the starting point for analysis and design of energy conversion devices.

1.2

MAGNETIC CIRCUITS

The magnetic circuit concept is helpful in understanding the behavior of many practical electromagnetic devices. A simple magnetic circuit is illustrated in Figure 1.1. The magnetic circuit in Figure 1.1 consists of a core having a rectangular cross section wrapped with N turns of wire. We will use this relatively simple device to review some fundamental relationships and definitions.

We start by recalling Ampère's law:†

$$\oint \overline{H} \cdot d\overline{\ell} = \int_s \overline{J} \cdot \overline{ds} \tag{1.1}$$

---

†Displacement current terms are negligible in most electromechanical energy conversion devices and transformers because time variation is at relatively low frequencies.

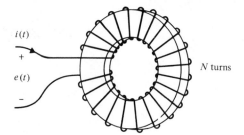

FIGURE 1.1 **Simple Magnetic Circuit**

In words, the line integral of magnetic field intensity $\overline{H}$ around any closed path is equal to the surface integral of current density $\overline{J}$ over the surface bounded by the closed path. The vector quantities involved are illustrated in Figure 1.2. In SI units, $\overline{H}$ is in amperes per meter, $\overline{J}$ is in amperes per square meter, $\overline{\ell}$ is path length in meters, and $\overline{s}$ is surface area in square meters. All quantities are vector quantities having direction as well as magnitude. Note that the integral on the right-hand side of (1.1) gives the total current through the surface bounded by the path of integration for $\overline{H}$ and it is a scalar quantity.

The magnetic circuit depicted in Figure 1.1 is a symmetrical configuration and permits a straightforward application of Ampère's law. Figure 1.3 is a cross-sectional view of the magnetic circuit showing appropriate dimensions.

A cylindrical coordinate system is chosen and unit vectors $\overline{a}_r$, $\overline{a}_\theta$, and $\overline{a}_z$ are shown in Figure 1.3 at radius r and at an arbitrary angle θ. A current direction is assumed in the winding such that for $r < r_i$ current is directed into the page and for $r > r_o$ current is directed out of the page. We use "×" and "·" symbols to indicate current direction into and out of the page, respectively. If the winding consists of many turns of relatively fine wire where the turns of the wire are spaced closely together, we can assume uniform current densities for those spaces actually occupied by winding material. This assumption results in a symmetrical configuration with respect to the angle θ. The magnetic field intensity in the azimuthal (θ) direction inside the winding (totally within the core) is obtained by direct application of Ampère's law. For a circular path of integration at radius r we have on the left-hand side of (1.1):

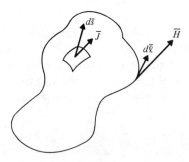

FIGURE 1.2 **Vector Quantities Associated with Ampère's Law**

FUNDAMENTAL CONCEPTS

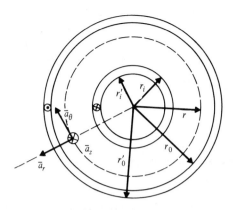

FIGURE 1.3 Cross-Sectional View of Core

$$\oint \bar{H} \cdot d\bar{\ell} = \oint (H_r \bar{a}_r + H_\theta \bar{a}_\theta + H_z \bar{a}_z) \cdot (r \, d\theta \, \bar{a}_\theta) \qquad (1.2)$$

$$= \int_0^{2\pi} H_\theta r \, d\theta = 2\pi r H_\theta$$

where $\bar{a}_r \cdot \bar{a}_\theta = \bar{a}_z \cdot \bar{a}_\theta = 0$, $\bar{a}_\theta \cdot \bar{a}_\theta = 1$, and H is invariant with θ based on symmetry considerations. On the right-hand side of (1.1), we have for a path of integration within the core:

$$\int \bar{J} \cdot d\bar{s} = Ni \qquad (1.3)$$

If we drop the θ subscript, the magnetic field intensity in the θ direction within the core must be

$$H = \frac{Ni}{2\pi r} \qquad (1.4)$$

By choosing paths of integration at various radii, we can similarly determine H in the θ direction for the other regions illustrated in Figure 1.3. If we exclude the end-turn regions, the current density in the region $r_i' \le r \le r_i$ is $Ni/[\pi(r_i^2 - r_i'^2)]$ directed into the page. Similarly, the current density in the region $r_o \le r \le r_o'$ is $Ni/[\pi(r_o'^2 - r_o^2)]$ directed out of the page. Thus, excluding the end-turn regions, we find

$$H = 0 \qquad\qquad\qquad\qquad r \le r_i'$$

$$H = \frac{Ni}{2\pi r} \frac{r^2 - r_i'^2}{r_i^2 - r_i'^2} \qquad\qquad r_i' \le r \le r_i$$

$$H = \frac{Ni}{2\pi r} \qquad\qquad\qquad r_i \le r \le r_o \qquad (1.5)$$

$$H = \frac{Ni}{2\pi r} \left(1 - \frac{r^2 - r_o^2}{r_o'^2 - r_o^2} \right) \qquad r_o \le r \le r_o'$$

$$H = 0 \qquad\qquad\qquad\qquad r \ge r_o'$$

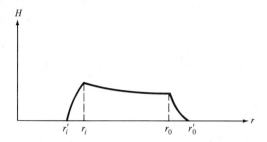

FIGURE 1.4 Variation of Magnetic Field Intensity with Radius

The H field in the θ direction excluding the end-turn regions might appear as shown in Figure 1.4.

The behavior internal to the end-turn regions is somewhat more complicated in that the H field in the θ direction is a function of both z and r. It does, however, reduce to zero outside the winding. This can be surmised by applying Ampère's law for appropriately chosen circular paths of integration outside the winding.

Magnetic flux density $\overline{B}$ in webers per square meter or teslas (Wb/m² or T; $1.0 \text{ Wb/m}^2 = 1.0 \text{ T}$) is related to magnetic field intensity $\overline{H}$ by the permeability of the material:

$$\overline{B} = \mu\overline{H} \tag{1.6}$$

where

$$\mu = \mu_r\mu_0 \tag{1.7}$$

In (1.7), μ_0 is the permeability of free space, $4\pi \times 10^{-7}$ henry per meter (H/m), and μ_r is the relative permeability. We are assuming here that the material is isotropic, which means that it has the same properties in all directions. The value of μ_r for ferromagnetic materials used in machines and transformers is high, say on the order of 10^3 to 10^5. If the core material in our magnetic circuit is ferromagnetic, flux densities within the winding material (regions where $r_i' < r < r_i$ and $r_o < r < r_o'$ and internal to the end-turn regions) will be negligible compared to flux densities within the core. This is illustrated in Figure 1.5.

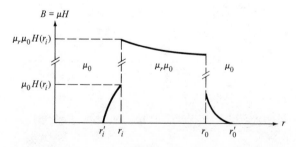

FIGURE 1.5 Variation of Magnetic Flux Density with Radius

Ferromagnetic materials in reality exhibit nonlinear characteristics, where the value of μ_r depends on the magnitude of H and its past behavior. We will examine some aspects of this behavior later, but as we will show, a relatively high value of μ_r permits us to assume linear behavior in many applications.

Magnetic flux φ in webers is obtained by integrating magnetic flux density $\bar{B}$ over some particular surface area.

$$\varphi = \int_s \bar{B} \cdot d\bar{s} \tag{1.8}$$

The vector quantities involved are illustrated in Figure 1.6. Note that magnetic flux is a scalar quantity.

To obtain flux in the core of our magnetic circuit example, we must integrate flux density in the core over the rectangular cross section. If we evaluate H at $r = (r_i + r_o)/2$ and assume this value over the entire cross section, we do not incur significant error (see Problem 1.2). Defining this value as H_c, (1.5) yields

$$H_c = \frac{Ni}{2\pi\left(\dfrac{r_i + r_o}{2}\right)} = \frac{Ni}{\ell_c} \tag{1.9}$$

where ℓ_c is the distance around the core at its midpoint. Then flux density is given by

$$B_c = \mu H_c = \frac{\mu Ni}{\ell_c} \tag{1.10}$$

and the flux is

$$\varphi = \int_s \bar{B} \cdot d\bar{s} = B_c A_c = \frac{\mu Ni A_c}{\ell_c} = \frac{Ni}{\ell_c/\mu A_c} \tag{1.11}$$

where A_c is the cross-sectional area of the core. We can view (1.11) as a cause-and-effect relationship where an exciting current i causes a flux φ in the core.

The quantity Ni in (1.11) is sometimes referred to as *magnetomotive force* (mmf). The units of mmf are ampere-turns (A · t) or, alternatively, amperes if N is considered dimensionless. The ratio of mmf to flux is defined as *reluctance:*

$$\mathcal{R} = \frac{Ni}{\varphi} \qquad \text{A · t/Wb} \tag{1.12}$$

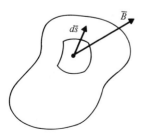

FIGURE 1.6 Vector Quantities Associated with Equation (1.8)

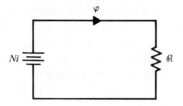

FIGURE 1.7 An Analogous Electrical Circuit to the Magnetic Circuit of Figure 1.1

The reciprocal of reluctance is defined as *permeance:*

$$\mathcal{P} = \frac{1}{\mathcal{R}} = \frac{\varphi}{Ni} \qquad \text{Wb/A} \cdot \text{t} \tag{1.13}$$

For the magnetic circuit we are considering we see from (1.11) that

$$\mathcal{R} = \frac{\ell_c}{\mu A_c} \tag{1.14}$$

This expression is similar to the expression for the resistance of a length of wire in terms of its resistivity ρ, its length ℓ, and its cross-sectional area A:

$$R = \frac{\rho \ell}{A} \tag{1.15}$$

An analogy is suggested between our magnetic circuit and a simple direct-current (dc) circuit as shown in Figure 1.7.

The analogy is sometimes useful in conceptualizing a magnetic circuit. To see this, we will consider an example.

EXAMPLE 1.1 Find the mmf required to produce a flux density of 0.5 Wb/m² in the air gap of the magnetic circuit depicted in Figure 1.8. Dimensions and relative permeability are

ℓ_g	gap length is 0.1 cm
A_c	core cross-sectional area is 1 cm²
ℓ_c	mean length of core is 6 cm
μ_r	relative permeability of core is 10^4

The example differs considerably from the symmetrical situation we encountered with the magnetic circuit configuration of Figure 1.1. The high permeability of the core results in negligible flux external to the core when compared to flux within the core. (There will

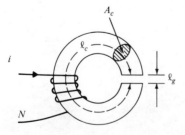

FIGURE 1.8 Magnetic Circuit For Example 1.1

be flux in the air gap.) If we neglect all flux except core flux and air-gap flux, the continuity law for magnetic flux;

$$\oint \overline{B} \cdot d\overline{s} = 0 \qquad (1.16)$$

requires that our core flux density B_c, assumed constant over any cross section, be the same at any point along the flux path within the core, since the core cross-sectional area is the same at any point along the path. That is, $\oint \overline{B} \cdot d\overline{s} = B_{c_1}A_c - B_{c_2}A_c = 0$, which requires that $B_{c_1} = B_{c_2}$ for any two points along the flux path ℓ_c within the core. This in turn requires that $H_c = B_c/\mu$ be constant along the length of the core. In the vicinity of the gap we will have some fringing, as illustrated in Figure 1.9. This can be accounted for (approximately) by increasing the effective cross-sectional area of the air gap over that of the core. We will assume that

$$A_g = 1.10A_c$$

If the effective cross-sectional area in the air gap A_g is assumed to apply for the entire length of the air gap ℓ_g, the continuity law for flux would require that flux density in the gap B_g be constant for the entire length of the air gap. Thus $H_g = B_g/\mu_0$ would be constant for the entire length of the air gap. Moreover, continuity of flux would require that $B_gA_g = B_cA_c = \varphi$. This implies a discontinuity in B at the transition from core to air gap since $A_g \neq A_c$. In reality, the normal component of B at a boundary must be continuous, and it will be in our magnetic circuit (see Figure 1.9). The use of an effective air-gap cross-sectional area A_g is obviously a simplifying assumption.

Now apply Ampère's law:

$$Ni = \oint \overline{H} \cdot d\overline{\ell}$$
$$= H_c\ell_c + H_g\ell_g$$

where

$$H_g = \frac{B_g}{\mu_0} = \frac{0.5}{4\pi \times 10^{-7}} = 3.98(10^5) \text{ A/m}$$

$$H_c = \frac{B_c}{\mu_r\mu_0} = \frac{\varphi}{A_c\mu_r\mu_0} = \frac{B_gA_g}{A_c\mu_r\mu_0}$$

$$= \frac{(0.5)(1.1)}{(10^4)(4\pi \times 10^{-7})} = 4.38(10^1) \text{ A/m}$$

Thus

$$Ni = 4.38(10)(0.06) + 3.98(10^5)(0.001)$$
$$= 2.63 + 3.98(10^2)$$
$$= 401 \text{ A or A} \cdot \text{t}$$

Thus a 100-turn coil carrying approximately 4 A could be used.

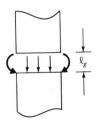

FIGURE 1.9 Typical Air-Gap Fringing

Note the relative magnitudes of H_g, the air-gap magnetic field intensity, and H_c, the core magnetic field intensity. Note also the relative magnitudes of $H_g\ell_g$ and $H_c\ell_c$. Ampère's law requires that these terms have the same units as the mmf Ni. $H_c\ell_c$ is sometimes referred to as the *mmf drop* in the core and $H_g\ell_g$ is referred to as the mmf drop in the air gap. The relative magnitudes of the quantities above suggest some approximations that will prove extremely useful in the future. To gain additional insights, we will solve the problem in a different way. Again starting with Ampère's law:

$$
\begin{aligned}
Ni &= \oint \overline{H} \cdot d\overline{\ell} \\
&= H_c\ell_c + H_g\ell_g \\
&= \frac{B_c}{\mu_c}\ell_c + \frac{B_g}{\mu_0}\ell_g \\
&= \varphi\frac{\ell_c}{\mu_c A_c} + \varphi\frac{\ell_g}{\mu_0 A_g} \\
&= \varphi[\mathcal{R}_c + \mathcal{R}_g]
\end{aligned}
$$

Solving yields

$$
\mathcal{R}_c = \frac{0.06}{(10^4)(4\pi \times 10^{-7})(10^{-4})} = 4.77(10^4) \text{ A} \cdot t/\text{Wb}
$$

$$
\mathcal{R}_g = \frac{0.001}{(4\pi \times 10^{-7})(1.1 \times 10^{-4})} = 7.23(10^6)
$$

$$
\varphi = B_g A_g = (0.5)(1.1 \times 10^{-4}) = 5.5(10^{-5}) \text{ Wb}
$$

Thus

$$
Ni = 5.5(10^{-5})[4.77(10^4) + 7.23(10^6)]
$$

$$
= 400 \text{ A} \cdot t
$$

which is the same as our previous result except for round-off differences. Note the relative magnitudes of air-gap reluctance $\mathcal{R}_g$ and core reluctance $\mathcal{R}_c$. The air-gap reluctance dominates. This is often the case. Errors we may have introduced by neglecting nonlinear effects of the core will be minimal if air-gap reluctance is significantly larger than core reluctance. Our use of the term "mmf drop" takes on added significance. It can easily be seen that

$$
H_c\ell_c = \varphi\mathcal{R}_c
$$

$$
H_g\ell_g = \varphi\mathcal{R}_g
$$

Other concepts suggested by the electrical circuit analog follow. The electrical circuit analog for our example is shown in Figure 1.10. A "voltage division" effect is suggested, with the major portion of the total mmf drop Ni appearing across the air-gap reluctance $\mathcal{R}_g$.

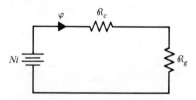

FIGURE 1.10 Electrical Circuit Analog

Several observations on Example 1.1 are in order. First, the magnetic circuit approach to the solution is an approximate technique. Its utility depends very much on relative orders of magnitude. If we replace the ferromagnetic core with some material whose permeability is close to that of free space (i.e., wood), the magnetic circuit solution approach is completely inappropriate. Referring to Figure 1.8, we see that if the coil were located on a small portion of a wood-core toroid, we would effectively have nothing more than an air-core coil in free space! Assuming that we do have a highly permeable core, the method may require use of rule-of-thumb factors to account for effects such as fringing in order to achieve the desired degree of accuracy. Although the electrical circuit analog provides some useful insights, in many cases we can solve magnetic circuit problems by direct application of Ampère's law and the continuity law for magnetic flux.

MAGNETIC MATERIALS

As discussed in the preceding section, the relationship between B and H is in fact nonlinear and multivalued for typical ferromagnetic materials used in machines and transformers. A B–H curve for a particular material is a means of illustrating this. Figure 1.11 shows a portion of a typical B–H curve. If H is increased from zero (i.e., point a in Figure 1.11), B will increase in a nonlinear manner in the first quadrant. Saturation typically occurs at a maximum value of 1.0 to 2.0 Wb/m^2. As H is decreased, a different path is followed. The path from b to c in the figure is an example. Note that at point c we have a nonzero value of B even though H has been reduced to zero. The value of B at this point is known as *remanence* and it is essentially a residual flux density. If we decrease H further, we eventually arrive at point d where B is again zero. The value of H required to reduce B to zero is defined as the *coercive force*. If we now continue to increase and decrease H to the same maximum and minimum values, we establish a B–H hysteresis loop as shown in Figure 1.12. The behavior is more complicated than we have indicated thus far. If we reverse direction at some point on the curve before H reaches its maximum (or minimum), we follow

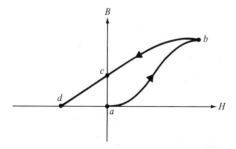

FIGURE 1.11 Typical Variation of B Versus H

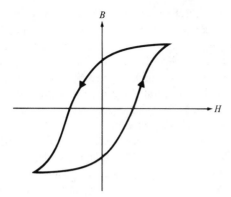

FIGURE 1.12 Typical Hysteresis Loop

a new curve not shown in Figure 1.12. Furthermore, the shape and size of the loop depend on the frequency of our excitation. During steady-state cyclic behavior we move on one of a family of hysteresis loops, depending on what the maximum and minimum values of H are. This is illustrated in Figure 1.13.

We will show in a later section that the energy loss per unit volume per cycle is directly proportional to the area inside the hysteresis loop. The dashed line connecting the tips of the loops in Figure 1.13 is the dc magnetization curve for the material. This is often referred to simply as the *magnetization curve*. A typical magnetization curve is shown in Figure 1.14. The dashed line in the figure indicates a linear approximation to the curve. Its slope is the permeability associated with the approximation. Obviously, this approximation would be poor if we move too far into the saturated region of the material or if we operate

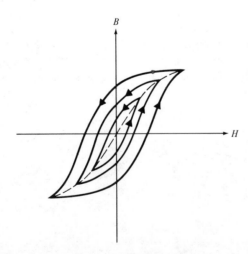

FIGURE 1.13 Family of Hysteresis Loops

FUNDAMENTAL CONCEPTS

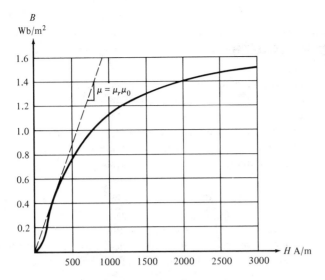

FIGURE 1.14 Typical Magnetization Curve

only with very low values of B and H (i.e., in the vicinity of the origin). It is clear that the appropriate value of μ_r depends entirely on what portion of the curve is used. It is possible to define permeability in a number of different ways. *Relative amplitude permeability* is defined as the ratio of B to $\mu_0 H$ at any point on the curve and it varies with the operating point. The maximum relative amplitude permeabilities for a number of different materials and their saturation flux densities are shown in Table 1.1.

In cases where operation will result in small displacements from a point on the magnetization curve, it will generally be preferable to use an incremental or differential permeability. This is defined as the slope of the curve at a particular point. Thus differential permeability at a point $H = H_0$ can be expressed as $dB/dH|_{H=H_0}$. Note that differential permeability at the origin, for the magnetization curve of Figure 1.14, would be less than the permeability associated with the dashed line.

Table 1.1 Characteristics of Soft Magnetic Materials[a]

Trade Name	Maximum Relative Amplitude Permeability	Saturation Flux Density (Wb/m²)
48 NI	200,000	1.25
Monimax	100,000	1.35
Permalloy	100,000	0.80
M-19	10,000	2.00

[a]Data excerpted from reference 1, Table 2.1, p. 32.

INDUCED VOLTAGES

Consider again the simple magnetic circuit with rectangular cross section that was analyzed in Section 1.2 and illustrated in Figure 1.1. If the core material is highly permeable, we can neglect flux external to the core and use magnetic circuit concepts, even if we concentrate the winding over a small portion of the core, as shown in Figure 1.15. To relate induced voltage at the terminals of the coil to current in the winding, we use Faraday's law:

$$\oint \overline{E} \cdot d\overline{\ell} = -\int_s \frac{\partial \overline{B}}{\partial t} \cdot d\overline{s} \tag{1.17}$$

In words, the line integral of the electric field around a closed path is equal to minus the partial derivative of the $\overline{B}$ field with respect to time integrated over any surface enclosed by the path. The terms in (1.17) are illustrated in Figure 1.16.

Each turn of the winding on the coil in Figure 1.15 constitutes a loop in which an electromotive force (emf) will be induced. Thus, for our configuration,

$$e_{turn} = \int \frac{\partial \overline{B}}{\partial t} \cdot d\overline{s} = \frac{d}{dt} \int \overline{B} \cdot d\overline{s} = \frac{d\varphi}{dt} \tag{1.18}$$

where e_{turn} is the voltage induced in each turn of the winding. There are N turns in series; thus the total coil voltage is given by

$$e(t) = N \frac{d\varphi}{dt} \tag{1.19}$$

where all of the flux φ is assumed to link all of the N turns. The minus sign (Lenz's law) in (1.17) has been correctly taken into account in (1.19) by the assignment of voltage polarity, winding sense, and flux direction shown in Figure 1.15. Current has been assigned such that positive current produces flux in the direction indicated in the figure.

Flux linkages are defined as the product of the number of turns and the flux linking the turns:

$$\lambda = N\varphi \tag{1.20}$$

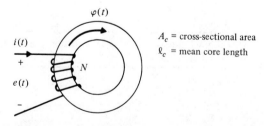

A_c = cross-sectional area
ℓ_c = mean core length

FIGURE 1.15

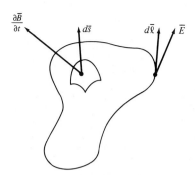

FIGURE 1.16 Illustration of Terms in Equation (1.17)

Faraday's law for a coil can thus be written

$$e = \frac{d\lambda}{dt} \qquad (1.21)$$

Inductance is defined as the ratio of flux linkages to current,

$$L = \frac{\lambda}{i} \qquad (1.22)$$

It should be clear that whatever problems we might have in defining a suitable value of relative permeability when we consider the nonlinearities of the iron core will also be present in our definition of inductance. Generally speaking, in discussions involving permeability μ or inductance L, an assumption of linearity is implied unless otherwise stated. Combining (1.21) and (1.22), we have

$$e = \frac{d\lambda}{dt} = \frac{d(Li)}{dt} \qquad (1.23)$$

If L is constant with respect to time (as it is in our present example), we produce the familiar *law of the inductor,* as expected:

$$e = L\frac{di}{dt} \qquad (1.24)$$

We will very shortly develop situations where L is not constant. This is typically due to relative motion between portions of the magnetic circuit. In such cases we must use the more general statement

$$e = \frac{d\lambda}{dt} = \frac{d(Li)}{dt} = \frac{Ldi}{dt} + \frac{idL}{dt} \qquad (1.25)$$

Equations (1.21), (1.23), and (1.25) are valid in general.

Inductance is related to the terms associated with the magnetic circuit in the following way:

$$L = \frac{\lambda}{i} = \frac{N\varphi}{i} = \frac{N}{i}\frac{Ni}{\mathcal{R}} = \frac{N^2}{\mathcal{R}} \qquad (1.26)$$

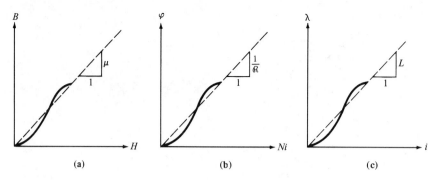

FIGURE 1.17

We see that inductance is directly proportional to the number of turns squared and inversely proportional to the magnetic circuit reluctance. In the present example, it is easy to relate inductance to dimensions in the magnetic circuit by substituting for the magnetic circuit reluctance:

$$L = \frac{N^2}{\mathcal{R}} = \frac{N^2 \mu A_c}{\ell_c} \tag{1.27}$$

where ℓ_c is the mean length of the core and A_c is its cross-sectional area. Equation (1.27) is useful for preliminary design estimates for inductors whose configurations are similar to that shown in Figure 1.15.

It may be helpful at this point to recall that we are essentially describing the same phenomenon in different ways. Consider Figure 1.17. If we start with a particular type of ferromagnetic core material characterized by the magnetization curve relating B and H in Figure 1.17(a) and construct a core with a single winding similar to the one shown in Figure 1.15, we can find a φ versus Ni curve as shown in Figure 1.17(b). These figures must have the same shape since multiplying B in Figure 1.17(a) by A_c gives us φ in Figure 1.17(b) and multiplying H in (a) by ℓ_c gives us Ni in (b). We have simply rescaled the vertical and horizontal axes. We can produce Figure 1.17(c) from (b) in a similar way by recalling the definition of flux linkages. Thus the shape of the B–H curve is maintained in the other two curves. The permeability μ, reluctance $\mathcal{R}$, and inductance L determine the slope of the respective straight lines where a linear characterization is appropriate.

1.5

ENERGY IN THE MAGNETIC FIELD

For convenience, Figure 1.15 is reproduced as Figure 1.18. The instantaneous power delivered to the magnetic circuit is given by

$$p(t) = i(t)e(t) \tag{1.28}$$

FUNDAMENTAL CONCEPTS

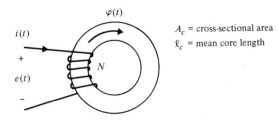

FIGURE 1.18

If the wire resistance is negligible, $e(t)$ is given by Faraday's law and

$$p(t) = i(t) \frac{d\lambda}{dt} \qquad (1.29)$$

If the current is expressed as a function of λ, we can separate variables and integrate to obtain the energy delivered to the device:

$$dW = p(t)\, dt = i(\lambda)\, d\lambda \qquad (1.30)$$

$$W = \int_{t_1}^{t_2} p(t)\, dt = \int_{\lambda_1}^{\lambda_2} i(\lambda)\, d\lambda \qquad (1.31)$$

Figure 1.19 illustrates graphically the integration for the case where λ_1 is zero. In Figure 1.19 the shaded area W corresponds to the energy delivered to the device as current i is increased from zero to i_2 and flux linkages λ increase from zero to λ_2. If we can assume linear behavior,

$$\lambda_2 = Li_2 \qquad (1.32)$$

consideration of Figure 1.19, modified for a linear relationship between λ and i, leads us to the familiar expression for stored energy in a linear inductor:

$$W = \tfrac{1}{2} Li_2^2 \qquad (1.33)$$

If we cannot assume linearity and if the curve is multivalued, returning the current to zero could result in the somewhat typical situation shown in Figure

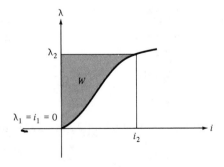

FIGURE 1.19 Energy Delivered to the Magnetic Circuit

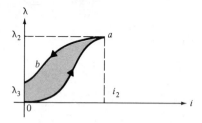

FIGURE 1.20 Example of Nonlinear Behavior

1.20. In this figure the shaded area represents energy not returned by the device. Essentially, as current is increased from zero to i_2 (curve $0a$), the energy delivered to the device is

$$W_{in} = \int_0^{\lambda_2} i(\lambda)\, d\lambda \qquad (1.34)$$

As current is decreased from i_2 to zero (curve ab), energy delivered to the device is

$$W'_{in} = \int_{\lambda_2}^{\lambda_3} i(\lambda)\, d\lambda = -\int_{\lambda_3}^{\lambda_2} i(\lambda)\, d\lambda \qquad (1.35)$$

The energy not returned by the device is

$$W_{in} + W'_{in} = \int_0^{\lambda_2} i(\lambda)\, d\lambda - \int_{\lambda_3}^{\lambda_2} i(\lambda)\, d\lambda \qquad (1.36)$$

which corresponds to the shaded area of Figure 1.20. The energy not returned is converted to heat due to behavior of the core material. It is generally regarded as a loss. During normal steady-state cyclic operation, the area inside a hysteresis loop corresponds to the energy loss for each cycle.

Equation (1.31) can be rewritten by changing the variable of integration ($\lambda = N\varphi$) as follows:

$$W = \int_{\lambda_1}^{\lambda_2} i(\lambda)\, d\lambda = \int_{\varphi_1}^{\varphi_2} i(\varphi)\, d(N\varphi) = \int_{\varphi_1}^{\varphi_2} Ni(\varphi)\, d\varphi \qquad (1.37)$$

This relates stored energy to variables associated with the magnetic circuit: mmf Ni and flux φ. If in the magnetic circuit of Figure 1.18 we assume that B and H are uniform over the cross section of the core,

$$d\varphi = A_c\, dB \qquad (1.38)$$

$$Ni = H\ell_c \qquad (1.39)$$

Changing the variable of integration and substituting for Ni in the last integral of (1.37) yields

$$W = \int_{B_1}^{B_2} H(B)\ell_c A_c\, dB = \ell_c A_c \int_{B_1}^{B_2} H(B)\, dB \qquad (1.40)$$

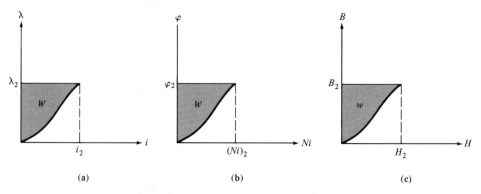

FIGURE 1.21 Graphical Depiction of Stored Energy Relationships for a Magnetic Circuit

Since, in (1.40), $\ell_c A_c$ is the volume of the core, the integral must be the energy density of the core or energy per unit volume.

We can summarize the foregoing concepts graphically as shown in Figure 1.21. In Figure 1.21(c), the lower case letter w indicates energy density rather than total energy.

1.6

ALTERNATING-CURRENT OPERATION

A very important mode of operation deserving special consideration is that of steady-state sinusoidal time variation. We again consider a simple magnetic circuit configuration as shown in Figure 1.22. If we assume negligible winding resistance, voltage $e(t)$ is given by Faraday's law:

$$e(t) = \frac{d\lambda}{dt} \tag{1.41}$$

If leakage flux is neglected, flux linkages are due only to the main core flux φ. This flux is sometimes referred to as *magnetizing flux*. Voltage $e(t)$ is given by

$$e(t) = \frac{d\lambda}{dt} = \frac{d(N\varphi)}{dt} = \frac{N\,d\varphi}{dt} \tag{1.42}$$

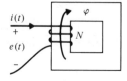

FIGURE 1.22

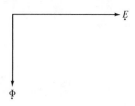

FIGURE 1.23

For steady-state sinusoidal time variation, $e(t)$ and $\varphi(t)$ can be represented by phasors:†

$$e(t) = \sqrt{2}\, E \cos(\omega t + \theta) \Leftrightarrow \dot{E} = E e^{j\theta} \qquad (1.43)$$

$$\varphi(t) = \Phi \cos(\omega t + \psi) \Leftrightarrow \dot{\Phi} = \frac{\Phi_{max} e^{j\psi}}{\sqrt{2}} \qquad (1.44)$$

For voltages and currents it is convenient to use root-mean-square (rms) or effective values as magnitudes of the respective phasor representations. The maximum value of flux Φ_{max} will generally be the quantity of interest. The appropriate relationship between phasors $\dot{E}$ and $\dot{\Phi}$ follows from (1.42):

$$\dot{E} = j\omega N \dot{\Phi} \qquad (1.45)$$

Equation (1.45) yields the angular relationship between $\dot{\Phi}$ and $\dot{E}$ shown in Figure 1.23. Equating magnitudes on both sides of (1.45), we have

$$E = \omega N \frac{\Phi_{max}}{\sqrt{2}} \qquad (1.46)$$

Thus the rms voltage E is related to the maximum flux Φ_{max} by (1.46). Since $\omega = 2\pi f$, (1.46) can be written as

$$E = \frac{2\pi}{\sqrt{2}} f N \Phi_{max} = 4.44 f N \Phi_{max} \qquad (1.47)$$

Equation (1.47) is useful for preliminary design estimates. If the flux density in a core is assumed uniform across the core cross section A_c,

$$\Phi_{max} = A_c B_{max} \qquad (1.48)$$

Recognizing that maximum flux densities in typical ferromagnetic materials are usually limited to 1 to 2 Wb/m², (1.47) and (1.48) can be used to determine the required core cross section A_c and number of turns N for a specified rms voltage E.

If we have linear behavior, the flux linkages are related to the current by the inductance and

$$\dot{\Phi} = \frac{L}{N} \dot{I} \qquad (1.49)$$

The angular relationship between $\dot{\Phi}$, $\dot{E}$, and $\dot{I}$ would then be as shown in Figure 1.24.

†See Appendix B for a concise review of the fundamentals of alternating-current (ac) steady-state analysis.

FUNDAMENTAL CONCEPTS

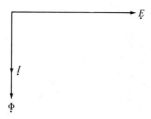

FIGURE 1.24

It was shown in Sections 1.3 and 1.5 that a nonzero area inside the B–H hysteresis loop associated with a given mode of steady-state ac operation implies energy losses within the core material. The relationship for the average power for steady-state sinusoidal time variation is†

$$P = EI \cos (\theta - \psi) \qquad (1.50)$$

where $\theta - \psi$ is the angle between E and I (the power-factor angle). For the phase relationship shown in Figure 1.24, the average power delivered to the magnetic circuit would be zero since $\theta - \psi = \pi/2$. It is important to remember that the relationship between E and I shown in Figure 1.24 is based on an assumption of linearity [Eq. (1.49)] that is not appropriate if we have "lossy" ferromagnetic material involved. Figure 1.24 does, however, show the correct 90° phase relationship between E and Φ regardless of the nature of the material. That is, if $e(t)$ is sinusoidal, $\varphi(t)$ must be sinusoidal and phase displaced from $e(t)$ by 90°. Similarly, if $\varphi(t)$ is sinusoidal, $e(t)$ must be sinusoidal and phase displaced from $\varphi(t)$ by 90°. This is a direct consequence of Faraday's law.

We can use a graphical development to determine the nature of the current when nonlinearities are taken into account. A B–H curve hysteresis loop has a corresponding φ versus Ni curve hysteresis loop. If $\varphi(t)$ is sinusoidal as shown in Figure 1.25, the waveform for $Ni(t)$ [and hence $i(t)$] can be determined graphically as shown in the figure.

The current waveform shown in Figure 1.25 is distorted but does possess half-wave symmetry; that is,

$$i(t) = -i\left(t \pm \frac{T}{2} \right) \qquad (1.51)$$

It therefore can be represented by a Fourier series containing a fundamental and only odd harmonics. Consideration of Figure 1.25 shows that the $Ni(t)$ [and hence $i(t)$] waveform leads the $\varphi(t)$ waveform by a small amount. The fundamental component of $i(t)$ will thus lead $\varphi(t)$. If we neglect harmonics and consider only the fundamental component of current, the phasor diagram in Figure 1.26 might be typical, where I is the phasor representation for the fundamental component of the current $i(t)$. The current I shown in Figure 1.26 can be resolved into components in phase with and in quadrature with E as shown in Figure 1.27. Referring to Figure 1.27, I_c is the core-loss component of current and I_m is the magnetizing component of current. The sum of the two $(I_m + I_c)$

†See Appendix C for a concise review of fundamental ac circuit power relationships.

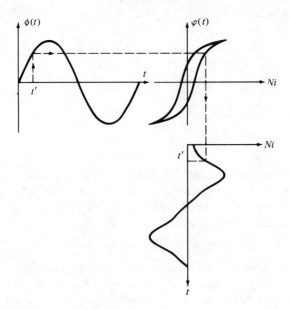

FIGURE 1.25

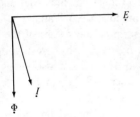

FIGURE 1.26

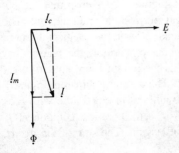

FIGURE 1.27

FUNDAMENTAL CONCEPTS

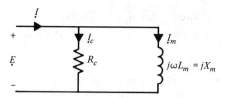

FIGURE 1.28

is referred to as *exciting current*. A simple linear equivalent circuit can be synthesized that yields the appropriate relationships between voltage and the *fundamental component* of current. Figure 1.27 suggests the simple equivalent circuit shown in Figure 1.28. Appropriate values of R_c and X_m can be determined by measurements made at the terminals of the device.

The effects of leakage flux and winding resistance can be accounted for by additional elements in the equivalent circuit. Reluctance associated with leakage flux paths depends primarily on the reluctance associated with the portion of the flux path through air or through the winding. Flux linkages due to leakage flux are therefore often assumed to be linearly related to current:

$$\lambda_\ell = L_\ell i \tag{1.52}$$

If distributed winding resistance is accounted for by a lumped resistance r, the terminal voltage is

$$v(t) = ri(t) + \frac{d\lambda_\ell}{dt} + e(t) \tag{1.53}$$

where $e(t)$ is determined as above using Faraday's law, neglecting winding resistance and leakage flux. Thus

$$v(t) = ri(t) + L_\ell \frac{di}{dt} + e(t) \tag{1.54}$$

and the corresponding phasor relationship is

$$\underset{\sim}{V} = r\underset{\sim}{I} + j\omega L_\ell \underset{\sim}{I} + \underset{\sim}{E} \tag{1.55}$$

We can modify the equivalent circuit shown in Figure 1.28 to obtain the equivalent circuit shown in Figure 1.29. Figure 1.29 is an equivalent circuit at the terminals for the magnetic circuit shown in Figure 1.22. The winding resistance r and leakage reactance X_ℓ are often negligible. The leakage reactance X_ℓ takes on a more significant role in the performance of power transformers, which we

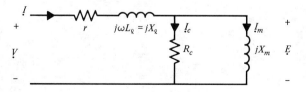

FIGURE 1.29

will consider shortly. The core-loss resistance R_c and magnetizing reactance X_m are valid only for operation near the frequency and flux magnitude for which they are determined.

EXAMPLE 1.2 Measurements made at the terminals of the magnetic circuit shown in Figure 1.22 under steady-state ac conditions (60 Hz) are as follows:

$$V = 120 \text{ V rms}$$

$$I = 1.8 \text{ A rms}$$

$$P = 75 \text{ W}$$

Find an equivalent circuit for the device.

The leakage reactance and winding resistance are generally negligible for such configurations. We thus find R_c and X_m for the circuit of Figure 1.28.

$$P = \frac{V^2}{R_c}$$

Therefore,

$$R_c = \frac{V^2}{P} = \frac{(120)^2}{75} = 192 \ \Omega$$

$$P = VI \cos \theta$$

$$\cos \theta = \frac{P}{VI} = \frac{75}{(120)(1.8)} = 0.347$$

Therefore,

$$\theta = 69.7^0$$

Figure 1.30 shows E and I plotted on a phasor diagram. Referring to Figure 1.30, we find

$$I_m = 1.8 \sin 69.7^0 = 1.69 \text{ A}$$

Therefore,

$$X_m = \frac{V}{I_m} = \frac{120}{1.69} = 71 \ \Omega$$

We turn our attention now to the nature of the losses in the material itself. Losses are due to induced eddy currents in the ferromagnetic material and hysteresis. Eddy-current losses are ohmic in nature. Hysteresis losses are due to energy converted to heat in the process of aligning and realigning material magnetic domains.

Lamination of the iron material can be used to reduce eddy-current losses.

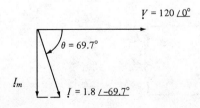

$V = 120 \angle 0^\circ$

$\theta = 69.7^\circ$

I_m

$I = 1.8 \angle{-69.7^\circ}$

FIGURE 1.30

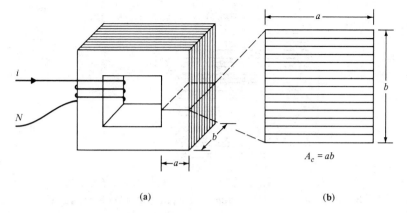

FIGURE 1.31 Magnetic Circuit with Laminated Material

Figure 1.31 shows a magnetic circuit and its associated cross section with material laminated to reduce losses. The core can be assembled by stacking thin sheets of magnetic material as shown in the figure. In power applications, the lamination thickness might be on the order of 0.5 to 5 mm. Ideally, the laminations are separated by thin sheets of highly resistive material. One approach is to permit a surface oxide to form on either side of each of the laminations and apply a coat of varnish or shellac in addition. The ratio of actual magnetic material cross-sectional area to gross cross-sectional area [a times b in Figure 1.31(b)] is defined as the *stacking factor*. Some typical lamination thicknesses and stacking factors are shown in Table 1.2.

The basis of eddy-current action and the effects of lamination can be explained approximately in the following way (see reference 2, pp. 22–27, for an alternative development). Consider Figure 1.32, which shows the general nature of induced eddy currents in our core cross section for a flux directed into the page and increasing with time. Note that the eddy currents tend to establish flux in opposition to our magnetizing flux in accordance with Lenz's law. We will neglect flux induced by the eddy currents themselves and assume a uniform flux density across the core cross-sectional area as before. We apply Faraday's law to determine the nature of the induced electric field in a lamination:

$$\oint \overline{E} \cdot d\overline{\ell} = \int \frac{\partial \overline{B}}{\partial t} \cdot d\overline{s} \qquad (1.56)$$

Table 1.2[a]

Lamination Thickness (mm)	Stacking Factor
0.0127	0.50
0.0254	0.75
0.0508	0.85
0.1 –0.25	0.90
0.27–0.36	0.95

[a]Reproduced from reference 1, Table 2.2, p. 34.

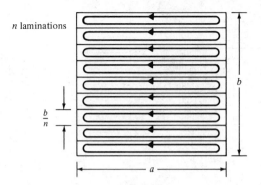

n laminations

$\frac{b}{n}$

b

a

FIGURE 1.32 Eddy Current Paths and Directions for Flux Increasing into the Page

(We omit the minus sign by taking $\overline{E}$ in the direction of eddy-current flow.) If we choose a path of integration at the periphery of a lamination and assume that $a \gg b/n$, the integral on the left becomes (approximately)

$$\oint \overline{E} \cdot d\overline{\ell} = E(2a) \tag{1.57}$$

where the magnitude of the electric field E is assumed constant along the length a. The integral on the right becomes

$$\int \frac{\partial \overline{B}}{\partial t} \cdot d\overline{s} = \frac{dB}{dt} a \frac{b}{n} \tag{1.58}$$

where we assume uniform flux density B over the lamination cross section. From (1.56), (1.57), and (1.58) we have

$$E = \frac{dB}{dt} \frac{b}{2n} \tag{1.59}$$

E varies linearly from zero at the center of a lamination to the value given by (1.59) at the periphery of the lamination. Since the power loss per unit of volume is σE^2, where σ is conductivity of the material, the average value of E^2 over the volume is of interest. This is given by (see Problem 1.15)

$$(E^2)_{\text{avg}} = \frac{1}{3} \left(\frac{dB}{dt} \right)^2 \left(\frac{b}{2n} \right)^2 \tag{1.60}$$

Now if the magnetic flux density B varies sinusoidally with time,

$$B = B_{\text{max}} \sin(\omega t + \psi) \tag{1.61}$$

then

$$\frac{dB}{dt} = 2\pi f B_{\text{max}} \cos(\omega t + \psi) \tag{1.62}$$

$$\left(\frac{dB}{dt} \right)^2 = \frac{(2\pi)^2 f^2 B_{\text{max}}^2}{2} [1 + \cos(2\omega t + 2\psi)] \tag{1.63}$$

Substituting (1.63) into (1.60), multiplying by σ, and taking the time average, we find the eddy-current power loss per unit of volume:

$$p_e = \frac{\sigma}{6}\,\pi^2 f^2 B_{max}^2 \left(\frac{b}{n}\right)^2 \qquad \text{W/m}^3 \qquad (1.64)$$

Since b/n is lamination thickness, (1.64) can be rewritten as

$$p_e = K_e(f B_{max} t_\ell)^2 \qquad \text{W/m}^3 \qquad (1.65)$$

which shows how p_e depends on the frequency f, maximum flux density B_{max}, and lamination thickness t_ℓ. Note that p_e does not depend on the core volume or geometry if the flux density is uniform throughout the core.

The dependence of p_e on frequency squared is of interest. If we start with (1.65), we can find the energy loss in each cycle per unit of volume due to eddy currents as

$$w_e = \frac{p_e}{f} = K_e f (B_{max} t_\ell)^2 \qquad \text{J/m}^3 \qquad (1.66)$$

Since the energy loss in each cycle is proportional to the area inside the B–H hysteresis loop, (1.66) suggests that the energy loss, and hence the shape of the B–H loop, will change if we operate at different frequencies. This type of variation is illustrated in Figure 1.33.

Losses due to hysteresis are more difficult to deal with analytically and empirically derived relationships are often used. If, for example, we can determine at low frequency that the area inside a B–H hysteresis loop can be taken as proportional to $(B_{max})^n$, then the energy dissipated due to hysteresis loss in each cycle per unit of volume is

$$w_h = K_h(B_{max})^n \qquad \text{J/m}^3 \qquad (1.67)$$

and the power loss per unit of volume due to hysteresis is

$$p_h = f w_h = K_h f (B_{max})^n \qquad \text{W/m}^3 \qquad (1.68)$$

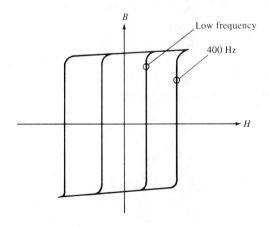

FIGURE 1.33 B versus H at Various Frequencies

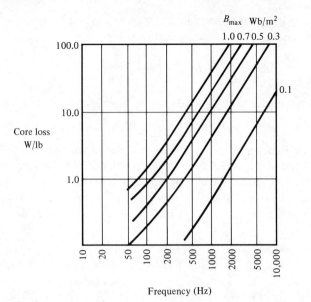

FIGURE 1.34 Typical Core Losses for M-19, 29-Gauge Steel

The constant n ranges typically from 1.5 to 2.5. The total core losses, then, per unit of volume, are the sum of the hysteresis and eddy-current losses:

$$p = p_h + p_e \quad W/m^3 \tag{1.69}$$

Manufacturers of magnetic materials frequently publish experimentally determined curves giving the total core loss per unit of volume (or per unit of weight) of magnetic material. Losses may be given as a function of the maximum flux density with the frequency as a parameter or as a function of frequency with the maximum flux density as a parameter. Exciting volt-amperes can be experimentally determined similarly. Figures 1.34 and 1.35 are examples.

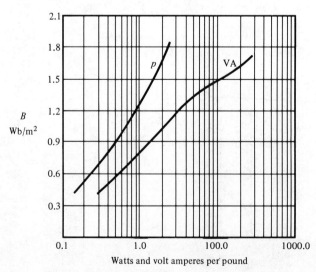

FIGURE 1.35 Typical Core Losses for M-19, 29-Gauge Steel at 60 Hz

EXAMPLE 1.3 The magnetic circuit shown in Figure 1.36 is characterized by the loss data shown in Figure 1.35. Find an equivalent linear circuit consisting of a resistance in parallel with a reactance that might be used to describe 60-Hz operation at maximum flux densities near 1.2 Wb/m².

The voltage induced is given by Faraday's law:

$$e(t) = \frac{N \, d\varphi}{dt} = NA_c \frac{dB}{dt}$$

Therefore, the maximum voltage would be

$$E_{max} = NA_c B_{max} \omega$$

The effective or rms voltage would be

$$E = \frac{E_{max}}{\sqrt{2}} = \frac{NA_c B_{max} \omega}{\sqrt{2}}$$

$$= \frac{200(5)^2(10^{-4})(1.2)(377)}{\sqrt{2}}$$

$$= 160 \text{ V}$$

From Figure 1.35,

$$p = 0.9 \text{ W/lb}$$

The total losses are

$$p = 0.9 \, \frac{W}{lb} \times 25 \text{ cm}^2 \times 60 \text{ cm} \times \frac{0.017 \text{ lb}}{cm^3}$$

$$= 22.95 \text{ W}$$

The appropriate value of resistance is then

$$R = \frac{E^2}{P} = \frac{(160)^2}{22.95} = 1115 \ \Omega$$

From Figure 1.35, volt-amperes per pound will be

$$3.0 \text{ VA/lb}$$

The rms current is then

$$I = \frac{VA}{E} = \frac{(3.0)(25)(60)(0.017)}{160} = 0.478 \text{ A}$$

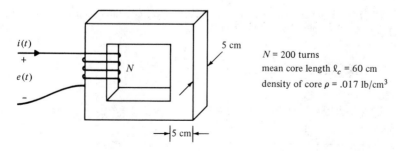

$i(t)$

$+$

$e(t)$

$-$

N

5 cm

5 cm

N = 200 turns

mean core length ℓ_c = 60 cm

density of core ρ = .017 lb/cm³

FIGURE 1.36

The magnetizing current is

$$I_m = \sqrt{I^2 - I_c^2}$$

$$= \sqrt{I^2 - \left(\frac{E}{R}\right)^2}$$

$$= \sqrt{(0.478)^2 - \left(\frac{160}{1115}\right)^2} = 0.456 \text{ A}$$

Thus the magnetizing reactance will be

$$X_m = \frac{E}{I_m} = \frac{160}{0.456} = 351 \ \Omega$$

and our circuit is as shown in Figure 1.37.

EXAMPLE 1.4   Determine the nature of the λ versus i characteristic we must have if the magnetic circuit of Figure 1.36 is represented by the equivalent circuit shown in Figure 1.37. Assume operation at 60 Hz under the set of conditions described in Example 1.3.

The total current in the circuit of Figure 1.37 is given by

$$i = i_R + i_L$$

$$= \frac{1}{R}\frac{d\lambda}{dt} + \frac{\lambda}{L}$$

Without loss of generality assume that

$$\lambda = \sqrt{2}\frac{E}{\omega}\sin \omega t$$

Then

$$\frac{d\lambda}{dt} = \sqrt{2}\, E \cos \omega t$$

$$i_R = \sqrt{2}\frac{E}{R}\cos \omega t$$

Thus

$$\left(\frac{\omega\lambda}{\sqrt{2}\, E}\right)^2 + \left(\frac{Ri_R}{\sqrt{2}\, E}\right)^2 = \sin^2 \omega t + \cos^2 \omega t = 1$$

and λ and i_R satisfy the equation for an ellipse. The current i_L is related to flux linkages λ by the inductance L:

$$i_L = \frac{\lambda}{L} = \frac{\omega}{X_m}\lambda$$

which is a linear relationship. The relationship between λ and $i = i_R + i_L$ can be developed graphically as shown in Figure 1.38.

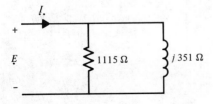

FIGURE 1.37

FUNDAMENTAL CONCEPTS

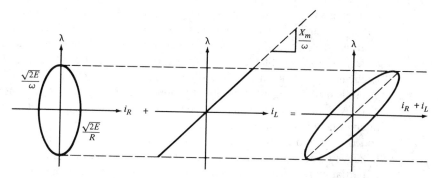

FIGURE 1.38

The λ versus i characteristics for the linear circuit of Figure 1.37 should be compared to some of our B–H curves for ferromagnetic cores illustrated previously. It should be obvious that the λ versus i curve, for our linear model, is only a reasonable approximation of the actual core characteristics.

1.7

COUPLED COILS

We have considered single-winding systems thus far. We now turn our attention to systems containing multiple windings and extend our definitions to include them. The transformer is one example and its multiple windings constitute a coupled-coil system. Most machine configurations involve interactions between windings and thus exhibit coupled-coil behavior.

A portion of a coupled-coil system consisting of three windings is shown in Figure 1.39. Fluxes indicated in the figure are shown conceptually, with no

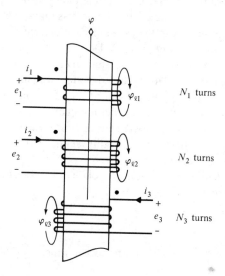

FIGURE 1.39 Coupled-Coil System

attempt made to show behavior in detail. The flux denoted φ is flux shown linking all turns of each winding in the system. If the core is highly permeable, such as what we might have in a power transformer, it is referred to as *magnetizing flux*. We define leakage flux as flux other than flux φ. *Leakage flux* due to current i_1 is denoted $\varphi_{\ell 1}$. It is shown linking some turns in winding 1 and in general must be regarded as linking some turns in the other two windings as well. A similar statement can be made about $\varphi_{\ell 2}$ and $\varphi_{\ell 3}$. The total flux linkages in any one winding will be the sum of the flux linkages caused by φ, $\varphi_{\ell 1}$, $\varphi_{\ell 2}$, and $\varphi_{\ell 3}$. In most systems a direct attack on the problem of determining flux linkages using the definition of flux linkages ($\lambda = N\varphi$) requires consideration of partial flux linkages. Except in relatively simple configurations, other methods are generally less troublesome. For the moment we will avoid the issue and assume that we have a linear system such that the flux linkages in any one winding can be expressed as a simple linear combination of currents:[†]

$$\overline{\lambda} = \overline{L}\,\overline{i} \tag{1.70}$$

In (1.70) all quantities are matrix quantities of the appropriate dimension:

$$\overline{\lambda} = \begin{bmatrix} \lambda_1 \\ \lambda_2 \\ \lambda_3 \end{bmatrix} \tag{1.71}$$

$$\overline{i} = \begin{bmatrix} i_1 \\ i_2 \\ i_3 \end{bmatrix} \tag{1.72}$$

$$\overline{L} = \begin{bmatrix} L_{11} & L_{12} & L_{13} \\ & L_{22} & L_{23} \\ & & L_{33} \end{bmatrix} \tag{1.73}$$

In coupled-coil systems the assignment of current direction and flux direction is important. Equation (1.70) implies that positive current in any winding gives rise to positive flux linkages in any winding of the system, assuming that all L_{ij} are ≥ 0. Equation (1.70) is consistent with the assignment of current directions and indicated winding senses shown in Figure 1.39. Dots are often used when direct determination of the winding sense is not possible. The convention used may be stated as: "Currents entering dotted (undotted) terminals establish flux in the same direction in the magnetic circuit." It can be seen from Figure 1.39 that, if we move the dots on each winding to the other terminal, (1.70) will still hold. The statement of the dot convention used here is consistent with other statements of the dot convention perhaps more readily applied when we are interested only in electrical-circuit conditions.

Lenz's law can be used to determine the voltage polarity. In particular, for the assignment of polarities shown in Figure 1.39,

$$\overline{e} = \frac{d\overline{\lambda}}{dt} = \overline{L}\,\frac{d\overline{i}}{dt} \tag{1.74}$$

where all inductances are assumed constant with respect to time.

[†]See Appendix F for a concise review of elementary matrix operations.

Note that in the inductance matrix shown in (1.73) only the main diagonal elements and elements above the diagonal are shown. Elements below the diagonal elements are related by

$$L_{ij} = L_{ji} \tag{1.75}$$

In other words, the matrix is symmetric:

$$\overline{L} = \overline{L}^T \tag{1.76}$$

where the superscript T indicates transpose of a matrix. It is sometimes helpful to recognize that each inductance has a physical significance:

$$L_{jk} = \lambda_j \Big|_{\substack{i_k = 1 \\ i_n = 0;\ \text{all } n \neq k}} \tag{1.77}$$

In words, L_{jk} is equal to the flux linkages we would have in winding j for a 1A current in winding k if all other winding currents were zero.

The stored energy in a system of coupled coils can be found by adding total energy delivered to each coil in the system.

Instantaneous power delivered to the first winding is

$$p_1(t) = i_1(t)e_1(t) = i_1(t)\frac{d\lambda_1(t)}{dt} \tag{1.78}$$

If currents can be expressed as functions of flux linkages, we can separate variables:

$$dW_1 = p_1(t)\ dt = i_1(\lambda_1, \lambda_2, \lambda_3)\ d\lambda_1 \tag{1.79}$$

For our linear system

$$\lambda_1 = L_{11}i_1 + L_{12}i_2 + L_{13}i_3 \tag{1.80}$$

Thus

$$d\lambda_1 = L_{11}\ di_1 + L_{12}\ di_2 + L_{13}\ di_3 \tag{1.81}$$

Substituting in (1.79), we have

$$dW_1 = L_{11}i_1\ di_1 + L_{12}i_1\ di_2 + L_{13}i_1\ di_3 \tag{1.82}$$

The energy delivered to the system at the terminals of the first winding as currents are raised from zero to final values I_1, I_2, and I_3 will be

$$W_1 = L_{11}\int_0^{I_1} i_1\ di_1 + L_{12}\int_0^{I_2} i_1\ di_2 + L_{13}\int_0^{I_3} i_1\ di_3 \tag{1.83}$$

Similarly, the energy delivered at the terminals of the second and third windings will be given by

$$W_2 = L_{21}\int_0^{I_1} i_2\ di_1 + L_{22}\int_0^{I_2} i_2\ di_2 + L_{23}\int_0^{I_3} i_2\ di_3 \tag{1.84}$$

$$W_3 = L_{31}\int_0^{I_1} i_3\ di_1 + L_{32}\int_0^{I_2} i_3\ di_2 + L_{33}\int_0^{I_3} i_3\ di_3 \tag{1.85}$$

The total stored energy in general will be the sum of the energies delivered at the terminals of each of the three windings.

Now recognize that the stored energy will be independent of the manner in which the three currents are raised to their final values I_1, I_2, and I_3 since the currents constitute a set of state variables for the system. If we raise i_1 to I_1 while keeping $i_2 = i_3 = 0$, we have from (1.83) (since $di_2 = di_3 = 0$)

$$\Delta W_1 = \tfrac{1}{2} L_{11} I_1^2 \tag{1.86}$$

Now raise i_2 to I_2 while keeping $i_1 = I_1$ and $i_3 = 0$. From (1.83) we obtain

$$\Delta W_1' = L_{12} I_1 I_2 \tag{1.87}$$

and from (1.84) we obtain

$$\Delta W_2 = \tfrac{1}{2} L_{22} I_2^2 \tag{1.88}$$

Now raise i_3 to I_3 while keeping $i_1 = I_1$ and $i_2 = I_2$. From (1.83) we obtain

$$\Delta W_1'' = L_{13} I_1 I_3 \tag{1.89}$$

From (1.84) we obtain

$$\Delta W_2' = L_{23} I_2 I_3 \tag{1.90}$$

and from (1.85) we obtain

$$\Delta W_3 = \tfrac{1}{2} L_{33} I_3^2 \tag{1.91}$$

The total energy delivered to the system must be the sum of the energies delivered at the terminals as we successively raised currents to their final values:

$$\begin{aligned} W &= \Delta W_1 + \Delta W_1' + \Delta W_2 + \Delta W_1'' + \Delta W_2' + \Delta W_3 \\ &= \tfrac{1}{2} L_{11} I_1^2 + L_{12} I_1 I_2 + \tfrac{1}{2} L_{22} I_2^2 + L_{13} I_1 I_3 + L_{23} I_2 I_3 + \tfrac{1}{2} L_{33} I_3^2 \end{aligned} \tag{1.92}$$

This result can be generalized to an arbitrary number of windings and written in the following ways:

$$W = \tfrac{1}{2} \sum_i \sum_j L_{ij} I_i I_j \tag{1.93}$$

or $\hspace{11cm}$ (1.94)

$$W = \tfrac{1}{2} \bar{I}^T \bar{L}\, \bar{I}$$

because of the symmetry of the inductance matrix.

EXAMPLE 1.5 Consider a two-winding linear system

$$\lambda_1 = L_{11} i_1 + L_{12} i_2$$
$$\lambda_2 = L_{21} i_1 + L_{22} i_2$$

Verify the symmetry condition

$$L_{12} = L_{21}$$

using stored energy concepts.

We must have the same stored energy in the coupled-coil system for a final value of i_1 and i_2 regardless of the manner in which i_1 and i_2 reach their final values if i_1 and i_2

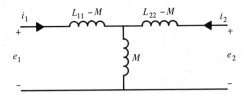

FIGURE 1.40

are state variables. If we repeat the development for our two-coil system described in (1.83) through (1.92) for the case where i_1 is raised to final value I_1 while i_2 is held at zero followed by raising i_2 to final value I_2 while i_1 is fixed at I_1, we have for the total stored energy

$$W = \tfrac{1}{2} L_{11}I_1^2 + L_{12}I_1I_2 + \tfrac{1}{2} L_{22}I_2^2$$

If we repeat the exercise but reverse the order of raising currents to their final values (i.e., first raise i_2 to final value I_2 holding i_1 at zero then raise i_1 to final value I_1 holding i_2 at I_2), we have for the total stored energy

$$W' = \tfrac{1}{2} L_{22}I_2^2 + L_{21}I_2I_1 + \tfrac{1}{2} L_{11}I_1^2$$

Comparing the two expressions, we see that we must have

$$L_{21} = L_{12}$$

if $W' = W$ for the same final current values I_1 and I_2. This common value of inductance is referred to as *mutual inductance*.

EXAMPLE 1.6   Consider a two-winding linear system

$$\lambda_1 = L_{11}i_1 + Mi_2$$
$$\lambda_2 = Mi_1 + L_{22}i_2$$

Find an equivalent circuit relating voltage and current at each of the windings.
    By Faraday's law,

$$e_1 = \frac{d\lambda_1}{dt} = L_{11}\frac{di_1}{dt} + M\frac{di_2}{dt}$$

$$e_2 = \frac{d\lambda_2}{dt} = M\frac{di_1}{dt} + L_{22}\frac{di_2}{dt}$$

The circuit in Figure 1.40 yields the correct current–voltage relationships. The circuit is not unique in that other circuits can be found that will yield the correct terminal voltage–current relationships.

1.8

THE POWER TRANSFORMER

We begin our discussion of the power transformer by reviewing the relationships associated with the ideal two-winding transformer. Well-designed power transformers do not depart significantly from ideal transformer behavior, and approximate power transformer behavior can often be reasonably determined by simply assuming the ideal transformer relationships. Consider the two-winding transformer shown in Figure 1.41. The ideal transformer relationships follow if we assume:

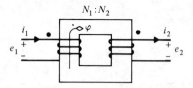

FIGURE 1.41

1. Zero leakage flux.
2. Zero winding resistance.
3. Infinite permeability of the core.

If we have zero leakage flux and zero winding resistance, voltages on either side of the transformer are determined by Faraday's law:

$$e_1 = \frac{d\lambda_1}{dt} = N_1 \frac{d\varphi}{dt} \tag{1.95}$$

$$e_2 = \frac{d\lambda_2}{dt} = N_2 \frac{d\varphi}{dt} \tag{1.96}$$

From (1.95) and (1.96),

$$\frac{d\varphi}{dt} = \frac{e_1}{N_1} = \frac{e_2}{N_2} \tag{1.97}$$

or

$$\frac{e_1}{e_2} = \frac{N_1}{N_2} \tag{1.98}$$

Current i_2 has been assigned in the direction of induced current (Lenz's law). It tends to establish the flux opposite to the direction of the magnetizing flux φ shown in Figure 1.41. This assignment of i_2 is typical for power transformer analysis and generally results in fewer minus signs. We now apply Ampère's law, taking into account our third assumption. If we have infinite permeability and if the flux density in the core is finite, it follows that within the core we must have $H = 0$. Thus for any path of integration within the core,

$$\int \overline{H} \cdot d\overline{\ell} = 0 = N_1 i_1 - N_2 i_2 \tag{1.99}$$

$$\frac{i_1}{i_2} = \frac{N_2}{N_1} \tag{1.100}$$

We can show from (1.98) and (1.100) that

$$e_1 i_1 = e_2 i_2 \tag{1.101}$$

Thus the instantaneous power into the transformer is equal to the instantaneous power out of the transformer, implying lossless operation. This is to be expected, in that we have assumed that all sources of losses are zero.

FUNDAMENTAL CONCEPTS

EXAMPLE 1.7 At time $t = 0$ a constant (dc) voltage of V volts is applied to the primary of the transformer shown in Figure 1.41. The secondary is left an open circuit. Assume ideal transformer relationships and determine i_1, i_2, e_2, and φ.

Since the secondary is an open circuit, $i_2 = 0$. The ideal transformer relationships would thus require that $i_1 = 0$. Since $e_1 = V$ for $t \geq 0$, the ideal transformer relationships would require that $e_2 = N_2/N_1$ V. We must find φ by Faraday's law:

$$\varphi(t) = \frac{1}{N_1} \int_0^t V \, d\tau + \varphi(0)$$

Assuming zero flux in the core at time $t = 0$, we have

$$\varphi(t) = \frac{V}{N_1} t$$

Thus magnetizing flux increases linearly with respect to time as long as $e_1 = V$. An obvious practical limitation presents itself. Magnetic materials eventually saturate and φ would approach a constant value. Time rate of change of flux would approach zero and Faraday's law would require that e_1 (and e_2) be zero under that condition. If we were to connect a well-regulated dc source to a real power transformer in an attempt to maintain constant voltage across a winding, the entire applied dc voltage would eventually have to be countered by the winding resistance voltage drop. For a large applied voltage and low winding resistance our transformer would become more of a heating device than an effective power transformer!

Equations (1.98) and (1.100) imply simple linear relationships between primary and secondary currents and primary and secondary voltages. Thus we can consider the special case of steady-state sinusoidal time variation by simply replacing appropriate time-varying quantities with their phasor representations in (1.98) and (1.100):

$$\frac{\underline{E}_1}{\underline{E}_2} = \frac{N_1}{N_2} \tag{1.102}$$

$$\frac{\underline{I}_1}{\underline{I}_2} = \frac{N_2}{N_1} \tag{1.103}$$

Figure 1.42 shows a source feeding a load impedance through an ideal transformer. In the figure we use the typical ideal transformer symbol associated with circuit analysis. Equations (1.102) and (1.103) are sufficient to characterize the ideal transformer, but it is often helpful to recognize that the effect of the equations is to cause quantities on one side of the transformer to appear as "magnified" by the appropriate turns ratio or ratio squared when viewed from the other side of the transformer. For example, (1.102) and (1.103) can be used to show that for the circuit of Figure 1.42,

$$\frac{\underline{E}_1}{\underline{I}_1} = \left(\frac{N_1}{N_2}\right)^2 \frac{\underline{E}_2}{\underline{I}_2} = \left(\frac{N_1}{N_2}\right)^2 Z_L \tag{1.104}$$

Thus for analysis we could replace the circuit of Figure 1.42 with the circuit of Figure 1.43. Similarly, we could refer quantities to the other side of the circuit, as shown in Figure 1.44.

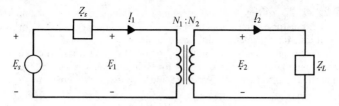

FIGURE 1.42

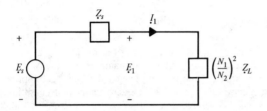

FIGURE 1.43

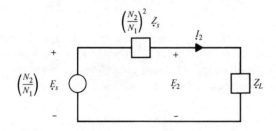

FIGURE 1.44

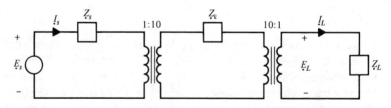

FIGURE 1.45

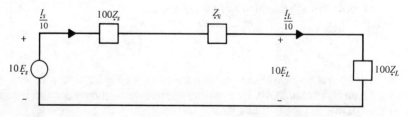

FIGURE 1.46

FUNDAMENTAL CONCEPTS

EXAMPLE 1.8   Consider the two-transformer circuit shown in Figure 1.45. Refer all quantities to the high-voltage circuit assuming ideal transformer relationships.

The circuit with quantities referred is shown in Figure 1.46. For given values of E_s, Z_{ts}, Z_{te}, and Z_L, a solution would be trivial. If we compare the circuit of Figure 1.46 with what we would have without the transformers, the fundamental benefit of the power transformers is evident: reduced line losses and line voltage drop in the high-tension circuit for a given high-tension circuit conductor size.

We can find an improved representation for the power transformer by simply relaxing the assumptions associated with the ideal transformer. A good starting point is to consider behavior under a no-load condition, that is, with the secondary an open circuit. With $I_2 = 0$ we should have $I_1 = 0$ by the ideal transformer relationship regardless of the value of E_1. In reality we physically have an inductor consisting of N_1 turns wrapped around a core of relatively high, but not infinite permeability (see Figure 1.41). If we relax the assumption of infinite permeability and allow for some core loss, we have the same configuration that was analyzed in section 1.6. Accordingly, the equivalent circuit shown in Figure 1.47 is appropriate. If the behavior of the transformer is close to that of an ideal transformer, we would expect the parallel combination of R_c and jX_m to constitute a relatively large impedance. Since the circuit is predominately inductive we would expect I_c to be small in magnitude compared to I_m.

Now consider the transformer with the secondary short circuited (i.e., with $E_2 = 0$). Ideal transformer relationships require that $E_1 = 0$. Thus the transformer should appear as a short circuit when viewed from the primary side. Note that we could arrive at the same conclusion by considering impedance at the secondary terminals referred to the primary of the transformer. Now if we relax the assumption of zero leakage flux and zero winding resistance, we expect to see a small but nonzero impedance at the terminals of the primary. We thus might consider the circuit shown in Figure 1.48. In the figure we have combined the effects of leakage flux and winding resistance for both the primary and secondary windings into a single leakage impedance $Z_{eq} = R_{eq} + jX_{eq}$.

In Figure 1.48 we have neglected the effects of R_c and jX_m since we expect their equivalent impedance to be large compared to the Z_{eq} we see with the secondary short circuited. Similarly, in Figure 1.47 we have neglected the effects of Z_{eq} when the secondary is an open circuit. Finally, we recognize that when the secondary is short circuited, the equivalent leakage impedance Z_{eq} we see at the primary must consist of primary effects and secondary effects referred to the primary. Thus we surmise that

$$R_{eq} = r_1 + \left(\frac{N_1}{N_2}\right)^2 r_2 \qquad (1.105)$$

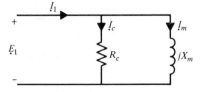

FIGURE 1.47 Equivalent Circuit for a Power Transformer with $I_2 = 0$

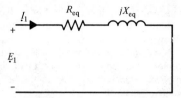

FIGURE 1.48 Equivalent Circuit for a Power Transformer with $E_2 = 0$

where r_1 and r_2 are the primary and secondary winding resistances, respectively, and that

$$X_{eq} = X_1 + \left(\frac{N_1}{N_2}\right)^2 X_2 \tag{1.106}$$

where X_1 and X_2 are the primary and secondary winding equivalent leakage reactances, respectively. An equivalent circuit that reproduces the correct no-load (approximately) and short-circuited conditions is shown in Figure 1.49.

From what we have said thus far, there is no particular reason to put the leakage impedance on one side or the other of the parallel combination of R_c and jX_m. Because of the relative magnitudes of the various parameters, it makes little difference. In the next section we will show in a more rigorous manner that the circuit shown in Figure 1.50 yields a more accurate description of actual behavior. In most situations the circuit shown in Figure 1.49 is considerably easier to work with, and it is adequate for most analyses.

EXAMPLE 1.9   Tests are conducted on a 10-kVA 120/480-V transformer with results as follows:

Open-circuit test (data taken on the low-voltage side):

$$V = 120 \text{ V} \quad \text{(rated low-side voltage)}$$
$$I = 4 \text{ A}$$
$$P = 75 \text{ W}$$

Short-circuit test (data taken on the high-voltage side):

$$V = 48 \text{ V}$$
$$I = 20.83 \text{ A} \quad \text{(rated high-side current)}$$
$$P = 500 \text{ W}$$

Find an equivalent circuit for the transformer referred to the high-voltage winding.

From the open-circuit test we can determine R_c and X_m. Neglecting leakage impedance in the equivalent circuit of Figure 1.49, we have the situation depicted in Figure 1.51.

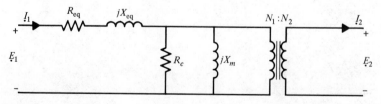

FIGURE 1.49

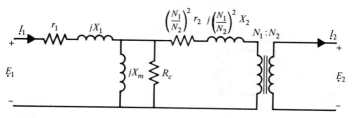

FIGURE 1.50

Thus

$$P = \frac{V^2}{R_c}$$

and

$$R_c = \frac{(120)^2}{75} = 192 \ \Omega$$

$$I_c = \frac{V}{R_c} = \frac{120}{192} = 0.625 \ \text{A}$$

$$I_m = \sqrt{I^2 - I_c^2} = \sqrt{(4)^2 - (0.625)^2} = 3.95 \ \text{A}$$

$$X_m = \frac{V}{I_m} = \frac{120}{3.95} = 30.4 \ \Omega$$

Referring R_c and X_m to the high side, we have

$$R_c' = \left(\frac{N_2}{N_2}\right)^2 R_c = \left(\frac{480}{120}\right)^2 (192) = 3072 \ \Omega$$

$$X_m' = \left(\frac{N_2}{N_1}\right)^2 X_m = \left(\frac{480}{120}\right)^2 (30.4) = 486 \ \Omega$$

From the short-circuit test we can determine R_{eq} and X_{eq}. During the short-circuit test we have the situation depicted in Figure 1.52. Thus

$$P = I^2 R_{eq}$$

and

$$R_{eq} = \frac{P}{I^2} = \frac{500}{(20.83)^2} = 1.15 \ \Omega$$

$$X_{eq} = \sqrt{Z_{eq}^2 - R_{eq}^2} = \sqrt{\left(\frac{48}{20.83}\right)^2 - (1.15)^2} = 1.99 \ \Omega$$

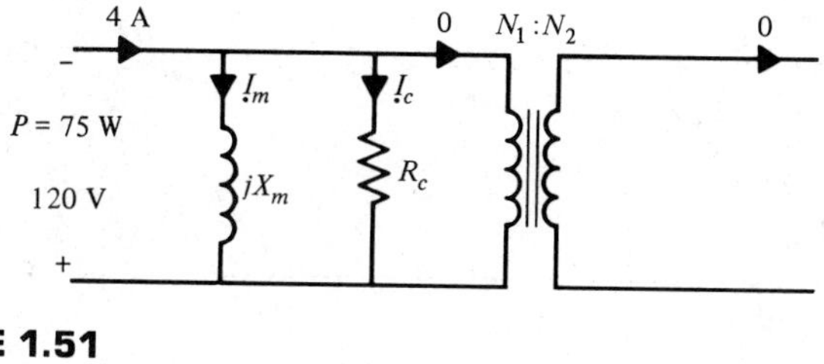

FIGURE 1.51

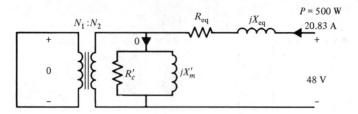

FIGURE 1.52

The equivalent circuit with all parameters referred to the high side is shown in Figure 1.53.

The tests described in this example are commonly used to determine parameters for an equivalent circuit for the power transformer.

1.9

POWER TRANSFORMER DESIGN CONSIDERATIONS

In Section 1.8 we relaxed the assumptions associated with the ideal transformer in order to develop a reasonable equivalent circuit structure. Example 1.9 showed that relatively simple measurements made at the transformer terminals could be used to establish parameter values. From a design viewpoint the power transformer designer must choose appropriate materials, geometric configurations, and dimensions to achieve a workable product. Generally, we would like to approach ideal transformer behavior. Thus we would like to minimize leakage impedance Z_{eq} and maximize core-loss impedance (the parallel combination of R_c and jX_m). Our objective in this section will be to develop some appreciation for the relationship between transformer equivalent-circuit parameter values and the dimensions and material properties of the device itself.

Figure 1.54 shows symbolically a two-winding core form transformer similar to the configurations discussed thus far. Figure 1.54 shows current directions into and out of the page and magnetizing flux φ assigned as before. Figure 1.55 shows dimensions of the transformer for both side and top views. Note that the winding material from windings 1 and 2 completely fills the "window" of the core. This window has area A_w where

$$A_w = h_w w_w \qquad (1.107)$$

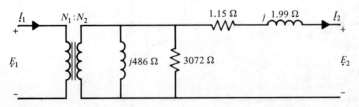

FIGURE 1.53

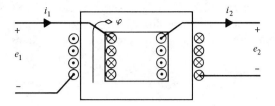

FIGURE 1.54

We can relate other dimensions in the figure to dimensions we have previously associated with similar core geometries. The cross-sectional area of the core A_c is

$$A_c = h_c w_c \tag{1.108}$$

The mean flux path length ℓ_c is

$$\ell_c = 2(w_w + w_c + h_w + w_c) \tag{1.109}$$

$$= 2w_w + 2h_w + 4w_c$$

The core volume V_c is

$$V_c = A_c \ell_c \tag{1.110}$$

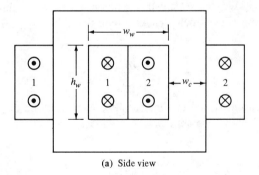

(a) Side view

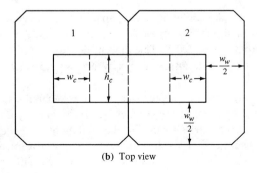

(b) Top view

FIGURE 1.55

We can use Faraday's law to establish a relationship between induced voltage, core cross-sectional area, and maximum flux density. Neglecting leakage flux and assuming a uniform flux density over the core cross section yields

$$e_1 = \frac{d\lambda_1}{dt} = N_1 \frac{d\varphi}{dt} = N_1 A_c \frac{dB}{dt} \tag{1.111}$$

If we are dealing with steady-state sinusoidal time variation, the rms voltage E_1 is given by

$$E_1 = \omega N_1 A_c \frac{B_{max}}{\sqrt{2}}$$
$$= 4.44\, f N_1 A_c B_{max} \tag{1.112}$$

[The reader will recall that this is nothing more than the same development we used in Section 1.6; see (1.47) and (1.48).] Similarly, the rms voltage E_2 is given by

$$E_2 = 4.44 f N_2 A_c B_{max} \tag{1.113}$$

Note that the ratio of voltage to turns is the same on either side of the transformer:

$$\frac{E_1}{N_1} = \frac{E_2}{N_2} = 4.44 f A_c B_{max} \tag{1.114}$$

Referring to Figure 1.55, we can establish a relationship between mmf NI, window area, and winding current density. Assuming that $J_1 = J_2 = J$ gives us

$$N_1 I_1 = \frac{J h_w w_w}{2} = \frac{J A_w}{2} \tag{1.115}$$

Note that, in (1.115), current I_1 and current density J must both be expressed as rms values. Note further that we anticipate, if the design is reasonable, that $N_1 I_1 \simeq N_2 I_2$. Thus

$$N_1 I_1 = N_2 I_2 = \frac{J A_w}{2} \tag{1.116}$$

Equations (1.114) and (1.116) are similar in that the maximum flux density B_{max} and current density J will be limited by physical considerations. B_{max} is limited by saturation to values from 1.0 to 2.0 T. The current density J is limited by our capability to dissipate heat due to losses. Typically, we might have J limited to 1.0 to 8.0 A/mm^2 rms.

The range of values is large because sophisticated cooling techniques can be used to dissipate the heat. Large power transformers generally employ circulating oil or water and a heat exchanger system. For the configuration depicted in Figure 1.55 we would expect to be on the low end of the range, since we have allowed no space for a coolant. Combining (1.114) and (1.116), the volt-ampere rating of the transformer will be

$$VA = E_1I_1 = E_2I_2 = \frac{4.44fA_cA_wB_{max}J}{2} \qquad (1.117)$$

EXAMPLE 1.10 Determine the volt-ampere rating of a 60-Hz power transformer whose dimensions are

$$w_c = 0.4 \text{ m}$$
$$h_c = 0.4 \text{ m}$$
$$w_w = 0.6 \text{ m}$$
$$h_w = 0.6 \text{ m}$$

Assume that the current density and flux density are limited to

$$J = 1.0 \text{ A/mm}^2$$
$$B_{max} = 1.5 \text{ T}$$

Applying (1.117), we obtain

$$VA = \frac{(4.44)(60)(0.4)^2(0.6)^2(1.5)(1.0 \times 10^6)}{2}$$
$$= 11.5(10^6) \text{ VA}$$
$$= 11.5 \text{ MVA}$$

The actual capacity will be somewhat less in that the effective cross-sectional areas A_c and A_w will be reduced by practical considerations. For example, some space must be allowed for winding insulation and spacing between high- and low-voltage windings. This will reduce A_W.

The design problem is inherently more difficult. That is, given a desired rating, the problem of finding core and window dimensions presents us with a large range of alternatives. Rules of thumb, experience, and empirical design formulas must generally be relied on. References 3–5 are helpful in this regard.

A transformer of the volt-ampere rating we have calculated might typically be one of three single-phase transformers used in a three-phase bank of transformers. Typical voltages might be 69 kV/$\sqrt{3}$ = 39.8 kV on the high side and 13.8 kV/$\sqrt{3}$ = 7.97 kV on the low side.†

EXAMPLE 1.11 For an assumed resistivity of $\rho = 1.724(10^{-8})$ $\Omega{\cdot}$m at 20°C, find the winding ohmic losses for the transformer shown in Figure 1.55. Assume the dimensions and values specified for Example 1.10.

A typical transformer operating temperature might be 80°C. Correcting resistivity for this temperature,

$$\rho = 1.724[1 + \alpha(80° - 20°)](10^{-8})$$
$$= 2.13(10^{-8}) \text{ }\Omega{\cdot}\text{m}$$

where the temperature coefficient of resistance α for copper is 0.00393. The losses per unit of volume in the winding material will be

$$p = \rho J^2 = 2.13(10^{-8})(10^6)^2 = 2.13(10^4) \text{ W/m}^3$$

---

†The voltages 69 kV and 13.8 kV are standard line-to-line voltages for power transmission. If the transformers are Y-connected on both high- and low-voltage sides, the line-to-neutral voltages are lower by a factor of $1/\sqrt{3}$. Appendixes D and E provide a review of three-phase circuit fundamentals.

If we refer to Figure 1.55, we see that the volume of the winding material V_w is

$$V_w = 2h_w \frac{w_w}{2}\left(h_c + \frac{w_w}{2} + w_c + \frac{w_w}{2}\right)(2) = 2A_w(h_c + w_c + w_w)$$

$$= 2(0.36)(0.4 + 0.4 + 0.6) = 1.008 \text{ m}^3$$

Thus the total winding losses would be

$$p = pV_w = 2.13(10^4)(1.008) = 21.47 \text{ kW}$$

Knowing the winding losses, we can find values of lumped resistance for each winding to account for these losses (see Problem 1.27). It is clear that the winding losses in our example are shared equally by each winding. Since the secondary current referred to the primary is approximately the same as primary current, it must be true that the secondary winding resistance referred to the primary will be approximately equal to the primary winding resistance, and vice versa (see Figure 1.50 to clarify this). In our example the winding losses are actually smaller than what we might expect in an 11.5-MVA transformer due to our assumed low current density. We have somewhat improved efficiency at the expense of transformer capacity.

The core-loss parameters R_c and X_m can be found as described in Section 1.6. Since we know the maximum flux density and frequency, we can find the core losses and rms volt-amperes per unit of volume (or weight) from manufacturers' curves. Knowing the core volume (or weight), we can find the total core losses and determine appropriate values of R_c and X_m. Example 1.3 illustrated the procedure.

We have thus far related winding resistances, r_1 and r_2, and core-loss parameters X_m and R_c, to physical properties of the transformer. We now turn our attention to the approximate determination of leakage reactances. As we will see, the leakage reactance of the transformer plays a major role in the choice of winding configuration and core structure. In fact, problems associated with leakage reactance are the main reason the configuration shown in Figure 1.55 would be a poor design for many applications.

We will find appropriate values of leakage reactance (or equivalently leakage inductance) by determining the stored energy associated with the leakage fields and equating it to the energy associated with the leakage inductances we assign to the transformer equivalent circuit. To do this we start by viewing the transformer as a set of coupled coils and write the following flux linkage relationships for the transformer of Figures 1.54 and 1.55:

$$\lambda_1 = \lambda_{\ell 11} + \lambda_{\ell 12} + N_1\varphi \tag{1.118}$$

$$\lambda_2 = \lambda_{\ell 21} + \lambda_{\ell 22} + N_2\varphi \tag{1.119}$$

In (1.118) and (1.119), λ_1 and λ_2 are the total flux linkages in windings 1 and 2, respectively. The terms on the right-hand side are the contributions to the total flux linkages. $\lambda_{\ell 11}$ represents the flux linkages in winding 1 due to leakage flux caused by the current in winding 1. $\lambda_{\ell 12}$ represents the flux linkages in winding 1 due to leakage flux caused by the current in winding 2. $\lambda_{\ell 21}$ and $\lambda_{\ell 22}$ are defined similarly. φ is the magnetizing flux confined to the core assumed to link all turns of both windings. Thus the flux linkages in winding 1 due to φ are given by $N_1φ$ and the flux linkages in winding 2 due to φ are given by $N_2φ$. If we assume linear behavior, we can write

$$\lambda_1 = L_{\ell 11}i_1 - L_{\ell 12}i_2 + N_1 \frac{N_1 i_1 - N_2 i_2}{R_m} \tag{1.120}$$

$$\lambda_2 = L_{\ell 21}i_1 - L_{\ell 22}i_2 + N_2 \frac{N_1 i_1 - N_2 i_2}{R_m} \tag{1.121}$$

where R_m is the reluctance of the core itself. The minus sign appears due to our assignment of current directions in accordance with power transformer convention.

EXAMPLE 1.12   Show that in (1.120) and (1.121),

$$L_{\ell 12} = L_{\ell 21}$$

Rearranging and collecting terms, we can rewrite (1.120) and (1.121) as

$$\lambda_1 = \left[L_{\ell 11} + \frac{N_1^2}{R_m} \right] i_1 - \left[L_{\ell 12} + \frac{N_1 N_2}{R_m} \right] i_2$$

$$\lambda_2 = \left[L_{\ell 21} + \frac{N_2 N_1}{R_m} \right] i_1 - \left[L_{\ell 22} + \frac{N_2^2}{R_m} \right] i_2$$

This is in the form

$$\lambda_1 = L_{11}i_1 + L_{12}i_2$$

$$\lambda_2 = L_{21}i_1 + L_{22}i_2$$

But as shown in Example 1.5, we must have $L_{12} = L_{21}$. Thus

$$L_{\ell 12} + \frac{N_1 N_2}{R_m} = L_{\ell 21} + \frac{N_1 N_2}{R_m}$$

$$L_{\ell 12} = L_{\ell 21}$$

Defining $L_{\ell 12} = L_{\ell 21} = M_\ell$, we can rewrite (1.120 and 1.121) as

$$\lambda_1 = L_{\ell 11}i_1 - M_\ell i_2 + N_1 \frac{N_1 i_1 - N_2 i_2}{R_m} \tag{1.122}$$

$$\lambda_2 = M_\ell i_1 - L_{\ell 22}i_2 + N_2 \frac{N_1 i_1 - N_2 i_2}{R_m} \tag{1.123}$$

Now refer variables from winding 2 to winding 1 in the following way:

$$i_2' = \frac{N_2}{N_1} i_2 \tag{1.124}$$

$$\lambda_2' = \frac{N_1}{N_2} \lambda_2 \tag{1.125}$$

Substituting from (1.124) and (1.125) into (1.122) and (1.123), we have

$$\lambda_1 = L_{\ell 11} i_1 - \left(\frac{N_1}{N_2} M_\ell\right) i_2' + N_1 \frac{N_1 i_1 - N_1 i_2'}{R_m} \tag{1.126}$$

$$\frac{N_2}{N_1} \lambda_2' = M_\ell i_1 - \left(\frac{N_1}{N_2} L_{\ell 22}\right) i_2' + N_2 \frac{N_1 i_1 - N_1 i_2'}{R_m} \tag{1.127}$$

Now define $L_{\ell 22}' = (N_1/N_2)^2 L_{\ell 22}$ and simply rearrange (1.126) and (1.127):

$$\lambda_1 = L_{\ell 11} i_1 + \frac{N_1^2}{R_m}(i_1 - i_2') - \left(\frac{N_1}{N_2} M_\ell\right) i_2' \tag{1.128}$$

$$\lambda_2' = \left(\frac{N_1}{N_2} M_\ell\right) i_1 - L_{\ell 22}' i_2' + \frac{N_1^2}{R_m}(i_1 - i_2') \tag{1.129}$$

The simple equivalent circuit in Figure 1.56 relates the variables shown in (1.128) and (1.129). [It may be helpful to differentiate both sides of (1.128) and (1.129) with respect to time to see this.] The self-leakage inductance of winding 2 referred to winding 1 (i.e., $L_{\ell 22}'$), will be approximately equal to the self-leakage inductance of winding 1, $L_{\ell 11}$. We may thus simplify the circuit of Figure 1.56 by defining

$$L_\ell = L_{\ell 11} - \frac{N_1}{N_2} M_\ell \approx L_{\ell 22}' - \frac{N_1}{N_2} M_\ell \tag{1.130}$$

We finally arrive at the equivalent circuit shown in Figure 1.57.

Consideration of the equivalent circuit permits some interesting observations. In a power transformer R_m will be small. We would expect that

$$\frac{N_1^2}{R_m} >> \frac{N_1}{N_2} M_\ell$$

If we have infinite permeability, $R_m = \ell_c/\mu A_c$ would be zero and the center leg of our equivalent circuit would present an infinitely large impedance for steady-state ac operation. If the leakage flux is zero, $L_\ell = 0$, and we arrive back at our ideal transformer. The equivalent circuit of Figure 1.57 can be improved if we take into account winding resistances and account for core losses with a resistance R_c. The reader should compare the equivalent circuit of Figure 1.57 with the one we developed heuristically in Figure 1.50 for steady-state ac operation. The leakage inductance L_ℓ is an equivalent leakage inductance that takes into account mutual coupling associated with the leakage fields. It is

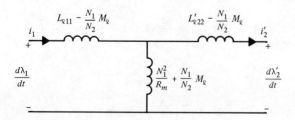

FIGURE 1.56

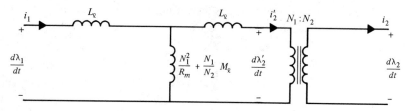

FIGURE 1.57

reasonable to expect that $i_1 \approx i_2'$ and stored energy associated with the leakage inductance term in the center leg of the transformer would be negligible. That is,

$$\frac{1}{2}\left(\frac{N_1}{N_2} M_\ell\right)(i_1 - i_2')^2 \approx 0$$

We could thus state that stored energy W_ℓ associated with the leakage inductance in the equivalent circuit is approximately

$$W_\ell \approx \tfrac{1}{2} L_\ell i_1^2 + \tfrac{1}{2} L_\ell i_2'^2 \tag{1.131a}$$

$$\approx L_\ell i_1^2$$

To relate the inductance L_ℓ to physical dimensions we will find stored energy by integrating the energy density associated with the leakage fields over the volume occupied by the leakage fields:

$$W_\ell = \int_v \frac{u_0 H_\ell^2 \, dv}{2} \tag{1.131b}$$

If we neglect fields external to the transformer and concentrate our attention on the leakage flux in the window of the transformer, we can evaluate (1.131b) in an approximate way. The top half of Figure 1.55 is reproduced as Figure 1.58. We can apply Ampère's law to find the H field in the window.

$$\oint \overline{H} \cdot \overline{d\ell} = \int_s \overline{J} \cdot \overline{ds} \tag{1.132}$$

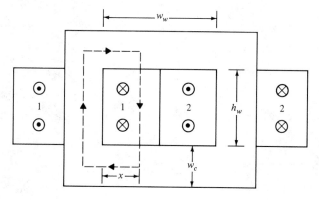

FIGURE 1.58

Choose a path of integration as shown in Figure 1.58 and assume that $H \approx 0$ in the core and constant in the window. Then the left-hand side of (1.132) becomes

$$\oint \overline{H} \cdot \overline{d\ell} = H_\ell h_w \tag{1.133}$$

The right-hand side of (1.132) is for $0 \le x \le w_w/2$,

$$\int_s \overline{J} \cdot \overline{ds} = J h_w x \tag{1.134}$$

For $w_w/2 \le x \le w_w$ the right-hand side of (1.132) is

$$\int_s \overline{J} \cdot \overline{ds} = \frac{J h_w w_w}{2} - J h_w \left(x - \frac{w_w}{2} \right) \tag{1.135}$$

By equating (1.133) and (1.134) or (1.135), we find H_ℓ in the window:

$$H_\ell = Jx \qquad\qquad 0 \le x \le \frac{w_w}{2} \tag{1.136}$$

$$H_\ell = J(w_w - x) \qquad \frac{w_w}{2} < x \le w_w \tag{1.137}$$

Figure 1.59 shows the relationship between H_ℓ and x within the window. Since $Jw_w h_w/2 = N_1 i_1$, we see that the maximum value of H_ℓ is

$$H_{\ell max} = \frac{N_1 i_1}{h_w} \tag{1.138}$$

We can now evaluate the stored energy associated with the window area by (1.131b):

$$W_\ell = (2) \int_0^{w_w/2} \frac{\mu_0}{2} \left(H_{\ell max} \frac{2x}{w_w} \right)^2 h_c h_w \, dx$$
$$= \frac{\mu_0 H_{\ell max}^2 h_c h_w w_w}{6} \tag{1.139}$$

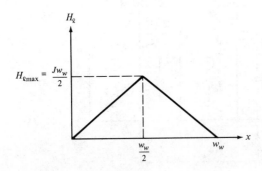

FIGURE 1.59

Now substitute for $H_{\ell max}$ using (1.138):

$$W_\ell = \frac{\mu_0}{6} \frac{h_c}{h_w} w_w (N_1 i_1)^2 \tag{1.140}$$

Equating (1.131a) and (1.140), we can solve for L_ℓ:

$$L_\ell = N_1^2 \frac{\mu_0}{6} \frac{h_c w_w}{h_w} \tag{1.141}$$

Leakage fields external to the window of the transformer can significantly increase the leakage inductance L_ℓ. Equation (1.141) must therefore be regarded as a lower limit on the value of the leakage inductance. More detailed procedures for estimating leakage inductance are given in references 3–5.

EXAMPLE 1.13 Determine the rms voltage that would be developed across the total leakage reactance of the transformer during normal operation for the transformer of Figure 1.55. Express this answer as a percentage of the rated rms voltage of the transformer. Evaluate the result numerically for the data specified in Example 1.10.

Making reasonable simplifying assumptions, we modify the equivalent circuit shown in Figure 1.57 to obtain (for steady-state ac operation) the circuit shown in Figure 1.60. We see from the figure that the total leakage reactance X_{eq} is $2\omega L_\ell$. The result we seek expressed as a ratio is

$$\frac{X_{eq} I_{1rated}}{E_{1rated}}$$

This can be rewritten as

$$\frac{X_{eq}}{E_{1rated}/I_{1rated}} = \frac{X_{eq}}{Z_{base}}$$

which is the per unit transformer reactance on the transformer rating as base.

Substituting using (1.114) and (1.116) to determine rated values and (1.141) for L_ℓ, we obtain

$$\frac{X_{eq} I_{1rated}}{E_{1rated}} = \left(2\omega N_1^2 \frac{\mu_0}{6} \frac{h_c w_w}{h_w}\right) \frac{J A_w}{2N_1} \frac{1}{4.44 N_1 f A_c B_{max}}$$

$$= \frac{\mu_0 w_w^2}{3\sqrt{2}} \frac{J}{w_c} \frac{J}{B_{max}}$$

Now we evaluate for the data given in Example 1.10:

$$\frac{X_{eq} I_{1rated}}{E_{1rated}} = \frac{(4\pi \times 10^{-7})(0.6)^2}{3\sqrt{2}(0.4)} \frac{10^6}{1.5}$$

$$= 0.178$$

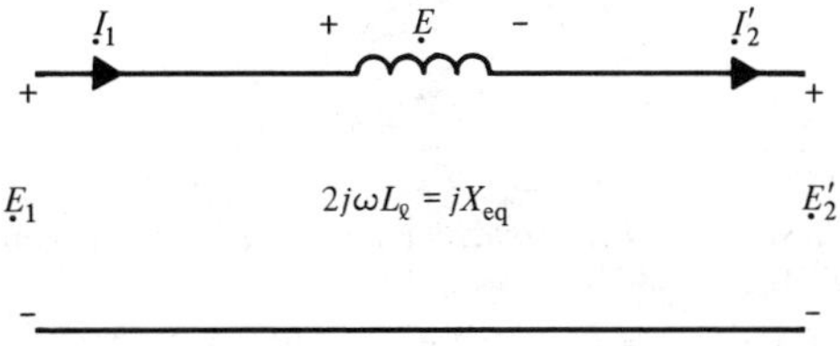

FIGURE 1.60

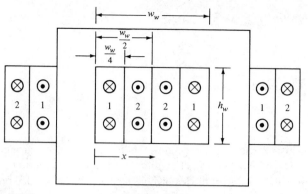

FIGURE 1.61

Thus the rms voltage developed across the leakage reactance is 17.8% of the rated transformer rms voltage. (The per unit leakage reactance is 17.8%.) If leakage fields external to the window are included, the reactance can be shown to be roughly three times higher than what we have calculated for this particular set of dimensions. This implies a reactance roughly five times as high as what would be expected for a well-designed transformer. The leakage reactance can be reduced significantly by increasing h_w [see (1.141)] and by altering the winding arrangement. By putting half of each of the winding turns on each of the two legs we can reduce considerably the average energy density in the window. Figure 1.61 shows this type of winding arrangement for the same core structure as that shown in Figures 1.55 and 1.58. Now by repeating the development for H_ℓ in the window we can show that H_ℓ as a function of x will be as shown in Figure 1.62. $H_{\ell max}$ can be shown to be

$$H_{\ell max} = \frac{N_1 i_1}{2h_w}$$

This winding arrangement reduces the energy associated with the leakage fields in the window and hence the total leakage reactance by a factor of 4 (see Problem 1.32).

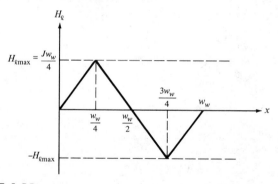

FIGURE 1.62

REFERENCES

[1] S. A. NASAR and L. E. UNNEWEHR, *Electromechanics and Electric Machines.* New York: John Wiley & Sons, Inc., 1983.

[2] G. R. SLEMON and A. STRAUGHEN, *Electric Machines.* Reading, Mass.: Addison-Wesley Publishing Co., Inc., 1980.

[3] DONALD G. FINK and H. WAYNE BEATY, *Standard Handbook for Electrical Engineers,* 11th ed. New York: McGraw-Hill Book Company, 1978.

[4] L. F. BLUME et al., *Transformer Engineering,* 2nd ed. New York: John Wiley & Sons, Inc., 1951.

[5] A. Still, *Elements of Electrical Design,* 2nd ed. New York: McGraw-Hill Book Company, 1932.

PROBLEMS

1.1   Find an expression for the magnetic field intensity H external to an infinitely long wire carrying current i. The expression should be in terms of distance from the centerline of the wire. Submit a simple sketch to show the relative directions and orientations of current and the magnetic field intensity.

1.2   Suppose that, in Figure 1.3, $r_o - r_i = 0.2r_i$. Determine the percent error incurred by calculating φ in the magnetic circuit by (1.11), which assumes a uniform flux density across the core cross section.

1.3   An infinitely long wire carrying a current of 1.0 A establishes a flux in the rectangular cross section of the core surrounding the wire as shown in the figure. If the relative permeability of the core material is 10^3, find the total core flux.

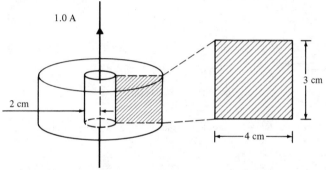

1.4   Consider the following magnetic circuit and magnetization curve for its core material. Make reasonable simplifying assumptions and determine the current required to establish a flux of $3.0(10^{-4})$ Wb in the core if the winding has 400 turns.

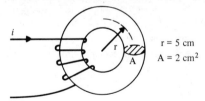

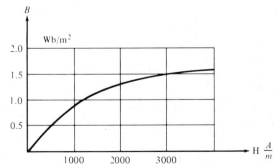

1.5 Consider the magnetic circuit and magnetization curve for Problem 1.4. If the magnetic material is replaced with material that has a linear B–H relationship where relative permeability μ_r is 500, find the current required to establish a flux of $3.0(10^{-4})$ Wb.

1.6 Neglect fringing and leakage flux and determine the mmf required to establish a flux density of 1.5 Wb/m² in the center leg of the magnetic circuit shown in the figure. The soft-iron core material is characterized by the magnetization curve of Figure 1.14. The depth of the core (into the page) is the same over the cross section.

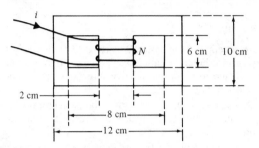

1.7 Consider the following magnetic circuit.

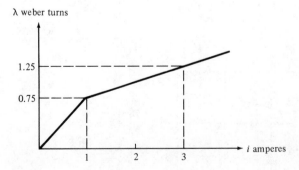

$A_c = 4$ cm²
$l_c = 25$ cm
$N = 100$ turns
$u_r = 10^3$
$u_o = 4\pi \times 10^{-7}\ \frac{H}{m}$

Neglect the leakage flux and winding resistance and assume linear core material. Steady-state current is $i(t) = 5 \cos 100t$ amperes.
a. Find flux $\varphi(t)$ in the core.
b. Find flux linkages $\lambda(t)$.
c. What is the inductance of the device?
d. Find voltage $e(t)$.

1.8 Consider the magnetic circuit shown in Figure 1.15. Assume that the core material has relative permeability $\mu_r = 3000$. Core cross-sectional area A_c is 2 cm² and the mean core length ℓ_c is 5 cm. The coil has 200 turns. If $e(t) = 10 \cos 100t$ V, find an expression for the steady-state flux $\varphi(t)$ and the steady-state current $i(t)$.

1.9 The λ versus i curve for a certain nonlinear inductor can be approximated by the following piecewise linear curve.

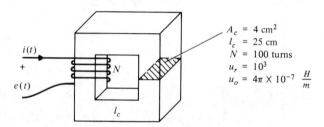

If the current is constant at $i = 2.0$ amperes, what is the stored energy in the inductor?

1.10 The relationship between flux linkages λ and current i for a certain nonlinear inductor is to be approximated by

$$\lambda = 3(1 - e^{-2i}) \qquad \text{Wb} \cdot \text{t}$$

Find the stored energy in the inductor for a current i of 0.5 A.

1.11 Consider the magnetic circuit configuration shown in the figure. Assume linear iron behavior where the iron has high but not infinite permeability. Neglect fringing and leakage flux. Find an expression for the ratio of energy stored in the air gap to energy stored in the core material. Evaluate (numerically) your result for the magnetic circuit of Example 1.1.

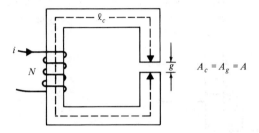

$A_c = A_g = A$

1.12 Refer to Figure 1.21, which presents a graphical depiction of stored energy relationships for a magnetic circuit. Assume linear behavior of the core material, that is, constant L, μ_r, and so on. Figure 1.21(a) relates variables commonly associated with an electric circuit. Find an expression for stored energy in terms of inductance and current. Find a second expression in terms of inductance and flux linkages. Repeat the exercise for Figure 1.21(b) and (c) using reluctance and permeability, respectively, as parameters.

1.13 Measurements made at the terminals of the circuit shown yield:

$$I = 2 \text{ amperes rms}$$
$$V = 120 \text{ volts rms}$$
$$P = 100 \text{ watts}$$

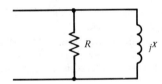

where current and voltage are sinusoidal at 60 Hz and P is the average real power. Find R and X.

1.14 The applied voltage at the terminals of the circuit in the figure is $e(t) = \sqrt{2}\,(117) \cos 377t$ V. Find:

a. The real power P consumed by the circuit.
b. The reactive power Q consumed by the circuit.
c. The power factor.
d. An expression for steady-state current $i(t)$.

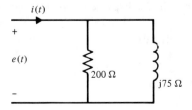

1.15   Suppose that the magnitude of the electric field intensity E varies linearly from zero at the center of a lamination (see Figure 1.32) to a maximum value of $E = (b/2n)(dB/dt)$ at the periphery of the lamination; that is,

$$E(y) = \frac{dB}{dt} y \qquad 0 \le y \le \frac{b}{2n}$$

Show that the average value of $E^2(y)$ over this region of space is given by (1.60).

1.16   A magnetic circuit is constructed using the laminated material characterized by Figure 1.34. The core cross-sectional area is 9 cm², the core length is 15 cm, and the winding has 300 turns. If a sinusoidal voltage $e(t) = \sqrt{2}\,(50) \cos 377t$ V exists at the terminals of the coil, find the core losses. (Assume a density for the material of 7500 kg/m³.)

1.17   A three-winding transformer configuration is shown in the accompanying figure.

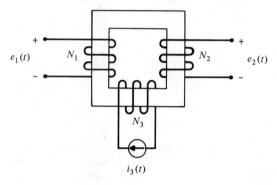

$N_1 = 100$ turns

$N_2 = 200$ turns

$N_3 = 50$ turns

$\mathcal{R} = $ reluctance of the core $= 10^5$ amp turns/weber

Find expressions for the induced voltages $e_1(t)$ and $e_2(t)$ if $i_3(t) = \sqrt{2}\,(10) \sin 10t$ amps and $i_1(t) = i_2(t) = 0$, that is, coils 1 and 2 are open circuits. Neglect the leakage flux and winding resistances.

1.18   Consider the coupled coil configuration shown in the figure, where core material has high relative permeability. Neglect the leakage flux and find all possible values of inductance that might be realized by connecting the two windings in series. Express your answer(s) in terms of N_1, N_2, and core reluctance $\mathcal{R}$.

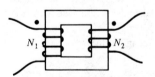

1.19   Consider the two-winding linear coupled-coil system of Example 1.6. Find the stored energy in the system for $i_1 = I_1$ and $i_2 = I_2$ by finding the stored energy associated with each of the inductances of the equivalent circuit shown in Figure 1.40.

1.20   Consider the two-winding linear system described in Example 1.6. Assume the following parameter values:

$$L_{11} = 10 \text{ mH}$$

$$L_{22} = 16 \text{ mH}$$

$$M = 3 \text{ mH}$$

One winding of the system is left an open circuit such that $i_2 = 0$. A 10-V rms 1000-Hz sinusoidal voltage is applied to the other winding. Find the rms current in this winding and the rms voltage across the terminals of the open-circuited winding.

1.21 A certain power transformer is known to have no-load losses of 1200 watts when operated under rated conditions. Suppose that the power company reduces the voltage by 3%. Find the no-load losses under this reduced voltage condition.

1.22 Manufacturer's data for a single-phase 5-kVA, 480-V primary, 120-V secondary transformer is as follows:

$$\text{No-load losses} \quad 40 \text{ W}$$
$$\text{Full-load losses} \quad 200 \text{ W}$$

The full-load losses are equal to load losses plus no-load losses.

Use the loss data furnished to establish values for resistances of an equivalent circuit that could be used to model the transformer for operation close to its rating. Sketch your equivalent circuit showing the unknown reactances and your numerical resistance values referred to the high side of the transformer.

1.23 A 10-kVA single-phase transformer has a rated primary voltage of 480 V and a rated secondary voltage of 120 V. The transformer is to be represented by the equivalent circuit shown, where parameters are referred to the primary winding.

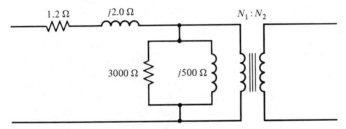

Make reasonable assumptions about transformer operation and determine:
a. The turns ratio N_1/N_2.
b. Rated secondary current.
c. Transformer no-load losses.
d. Transformer winding losses at full load.
e. Transformer efficiency at half load.

1.24 The transformer of Example 1.9 delivers rated kVA at rated voltage and at unity power factor to a load connected to its secondary. Find the transformer efficiency.

1.25 The transformer of Example 1.9 delivers rated kVA at rated voltage and at unity power factor to a load connected to its secondary (the low-voltage winding). Find the rms voltage at the terminals of the primary windings.

1.26 The manufacturer's data for a 1000-kVA 480-V secondary transformer is as follows:

$$\text{No-load losses} \quad 2,700 \text{ W}$$
$$\text{Full-load losses} \quad 17,000 \text{ W}$$

Suppose that the transformer operates at 90% capacity 16 hours a day, 5 days a week, and at 5% capacity at all other times. Estimate the annual losses in kilowatt-hours. (Assume that full-load losses are equal to load losses plus no-load losses.)

1.27 For the transformer configurations described in Section 1.9 (i.e., Figure 1.55), find values of lumped resistance for each winding which would account for the winding ohmic losses determined in Examples 1.10 and 1.11. Assume that the primary rated voltage is 39.8 kV and the secondary rated voltage is 7.97 kV.

1.28 For the transformer configurations described in Section 1.9 (i.e., Figure 1.55), find values of R_c and X_m using the loss data shown in Figure 1.35. Assume that the primary rated voltage is 39.8 kV and the secondary rated voltage is 7.97 kV. Refer parameters to the primary winding. Use the dimensions and values specified for Example 1.10. The core density is 0.017 lb/cm^3.

1.29 For the transformer configuration described in Section 1.9 (i.e., Figure 1.55), the primary rated voltage is 39.8 kV and the secondary rated voltage is 7.97 kV. Use the dimensions and values specified in Example 1.10 and determine the number of turns N_1 on the primary and the number of turns N_2 on the secondary.

1.30 The transformer equivalent circuit shown in Figure 1.57 has all parameters referred to winding 1. Find a similar equivalent circuit with all parameters referred to winding 2.

1.31 In Example 1.13 an expression is developed for the ratio of the rms voltage developed across the total leakage reactance to the rated rms transformer voltage:

$$\frac{X_{eq}I_{1rated}}{E_{1rated}} = \frac{\mu_0 w_w^2}{3\sqrt{2}\, w_c} \frac{J}{B_{max}}$$

Verify that the ratio is a dimensionless quantity.

1.32 Find an expression for total leakage reactance $X_{eq} = 2\omega L_\ell$ for the winding configuration shown in Figure 1.61. The expression should be referred to winding 1 and should be in terms of the dimensions h_w and w_w shown on the figure and an assumed core depth h_c.

CHAPTER 2

Fundamentals of Electromechanical Energy Conversion

INTRODUCTION

Relative motion between portions of a magnetic circuit or relative motion between current-carrying conductors and magnetic fields invariably results in an energy conversion process. In this chapter we extend the concepts of Chapter 1 to account for relative motion. We review the nature of electromagnetic forces and torques and the basic equations of motion governing mechanical systems. Incorporation of the effects of motion in our circuit equations then yields a mathematical model for analysis of rudimentary electromechanical devices.

FORCES AND TORQUES—
THE MICROSCOPIC VIEWPOINT

The force on a point charge moving in a magnetic field is given by the Lorentz force law:

$$\overline{F} = q(\overline{E} + \overline{v} \times \overline{B})$$

if $\overline{E}$ is zero,

$$\overline{F} = q(\overline{v} \times \overline{B}) \quad \text{N} \tag{2.1}$$

where $\overline{F}$ is the force in newtons, q is the charge in coulombs, $\overline{v}$ the velocity in meters per second, and $\overline{B}$ the magnetic flux density in webers per square meter. The vector quantities involved are illustrated in Figure 2.1. The direction of $\overline{F}$ is determined by the indicated vector cross product in (2.1). By the rules for vector cross products, the magnitudes are related by

$$F = qvB \sin \theta \quad \text{N} \tag{2.2}$$

where θ is the angle between $\overline{B}$ and $\overline{v}$ as shown in Figure 2.1. The concept can be extended to determine the force on a conductor or portion of a conductor. A simplified model of metallic conducting material treats the material as a lattice-like structure of positive ions essentially fixed in the material. This positively charged fixed lattice structure is electrically balanced by a negatively charged "gas" of free, or conduction electrons which can drift with respect to the lattice structure under the influence of an electric field. The net charge of any particular volume element of conducting material is effectively zero. In accordance with (2.1), a force will be developed on a negatively charged electron which moves with respect to a magnetic field. Coulomb forces of attraction between the free negatively charged electrons and the fixed positive ions cause forces acting on the electrons to be transmitted to the lattice structure of the conducting material itself. Current is defined by convention in terms of the motion of positive charge. It is therefore convenient to describe behavior in terms of an equivalent positive line charge density, or volume charge density, equal in magnitude to the respective negative free electron charge density, but having a drift velocity at any point opposite in direction to that of the free electrons. We will adopt this convention in describing forces on conducting materials. It may be helpful to note some rough orders of magnitude. The number of conducting electrons in copper is on the order of 10^{29} electrons per cubic meter. Multiplying by the charge on an electron yields a free electron charge density on the order of -10^{10} coulombs (c) per cubic meter. For typical current levels in copper, drift velocities are on the order of 10^{-4} m/s.

Consider the current-carrying conductor shown in Figure 2.2. Assume that the conductor has negligible cross-sectional area and line charge density ρ in coulombs per meter. Current must be related to ρ and charge velocity magnitude v by

$$i = \rho v \quad \text{A} \quad \text{(a scalar)} \tag{2.3}$$

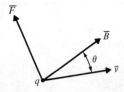

FIGURE 2.1 Vector Quantities Associated with Equation (2.1)

FIGURE 2.2 Current-Carrying Conductor in a Magnetic Field

Charge q on a differential element of the conductor will be

$$q = \rho \, d\ell \qquad C \tag{2.4}$$

We find the force $d\overline{F}$ acting on a differential element of the conductor as follows:

$$\begin{aligned}
d\overline{F} &= q(\overline{v} \times \overline{B}) \\
&= \rho \, d\ell(\overline{v} \times \overline{B}) \\
&= i \, d\ell\left(\frac{\overline{v}}{v} \times \overline{B}\right) \\
&= i \, d\overline{\ell} \times \overline{B}) \qquad N
\end{aligned} \tag{2.5}$$

since $d\overline{\ell}$ and $\overline{v}$ have the same direction at any point on the conductor. An integration over the length of the conductor yields a total vector force $\overline{F}$ acting on the conductor. For a straight length of conductor L perpendicular to a uniform field of flux density B, (2.5) yields the familiar

$$F = BLi \qquad N \tag{2.6}$$

where the direction $\overline{F}$ can be determined by consideration of (2.5). If conductor cross-sectional area is not negligible, a more general approach valid for any medium with current density $\overline{J}$ can be used. The current density is related to the charge density σ in coulombs per cubic meter and $\overline{v}$ by

$$\overline{J} = \sigma\overline{v} \qquad A/m^2 \tag{2.7}$$

We find the force $d\overline{F}$ acting on a differential volume element dV having total charge $q = \sigma \, dV$ as follows:

$$\begin{aligned}
d\overline{F} &= q(\overline{v} \times \overline{B}) \\
&= \sigma \, dV(\overline{v} \times \overline{B}) \\
&= dV(\sigma\overline{v} \times \overline{B}) \\
&= dV(\overline{J} \times \overline{B}) \qquad N
\end{aligned} \tag{2.8}$$

Thus the force density at any point must be

$$\bar{f} = \frac{d\bar{F}}{dV} = \bar{J} \times \bar{B} \qquad \text{N/m}^3 \qquad (2.9)$$

When applying (2.5), (2.8), or (2.9), we are generally faced with the problem of summing forces (or torques). The vector sum of the force is applied at the center of mass to determine the motion of the center of mass. The vector sum of the torques about a particular axis is used to determine the rotational motion about that axis.

EXAMPLE 2.1 Find an expression for the force per unit of length developed between the two conductors of a circuit having the configuration shown in Figure 2.3.

Assuming that $r << d$ and that the total circuit length $L >> d$, the field in the vicinity of one conductor due to current in the other conductor can be found from Ampère's circuital law:

$$\oint \bar{H} \cdot d\bar{\ell} = i$$

$$H(2\pi d) = i$$

$$H = \frac{i}{2\pi d}$$

$$B = \mu_0 H = \frac{\mu_0 i}{2\pi d} \qquad \text{Wb/m}^2$$

Thus

$$F = BLi = \frac{L\mu_0 i^2}{2\pi d}$$

and the force per unit of length will be

$$f = \frac{F}{L} = \frac{\mu_0 i^2}{2\pi d} \qquad \text{N/m}$$

We can determine the direction of force by applying (2.1), (2.5), (2.8), or (2.9). For currents oppositely directed (typical of a circuit), the force tends to move the conductors apart.

A more interesting and practical problem is the determination of forces on rigid bus structures under fault (short-circuit) conditions. These structures must be braced to withstand the forces that develop during a short circuit. For example, a bus that is rated at 600 A continuous current might typically be braced to withstand fault conditions of 22,000 A rms symmetrical.

EXAMPLE 2.2 Find an expression for the force per unit of length developed between the two rigid copper buses shown in Figure 2.4. Assume that the buses constitute the two conductors of a single-phase circuit carrying current i.

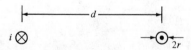

FIGURE 2.3

FIGURE 2.4

As shown in Figure 2.5, a force is developed between differential elements of each of the two buses. If we find the magnetic flux density B at a point in the right bus due to a current at a point in the left bus, we can determine a differential force dF per unit of length between the buses. Assume that the thickness t is small compared to d. Then if current density over the bus cross section is uniform, the current in a differential element of the bus on the left will be $(i/h) dy_1$. The magnetic field intensity H at a point y_2 on the right bus due to a differential element at y_1 can be found using Ampère's circuital law:

$$\oint \overline{H} \cdot d\overline{\ell} = I$$

$$H 2\pi \sqrt{d^2 + (y_1 - y_2)^2} = \frac{i}{h} dy_1$$

Thus

$$B = \mu_0 H = \frac{\mu_0}{2\pi} \frac{i}{h} \frac{dy_1}{\sqrt{d^2 + (y_1 - y_2)^2}} \qquad \text{Wb/m}^2$$

The horizontal component of force per unit of length at a point y_2 can be found by applying (2.6):

$$dF = BL\left(\frac{i}{h} dy_2\right)\frac{d}{\sqrt{d^2 + (y_2 - y_1)^2}} \qquad \text{N}$$

$$df = \frac{dF}{L} = \frac{\mu_0 d}{2\pi}\left(\frac{i}{h}\right)^2 \frac{1}{d^2 + (y_2 - y_1)^2} dy_1 \, dy_2 \qquad \text{N/m}$$

Then the total horizontal component of force per unit of length will be given by

$$f = \frac{\mu_0 d}{2\pi}\left(\frac{i}{h}\right)^2 \int_0^h \int_0^h \frac{1}{d^2 + (y_2 - y_1)^2} dy_1 \, dy_2 \qquad \text{N/m}$$

The integration is somewhat tedious but yields

$$f = \frac{\mu_0}{2\pi}\left(\frac{i}{h}\right)^2 \left[2h \tan^{-1}\frac{h}{d} - d \ln\left(\frac{d^2 + h^2}{d^2}\right)\right] \qquad \text{N/m}$$

(This final expression appears in reference 1 with some additional discussion.)

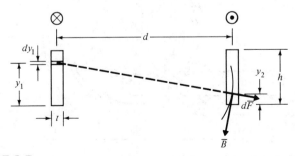

FIGURE 2.5

If $h = 4$ cm, $d = 10$ cm, and current is 22,000 A rms, substitution in our expression yields

$$f(t) = 943.8(1 + \cos [(2)(377)t] \text{ N/m}$$

where $i(t)$ is taken to be $\sqrt{2}(22)(10^3) \cos 377t$. Thus the time-average force is 943.8 N/m (≈ 65 lb/ft). It should be noted that switching transients can play a significant role here. We have assumed a symmetrical sinusoidal waveform for purposes of illustration.

EXAMPLE 2.3
Find an expression for torque developed on the rotor of the simplified four-pole dc machine shown in Figure 2.6. Generalize the expression for an arbitrary number of poles.

Figure 2.6 illustrates the fundamental aspects of conventional dc machine operation. The field circuit of the machine is on the stator. It produces a stationary magnetic field whose radial component of magnetic flux density at radius r is a periodic function of the angle θ defined in Figure 2.6. This is illustrated in Figure 2.7. In general, for a machine with P poles we would have $B(\theta) = B(\theta \pm 4\pi/P) = -B(\theta \pm 2\pi/P)$, that is, a periodic function of θ with half-wave symmetry.

The armature winding of a machine is the winding that has the major voltage induced in it. In the case of a dc machine, the armature winding is on the rotor. (Most ac machines have their armature windings on the stator.) The rotor (armature) winding of a dc machine is typically distributed in slots around the periphery of the rotor. Stationary brushes supply armature current i_a to the moving conductors of the armature winding in such a way that the pattern of armature currents into and out of the page as shown in Figure 2.6 is stationary. By commutator action, the armature retains this stationary current distribution regardless of the position of the rotor with respect to the stator. This is the essence of conventional dc machine operation.

We will assume that the axial length of the machine is sufficient to permit a two-dimensional analysis (i.e., end effects will be neglected). The mmf due to armature current i_a distorts the flux density distribution of Figure 2.7. In a well-designed machine this effect can be minimized and it will be neglected here. Armature slot effects will also be neglected, and we will treat the winding as a belt of conductors of uniform thickness t around the periphery of a perfectly round rotor.

By assuming that the thickness t of the belt of armature conductors is small compared to rotor radius r, we can develop the desired result using (2.6). If the total number of uniformly distributed armature conductors is Z, the number in a differential element of

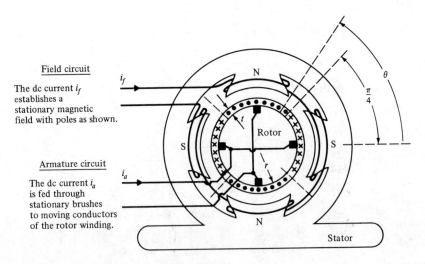

Field circuit

The dc current i_f establishes a stationary magnetic field with poles as shown.

Armature circuit

The dc current i_a is fed through stationary brushes to moving conductors of the rotor winding.

FIGURE 2.6 Simplified Four-Pole DC Machine

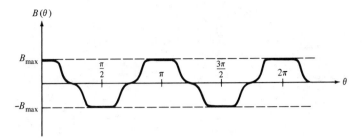

FIGURE 2.7 Radial Component of Air-Gap Magnetic Flux Density due to the Field Current of a Four-Pole DC Machine

arc length will be $Zr \, d\theta/2\pi r = (Z/2\pi) \, d\theta$. If each conductor carries current i_c, the contribution to total torque by a differential element of arc length at position θ will be

$$dT = rBLi$$

$$= rB(\theta)Li_c \frac{Z}{2\pi} d\theta \qquad \text{N} \cdot \text{m}$$

where L is the axial length of the rotor. [We assume that $B(\theta)$ is the same over the length L; that is, end effects are neglected.] We can obtain the torque contribution for those conductors under one pole by integrating from $\theta = -\pi/4$ to $\theta = +\pi/4$ in this case. In general, for a P-pole machine, integration must be from $-\pi/P$ to $+\pi/P$. Thus torque per pole will be

$$T_p = \frac{Z}{2\pi} i_c \int_{-\pi/P}^{+\pi/P} B(\theta)Lr \, d\theta \qquad \text{N} \cdot \text{m}$$

Now for any number of poles, the integral in the expression yields the total flux per pole:

$$\Phi = \int_{-\pi/P}^{+\pi/P} B(\theta)Lr \, d\theta \qquad \text{Wb}$$

Thus the total torque will be

$$T = PT_p$$

$$= \frac{PZ}{2\pi} i_c \Phi \qquad \text{N} \cdot \text{m}$$

where i_c is the current in one conductor. The actual winding of a dc armature generally involves a number of parallel paths. Thus $i_c = i_a/a$, where i_a is armature terminal current and a is the number of parallel paths through the winding. Thus torque is generally expressed as

$$T = \frac{PZ}{2\pi a} i_a \Phi \qquad \text{N} \cdot \text{m}$$

For a given machine our final expression tells us that the torque is directly proportional to the product of armature current and field flux. The result is not surprising when we consider (2.6) as the starting point in the development.

Equation (2.6) can be used to develop a familiar result for motional emf induced if a conductor moves through a magnetic field. Consider the situation depicted in Figure 2.8. A straight conductor of length L moves through a magnetic field of flux density $\overline{B}$, where $\overline{B}$, $\overline{v}$, and L are mutually perpendicular. A positive charge q at any point on the conductor experiences a force in accordance

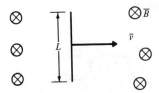

FIGURE 2.8

with (2.1) (directed upward in this case). The work that would be done in moving this charge q from one end to the other would be given by the product of force and distance:

$$W = qvBL \qquad J$$

Thus the emf or work per unit charge will be

$$e = \frac{W}{q} = BLv \qquad V \qquad (2.10)$$

We can determine the polarity of this emf quite simply by using (2.1).

EXAMPLE 2.4   Find an expression for the open-circuit voltage induced in a single rectangular loop of wire rotated in a magnetic field as shown in Figure 2.9.

The field produces an air-gap flux whose radial component of magnetic flux density at radius r is given by $B(\theta)$. $B(0) = B_{max}$ and $B(\theta) = B(\theta \pm 2\pi) = -B(\theta \pm \pi)$, where θ is an arbitrary angle measured from the horizontal axis as shown in Figure 2.9. The field is similar to the field described in Example 2.3, Figure 2.7, but is a two-pole field.

The axial length of the rotor is L and the rotor turns at constant speed ω. Assume that $B(\theta)$ is the same over the axial length L. We can use (2.10) to achieve the desired result. The velocity of one side of the loop with respect to the field will be ωr. Thus the voltage induced in one side of the loop at angle θ_r will be

$$e_s = B(\theta_r)Lv$$

$$= B(\theta_r)L\omega r \qquad V$$

The two sides of the loop are in series; thus the total induced voltage will be

$$e = 2e_s = 2L\omega rB(\theta_r) \qquad V$$

If the loop is turned at constant speed ω, then θ_r as a function of time is $\theta_r(t) = \omega t + \theta_0$, where θ_0 is the angle at time $t = 0$. We thus have

$$e(t) = 2 L\omega rB(\omega t + \theta_0) \qquad V$$

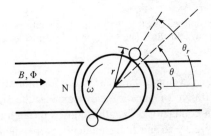

FIGURE 2.9

By observing the waveform $e(t)$ on an oscilloscope, we can indirectly determine the variation of field flux density B with angle θ. A coil used in this manner is referred to as a *search coil*.

We can also use Faraday's law to achieve the desired result:

$$e(t) = \frac{d\lambda}{dt} = \frac{N \, d\varphi}{dt} = \frac{d\varphi}{dt} \quad \text{V}$$

where $N = 1$.

We need an expression for flux as a function of time. Choose reference axes as shown in Figure 2.10. The surface of integration for the flux calculation is taken as the surface shown by the dashed line in Figure 2.10. $B(\theta)$ is the radial component of the magnetic flux density and it will be normal to this surface at every point. Thus for flux linking the loop we will have

$$\varphi(t) = \int_{\omega t + \theta_0 - \pi}^{\omega t + \theta_0} B(\theta) Lr \, d\theta \quad \text{Wb}$$

To differentiate with respect to time, apply (fundamental theorem of calculus and the chain rule for differentiation)

$$\frac{d}{dt} \int_{a(t)}^{b(t)} f(x) \, dx = f(b(t)) \frac{db(t)}{dt} - f(a(t)) \frac{da(t)}{dt}$$

Thus

$$\frac{d\varphi}{dt} = B(\omega t + \theta_0) Lr\omega - B(\omega t + \theta_0 - \pi) Lr\omega$$

But

$$B(\omega t + \theta_0 - \pi) = -B(\omega t + \theta_0)$$

Thus

$$e(t) = \frac{d\varphi}{dt} = 2L\omega r B(\omega t + \theta_0) \quad \text{V}$$

which is the same as our previous result.

Note that the nature of the flux density distribution $B(\theta)$ determines the nature of the voltage waveform $e(t)$. In particular, if we want a sinusoidal voltage waveform, we must have a sinusoidal flux density distribution.

To completely describe a given system, we must write the equation(s) of motion and appropriate circuit equations, taking into account relative motion.

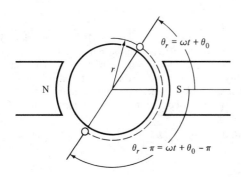

FIGURE 2.10

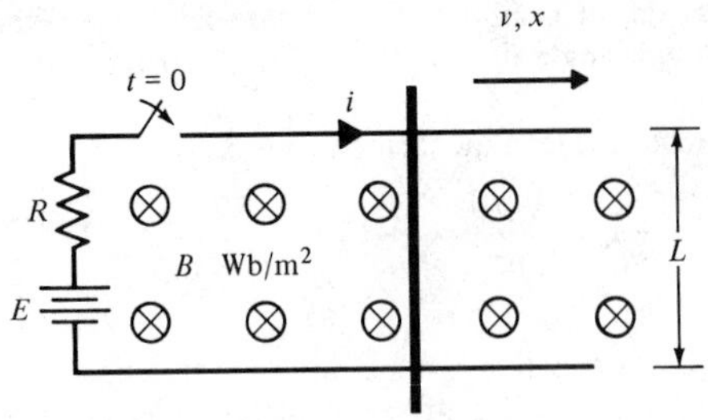

FIGURE 2.11

EXAMPLE 2.5 Write a set of equations in state-variable form to describe the behavior of the rail system depicted in Figure 2.11 for $t \geq 0$. Assume that the bar is free to slide in either direction with a retarding force due to friction which is directly proportional to velocity of the bar. The magnetic field is uniform and constant.

The electromagnetic force developed on the bar may be found from (2.6)

$$F = BLi$$

In reality there will be a second component of force due to the variation of the self-inductance of the circuit with position of the bar. We will examine the nature of this force in the next section. For now we assume that the BLi force is the significant force. We can write an equation of motion for the bar by setting the product of the mass m and acceleration dv/dt equal to the sum of all externally applied forces in the direction of motion:

$$m \frac{dv}{dt} = F - bv$$

$$= BLi - bv$$

In the equations above, the force F, velocity v, and acceleration dv/dt are all defined positive in the same direction (to the right in Figure 2.11). Since the current is changing with time, we account for a self-inductance term $L_a(x)$ in the circuit equation:

$$E = iR + L_a(x) \frac{di}{dt} + i \frac{dL_a(x)}{dt} + e$$

$$= iR + L_a(x) \frac{di}{dt} + i \frac{dL_a(x)}{dx} \frac{dx}{dt} + e$$

$$= iR + L_a(x) \frac{di}{dt} + iv \frac{dL_a(x)}{dx} + e$$

where e is determined from (2.10):

$$e = BLv$$

The inductance $L_a(x)$ increases with increasing values of x. If $L_a(x)$ increases linearly with x, then $dL_a(x)/dx$ is simply the inductance per unit of rail circuit length. This value can be kept small by suitably designing the device. If we restrict the problem and consider relatively low values of velocity v and operation over a limited distance x, it is not unreasonable to neglect the $iv[dL_a(x)/dx]$ term in the circuit equation and take $L_a(x)$ as a constant. Then combining the equations above and rewriting in state-variable form, we have

$$\frac{dv}{dt} = -\frac{b}{m} v + \frac{BL}{m} i \qquad (2.5a)$$

$$\frac{di}{dt} = -\frac{BL}{L_a} v - \frac{R}{L_a} i + \frac{E}{L_a} \qquad (2.5b)$$

The system is thus characterized by two linear first-order time-invariant ordinary differential equations with constant forcing function. Assumption of a set of values for the parameters and recognition of the relevant initial conditions

$$i(0) = v(0) = 0$$

would permit a straightforward solution. Generally, the field will be a function of some current which itself is a state variable. It is easy to see that if $B = ki_f$ and i_f is a state variable, we will have products of state variables in the equations, rendering them nonlinear. If inductive coupling exists between the field circuit and the rail circuit, we add another complication. The assumption of a constant B field has avoided these difficulties, and by choosing a suitable structure, it is not unrealistic.

To gain some additional insights, we will examine (2.5a) and (2.5b) in more detail. Multiply (2.5a) by mv and rearrange to obtain

$$mv \frac{dv}{dt} + bv^2 = (BLv)i$$

This can be rewritten as

$$\frac{d}{dt}\left(\frac{1}{2} mv^2\right) + bv^2 = ei$$

The first term on the left is the time rate of change of the kinetic energy of the bar. The second term on the left is the product of the frictional retarding force bv and velocity v. It is the instantaneous power loss due to friction. The term in the right is the instantaneous electrical power delivered to the bar, assuming that any bar resistance has been accounted for elsewhere. Thus the product of motional emf and current is power which is being converted into an increase of the kinetic energy of the bar and into heat due to friction.

Now multiply the second equation by $L_a i$ and rearrange to obtain

$$L_a i \frac{di}{dt} + (BLv)i + Ri^2 = Ei$$

This can be rewritten as

$$\frac{d}{dt}\left(\frac{1}{2} L_a i^2\right) + ei + Ri^2 = Ei$$

Thus the instantaneous total power Ei is being converted as resistive losses, time rate of change of energy stored in the magnetic field, and in the ei term described previously. Note finally that

$$(BLv)i = ei$$
$$(BLi)v = Fv$$

Thus

$$ei = Fv$$

which directly relates the electrical power ei to the mechanical power Fv.

2.3

FORCES AND TORQUES— THE MACROSCOPIC VIEWPOINT

Consider the lossless system shown in Figure 2.12. The system may be regarded as ideal in that it does not have the usual causes of losses, such as electrical resistance and mechanical friction. For actual systems we will account for these

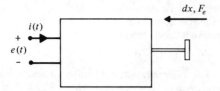

FIGURE 2.12 Lossless Electromechanical System

effects separately. The device is assumed to be electromagnetic in nature such that $e(t)$ is determined by Faraday's law:

$$e(t) = \frac{d\lambda}{dt} \tag{2.11}$$

Instantaneous electrical power delivered to the device is

$$p(t) = e(t)i(t) = i(t)\frac{d\lambda}{dt} \tag{2.12}$$

If the current i can be expressed as a function of the flux linkages λ, we can separate variables and write

$$dW_e = p(t)\,dt = i\,d\lambda \tag{2.13}$$

where dW_e is a differential amount of electrical input energy to the system.

If F_e is the developed electrical force and a displacement dx occurs in the same direction as the force (note that we have defined x and hence dx in the same direction as F_e in Figure 2.12), the system does a differential amount of mechanical work:

$$dW_m = F_e\,dx \tag{2.14}$$

The net differential energy increase in the system will be

$$dW = dW_e - dW_m \tag{2.15}$$
$$= i\,d\lambda - F_e\,dx$$

Now if λ and x are chosen as independent variables, $W = W(\lambda, x)$, $i = i(\lambda, x)$, and $F_e = F_e(\lambda, x)$. We must have

$$dW(\lambda, x) = \frac{\partial W(\lambda, x)}{\partial \lambda}\,d\lambda + \frac{\partial W(\lambda, x)}{\partial x}\,dx \tag{2.16}$$

By comparing (2.16) and (2.15), we see that

$$i = i(\lambda, x) = \frac{\partial W(\lambda, x)}{\partial \lambda} \tag{2.17}$$

$$F_e = F_e(\lambda, x) = -\frac{\partial W(\lambda, x)}{\partial x} \tag{2.18}$$

Equation (2.18) yields an important result. If we can find stored energy in a system as a function of λ and x, we can find an expression for the electrical force developed by performing the partial differentiation indicated in (2.18).

It should be noted that the same development could be made for rotational motion. That is, if the developed electrical torque T_e is of interest, (2.14) is replaced with

$$dW_m = T_e \, d\theta \tag{2.19}$$

If λ and θ are then chosen as the independent variables, we find by the same reasoning that

$$T_e = T_e(\lambda, \theta) = -\frac{\partial W(\lambda, \theta)}{\partial \theta} \tag{2.20}$$

EXAMPLE 2.6 Find an expression for the electrical force developed between the members of the magnetic circuit shown in Figure 2.13.

We start by finding an expression for the stored energy W as a function of λ and x. If we assume "linear" iron, the relationship between λ and i for various values of x might appear as shown in Figure 2.14. The assumption of linearity simplifies the example. The concepts we are applying are valid if the magnetic circuit is nonlinear as long as λ is a single-valued function of i, and vice versa. As x decreases, the magnetic circuit reluctance decreases. For a given current, decreasing the value of x results in larger values of flux, and hence flux linkages λ. Thus inductance increases as x decreases, as shown in Figure 2.14. If we express the current i as a function of λ and x (the independent variables), we can write

$$W(\lambda, x) = \frac{\lambda i(\lambda, x)}{2} = \frac{\lambda^2}{2L(x)}$$

where the inductance L is a function of x. If R_{core} is the reluctance of the magnetic circuit when $x = 0$, then

$$R(x) = R_{core} + R_{gap}(x)$$

$$= R_{core} + \frac{2x}{\mu_0 A}$$

where we have neglected leakage and fringing effects. Then

$$L(x) = \frac{N^2}{R(x)}$$

$$W(\lambda, x) = \frac{\lambda^2}{2N^2}\left(R_{core} + \frac{2x}{\mu_0 A}\right)$$

Now $F_e(\lambda, x)$ can be found from (2.18) as

$$F_e(\lambda, x) = -\frac{\partial W(\lambda, x)}{\partial x} = -\frac{\lambda^2}{N^2 \mu_0 A}$$

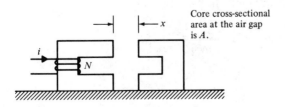

Core cross-sectional area at the air gap is A.

FIGURE 2.13

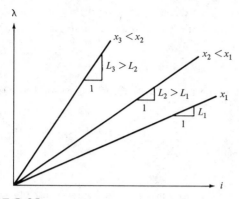

FIGURE 2.14

The reader will recall that in our development of (2.18) we assumed that F_e was defined as positive in the same direction as positive x and dx. The minus sign in our result tells us that the direction of F_e is opposite to the direction of increasing x (increasing air gap). The direction of developed force is not surprising. It tends to pull the magnetic circuit together. We will see that force (torque) is always developed in such a direction that if it could cause a displacement (rotation), the magnetic circuit reluctance would decrease.

Our result for force $F_e(\lambda, x)$ shows that F_e is independent of x, assuming that λ is constant. (Obviously, we must consider the limiting effects of our assumptions of zero leakage flux and fringing.) Now if λ is held constant, there can be no electrical input to the system since $dW_e = i\, d\lambda = 0$. Thus, if λ is held constant and a displacement occurs in the direction of the developed force, the mechanical work done must be equal to a decrease in the stored energy of the system. For example, suppose that the initial value of x is x_1 and a displacement at constant λ occurs to a final value of $x_2 < x_1$. The change (increase) in stored energy is

$$\Delta W = W(\lambda, x_2) - W(\lambda, x_1)$$

$$= \frac{\lambda^2}{2N^2}\left(R_{\text{core}} + \frac{2x_2}{\mu_0 A}\right) - \frac{\lambda^2}{2N^2}\left(R_{\text{core}} + \frac{2x_1}{\mu_0 A}\right)$$

$$= \frac{\lambda^2}{N^2}\frac{x_2 - x_1}{\mu_0 A} = -\frac{\lambda^2(x_1 - x_2)}{N^2\mu_0 A} < 0$$

since $x_1 > x_2$. Thus the stored energy has decreased. The mechanical work done by the system is

$$\Delta W_m = \int_{x_1}^{x_2} F_e\, dx$$

$$= -\frac{\lambda^2}{N^2\mu_0 A}\int_{x_1}^{x_2} dx$$

$$= +\frac{\lambda^2}{N^2\mu_0 A}(x_1 - x_2) > 0$$

$$= -\Delta W$$

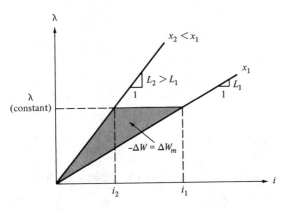

FIGURE 2.15

The decrease in stored energy $-\Delta W = \Delta W_m$ is illustrated graphically in Figure 2.15. (The reader should determine the nature of the electrical driving circuit that must be connected to the winding to realize a constant value of λ.)

Often it is more convenient to choose current rather than flux linkages as one of the independent variables. This can be done simply, provided that we keep clearly in mind our choice of independent variables. From our previous development, the net differential energy increase in the system is given by (2.15), repeated here for convenience:

$$dW = dW_e - dW_m \tag{2.15}$$

$$= i\, d\lambda - F_e\, dx$$

If i and x are chosen as the independent variables, λ must be expressed as a function of i and x, that is, $\lambda = \lambda(i, x)$. Similarly, $W = W(i, x)$ and $F_e = F_e(i, x)$. Then we must have

$$d\lambda(i, x) = \frac{\partial \lambda(i, x)}{\partial i}\, di + \frac{\partial \lambda(i, x)}{\partial x}\, dx \tag{2.21}$$

Substituting (2.21) in (2.20) yields

$$dW(i, x) = i\left[\frac{\partial \lambda(i, x)}{\partial i}\, di + \frac{\partial \lambda(i, x)}{\partial x}\, dx\right] - F_e(i, x)\, dx \tag{2.22}$$

$$= \frac{i\, \partial \lambda(i, x)}{\partial i}\, di + \left[\frac{i\, \partial \lambda(i, x)}{\partial x} - F_e(i, x)\right] dx$$

We must also have

$$dW(i, x) = \frac{\partial W(i, x)}{\partial i}\, di + \frac{\partial W(i, x)}{\partial x}\, dx \tag{2.23}$$

By comparing (2.22) and (2.23), we see that

$$\frac{\partial W(i, x)}{\partial i} = i\,\frac{\partial \lambda(i, x)}{\partial i} \tag{2.24}$$

$$\frac{\partial W(i, x)}{\partial x} = i\,\frac{\partial \lambda(i, x)}{\partial x} - F_e(i, x) \tag{2.25}$$

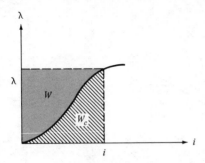

FIGURE 2.16 Graphical Depiction of Stored Energy and Coenergy Relationships

From (2.25),

$$F_e(i, x) = i \frac{\partial \lambda(i, x)}{\partial x} - \frac{\partial W(i, x)}{\partial x}$$ (2.26)

$$= \frac{\partial}{\partial x} [i\lambda(i, x) - W(i, x)]$$

By defining coenergy $W_c = W_c(i, x)$ as

$$W_c(i, x) = i\lambda(i, x) - W(i, x)$$ (2.27)

We can write

$$F_e(i, x) = + \frac{\partial W_c(i, x)}{\partial x}$$ (2.28)

A similar development can be made for rotational motion, resulting in

$$T_e = + \frac{\partial W_c(i, \theta)}{\partial \theta}$$ (2.29)

Equation (2.27) has a simple geometric interpretation, as shown in Figure 2.16. Coenergy, as defined by (2.27) and illustrated in Figure 2.16, may be regarded simply as a function defined for the sake of convenience. Its usefulness will be demonstrated in the examples that follow.

We have developed our force (torque) relationships [Eqs. (2.18), (2.20), (2.28), and (2.29)] by treating quantities most conveniently related to electrical circuit quantities, namely flux linkages λ and current i. We could just as easily have done the entire development in terms of magnetic circuit quantities: flux φ and magnetomotive force Ni. Similarly, we could have completed the development in terms of energy and coenergy densities, starting with magnetic flux density B and magnetic field intensity H. The reader should review Sections 1.4 and 1.5, in particular Figures 1.17 and 1.21, to fix these ideas.

EXAMPLE 2.7   Find an expression for the electrical force developed between the members of the magnetic circuit shown in Figure 2.13 (i.e., Example 2.6) by treating current i and displacement x as the independent variables.

We again assume "linear" iron such that the relationship between λ and i for various values of x will be as shown in Figure 2.14. For any value of i and x (the independent variables), we can write

$$W_c = W = \frac{\lambda i}{2}$$

where, because of our assumption of linearity, W and W_c are numerically equal. Since i and x have been chosen (arbitrarily) as independent variables, it is convenient to work with coenergy:

$$W_c(i, x) = \frac{\lambda(i, x)i}{2} = \frac{1}{2} L(x)i^2$$

and with (2.28):

$$F_e(i, x) = + \frac{\partial W_c(i, x)}{\partial x}$$

The inductance $L(x)$ was found in Example 2.6 to be (neglecting leakage and fringing)

$$L(x) = \frac{N^2}{R(x)} = \frac{N^2}{R_{core} + 2x/\mu_0 A}$$

Thus

$$W_c(i, x) = \frac{1}{2} \frac{N^2}{R_{core} + 2x/\mu_0 A} i^2$$

$$F_e(i, x) = + \frac{\partial W_c(i, x)}{\partial x}$$

$$= -\frac{(Ni)^2}{\mu_0 A} \left(\frac{1}{R_{core} + 2x/\mu_0 A} \right)^2$$

As before, the minus sign tells us that the direction of F_e is opposite to the direction of increasing x (increasing air gap). Our result for F_e appears different from the result reached in Example 2.6. The results are the same but they are expressed in terms of different variables (see Problem 2.9).

If the current is held constant and a displacement occurs in the direction of the developed force, mechanical work will be done by the system. The displacement in the direction of the developed force decreases the magnetic circuit reluctance. Since the current is held constant, the flux, and hence flux linkages, will increase, resulting in an electrical energy input to the system. For example, suppose that the initial value of x is x_1 and a displacement at constant current occurs to a final value of $x = x_2 < x_1$. The electrical energy input to the system will be

$$\Delta W_e = \int_{\lambda_1}^{\lambda_2} i \, d\lambda = i(\lambda_2 - \lambda_1) > 0$$

since $\lambda_2 > \lambda_1$. The change in stored energy will be

$$\Delta W = W(i, x_2) - W(i, x_1)$$

$$= \tfrac{1}{2} i\lambda_2 - \tfrac{1}{2} i\lambda_1 \qquad \text{(because of our assumption of linearity)}$$

$$= \tfrac{1}{2} i(\lambda_2 - \lambda_1)$$

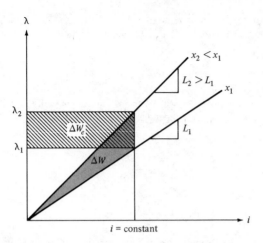

FIGURE 2.17

which is one-half of the electrical energy input ΔW_e. The remaining half must have been converted to mechanical energy. This can be verified (by the reader) by performing the following integration:

$$\Delta W_m = \int_{x_1}^{x_2} F_e(i, x) \, dx$$

These concepts are illustrated graphically in Figure 2.17.

In general, in a linear system, a displacement at constant current results in a change in stored energy equal to the mechanical work done. In our example, one-half of the electrical input went into an increase in stored energy and the other half was converted to output mechanical work. If the process is reversed such that mechanical work is input and electrical energy is output, we would find that the system stored energy would decrease by an amount equal to the mechanical work done. The electrical output energy would be twice the mechanical input energy.

EXAMPLE 2.8 Write a set of equations in state-variable form to describe the behavior of the system shown in Figure 2.18.

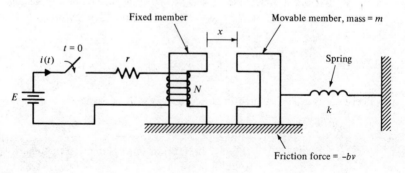

FIGURE 2.18

We will assume that the right member of the magnetic circuit does not come in contact with the left member (i.e., x remains > 0). Assuming that the system is initially at rest, the right member will have moved to a position $x = x_0$ where the spring force is zero. The initial conditions must then be

$$x(0) = x_0$$

$$i(0) = \lambda(0) = v(0) = 0$$

We must find an expression for the electrical force developed, then write an equation of motion for the right member, and finally a circuit equation to describe the system completely. We arbitrarily choose flux linkages and displacement x as independent variables. As in Examples 2.6 and 2.7, we will assume a linear magnetic circuit (just to simplify the example) and neglect leakage flux and fringing. The relevant results from these examples are

$$L(x) = \frac{N^2}{R(x)} = \frac{N^2}{R_{core} + 2x/\mu_0 A}$$

$$F_e(\lambda, x) = -\frac{\lambda^2}{N^2 \mu_0 A}$$

We write an equation of motion by setting the mass m times the acceleration dv/dt equal to the sum of forces acting in the direction of increasing x:

$$\frac{m \, dv}{dt} = F_e(\lambda, x) - bv - k(x - x_0)$$

The velocity v and displacement x are related by

$$\frac{dx}{dt} = v$$

Our circuit equation is, for $t \geq 0$,

$$E = ri + \frac{d\lambda}{dt}$$

$$= \frac{r}{L(x)} \lambda + \frac{d\lambda}{dt}$$

Substituting for F_e and $L(x)$ and rearranging, we find

$$\frac{dv}{dt} = \frac{\lambda^2}{mN^2 \mu_0 A} - \frac{b}{m} v - \frac{k}{m}(x - x_0)$$

$$\frac{dx}{dt} = v$$

$$\frac{d\lambda}{dt} = \frac{rR_{core}}{N^2} \lambda - \frac{2r}{N^2 \mu_0 A} x\lambda + E$$

The equations are nonlinear due to the λ^2 and $x\lambda$ terms. Solution by numerical integration using the computer can easily be accomplished for a known set of parameters and initial conditions (see Appendix G). Equilibrium points may be determined by setting time derivatives to zero and solving for λ, v, and x. The nature of system damping about these points may be examined by writing linear differential equations describing small displacements of state variables from these equilibrium points.

An analysis of equilibrium points yields additional insights on system behavior. At an equilibrium point

$$\frac{dv}{dt} = \frac{dx}{dt} = \frac{d\lambda}{dt} = 0$$

We see from the first and second state-variable equations that, at an equilibrium point, the spring restraining force $k(x - x_0)$ exactly balances the developed electrical force F_e. Obviously, an equilibrium point is one where the movable member is stationary, and our second state-variable equation reflects this. The third state-variable equation for equilibrium conditions is solved perhaps more simply by reexamining the figure. If $d\lambda/dt = 0$, Kirchhoff's voltage law requires that $E = ir$ at an equilibrium point. Eliminate flux linkages λ using the definition of flux linkages:

$$\lambda = Li$$

$$= \frac{N^2}{R_{\text{core}} + 2x/\mu_0 A} \frac{E}{r}$$

Now substitute in the expression for electrical force:

$$F_e = -\frac{\lambda^2}{N^2 \mu_0 A}$$

$$= -\left(\frac{E}{r}\right)^2 \frac{1}{\mu_0 A} \left(\frac{N}{R_{\text{core}} + 2x/\mu_0 A}\right)^2$$

The electrical force F_e, as given above, is the force on the movable member in the direction of increasing x (i.e., to the right). If the electrical force to the left (i.e., $-F_e$) and the spring restraining force to the right [i.e., $-k(x - x_0)$] are plotted on the same graph, equilibrium points occur where the curves intersect. Figure 2.19 illustrates this for various values of the spring constant k. For a relatively weak spring constant (k_1 in the figure), there can be no equilibrium points in the interval $0 < x \le x_0$. We would expect that the movable member would eventually come to rest at $x = 0$. For a spring constant $k = k_2$, we have two potential equilibrium points in the interval $0 < x \le x_0$, identified as "1" and "2" on the figure. It can be seen from examination of Figure 2.19 that point 1 is a stable equilibrium point and point 2 is an unstable equilibrium point. To see this, consider point 2. At the precise point where the curves intersect, the electrical force and spring force are balanced. If a small displacement from this equilibrium point occurs in the direction of decreasing x, the net accelerating force will be in the direction of decreasing x, which is away from point 2. Similarly, a small displacement in the direction of increasing x yields a net accelerating force away from point 2. The converse is true for point 1. For the value $k = k_3$, we have one equilibrium point identified as

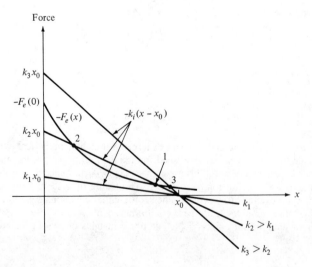

FIGURE 2.19

FIGURE 2.20 Large Solenoid-Actuated Brake for Industrial Applications. The Cylindrical-Shaped Enclosure Houses the Electromagnetic Actuator. (Courtesy of the Square D Company)

point 3 on the figure and it is stable. The graphical approach illustrated here is useful for an intuitive appreciation of system behavior as well as for preliminary design considerations. Solution of the state-variable equations is then necessary to quantify system behavior as a function of time and to incorporate the effects of damping and mass. Many practical devices, such as contactors, relays, and electromagnetically activated clutches and brakes, can be analyzed in this manner.

2.4

FORCES AND TORQUES IN MULTIPLY EXCITED SYSTEMS

We have restricted our attention thus far to systems that have a single winding. Most rotating machine configurations have multiple windings and may thus be regarded as multiply excited. A lossless multiply excited system is depicted in Figure 2.21. Note that we could replace the force F_e and differential displacement dx with torque T_e and differential angular displacement $d\theta$ in order to consider a rotational system. Although it is possible to consider multiple mechanical inputs (outputs), we will restrict our attention to devices that can be characterized by a single translational or rotational displacement. Our previous

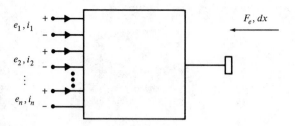

FIGURE 2.21

development for the determination of force (torque) relationships in singly excited systems can be extended simply. A differential amount of electrical energy
input to the system will be

$$dW_e = i_1 \, d\lambda_1 + i_2 \, d\lambda_2 + \cdots + i_n \, d\lambda_n \tag{2.30}$$

If F_e is a developed electrical force and a displacement dx occurs in the same
direction as this force, the system does a differential amount of mechanical
work:

$$dW_m = F_e \, dx \tag{2.31}$$

The net differential energy increase in the system will be

$$dW = dW_e - dW_m \tag{2.32}$$
$$= i_1 \, d\lambda_1 + i_2 \, d\lambda_2 + \cdots + i_n \, d\lambda_n - F_e \, dx$$

If independent variables are chosen as $\lambda_1, \lambda_2, \ldots, \lambda_n$ and x, then

$$W = W(\lambda_1, \lambda_2, \ldots, \lambda_n, x)$$

and we must have

$$dW(\lambda_1, \lambda_2, \ldots, \lambda_n, x) = \frac{\partial W(\lambda_1, \lambda_2, \ldots, \lambda_n, x)}{\partial \lambda_1} \, d\lambda_1$$

$$+ \frac{\partial W(\lambda_1, \lambda_2, \ldots, \lambda_n, x)}{\partial \lambda_2} \, d\lambda_2$$

$$\vdots \tag{2.33}$$

$$+ \frac{\partial W(\lambda_1, \lambda_2, \ldots, \lambda_n, x)}{\partial \lambda_n} \, d\lambda_n$$

$$+ \frac{\partial W(\lambda_1, \lambda_2, \ldots, \lambda_n, x)}{\partial x} \, dx$$

By comparing (2.32) and (2.33), we see that

$$F_e = F_e(\lambda_1, \lambda_2, \ldots, \lambda_n, x) = -\frac{\partial W(\lambda_1, \lambda_2, \ldots, \lambda_n, x)}{\partial x} \tag{2.34}$$

For a rotational system we would have

$$T_e = T_e(\lambda_1, \lambda_2, \ldots, \lambda_n, \theta) = -\frac{\partial W(\lambda_1, \lambda_2, \ldots, \lambda_n, \theta)}{\partial \theta} \tag{2.35}$$

The reader should compare (2.34) and (2.35) to the expressions for a singly excited system developed previously, that is, (2.18) and (2.20).

If currents $i_1, i_2, \ldots, i_n$ and displacement x are chosen as independent variables, flux linkages $\lambda_1, \lambda_2, \ldots, \lambda_n$ must be expressed as functions of the independent variables chosen. That is, $\lambda_i = \lambda_i(i_1, i_2, \ldots, i_n, x)$ for all λ_i, $W = W(i_1, i_2, \ldots, i_n, x)$, and $F_e = F_e(i_1, i_2, \ldots, i_n, x)$. Then we must have

$$
\begin{aligned}
d\lambda_i(i_1, i_2, \ldots, i_n, x) &= \frac{\partial \lambda_i(i_1, i_2, \ldots, i_n, x)}{\partial i_1} di_1 \\
&+ \frac{\partial \lambda_i(i_1, i_2, \ldots, i_n, x)}{\partial i_2} di_2 \\
&\quad\quad\quad \vdots \\
&+ \frac{\partial \lambda_i(i_1, i_2, \ldots, i_n, x)}{\partial i_n} di_n \\
&+ \frac{\partial \lambda_i(i_1, i_2, \ldots, i_n, x)}{\partial x} dx \\
&= \frac{\partial \lambda_i(i_1, i_2, \ldots, i_n, x)}{\partial x} dx \\
&+ \sum_{j=1}^{n} \frac{\partial \lambda_i(i_1, i_2, \ldots, i_n, x)}{\partial i_j} di_j
\end{aligned}
\tag{2.36}
$$

Similarly, we must have

$$
dW(i_1, i_2, \ldots, i_n, x) = \sum_{j=1}^{n} \frac{\partial W(i_1, i_2, \ldots, i_n, x)}{\partial i_j} di_j + \frac{\partial W(i_1, i_2, \ldots, i_n, x)}{\partial x} dx
\tag{2.37}
$$

Now substitute (2.36) into (2.32):

$$
\begin{aligned}
dW(i_1, i_2, \ldots, i_n, x) \\
= \sum_{i=1}^{n} i_i \frac{\partial \lambda_i(i_1, i_2, \ldots, i_n, x)}{\partial x} dx \\
+ \sum_{i=1}^{n} \sum_{j=1}^{n} i_i \frac{\partial \lambda_i(i_1, i_2, \ldots, i_n, x)}{\partial i_j} di_j - F_e(i_1, i_2, \ldots, i_n, x) dx \\
= \left(-F_e(i_1, i_2, \ldots, i_n, x) + \sum_{i=1}^{n} i_i \frac{\partial \lambda_i(i_1, i_2, \ldots, i_n, x)}{\partial x} \right) dx \\
+ \sum_{i=1}^{n} \sum_{j=1}^{n} i_i \frac{\partial \lambda_i(i_1, i_2, \ldots, i_n, x)}{\partial i_j} di_j
\end{aligned}
\tag{2.38}
$$

By comparing (2.37) and (2.38), we can write

$$F_e(i_1, i_2, \ldots, i_n, x)$$

$$= \sum_{i=1}^{n} i_i \frac{\partial \lambda_i(i_1, i_2, \ldots, i_n, x)}{\partial x} - \frac{\partial W(i_1, i_2, \ldots, i_n, x)}{\partial x} \qquad (2.39)$$

$$= \frac{\partial}{\partial x} \left[\sum_{i=1}^{n} i_i \lambda_i(i_1, i_2, \ldots, i_n, x) - W(i_1, i_2, \ldots, i_n, x) \right]$$

By defining coenergy $W_c = W_c(i_1, i_2, \ldots, i_n, x)$ as

$$W_c(i_1, i_2, \ldots, i_n, x) = \sum_{i=1}^{n} i_i \lambda_i(i_1, i_2, \ldots, i_n, x)$$

$$- W(i_1, i_2, \ldots, i_n, x) \qquad (2.40)$$

we can say

$$F_e(i_1, i_2, \ldots, i_n, x) = + \frac{\partial W_c(i_1, i_2, \ldots, i_n, x)}{\partial x} \qquad (2.41)$$

The reader should compare (2.39), (2.40), and (2.41) with the expressions for a singly excited system developed previously [i.e., (2.26), (2.27), and (2.28)].

A similar development can be made for rotational motion resulting in

$$T_e(i_1, i_2, \ldots, i_n, \theta) = + \frac{\partial W_c(i_1, i_2, \ldots, i_n, \theta)}{\partial \theta} \qquad (2.42)$$

Equations (2.34), (2.35), (2.41), and (2.42) essentially summarize the macroscopic viewpoint. Force and torque expressions for singly excited systems follow by simply recognizing that $n = 1$ in such systems. We summarize relevant relationships in more compact form by defining

$$\bar{i} = \begin{bmatrix} i_1 \\ i_2 \\ \cdot \\ \cdot \\ \cdot \\ i_n \end{bmatrix} \qquad (2.43a)$$

$$\bar{\lambda} = \begin{bmatrix} \lambda_1 \\ \lambda_2 \\ \cdot \\ \cdot \\ \cdot \\ \lambda_n \end{bmatrix} \qquad (2.43b)$$

Thus

$$F_e(\bar{\lambda}, x) = - \frac{\partial W(\bar{\lambda}, x)}{\partial x} \qquad (2.44)$$

$$F_e(\bar{i}, x) = +\frac{\partial W_c(\bar{i}, x)}{\partial x} \tag{2.45}$$

$$T_e(\bar{\lambda}, \theta) = -\frac{\partial W(\bar{\lambda}, \theta)}{\partial \theta} \tag{2.46}$$

$$T_e(\bar{i}, \theta) = +\frac{\partial W_c(\bar{i}, \theta)}{\partial \theta} \tag{2.47}$$

Our relationships are valid regardless of whether or not we have linear relationships between flux linkages and currents. If linearity is assumed, flux linkages in any one winding can be expressed as a simple linear combination of currents:

$$\bar{\lambda} = \bar{L}i$$
$$= \bar{L}(x)\bar{i} \tag{2.48}$$

or

$$= \bar{L}(\theta)\bar{i}$$

where $\bar{L}$ is an $n \times n$ matrix of inductance coefficients, which we now recognize will, in general, be functions of x or θ. In Section 1.7 we showed that the total stored energy in such a linear system could be expressed as

$$W(\bar{i}) = \tfrac{1}{2} \sum_i \sum_j L_{ij} i_i i_j$$
$$= \tfrac{1}{2} \bar{i}^T \bar{L}\bar{i} \tag{2.49}$$

Now recognize that

$$\sum_{i=1}^{n} i_i \lambda_i = \bar{i}^T \bar{\lambda} = \bar{i}^T \bar{L}\bar{i} \tag{2.50}$$

Now let $\bar{L} = \bar{L}(x)$ and substitute (2.49) and (2.50) in (2.40):

$$W_c(i_1, i_2, \ldots, i_n, x) = \bar{i}^T \bar{L}(x)\bar{i} - \tfrac{1}{2} \bar{i}^T \bar{L}\bar{i}$$
$$= \tfrac{1}{2} \bar{i}^T \bar{L}(x)\bar{i} \tag{2.51}$$
$$= W(i_1, i_2, \ldots, i_n, x)$$

Thus *for a linear system,*

$$W_c(\bar{i}, x) = W(\bar{i}, x) = \tfrac{1}{2} \bar{i}^T \bar{L}(x)\bar{i} \tag{2.52}$$

and similarly,

$$W_c(\bar{i}, \theta) = W(\bar{i}, \theta) = \tfrac{1}{2} \bar{i}^T \bar{L}(\theta)\bar{i} \tag{2.53}$$

EXAMPLE 2.9 Find an expression for electrical torque T_e in terms of i_1, i_2, and θ for the system shown in Figure 2.22.

If we assume linearity, (2.47) and (2.53) can be used to achieve the desired result. The flux linkages in windings 1 and 2 can be expressed as

$$\lambda_1 = L_{11}(\theta)i_1 + M(\theta)i_2$$
$$\lambda_2 = M(\theta)i_1 + L_{22}(\theta)i_2$$

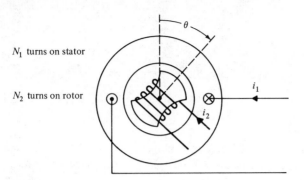

N_1 turns on stator

N_2 turns on rotor

FIGURE 2.22

We examine the nature of $L_{11}(\theta)$ by recalling that

$$L_{11}(\theta) = \left.\frac{\lambda_1}{i_1}\right|_{i_2=0}$$

$$= \frac{N_1^2}{\mathcal{R}(\theta)}$$

It is clear that the magnetic circuit reluctance will be at a minimum when $\theta = 0$, π, 2π, ... , $n\pi$. It will be at a maximum when $\theta = \pi/2$, $3\pi/2$, $5\pi/2$, ... , $\pi/2 + n\pi$. The inductance $L_{11}(\theta)$ is inversely proportional to $\mathcal{R}(\theta)$ and peaks when $\mathcal{R}(\theta)$ is at a minimum, and vice versa. It is thus a periodic function of 2θ. To simplify the example, we will assume that the inductance $L_{11}(\theta)$ varies sinusoidally, that is,

$$L_{11}(\theta) = L_s + \Delta L_s \cos 2\theta$$

where $L_s + \Delta L_s$ and $L_s - \Delta L_s$ are maximum and minimum values, respectively, of the stator self-inductance. The self-inductance of winding 2 is independent of rotor position θ. Thus

$$L_{22}(\theta) = L_r$$

The mutual inductance $M(\theta)$ can be examined by recalling that

$$M(\theta) = \left.\frac{\lambda_1}{i_2}\right|_{i_1=0}$$

For a fixed current i_2, flux linkages λ_1 in winding 1 will peak when $\theta = 0$, 2π, ... , $2n\pi$. They will be zero at $\theta = \pi/2$, $3\pi/2$, ... , $\pi/2 + n\pi$. They will peak in the negative direction at $\theta = \pi$, 3π, 5π, ... , $\pi + 2n\pi$. If we assume sinusoidal variation we will have

$$M(\theta) = L_m \cos \theta$$

where L_m is the maximum value of stator–rotor mutual inductance. Now from (2.53)

$$W_c(i_1, i_2, \theta) = \tfrac{1}{2} L_{11}(\theta)i_1^2 + M(\theta)i_1 i_2 + \tfrac{1}{2} L_{22}(\theta)i_2^2$$

and from (2.47)

$$T_e(i_1, i_2, \theta) = + \frac{\partial W_c(i_1, i_2, \theta)}{\partial \theta}$$

$$= \frac{i_1^2}{2}\frac{dL_{11}(\theta)}{d\theta} + i_1 i_2 \frac{dM(\theta)}{d\theta} + \frac{i_2^2}{2}\frac{dL_{22}(\theta)}{d\theta}$$

$$= -i_1^2 \Delta L_s \sin 2\theta - i_1 i_2 L_m \sin \theta$$

which is the desired result.

Examination of the expression for T_e reveals some interesting characteristics. If the unsymmetrical rotor shown in Figure 2.22 is replaced with a round rotor, as shown in Figure 2.23(b), there will be no variation of magnetic circuit reluctance with position of the rotor. That is, $\mathcal{R}(\theta) = \mathcal{R}$ a constant. The self-inductance term $L_{11}(\theta)$ will be independent of rotor position. Thus ΔL_s will be zero and we will have

$$T_e = - i_1 i_2 L_m \sin \theta$$

If the unsymmetrical rotor is used [Figure 2.23(a)], we can have nonzero torque even if i_2 is zero. For example, we might leave the rotor circuit as an open circuit (or remove the winding entirely) and have

$$T_e = - i_1^2 \, \Delta L_s \sin 2\theta$$

Torque developed in this manner is often referred to as *reluctance torque*. Any machine configuration that has a variation in magnetic circuit reluctance with rotor position will exhibit reluctance torque. By examining Figure 2.22 and recalling that our methods always yield torque in the direction of positive $d\theta$, we see that reluctance torque tends to align the rotor with the axis of the stator winding; that is, torque is developed in such a direction as to minimize magnetic circuit reluctance.

The special case of sinusoidal excitation is of interest. Consider the following case:

1. Machine with an unsymmetrical rotor (i.e., Figure 2.22).
2. $i_2 = 0$ (the rotor circuit an open circuit).
3. Rotor turning at constant speed ω_r:

$$\theta(t) = \omega_r t + \theta_0$$

4. Sinusoidal time variation of the stator current, that is,

$$i_1(t) = \sqrt{2}\, I \cos \omega_s t$$

Then we have for torque:

$$
\begin{aligned}
T_e &= -\left(\sqrt{2}\, I \cos \omega_s t\right)^2 \Delta L_s \sin\left(2\omega_r t + 2\theta_0\right) \\
&= -I^2\, \Delta L_s\, (1 + \cos 2\omega_s t)\, \sin\,(2\omega_r t + 2\theta_0) \\
&= -I^2\, \Delta L_s \left\{ \sin\,(2\omega_r t + 2\theta_0) + \tfrac{1}{2} \sin\left[2(\omega_s + \omega_r)t + 2\theta_0\right] \right. \\
&\qquad\qquad \left. + \tfrac{1}{2} \sin\left[2(\omega_r - \omega_s)t + 2\theta_0\right] \right\}
\end{aligned}
$$

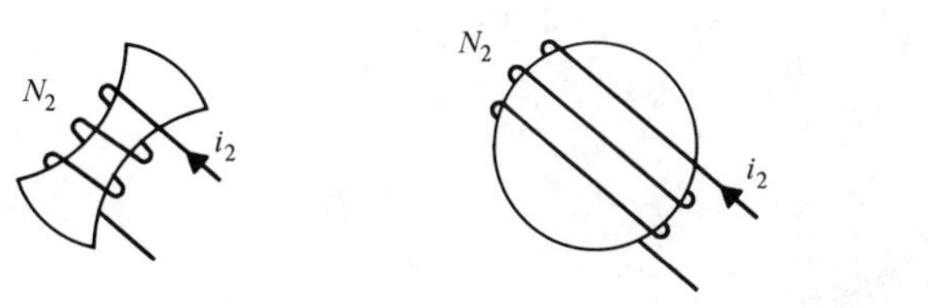

(a) Unsymmetrical rotor   (b) Round rotor (symmetrical rotor)

FIGURE 2.23

By examining the terms in braces we see that we can only have a nonzero time-average torque contribution from the third term and then only if $\omega_r = \omega_s$. Furthermore, the average torque would depend on θ_0, that is, the position of the rotor at time $t = 0$. Development of nonzero average torque at only one speed is a characteristic of synchronous machines. Even if, in our present example, we could shape the members of the magnetic circuit to achieve the sinusoidal inductance variation we have assumed, our machine would have significant undesirable characteristics. We would have a double-frequency torque component and a torque component at $4\omega_s$, both of large magnitude compared to our average torque term. We would also have harmonics in the induced stator and rotor voltage waveforms. These characteristics would be unacceptable in large rotating machinery. Fortunately, they can be eliminated by going to a polyphase machine.

2.5

TORQUE IN SMOOTH-AIR-GAP MACHINES

A smooth-air-gap machine is depicted in Figure 2.24. The stator of the machine has a single winding whose turns are distributed in slots as shown in the figure. The actual connections between turns of the stator winding need not concern us at the moment. We will simply assume that connections are made in such a way as to produce currents in the direction indicated by the "×" and "·" symbols shown in the figure. A single rotor winding is similarly distributed in slots about the periphery of the rotor.

The terminology "smooth air gap" is a misnomer. The slots in which rotor and stator windings are distributed prevent us from having a truly smooth, or uniform, air gap. We use the terminology to describe a situation where the

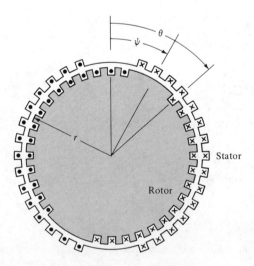

FIGURE 2.24 A Smooth-Air-Gap Machine

magnetic circuit reluctance does not change appreciably as the angle ψ (the angle between the main axis of the stator winding and the main axis of the rotor winding) is varied. If we were to replace our round rotor with an unsymmetrical rotor such as the one depicted in Figure 2.23(a), we would have a significant variation in magnetic circuit reluctance with changes in ψ and the assumption of a smooth air gap would not be appropriate. With a round (symmetrical) rotor [Figure 2.23(b)], the essential action of the machine can be treated by assuming a smooth gap. The effects of slots can be reduced in practice by "skewing" the slots of the rotor in the axial direction.

If the stator winding is suitably arranged in the slots of the stator, the magnetic flux density in the air gap due to stator current will be nearly sinusoidal in nature. That is, the radial component of air-gap magnetic flux density at radius r due to stator current can be expressed approximately as

$$B_s(\theta) = B_{smax} \cos \theta \tag{2.54}$$

In reality $B_s(\theta)$ will consist of a fundamental and odd harmonics. The harmonics produce undesirable effects and a major effort is made to minimize them in design. By making similar assumptions about the rotor winding distribution, we can express the radial component of the air-gap magnetic flux density at radius r due to rotor current as

$$B_r(\theta) = B_{rmax} \cos (\theta - \psi) \tag{2.55}$$

Figure 2.25 presents a conceptual version of the situation we have described. The angle ψ is the angle between the main axis of the stator and rotor windings. The angle θ is used to denote any arbitrary angular displacement from the main axis of the stator winding.

We can find an expression for torque developed between the rotor and stator by expressing energy or coenergy in terms of convenient variables and taking the partial derivative with respect to the angular displacement between the rotor and stator (i.e., with respect to ψ). In anticipation of the fact that currents will be a convenient choice, we elect to find an expression for coenergy. The radial components of the air-gap magnetic field intensities $H_s(\theta)$ and $H_r(\theta)$ are related

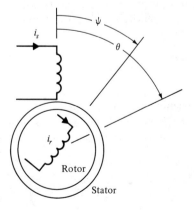

FIGURE 2.25

to the respective magnetic flux densities $B_s(\theta)$ and $B_r(\theta)$ by the permeability of free space μ_0:

$$H_s(\theta) = \frac{B_s(\theta)}{\mu_0} = \frac{B_{smax}}{\mu_0} \cos\theta = H_{smax} \cos\theta \tag{2.56}$$

$$H_r(\theta) = \frac{B_r(\theta)}{\mu_0} = \frac{B_{rmax}}{\mu_0} \cos(\theta - \psi) = H_{rmax} \cos(\theta - \psi) \tag{2.57}$$

The total radial component of the air-gap magnetic field intensity at radius r will thus be

$$H(\theta) = H_{smax} \cos\theta + H_{rmax} \cos(\theta - \psi) \tag{2.58}$$

If we assume that the stator and rotor iron has high relative permeability, the stored energy will be predominately associated with the air gap. (As $\mu_r \to \infty$, $H \to 0$ in the stator and rotor iron.) If the tangential component of the air-gap magnetic field intensity is assumed negligible compared to the radial component, the coenergy density in the gap can be expressed as

$$w_c(\theta) = \frac{\mu_0 H^2(\theta)}{2}$$

$$= \frac{\mu_0}{2}[H_{smax}^2 \cos^2\theta + 2H_{smax}H_{rmax} \cos\theta \cos(\theta - \psi)$$

$$+ H_{rmax}^2 \cos^2(\theta - \psi)] \tag{2.59}$$

$$= \frac{\mu_0}{2}\left\{ \frac{H_{smax}^2}{2}(1 + \cos 2\theta) + H_{smax}H_{rmax} \cos(2\theta - \psi) \right.$$

$$\left. + H_{smax}H_{rmax} \cos\psi + \frac{H_{rmax}^2}{2}[1 + \cos(2\theta - 2\psi)] \right\} \quad J/m^3$$

The total coenergy W_c can be found by integrating this coenergy density over the volume of the air gap. If we assume that the gap g is small compared to the radius r, we can neglect the variation of magnetic field intensity with respect to radius. For a given angle ψ the coenergy density is only a function of θ and the total coenergy will be

$$W_c = \int_0^{2\pi} w_c(\theta)g\ell r \, d\theta \tag{2.60}$$

where g is the gap width, ℓ is the machine axial length, and r may be taken as the radius of the rotor. Now substitute (2.59) into (2.60) and recognize that the integral over 2π will be zero for all sinusoidal terms in θ or 2θ. We find for the coenergy:

$$W_c = \pi\mu_0 g\ell r\left(\frac{H_{smax}^2}{2} + H_{smax}H_{rmax} \cos\psi + \frac{H_{rmax}^2}{2} \right) \tag{2.61}$$

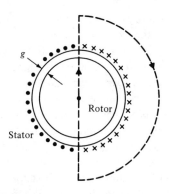

FIGURE 2.26

We can find H_{smax} in terms of the stator current i_s by applying Ampère's law:

$$\oint \overline{H} \cdot d\overline{\ell} = \int \overline{J} \cdot d\overline{s} \qquad (2.62)$$

Choose a path of integration as shown by the dashed line in Figure 2.26. Because of our assumption of high permeability in the stator and rotor iron, H is negligible except in the gap, where it is taken as a constant over the gap width. Thus the left-hand side of (2.62) becomes

$$\oint \overline{H} \cdot d\overline{\ell} = gH_{smax} \cos (0) - gH_{smax} \cos (\pi) = 2gH_{smax}$$

The right-hand side of (2.62) is $N_s i_s$, where N_s is the number of turns on the stator winding. Thus

$$H_{smax} = \frac{N_s i_s}{2g} \qquad (2.63)$$

Similarly,

$$H_{rmax} = \frac{N_r i_r}{2g} \qquad (2.64)$$

Substitution in (2.61) yields

$$W_c(i_s, i_r, \psi) = \frac{\pi \mu_0 \ell r}{g} \left[\frac{(N_s i_s)^2}{8} + \frac{N_s N_r i_s i_r}{4} \cos \psi + \frac{(N_r i_r)^2}{8} \right] \qquad (2.65)$$

Thus we can find the torque as

$$T_e(i_s, i_r, \psi) = + \frac{\partial W_c(i_s, i_r, \psi)}{\partial \psi}$$

$$= - \frac{\pi \mu_0 \ell r N_s N_r}{4g} i_s i_r \sin \psi \qquad (2.66)$$

Examination of (2.66) reveals that the torque is developed in such a direction as to tend to align the axes of the stator and rotor windings. The torque is proportional to the product of stator and rotor current and the sine of the angle between the axes of the stator and rotor windings. If the stator and rotor currents are constants (i.e., direct currents), the angle between the stator and rotor axes must be a constant if the developed torque is to be constant. If the stator and rotor currents are constants and the stator (and stator winding) rotates at constant speed ω_s, there is only one constant speed that the rotor can have in order to develop nonzero average torque: ω_s. That is, the two must rotate in synchronism. Machines that operate with a developed torque that is constant in time can be analyzed in terms of single equivalent stator and rotor windings carrying constant currents and having a fixed angle between their respective axes. Thus (2.66) is more generally applicable than what one might suspect based on the derivation presented. For a given set of machine physical dimensions and stator and rotor winding data, it can be used to determine an upper limit on the torque under idealized conditions. By substituting in (2.66) using (2.63) and (2.64), the torque can be expressed in terms of magnetic field intensities. It can also be expressed in terms of magnetic flux densities, total flux per pole, and so on. Recognizing that the resultant magnetic flux densities will typically be limited to approximately 1 T is also relevant. We will find that for particular machines, alternative means of determining the torque under steady-state conditions using equivalent circuits are in general simpler. Such methods, however, may yield less insight into the restrictions imposed by dimensions and material properties.

2.6

ROTATING FIELDS IN POLYPHASE AC MACHINES

If multiple windings displaced by a fixed angle in space are excited by currents appropriately displaced in time, it is possible to establish a magnetic field that rotates with respect to the windings. The nature of the field depends on the nature of the flux density distribution for each winding and on the nature of the excitation. Figure 2.27 shows the stator and rotor of a smooth-air-gap machine. Two windings are shown on the stator, whose main axes are displaced by $\pi/2$ radians. The angle θ denotes an arbitrary position whose angular displacement is θ radians from the main axis of the winding carrying current i_{as}. To simplify

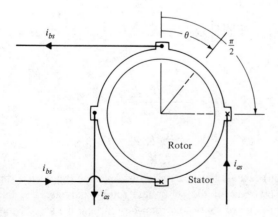

FIGURE 2.27

FUNDAMENTALS OF ELECTROMECHANICAL ENERGY CONVERSION

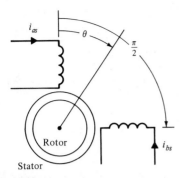

FIGURE 2.28

the figure, the windings are shown as concentrated windings. Generally, each winding would be distributed in slots about the stator as suggested in Figure 2.24. The subscripts *as* and *bs* are used to distinguish between the two stator windings, which are assumed to constitute two distinct circuits. Figure 2.28 presents a conceptual version of the situation depicted in Figure 2.27.

If we assume a sinusoidal distribution of the radial component of air-gap magnetic flux density (and field intensity) for each winding, we can write

$$H_{as}(\theta) = H_{asmax} \cos \theta \tag{2.67}$$

$$H_{bs}(\theta) = H_{bsmax} \cos \left(\theta - \frac{\pi}{2} \right) \tag{2.68}$$

where $H_{as}(\theta)$ and $H_{bs}(\theta)$ are the radial components of magnetic field intensity due to windings a and b, respectively. By applying Ampère's law (as in Section 2.5), we can express the maximum field intensities in terms of turns, current, and air-gap width:

$$H_{asmax} = \frac{N_{as}i_{as}}{2g} \tag{2.69}$$

$$H_{bsmax} = \frac{N_{bs}i_{bs}}{2g} \tag{2.70}$$

For a symmetrical two-phase machine we will have $N_{as} = N_{bs} = N_s$. Thus the resultant magnetic field intensity in the gap due to phases *a* and *b* can be expressed as

$$H_s(\theta) = H_{as}(\theta) + H_{bs}(\theta) \tag{2.71}$$

$$= \frac{N_s}{2g} \left[i_{as} \cos \theta + i_{bs} \cos \left(\theta - \frac{\pi}{2} \right) \right]$$

Now if i_{as} and i_{bs} constitute a set of balanced two-phase currents, we will have

$$i_{as} = \sqrt{2} I_s \cos (\omega_s t + \varphi_s) \tag{2.72}$$

$$i_{bs} = \sqrt{2} I_s \cos \left(\omega_s t + \varphi_s - \frac{\pi}{2} \right) \tag{2.73}$$

where φ_s is arbitrary. The currents are equal in magnitude but phase displaced in time by $\pi/2$ radians. We can take $\varphi_s = 0$ for convenience and substitute (2.72) and (2.73) into (2.71) to obtain

$$H_s(\theta, t) = \frac{\sqrt{2}\, I_s N_s}{2g} \left[\cos \omega_s t \cos \theta + \cos \left(\omega_s t - \frac{\pi}{2} \right) \cos \left(\theta - \frac{\pi}{2} \right) \right]$$

$$= \frac{\sqrt{2}\, I_s N_s}{2g} \cos (\omega_s t - \theta) \qquad\qquad (2.74)$$

Equation (2.74) is the equation of a sinusoidally distributed magnetic field intensity wave which rotates at speed ω_s. Its effect is the same as that which would be realized by a single winding of N_s turns carrying a direct current $i_s = \sqrt{2}\, I_s$ rotating at speed ω_s. The situation is depicted conceptually in Figure 2.29. This same result may be produced by other choices of polyphase systems, the most commonly used being the three-phase system. If a single winding on the rotor carries a direct current, we can satisfy the conditions for constant torque described in Section 2.5 provided that the rotor turns at speed ω_s. This is the essence of synchronous machine action, so named because of the requirement that the rotor turn in synchronism with the rotating field established by the stator.

It is not necessary that the rotor itself turn at synchronous speed ω_s. If the field established by the rotor currents rotates at speed ω_s, we can satisfy the conditions described in Section 2.5 for constant torque. A polyphase rotor winding carrying sinusoidal currents appropriately phase displaced can produce a field that rotates with respect to the rotor. Consider the rotor depicted conceptually in Figure 2.30. The figure shows a two-phase winding on the stator and a three-phase winding on the rotor. The rotor and its three-phase winding are assumed to turn at speed ω_r. If rotor currents must also flow in stationary circuits external to the machine, a *slip-ring* system is generally required. The actual number of phases on either the stator or rotor is irrelevant to the final result provided that both stator and rotor are capable of establishing fields that rotate in synchronism. If the stator windings are excited by the currents shown in (2.72) and (2.73), a rotating stator field is produced as depicted conceptually in

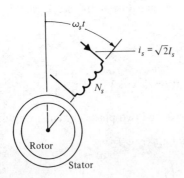

FIGURE 2.29

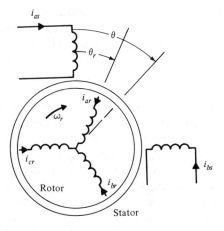

FIGURE 2.30

Figure 2.29. Now if the rotor windings each establish sinusoidally distributed magnetic field intensities, we can write

$$H_{ar}(\theta) = H_{ar\max} \cos (\theta - \theta_r) \tag{2.75}$$

$$H_{br}(\theta) = H_{br\max} \cos \left(\theta - \theta_r - \frac{2\pi}{3} \right) \tag{2.76}$$

$$H_{cr}(\theta) = H_{cr\max} \cos \left(\theta - \theta_r - \frac{4\pi}{3} \right) \tag{2.77}$$

By expressing maximum rotor field intensities in terms of rotor winding turns and rotor winding currents using Ampère's law and assuming an equal number of turns on each rotor winding (i.e., a symmetrical rotor), we can express the resultant rotor magnetic field intensity as

$$
\begin{aligned}
H_r(\theta) &= H_{ar}(\theta) + H_{br}(\theta) + H_{cr}(\theta) \\
&= \frac{N_r}{2g} \left[i_{ar} \cos (\theta - \theta_r) + i_{br} \cos \left(\theta - \theta_r - \frac{2\pi}{3} \right) \right. \\
&\quad \left. + i_{cr} \cos \left(\theta - \theta_r - \frac{4\pi}{3} \right) \right]
\end{aligned} \tag{2.78}
$$

Now if i_{ar}, i_{br}, and i_{cr} constitute a set of balanced three-phase currents at frequency $\omega_s - \omega_r$ (the reason for this choice will become apparent shortly), we can write

$$i_{ar} = \sqrt{2} \, I_r \cos [(\omega_s - \omega_r)t - \varphi_r] \tag{2.79}$$

$$i_{br} = \sqrt{2} \, I_r \cos \left[(\omega_s - \omega_r)t - \varphi_r - \frac{2\pi}{3} \right] \tag{2.80}$$

$$i_{cr} = \sqrt{2} \, I_r \cos \left[(\omega_s - \omega_r)t - \varphi_r - \frac{4\pi}{3} \right] \tag{2.81}$$

The arbitrary angle φ_r will be retained in what follows for the sake of generality. Now if the rotor turns at speed ω_r, then $\theta_r(t) = \omega_r t + \theta_r(0)$. Substitution of (2.79), (2.80), and (2.81) into (2.78) and simplification yields

$$H_r(\theta, t) = \frac{N_r \sqrt{2}\, I_r}{2g} \frac{3}{2} \cos [\omega_s t - \theta - \varphi_r + \theta_r(0)] \qquad (2.82)$$

Equation (2.82) is the equation of a sinusoidally distributed magnetic field intensity wave which rotates at speed ω_s with respect to the stator. It rotates at speed $\omega_s - \omega_r$ with respect to the rotor. Its effect is the same as that which would be realized by a single rotor winding of N_r turns carrying a direct current $i_r = \frac{3}{2} \sqrt{2}\, I_r$ and rotating at speed ω_s with respect to the stator. This wave would rotate in synchronism with the stator-induced wave of (2.74) and we could satisfy the conditions required for constant torque. In conventional induction machine action the rotor currents are induced in short-circuited rotor windings by a resultant magnetic field that rotates at speed $\omega_s - \omega_r$ with respect to the rotor. Our assumption about the frequency $\omega_s - \omega_r$ of induced rotor currents is thus representative of conventional induction machine action. Figure 2.31 summarizes conceptually our final result for induction machine action. The figure shows the fictitious stator and rotor windings carrying direct currents and rotating at speed ω_s. The phase "a" stator winding is shown in dashed line for the sake of clarity. All other actual windings carrying alternating currents have been omitted. Note that the rotor itself turns at speed ω_r and it is not necessary that $\omega_r = \omega_s$. The actual angular displacement between the fictitious rotor and stator windings $[-\varphi_r + \theta_r(0)$ in this case] can be established by analysis of an equivalent circuit, which we will develop shortly.

We have presented only two possibilities for the establishment of synchronous rotation of stator and rotor magnetic fields. A more general treatment of the conditions necessary for conversion of nonzero average power can be found in Chapter 4 of reference 2.

Our examples in this section have been based on stator and rotor windings

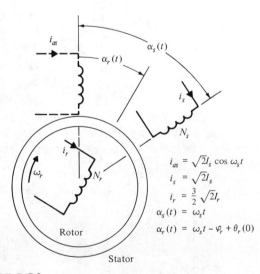

$$i_{as} = \sqrt{2}I_s \cos \omega_s t$$
$$i_s = \sqrt{2}I_s$$
$$i_r = \frac{3}{2}\sqrt{2}I_r$$
$$\alpha_s(t) = \omega_s t$$
$$\alpha_r(t) = \omega_s t - \varphi_r + \theta_r(0)$$

FIGURE 2.31

FUNDAMENTALS OF ELECTROMECHANICAL ENERGY
CONVERSION

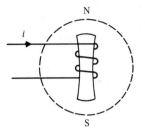

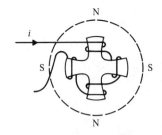

(a) Rotor wound for two poles   (b) Rotor wound for four poles

FIGURE 2.32 Rotor Windings for Different Numbers of Poles

which establish two-pole fields. Stator and rotor windings of a machine can be designed to develop an arbitrary even number of magnetic poles. Figure 2.32 illustrates rotor windings that would establish two- and four-pole magnetic fields. In a somewhat similar way, each phase winding on the stator of a machine can be designed to develop an arbitrary even number of magnetic poles. If each phase of a suitably designed polyphase stator winding is wound in such a way as to establish P magnetic poles, a rotating magnetic field produced by the winding will have P magnetic poles, and it will rotate at a speed $\omega_m = (2/P)\omega_s$, where ω_s is the radian frequency of the balanced polyphase winding currents. Thus the synchronous speed for induction machines and synchronous machines will be determined by the number of poles for which the machine is wound. For example, if the electrical frequency is 60 Hz, a two-pole synchronous or induction machine has a synchronous speed of 3600 rpm, a four-pole machine has a synchronous speed of 1800 rpm, and in general a P-pole machine will have a synchronous speed of $60f$ divided by the number of pole pairs:

$$n = \frac{60f}{P/2}$$

$$= \frac{120f}{P} \quad \text{rpm}$$

(2.83)

where f is the electrical frequency of the supply in hertz.

2.7

FIELD ORIENTATIONS IN CONVENTIONAL AC AND DC MACHINES

In Section 2.5 we derived an expression for developed torque between single-stator and single-rotor windings assuming a smooth air gap. Conditions for a constant torque in time were presented. In Section 2.6 we showed that a polyphase winding appropriately excited by sinusoidal currents will establish a rotating sinusoidally distributed magnetic field. A polyphase winding system thus produces the same effect as a single winding carrying direct current and rotating at a fixed speed. The actual orientation of stator- and rotor-established fields in ac machines is most easily determined by using equivalent circuits which we will develop shortly. In ac machines the resultant field, which is the sum of the

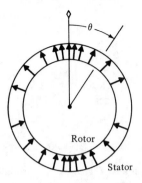

FIGURE 2.33 Space Vector and the Distribution It Represents

rotor-induced and stator-induced effects is of interest for two reasons. First, saturation effects essentially limit the resultant field. Second, induced voltages in stator and rotor windings must take into account the resultant field.

Generally, stator winding resistance is kept very low and stator terminal voltage can essentially be determined by applying Faraday's law if we use the resultant field. If a particular terminal voltage condition exists at the terminals of a stator winding, the resultant field magnitude, speed of rotation, *and* position with respect to a fixed reference angle (such as the angle of the main axis of the phase "a" stator winding) must all be established in accordance with Faraday's law. In other words, specification of the terminal voltage magnitude, frequency, and phase angle for any phase of the stator winding will essentially fix the magnitude, frequency, and position of the *resultant* rotating magnetic field. The rotor and stator winding fields must then be established in such a way to yield the appropriate resultant field. If currents in stator and rotor windings establish fields that are sinusoidally distributed around the gap, the resultant field will also be sinusoidally distributed around the gap (see Problem 2.20). In this case fields add in the manner of vectors and it is often convenient to discuss them in terms of a rotating space vector. In this context, Figure 2.33 illustrates a space vector and the field it represents. In the figure the nature of the distribution around the air gap is illustrated by lines drawn in a radial direction. The

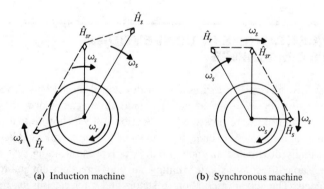

(a) Induction machine   (b) Synchronous machine

FIGURE 2.34 Typical Field Alignments in Induction and Synchronous Machines (Operating as Motors)

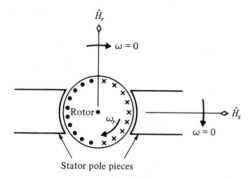

FIGURE 2.35 Typical Field Alignment in a DC Machine

relative "density" of the lines indicates the relative magnitude of the field at any point and the field at any point is taken to be in the direction shown by the lines. Thus if we are discussing the radial component of a magnetic field intensity H that is assumed to have a sinusoidal distribution, we have

$$\overline{H}(\theta) = H_{max} \cos \theta \overline{a}_r \qquad (2.84)$$

A space-vector representation of this field would simply be a vector equal in magnitude to H_{max} and it would be aligned with the $\theta = 0$ direction, that is, the direction in which $\overline{H}(\theta)$ is a positive maximum. We will denote such space vectors with the following notation: $\hat{H}$.

Typical induction machine and synchronous machine field alignments are shown in Figure 2.34. In a dc machine, commutator action yields an armature (rotor) field which is fixed in position regardless of the position or speed of the rotor. The field (stator) winding is stationary and carries a dc current, thus establishing a field that is stationary. Dc machines in general are not smooth-air-gap machines, and their rotor and stator fields are not in general sinusoidally distributed. Thus, although we might represent conceptually the stator and rotor fields with space vectors as shown in Figure 2.35 it is, strictly speaking, not correct unless we treat only the fundamental components of the respective fields.

Recognition of the basic nature of conventional machine action as depicted in Figures 2.34 and 2.35 will be of assistance as we proceed to analyze their behavior in greater detail in the following chapters.

REFERENCES

[1] DONALD G. FINK and H. WAYNE BEATY, *Standard Handbook for Electrical Engineers,* 11th ed. New York: McGraw-Hill Book Company, 1978.
[2] H. H. WOODSON and J. R. MELCHER, *Electromechanical Dynamics,* Part I: *Discrete Systems.* New York: John Wiley & Sons, Inc., 1968.

PROBLEMS

2.1 Find the force per unit of length developed between the two conductors of a single-phase circuit if the circuit supplies a 5-kW 120-V rms load whose power factor is 0.90 lagging. Conductors are spaced 15 cm apart. Find expressions for both instantaneous force and average force. Submit a simple sketch showing the instantaneous force and average force plotted as a function of time.

2.2 Find the average force per unit of length developed between the buses of Example 2.2 (using the same numerical data as used in Example 2.2) by using the expression for force per unit of length between two conductors which was developed in Example 2.1. Compare the result to the results of Example 2.2.

2.3 A straight conductor carrying current I rotates as shown in the figure, in a uniform magnetic field directed into the page. Find an expression for torque developed on the conductor and indicate its direction (i.e., clockwise or counterclockwise).

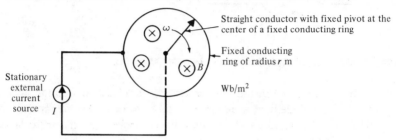

2.4 Refer to Problem 2.3, which shows a straight conductor rotating in a magnetic field. Find an expression for the voltage developed across the terminals of the stationary current source.

2.5 A four-turn coil arranged as shown in the figure replaces the single loop of wire considered in Example 2.4. If $B(\theta) = B_{max} \cos \theta$, where B is the radial component of the

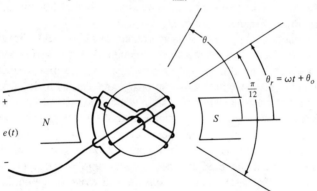

magnetic flux density (stationary with respect to the stator), find an expression for the open-circuit voltage $e(t)$ if the rotor turns at constant speed ω. Assume that $\theta_0 = \pi/24$. The axial length of the rotor is L. Assume that $B(\theta)$ is the same along the axial length L. The radius of the rotor is r.

2.6 Suppose that, in the magnetic circuit of Example 2.6 (Figure 2.13), the voltage at the terminals of the winding is constrained to be

$$e(t) = \sqrt{2}\,E \cos \omega t \quad V$$

Use the results of Example 2.6 to find an expression for developed force F_e as a function of time. [HINT: Apply Faraday's law.]

2.7 Suppose that the magnetic circuit of Example 2.6 (Figure 2.13) has a voltage at the terminals of the winding of

$$e(t) = \sqrt{2}\,E \cos \omega t \quad V$$

Assume steady-state conditions and use the results of Example 2.6 to find an expression for current as a function of time and displacement x.

2.8 A certain lossless magnetic circuit has the following relationship between the current i, flux linkages λ, and displacement x:

$$i = k_1 \lambda^3 + k_2 \lambda x$$

Sketch some typical curves relating λ and i with x as a parameter. Find an expression for F_e, the electrical force developed in the direction of increasing x. (Assume that $k_1 = 1.0$ A/Wb^3 and $k_2 = 1.0$ A/Wb $\cdot$ m.)

2.9 Compare expressions for force F_e developed in Examples 2.6 and 2.7. Show that the two yield the same result.

2.10 A coil oriented as shown in the figure has a movable (up and down) iron core. The inductance of the coil is found (experimentally) to be given by

$$L(y) = \frac{L_0}{1 + ky^2}$$

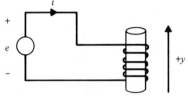

where $y = 0$ corresponds to the situation where the plunger is centered in the coil. Find an expression for the force F_e developed on the core as a function of the current i and displacement y.

2.11 Suppose that in Problem 2.10, $e(t) = \sqrt{2} E \cos \omega t$. Assume steady-state conditions, and find an expression for F_e as a function of the displacement y and time t. (Neglect the resistance of the wire.)

2.12 Suppose (as in Examples 2.6, 2.7, and 2.8) that the inductance of a magnetic circuit is given by

$$L(x) = \frac{N^2}{R_{\text{core}} + 2x/\mu_0 A}$$

Show that voltage across the terminals of the winding can be expressed as

$$e(t) = L(x) \frac{di}{dt} - \frac{2}{\mu_0 A} \frac{L^2(x)}{N^2} vi$$

where v is velocity (i.e., $v = dx/dt$).

2.13 The magnetic circuit shown has a reluctance given by:

$$\mathcal{R}(\theta) = \mathcal{R}_0 + \mathcal{R}_1 \theta$$

where θ is expressed in radians. The angle θ is assumed to be small.

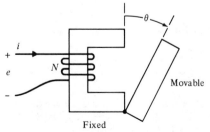

Find an expression for the electrical torque developed on the movable member of the magnetic circuit. Choose flux linkages and angular displacement as independent variables.

2.14 Consider the magnetic circuit of Problem 2.13. Find an expression for electrical torque developed on the movable member of the magnetic circuit if the current and angular displacement are chosen as independent variables.

2.15 Consider the magnetic circuit of Problem 2.13. Suppose that $i = I_0$, that is, a constant dc current. If the angle θ varies according to $\theta(t) = ct^3$, where c is a constant and t is the time in seconds, find the winding voltage $e(t)$. Neglect the winding resistance and leakage flux.

2.16 Write a set of equations in state-variable form to describe the system shown in Figure 2.18 (i.e., Example 2.8). Choose the current i and displacement x as independent variables. Make the same simplifying assumptions as were made in Example 2.8 and use the results of Problem 2.12 to simplify the answer.

2.17 Suppose that in Example 2.9 we use the round-rotor configuration, that is, the rotor shown in Figure 2.23(b). Find an expression for the time-average torque if

$$i_1 = \sqrt{2}\, I \cos \omega_s t$$

$$i_2 = I_2 \quad \text{(i.e., a direct current)}$$

$$\omega_r = \omega_s$$

$$\theta_0 = -\frac{\pi}{4} \quad \text{(i.e., the rotor angle at } t = 0)$$

2.18 The uniform-air-gap machine depicted in Figure 2.24 may be analyzed as a set of coupled coils. If linearity is assumed,

$$\lambda_s = L_{ss} i_s + M(\psi) i_r$$

$$\lambda_r = M(\psi) i_s + L_{rr} i_r$$

Find expressions for L_{ss}, $M(\psi)$, and L_{rr} in terms of the machine axial length ℓ, rotor radius r, gap width g, stator winding turns N_s, and rotor winding turns N_r. [HINT: Consider (2.65).]

2.19 A certain two-winding machine having a uniform air gap is operated in such a way that magnetic saturation does not occur. The flux linkages are linearly related to currents by

$$\lambda_s = L_{ss} i_s + M(\psi) i_r$$

$$\lambda_r = M(\psi) i_s + L_{rr} i_r$$

Because the flux density in the air gap is *not* sinusoidally distributed in space for this particular machine, the mutual inductances $M(\psi)$ is not a sinusoidal function of ψ. Find an expression for torque developed on the rotor of the machine if $M(\psi) = M_1 \cos \psi + M_3 \cos 3x + M_5 \cos 5\psi$.

2.20 Consider the machine configuration shown in Figure 2.24. Assumption of a sinusoidal distribution of the radial component of the magnetic field intensity around the gap permitted us to write [Eq. (2.58)]

$$H(\theta) = H_{smax} \cos \theta + H_{rmax} \cos (\theta - \psi)$$

This resultant magnetic field intensity $H(\theta)$ can also be expressed as

$$H(\theta) = H_{srmax} \cos (\theta - \alpha)$$

Find expressions for H_{srmax} and α in terms of H_{smax}, H_{rmax}, and ψ. (A sketch of the relevant geometry may be helpful.)

2.21 A certain machine consisting of a single-stator winding and single-rotor winding as shown in Figure 2.24 is designed to be operated in such a way that $N_s i_s \approx N_r i_r$. The resultant flux density in the air gap is limited to 1 T. Find the maximum possible torque the machine can develop if the axial length ℓ is 30 cm, the rotor radius r is 12 cm, and the gap width g is 0.30 cm. Assume that ψ must be $\leq \pi/2$.

FUNDAMENTALS OF ELECTROMECHANICAL ENERGY CONVERSION

2.22 In the magnetic circuit shown, the inductance is known to vary in an approximately *sinusoidal* manner as a function of the angle θ between the minimum and maximum values of L_{min} and L_{max}, respectively. Note that θ is specifically defined in the figure as the angle measured in a clockwise direction from the vertical to the axis of one of the rotor poles.

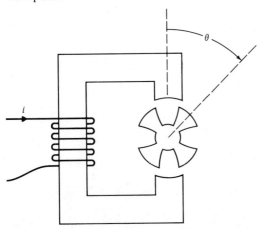

Find an expression for torque on the rotor in the clockwise direction. The answer should be in terms of parameters L_{min} and L_{max} and the independent variables chosen as the current i and angle θ.

(HINT: To start the problem, sketch a plot showing the inductance as a function of the angle θ assuming sinusoidal variation with θ; then find a suitable expression for the inductance as a function of θ.)

CHAPTER 3

The Induction Machine

INTRODUCTION

The three-phase induction (Figure 3.1) machine is widely regarded as the work-horse of industry. It is often the preferred choice over either a dc machine or a synchronous machine for an application requiring a fixed-speed drive motor. In a conventional polyphase induction machine, induced rotor currents establish a rotating magnetic field which rotates in synchronism with a rotating field due to armature (stator) currents. Because the rotor currents are induced, it is possible to operate the machine with no external connections to the rotor circuits. Thus brushes and slip rings can be eliminated. Most induction machines have *squirrel-cage rotors,* so named because of the resemblance of the rotor conductor assembly to a cage-like structure. The squirrel-cage induction machine is relatively simple in construction, simple in operation, rugged, maintenance free, and generally less expensive than either a dc machine or a synchronous machine of the same rating. A particularly attractive feature of the machine is its capability, in many applications, to be started from standstill using an *across-the-line* start. This means that with the rotor initially at rest we simply apply rated terminal voltage to its armature windings and allow the machine to come up to speed. This technique is generally not appropriate for synchronous or dc machines.

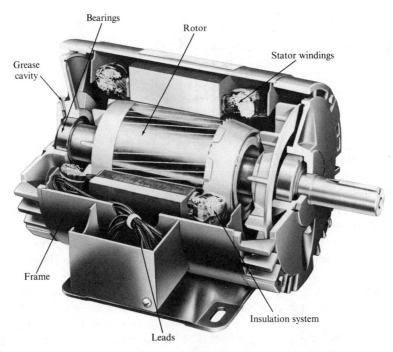

Bearings

Rotor

Stator windings

Grease cavity

Frame

Insulation system

Leads

FIGURE 3.1 Typical Three-Phase Induction Machine. (Courtesy of Marathon Electric)

The induction machine is essentially a fixed-speed drive. To operate efficiently, its rotor must turn at a speed near the synchronous speed, where the synchronous speed is determined by the frequency of the applied armature voltages and the number of poles for which the machine is wound. Efficient variable-speed operation essentially requires varying the frequency of the power supply. Advances in solid-state technology have permitted development of efficient variable-frequency power supplies and have thus significantly increased the range of applications for induction machines.

To analyze induction machine characteristics in a more comprehensive way, we start by developing an appropriate circuit model for steady-state operation of a polyphase induction machine. We then extend the model in a simple way to characterize performance for unbalanced operation, that is, when machine voltages and currents do not constitute balanced polyphase sets of variables. Performance of single-phase induction machines is treated as a special case of unbalanced operation.

3.2

THE STEADY-STATE EQUIVALENT CIRCUIT

The induction machine may be regarded as a smooth-air-gap machine and we can begin our treatment by extending the discussion presented in Sections 2.5, 2.6, and 2.7. As shown in Section 2.6, balanced polyphase currents in windings

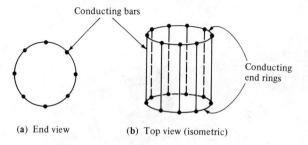

(a) End view   (b) Top view (isometric)

FIGURE 3.2 Squirrel-Cage Rotor

suitably displaced in space can establish a rotating magnetic field. The inter-action of rotating fields established by stator and rotor winding currents provides the essence of induction machine action. The reader will recall from Chapter 2 that we were able to establish such rotating fields with either a two-phase or a three-phase system. Other polyphase systems are possible. As shown in Chapter 2, the rotor and stator windings need not have the same number of phases. They must, however, establish fields having the same number of poles in order for the fields to interact properly when rotating in synchronism. A squirrel-cage rotor is interesting in this regard. It consists of conducting bars fitted in slots around the periphery of the rotor. The bars are effectively ''short circuited'' at each end of the rotor either by a ring or a plate of material having negligible resistance. Figure 3.2 illustrates conceptually a squirrel-cage rotor. It can be shown that a squirrel-cage rotor constitutes a rotor winding having a number of phases equal to the total number of rotor conducting bars divided by the number of stator winding pole pairs. It can also be shown that a squirrel-cage rotor will establish a magnetic field having the same number of poles as the magnetic field established by the stator currents. An induction machine that has actual windings on its rotor is referred to as a *wound-rotor induction machine,* and its rotor must be wound in a manner compatible with the stator windings.

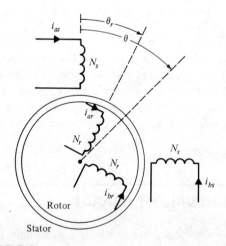

FIGURE 3.3

Without loss of generality, we can treat all cases of interest by considering an equivalent symmetrical two-pole two-phase induction machine having two stator windings and two rotor windings. This configuration is illustrated conceptually in Figure 3.3. The angle θ is an arbitrary angle measured from the axis of the phase a stator winding. The angle θ_r is the angle between the axes of the phase a stator winding and the phase a rotor winding. The machine is symmetrical in that we have an equal number of turns N_s on both stator windings and an equal number of turns N_r on both rotor windings. Note that the axes of the two stator (and rotor) windings are displaced by a fixed angle of $\pi/2$ radians.

We can analyze the configuration shown in Figure 3.3 by regarding stator and rotor windings as a set of coupled coils. Thus

$$\bar{\lambda} = \bar{L}(\theta_r)\bar{i} \tag{3.1}$$

where

$$\bar{\lambda} = \begin{bmatrix} \lambda_{as} \\ \lambda_{bs} \\ \lambda_{ar} \\ \lambda_{br} \end{bmatrix} \tag{3.2}$$

$$\bar{i} = \begin{bmatrix} i_{as} \\ i_{bs} \\ i_{ar} \\ i_{br} \end{bmatrix} \tag{3.3}$$

$$L(\theta_r) = \begin{bmatrix} L_{asas}(\theta_r) & L_{asbs}(\theta_r) & L_{asar}(\theta_r) & L_{asbr}(\theta_r) \\ & L_{bsbs}(\theta_r) & L_{bsar}(\theta_r) & L_{bsbr}(\theta_r) \\ & & L_{arar}(\theta_r) & L_{arbr}(\theta_r) \\ \text{(symmetric matrix)} & & & L_{brbr}(\theta_r) \end{bmatrix} \tag{3.4}$$

Consideration of Figure 3.3 suggests that some of the inductances in (3.4) will be independent of rotor angle θ_r and that some will be zero. We will examine these inductances in more detail. We will make the same assumptions as those made in Sections 2.5 and 2.6:

1. All windings establish magnetic fields whose radial components of air-gap magnetic flux density and magnetic field intensity are sinusoidal functions of the angle θ for any value of θ_r.
2. Stator and rotor iron is highly permeable such that stored energy is predominately associated with the air gap. Saturation effects are negligible; that is, the overall magnetic circuit may be characterized as linear.
3. Tangential components of air-gap magnetic flux density and field intensity are negligible compared to radial components.
4. The axial length ℓ of the rotor is sufficiently long that end effects are negligible.
5. The air gap g is small compared to the radius r of the rotor, so that magnetic flux density and field intensity are essentially invariant with radius in the gap.
6. The effects of slots are negligible.

Proceeding as in Sections 2.5 and 2.6, the radial components of air-gap magnetic field intensity for each of the windings will be

$$H_{as}(\theta) = H_{asmax} \cos \theta \qquad (3.5)$$

$$H_{bs}(\theta) = H_{bsmax} \cos \left(\theta - \frac{\pi}{2} \right) \qquad (3.6)$$

$$H_{ar}(\theta) = H_{armax} \cos (\theta - \theta_r) \qquad (3.7)$$

$$H_{br}(\theta) = H_{brmax} \cos \left(\theta - \theta_r - \frac{\pi}{2} \right) \qquad (3.8)$$

The maximum field intensities are related to currents by applying Ampère's law [see Figure 2.26 and Eqs. (2.63) and (2.64)]:

$$H_{asmax} = \frac{N_s i_{as}}{2g} \qquad (3.9)$$

$$H_{bsmax} = \frac{N_s i_{bs}}{2g} \qquad (3.10)$$

$$H_{armax} = \frac{N_r i_{ar}}{2g} \qquad (3.11)$$

$$H_{brmax} = \frac{N_r i_{br}}{2g} \qquad (3.12)$$

The total resultant magnetic field intensity in the gap at angle θ will be

$$H(\theta) = H_{as}(\theta) + H_{bs}(\theta) + H_{ar}(\theta) + H_{br}(\theta) \qquad (3.13)$$

The coenergy density in the gap will be

$$w_c(\theta) = \frac{\mu_0 H^2(\theta)}{2} \qquad (3.14)$$

Substitute (3.5), (3.6), (3.7) and (3.8) in (3.14) and simplify:

$$
\begin{aligned}
w_c(\theta) = \frac{\mu_0}{2} \Bigg\{ & \frac{H_{asmax}^2}{2} (1 + \cos 2\theta) + \frac{H_{bsmax}^2}{2} (1 - \cos 2\theta) \\
& + \frac{H_{armax}^2}{2} [1 + \cos 2(\theta - \theta_r)] + \frac{H_{brmax}^2}{2} [1 - \cos 2(\theta - \theta_r)] \\
& + H_{asmax} H_{bsmax} \sin 2\theta + H_{asmax} H_{armax} \cos (2\theta - \theta_r) \\
& + H_{asmax} H_{armax} \cos \theta_r + H_{asmax} H_{brmax} \sin (2\theta - \theta_r) \\
& - H_{asmax} H_{brmax} \sin \theta_r + H_{bsmax} H_{armax} \sin (2\theta - \theta_r) \\
& + H_{bsmax} H_{armax} \sin \theta_r - H_{bsmax} H_{brmax} \cos (2\theta - \theta_r) \\
& + H_{bsmax} H_{brmax} \cos \theta_r + H_{armax} H_{brmax} \sin (2\theta - 2\theta_r) \Bigg\}
\end{aligned}
$$

$$(3.15)$$

The total coenergy is obtained by integrating over the volume of the air gap:

$$W_c = \int_0^{2\pi} w_c(\theta) g \ell r \, d\theta \tag{3.16}$$

where g is the gap width, ℓ is the rotor axial length, and r may be taken as the radius of the rotor. Substitute (3.15) and (3.16) and recognize that the integral over 2π will be zero for all sinusoidal terms in θ or 2θ. Then substitute using (3.9), (3.10), (3.11), and (3.12) to yield

$$
\begin{aligned}
W_c = \frac{\pi \mu_0 \ell r}{4g} \Big(& \tfrac{1}{2}N_s^2 i_{as}^2 + \tfrac{1}{2}N_s^2 i_{bs}^2 + \tfrac{1}{2}N_r^2 i_{ar}^2 + \tfrac{1}{2}N_r^2 i_{br}^2 \\
& + N_s N_r i_{as} i_{as} \cos \theta_r - N_s N_r i_{as} i_{br} \sin \theta_r \\
& + N_s N_r i_{bs} i_{ar} \sin \theta_r + N_s N_r i_{bs} i_{br} \cos \theta_r \Big)
\end{aligned}
\tag{3.17}
$$

Now express coenergy W_c in terms of inductances:

$$
\begin{aligned}
W_c &= \tfrac{1}{2}\bar{i}^T \bar{L}(\theta_r) \bar{i} \\
&= \tfrac{1}{2}L_{asas} i_{as}^2 + \tfrac{1}{2}L_{bsbs} i_{bs}^2 + \tfrac{1}{2}L_{arar} i_{ar}^2 + \tfrac{1}{2}L_{brbr} i_{br}^2 \\
&\quad + L_{asbs} i_{as} i_{bs} + L_{asar} i_{as} i_{ar} + L_{asbr} i_{as} i_{br} \\
&\quad + L_{bsar} i_{bs} i_{ar} + L_{bsbr} i_{bs} i_{br} + L_{arbr} i_{ar} i_{br}
\end{aligned}
\tag{3.18}
$$

Now equate coefficients by comparing (3.17) and (3.18):

$$L_{asas} = L_{bsbs} = N_s^2 \frac{\pi \mu_0 \ell r}{4g} \tag{3.19}$$

$$L_{arar} = L_{brbr} = N_r^2 \frac{\pi \mu_0 \ell r}{4g} \tag{3.20}$$

$$L_{asbs} = L_{arbr} = 0 \tag{3.21}$$

$$L_{asar} = L_{bsbr} = N_s N_r \frac{\pi \mu_0 \ell r}{4g} \cos \theta_r \tag{3.22}$$

$$L_{asbr} = -N_s N_r \frac{\pi \mu_0 \ell r}{4g} \sin \theta_r \tag{3.23}$$

$$L_{bsar} = +N_s N_r \frac{\pi \mu_0 \ell r}{4g} \sin \theta_r \tag{3.24}$$

Note that $L_{\rho\sigma,\mu\nu} = L_{\mu\nu,\rho\sigma}$; that is, the inductance matrix $\bar{L}(\theta_r)$ is symmetric. We have neglected stator and rotor leakage inductances in our development.

The stator and rotor flux linkages due to leakage flux are independent of rotor position and can be expressed for either phase as

$$\lambda_s = L_{\ell s} i_s \tag{3.25}$$

$$\lambda_r = L_{\ell r} i_r \tag{3.26}$$

If we define M as

$$M = \frac{N_s^2 \pi \mu_0 \ell r}{4g} \tag{3.27}$$

and incorporate the leakage terms, we can summarize our results as

$$
\begin{bmatrix} \lambda_{as} \\ \lambda_{bs} \\ \lambda_{ar} \\ \lambda_{br} \end{bmatrix}
=
\begin{bmatrix}
L_{\ell s} + M & 0 & \dfrac{N_r}{N_s} M \cos \theta_r & -\dfrac{N_r}{N_s} M \sin \theta_r \\[2ex]
 & L_{\ell s} + M & \dfrac{N_r}{N_s} M \sin \theta_r & \dfrac{N_r}{N_s} M \cos \theta_r \\[2ex]
\text{(symmetric matrix)} & & L_{\ell r} + \left(\dfrac{N_r}{N_s}\right)^2 M & 0 \\[2ex]
 & & & L_{\ell r} + \left(\dfrac{N_r}{N_s}\right)^2 M
\end{bmatrix}
\begin{bmatrix} i_{as} \\ i_{bs} \\ i_{ar} \\ i_{br} \end{bmatrix}
\tag{3.28}
$$

We ultimately seek an equivalent circuit referred to the armature windings that are on the stator. Thus, in the manner of referring variables from one winding of a transformer to the other, we refer rotor variables to the stator circuits in the following way:

$$i'_{ar} = \frac{N_r}{N_s} i_{ar} \tag{3.29}$$

$$i'_{br} = \frac{N_r}{N_s} i_{br} \tag{3.30}$$

$$\lambda'_{ar} = \frac{N_s}{N_r} \lambda_{ar} \tag{3.31}$$

$$\lambda'_{br} = \frac{N_s}{N_r} \lambda_{br} \tag{3.32}$$

Rotor leakage inductance referred to the stator will be

$$L'_{\ell r} = \left(\frac{N_s}{N_r}\right)^2 L_{\ell r} \tag{3.33}$$

Substitution of (3.29) through (3.33) in (3.28) yields

$$
\begin{bmatrix} \lambda_{as} \\ \lambda_{bs} \\ \lambda'_{ar} \\ \lambda'_{br} \end{bmatrix}
=
\begin{bmatrix}
L_{\ell s} + M & 0 & M \cos \theta_r & -M \sin \theta_r \\
 & L_{\ell s} + M & M \sin \theta_r & M \cos \theta_r \\
\text{(symmetric matrix)} & & L'_{\ell r} + M & 0 \\
 & & & L'_{\ell r} + M
\end{bmatrix}
\begin{bmatrix} i_{as} \\ i_{bs} \\ i'_{ar} \\ i'_{br} \end{bmatrix}
\tag{3.34}
$$

Equation (3.34) is valid in general within the limits of our simplifying assumptions. We can now complete the development of an equivalent circuit for the special case of balanced operation under steady-state conditions. For balanced operation, stator currents must constitute a balanced two-phase set of currents. Let

$$i_{as} = \sqrt{2}\, I_s \cos \omega_s t \tag{3.35}$$

$$i_{bs} = \sqrt{2}\, I_s \cos \left(\omega_s t - \frac{\pi}{2} \right) \tag{3.36}$$

The slip s is defined as

$$s = \frac{\omega_s - \omega_r}{\omega_s} \tag{3.37}$$

where ω_r is the rotor speed.

Let the stator and rotor armature resistances be r_s and r_r, respectively. If the rotor windings are short circuited, we have

$$v'_{ar} = 0 = i'_{ar} r'_r + \frac{d\lambda'_{ar}}{dt} \tag{3.38}$$

$$v'_{br} = 0 = i'_{br} r'_r + \frac{d\lambda'_{br}}{dt} \tag{3.39}$$

The rotor angle θ_r is related to the rotor speed by

$$\theta_r = \omega_r t + \theta_r(0) \tag{3.40}$$

The angle $\theta_r(0)$ is the initial rotor angle, which may be taken as zero without loss of generality.

Now substitute in (3.38) and (3.39) using (3.34), (3.35), and (3.36). Simplify and rearrange to obtain

$$i'_{ar} r'_r + (L'_{\ell r} + M) \frac{d}{dt} i'_{ar} = (\omega_s - \omega_r) M \sqrt{2}\, I_s \cos \left[(\omega_s - \omega_r)t - \frac{\pi}{2} \right] \tag{3.41}$$

$$i'_{br} r'_r + (L'_{\ell r} + M) \frac{di'_{br}}{dt} = (\omega_s - \omega_r) M \sqrt{2}\, I_s \cos [(\omega_s - \omega_r)t - \pi] \tag{3.42}$$

Equations (3.41) and (3.42) are first-order linear differential equations whose forcing functions are sinusoidal time functions at frequency $\omega_s - \omega_r$. In the steady state, all rotor variables will be sinusoidal at this frequency, and we may use phasor concepts to determine steady-state currents i'_{ar} and i'_{br}. Note that i'_{br} will be the same as i'_{ar} but phase displaced by $\pi/2$ radians. Thus we need consider only the phase a rotor winding to characterize rotor conditions completely. Phasor representations for i_{as} and i'_{ar} may be chosen as

$$i_{as} \leftrightarrow I_s = I_s \underline{/0} \tag{3.43}$$

$$i'_{ar} \leftrightarrow I'_r = I'_r \underline{/-\varphi_r} \tag{3.44}$$

where frequency in the armature circuits is ω_s and frequency in the rotor circuits is $\omega_s - \omega_r$. From the definition of slip, we can express $\omega_s - \omega_r$ as $s\omega_s$. Then a corresponding phasor equation for (3.41) can be written as

$$\underline{I}'_r r'_r + js\omega_s (L'_{\ell r} + M)\underline{I}'_r = s\omega_s MI_s \underline{/-\pi/2}$$
$$= -js\omega_s M\underline{I}_s \tag{3.45}$$

We can express i'_{ar} and i'_{br} as

$$i'_{ar} = \sqrt{2}\, I'_r \cos(s\omega_s t - \varphi_r) \tag{3.46}$$

$$i'_{br} = \sqrt{2}\, I'_r \cos\left(s\omega_s t - \varphi_r - \frac{\pi}{2}\right) \tag{3.47}$$

The circuit equation for the phase a armature winding will be

$$v_{as} = i_{as} r_s + \frac{d\lambda_{as}}{dt} \tag{3.48}$$

Now substitute in (3.48) using (3.34), (3.35), (3.37), (3.46), and (3.47). Simplify and rearrange to obtain

$$v_{as} = i_{as} r_s + (L_{\ell s} + M)\frac{di_{as}}{dt} - \sqrt{2}\, I'_r \omega_s M \cos\left(\omega_s t - \varphi_r - \frac{\pi}{2}\right) \tag{3.49}$$

The corresponding phasor equation for (3.49) must be

$$\underline{V}_s = \underline{I}_s r_s + j\omega_s (L_{\ell s} + M)\underline{I}_s + j\omega_s M\underline{I}'_r \tag{3.50}$$

Now rewrite (3.45) as

$$0 = \frac{\underline{I}'_r r'_r}{s} + j\omega_s (L'_{\ell r} + M)\underline{I}'_r + j\omega_s M\underline{I}_s \tag{3.51}$$

Equations (3.50) and (3.51) can be treated as two equations in the two unknowns $\underline{I}'_r$ and $\underline{I}_s$ for specified values of $\underline{V}_s$ and slip s. $\underline{V}_s$ is the phasor representation for armature terminal voltage v_{as}. Generally, we simply assume that the machine is operated at or near rated voltage and choose the angle of the voltage as our reference. That is,

$$\underline{V} = V_s \underline{/0°} \tag{3.52}$$

For particular values of slip s we can solve for $\underline{I}'_r$ and $\underline{I}_s$ and determine all characteristics of interest. Although phasor $\underline{I}'_r$ represents i'_{ar}, which is the sinusoidal rotor current (referred to the stator by the turns ratio) in phase a of the rotor winding at frequency $\omega_s - \omega_r$, and phasor $\underline{I}_s$ represents i_{as}, which is the sinusoidal stator current in phase a of the stator winding at the frequency ω_s, we can easily synthesize a single equivalent circuit that is consistent with (3.50) and (3.51). The circuit shown in Figure 3.4 is generally regarded as the classical steady-state equivalent circuit for the induction machine.

It should be noted that the equivalent circuit is similar to the equivalent circuit for a transformer considered in Chapter 1. We have found an approximate value for the parameter M in terms of machine dimensions [Eq. (3.27)]. We can similarly find approximate expressions for the remaining parameters. As in trans-

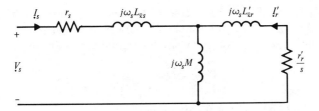

FIGURE 3.4 Steady-State Equivalent Circuit for the Polyphase Induction Machine

former analysis it is generally simpler to deal with machine reactances than inductances. Thus we typically will use

$$X_m = \omega_s M \tag{3.53}$$

$$x_{\ell s} = \omega_s L_{\ell s} \tag{3.54}$$

$$x'_{\ell r} = \omega_s L'_{\ell r} \tag{3.55}$$

The steady-state equivalent circuit of Figure 3.4 is also valid for three-phase machines. Armature voltage and current phasors must be based on appropriate phase quantities, as we will show in the examples that follow.

3.3

INDUCTION MACHINE
STEADY-STATE CHARACTERISTICS

We can determine steady-state characteristics of the induction machine by analyzing the steady-state equivalent circuit shown in Figure 3.4. The circuit is redrawn in Figure 3.5 using the reactances defined by (3.53), (3.54), and (3.55). Upon examination of the equivalent circuit we see that the real power can be converted only in the resistive elements r_s and r'_r/s. The resistance r_s is resistance of the armature winding. Thus the armature winding losses per phase must be

$$P_s = I_s^2 r_s \tag{3.56}$$

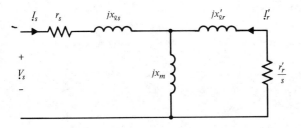

FIGURE 3.5

where $I_s = |\underline{I}_s|$, that is, the rms value of the armature current. The power converted in the resistance r'_r/s must be the total power per phase converted in the rotor circuit. It is the power per phase transferred across the air gap:

$$P_g = I_r'^2 \frac{r'_r}{s} \tag{3.57}$$

The rotor winding losses per phase are given by

$$P_r = I_r'^2 r'_r \tag{3.58}$$

The mechanical power converted per phase must be the difference between the total power per phase across the gap P_g and the rotor winding losses per phase P_r:

$$P_m = P_g - P_r$$

$$= I_r'^2 \frac{r'_r}{s} - I_r'^2 r'_r \tag{3.59}$$

$$= I_r'^2 r'_r \frac{1-s}{s}$$

The relationships above suggest that we might treat the rotor resistance r'_r/s as the sum of two resistances:

$$\frac{r'_r}{s} = r'_r + r'_r \frac{1-s}{s} \tag{3.60}$$

where r'_r determines rotor winding losses and $r'_r [(1 - s)/s]$ determines mechanical power converted. The equivalent circuit of Figure 3.5 can be simply modified as shown in Figure 3.6 to reflect these observations. The ratio of mechanical power converted P_m to total air-gap power P_g can be found directly from (3.57) and (3.59):

$$\frac{P_m}{P_g} = 1 - s \tag{3.61}$$

An important limitation is immediately apparent. We must operate at relatively low values of the slip s to achieve efficient operation. The slip will typically be 0.02 to 0.06 for operation at rated machine output for a well-designed machine.

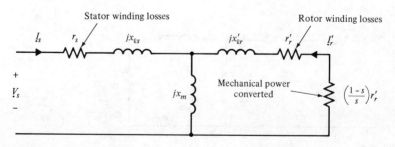

FIGURE 3.6

From the definition of slip given in (3.37) we can express the rotor speed ω_r in terms of the synchronous speed ω_s.†

$$\omega_r = (1 - s)\omega_s \tag{3.62}$$

The mechanical power converted is the product of ω_r and the developed electrical torque T_e. Thus, from (3.59) and (3.62),

$$T_e = \frac{P_m}{\omega_r} = \frac{I_r'^2 r_r'}{s\omega_s} \tag{3.63}$$

on a per phase basis.

The torque versus speed characteristics are of particular interest. We can find $I_r'^2$ as a function of slip by simplifying the equivalent circuit of Figure 3.5 as shown in Figure 3.7. In Figure 3.7, V_{eq} and Z_{eq} are the Thévenin equivalent voltage and impedance, respectively, for that portion of the equivalent circuit to the left of the boundary indicated in Figure 3.7. Thus

$$\underset{\sim}{V}_{eq} = \underset{\sim}{V}_s \frac{jx_m}{r_s + j(x_{\ell s} + x_m)} \tag{3.64}$$

$$\underset{\sim}{Z}_{eq} = \frac{jx_m(r_s + jx_{\ell s})}{r_s + j(x_{\ell s} + x_m)} \tag{3.65}$$

$$= r_{eq} + jx_{eq}$$

Now from the equivalent circuit of Figure 3.7 we have

$$-\underset{\sim}{I}_r' = \frac{V_{eq}}{r_{eq} + r_r'/s + j(x_{eq} + x_{\ell r}')} \tag{3.66}$$

Thus

$$I_r'^2 = \frac{V_{eq}^2}{(r_{eq} + r_r'/s)^2 + (x_{eq} + x_{\ell r}')^2} \tag{3.67}$$

†We assume here that we have a two-pole machine such that the synchronous speed for the rotor is ω_s.

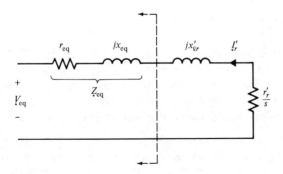

FIGURE 3.7

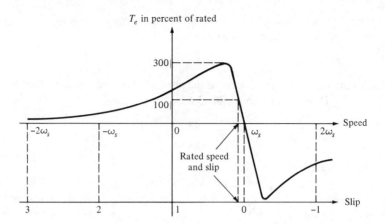

T_e in percent of rated

300

100

$-2\omega_s$ $-\omega_s$ 0 ω_s $2\omega_s$ Speed

Rated speed
and slip

Slip

3   2   1   0   -1

FIGURE 3.8

Substituting in (3.63) using (3.67), we can express the developed electrical torque per phase T_e as

$$T_e = \frac{V_{eq}^2 r_r'}{s\omega_s[(r_{eq} + r_r'/s)^2 + (x_{eq} + x_{\ell r}')^2]} \qquad (3.68)$$

Note that from (3.64), V_{eq} is directly proportional to V_s. Thus, for any value of the slip s, we see that the torque will vary as the square of the rms value of the armature terminal voltage. This is an extremely important consideration in machine applications.

For any given set of machine parameters we can easily determine r_{eq} and x_{eq} and evaluate (3.68) for various values of slip. The general shape of a torque versus slip curve is shown in Figure 3.8. The torque goes to zero at slip $s = 0$. At zero slip the rotor turns at synchronous speed, and we have no relative motion between the rotor and the rotating magnetic field due to armature currents. Thus we have no induced rotor currents and hence no developed torque. The reader should verify this from equivalent-circuit considerations as well. Figure 3.8 shows three modes of operation: braking, motoring, and generating. By recognizing that power per phase across the air gap P_g essentially is converted to rotor winding losses per phase P_r and to mechanical power per phase P_m,

$$P_g = P_m + P_r \qquad (3.69)$$

we can characterize the three modes of operation as shown in Figure 3.9. Figure 3.9 also reflects the locked-rotor (stall) condition and idling (no-load) condition. The various modes of operation and conditions described in Figure 3.9 can all be described in the steady state by our equivalent circuit or by simply considering (3.50) and (3.51), from which the equivalent circuit was synthesized.

EXAMPLE 3.1 In Section 2.6 we showed that induction machine action could be described in terms of fictitious single stator and rotor windings carrying direct currents and rotating in synchronism as shown in Figure 2.31. Figure 2.34 shows a typical rotating

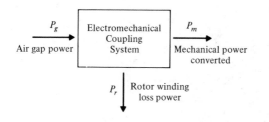

Mode of Operation	Speed/Slip	Power (Equations 3.57, 3.58, and 3.59)
Braking	$\omega_r < 0$ $s > 1$	$P_g > 0$ $P_m < 0$ $P_r > 0$
Motoring	$0 < \omega_r < \omega_s$ $1 > s > 0$	$P_g > 0$ $P_m > 0$ $P_r > 0$
Generating	$\omega_r > \omega_s$ $s < 0$	$P_g < 0$ $P_m < 0$ $P_r > 0$
Locked rotor condition	$\omega_r = 0$ $s = 1$	$P_g > 0$ $P_m = 0$ $P_r = P_g$
Idling† condition	$\omega_r = \omega_s$ $s = 0$	$P_g = 0$ $P_m = 0$ $P_r = 0$

†Friction and windage losses are neglected here.

FIGURE 3.9

field alignment. Find an expression for the fixed angle between the axis of the rotating field due to stator currents and the axis of the rotating field due to rotor currents.

Figure 2.31 was developed under the assumption of a three-phase rotor winding. For two-phase stator and rotor windings we will have as direct currents in the fictitious windings:

$$i_s = \sqrt{2}\, I_s$$

$$i_r = \sqrt{2}\, I_r$$

Consideration of Figure 2.31 and (2.72) with ($\varphi_s = 0$), (2.79), (3.35), and (3.46) leads us to the answer. The angle $\theta_r(0)$ may be arbitrarily chosen as zero, as discussed earlier. The angle between the rotating fields is thus φ_r, and it is found by solving for $I_r' = I_r' \underline{/-\varphi_r}$ using (3.51):

$$\varphi_r = \frac{\pi}{2} + \tan^{-1} \frac{s\omega_s(L_{\ell r}' + M)}{r_r'}$$

The angle between the axis of the resultant field due to combined effects of stator and rotor currents and the axis of the field due to rotor currents alone is also of interest. As shown in Figure 2.34(b), we can represent the rotating fields with rotating space vectors $\hat{H}_r$, $\hat{H}_s$, and $\hat{H}_{sr}$. We have thus far determined the

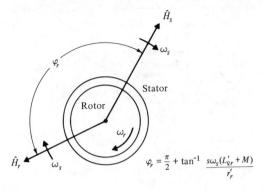

$$\varphi_r = \frac{\pi}{2} + \tan^{-1}\frac{s\omega_s(L'_{\ell r} + M)}{r'_r}$$

FIGURE 3.10

angle between $\hat{H}_s$ and $\hat{H}_r$ as shown in Figure 3.10. The magnitudes of $\hat{H}_s$ and $\hat{H}_r$ must be (from Ampère's law)

$$|\hat{H}_s| = H_{smax} = \frac{\sqrt{2}\, N_s I_s}{2g}$$

$$|\hat{H}_r| = H_{rmax} = \frac{\sqrt{2}\, N_r I_r}{2g}$$

In the expression for $|\hat{H}_r|$ above, I_r must be the rms value of the actual rotor current. It can be found by solving for $I'_r = I'_r \underline{/-\varphi_r}$ using (3.51) and referring the result to the rotor:

$$I_r = \frac{N_s}{N_r} I'_r = \frac{N_s}{N_r} I_s \frac{s\omega_s M}{\sqrt{r'^2_r + s^2\omega^2_s(L'_{\ell r} + M)^2}}$$

Thus

$$|\hat{H}_r| = \frac{\sqrt{2}\, N_s I_s}{2g} \frac{s\omega_s M}{\sqrt{r'^2_r + s^2\omega^2_s(L'_{\ell r} + M)^2}}$$

$$= |\hat{H}_s| \frac{s\omega_s M}{\sqrt{r'^2_r + s^2\omega^2_s(L'_{\ell r} + M)^2}}$$

and we have

$$\frac{|\hat{H}_r|}{|\hat{H}_s|} = \frac{s\omega_s M}{\sqrt{r'^2_r + s^2\omega^2_s(L'_{\ell r} + M)^2}}$$

Figure 3.11 illustrates the relevant geometry. The desired angle between $\hat{H}_{sr}$ and $\hat{H}_r$ is $\pi/2 + \tan^{-1}(b/a)$. From Figure 3.11,

$$\frac{b + |\hat{H}_r|}{|\hat{H}_s|} = \frac{b}{|\hat{H}_s|} + \frac{|\hat{H}_r|}{|\hat{H}_s|} = \cos\beta = \frac{s\omega_s(L'_{\ell r} + M)}{\sqrt{r'^2_r + s^2\omega^2_s(L'_{\ell r} + M)^2}}$$

$$\frac{a}{|\hat{H}_s|} = \sin\beta = \frac{r'_r}{\sqrt{r'^2_r + s^2\omega^2_s(L'_{\ell r} + M)^2}}$$

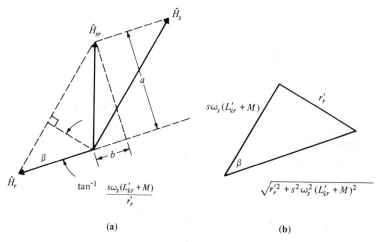

(a)   (b)

FIGURE 3.11

Thus

$$\frac{b}{a} = \frac{b/|\hat{H}_s|}{a/|\hat{H}_s|} = \frac{s\omega_s L'_{\ell r}}{r'_r} = \frac{s\omega_s L_{\ell r}}{r_r}$$

and our final result is

$$\varphi_{sr} = \frac{\pi}{2} + \tan^{-1} \frac{s\omega_s L_{\ell r}}{r_r}$$

This final result can be explained in a simple way. By Faraday's law, the voltage induced in a rotor winding by the resultant magnetic field will differ in phase by $\pi/2$ radians from the resultant flux linking its turns. The rotor current will lag this voltage by an additional amount based on the power-factor angle (imped-ance angle) of the rotor circuit, which is $\tan^{-1} (s\omega_s L_{\ell r}/r_r)$.

EXAMPLE 3.2 In Section 2.5 we found an expression for the torque developed between single stator and rotor windings for smooth-air-gap machines, that is, (2.66):

$$T_e = \frac{-\pi\mu_0 \ell r N_s N_r}{4g} i_s i_r \sin \psi \tag{2.66}$$

In Section 2.6 we showed that induction machine action could be described in terms of fictitious single stator and rotor windings carrying direct currents and rotating in syn-chronism (see Figures 2.31 and 2.34). Show that the torque as given by (2.66) is con-sistent with the torque determined on the basis of equivalent-circuit concepts.

The angle ψ is defined in Figure 2.24. It is the angle by which the field due to the rotor current leads the field due to the stator current. As we have seen in Example 3.1, for induction machines the field due to rotor current lags the field due to stator currents by the angle φ_r. Thus

$$\psi = -\varphi_r = -\frac{\pi}{2} - \tan^{-1}\frac{s\omega_s(L'_{\ell r} + M)}{r'_r}$$

$$\sin \psi = \sin(-\varphi_r) = -\sin \varphi_r = -\cos\left[\tan^{-1}\frac{s\omega_s(L'_{\ell r} + M)}{r'_r}\right]$$

$$= \frac{-r'_r}{\sqrt{r'^2_r + s^2\omega^2_s(L'_{\ell r} + M)^2}}$$

For two-phase stator and rotor windings, we will have as direct currents in the fictitious single windings:

$$i_s = \sqrt{2}\, I_s$$

$$i_r = \sqrt{2}\, I_r = \sqrt{2}\frac{N_s}{N_r} I'_r$$

Now express I_s in terms of I'_r by solving (3.51):

$$I_s = \frac{\sqrt{r'^2_r + s^2\omega^2_s(L'_{\ell r} + M)^2}}{s\omega_s M} I'_r$$

The mutual inductance M is given by (3.27)

$$M = \frac{N^2_s\pi\mu_0\ell r}{4g}$$

Now substitute in the expression for torque as given by (2.66):

$$T_e = -\frac{\pi\mu_0\ell r N_s N_r}{4g} i_s i_r \sin \psi$$

$$= \frac{-M}{N^2_s} N_s N_r (\sqrt{2}\, I_s)\left(\sqrt{2}\frac{N_s}{N_r} I'_r\right)\sin \psi$$

$$= +2MI'^2_r \frac{\sqrt{r'^2_r + s^2\omega^2_s(L'_{\ell r} + M)^2}}{s\omega_s M} \frac{r'_r}{\sqrt{r'^2_r + s^2\omega^2_s(L'_{\ell r} + M)^2}}$$

$$= \frac{2I'^2_r r'_r}{s\omega_s}$$

which is the result we would obtain using our equivalent-circuit concepts for a two-phase machine [see (3.63)].

The effects of the rotor resistance r'_r on torque versus slip characteristics are extremely important. The value of slip at which torque is a maximum can be found by consideration of (3.63), which shows that the torque T_e is proportional to the power delivered to the resistance r'_r/s. Applying circuit concepts to the equivalent circuit of Figure 3.7, we can show that the power delivered to this resistance will be maximized when the resistance is equal to the magnitude of the impedance between it and the source voltage V_{eq}. That is,

$$\frac{r'_r}{s_m} = \sqrt{r^2_{eq} + (x_{eq} + x'_{\ell r})^2} \tag{3.70}$$

Thus the slip at which torque will be a maximum is given by

$$s_m = \frac{r'_r}{\sqrt{r^2_{eq} + (x_{eq} + x'_{\ell r})^2}} \tag{3.71}$$

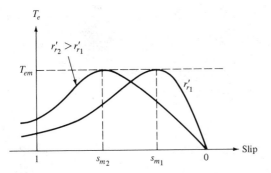

FIGURE 3.12 Effects of Rotor Resistance on Torque Versus Slip Curves

Now by substituting this value of slip into the expression for the torque T_e given by (3.68), we can find the value of maximum torque:

$$T_{em} = \frac{1}{2\omega_s} \frac{V_{eq}^2}{r_{eq} + \sqrt{r_{eq}^2 + (x_{eq} + x'_{\ell r})^2}} \tag{3.72}$$

Examination of (3.71) and (3.72) reveals that s_m is directly proportional to rotor resistance and that T_{em} is independent of rotor resistance. This is illustrated in Figure 3.12. If the terminals of the rotor windings of a wound-rotor machine are accessible through slip rings, the rotor resistance can be controlled in an external stationary circuit. The device required is sometimes referred to as a *secondary controller*. A wound-rotor induction machine and its secondary controller are illustrated conceptually in Figure 3.13. Typical operation entails setting the secondary resistance to a relatively high value for starting. This value is chosen based on the desired starting torque for the intended application. The relatively high rotor circuit resistance at starting also acts to reduce the starting current, sometimes referred to as *starting inrush*. As the machine comes up to speed, the secondary resistance is reduced. Any such external resistance should

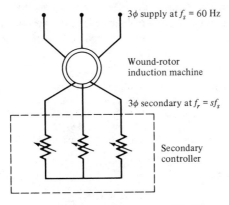

FIGURE 3.13 Schematic Diagram of a Three-Phase Wound-Rotor Induction Machine and Its Secondary Controller

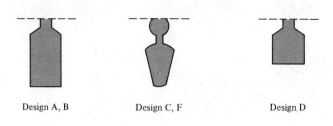

Design A, B   Design C, F   Design D

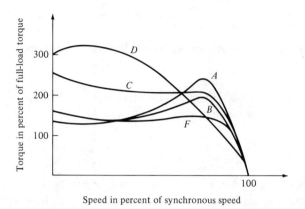

Speed in percent of synchronous speed

FIGURE 3.14 Effects of Rotor Bar Geometry on Torque Characteristics in Squir-rel-Cage Machines

be zero for operation at rated machine conditions so that rotor losses will be minimal.

This same effect is used in squirrel-cage machines in an indirect way to achieve desirable torque characteristics and to limit inrush current. Depending on the geometry of the rotor bars, as shown in Figure 3.14, the effective rotor resistance at standstill (60-Hz rotor currents) can be several times higher than the effective rotor resistance during normal running conditions (2- to 5-Hz rotor currents). When the frequency of the rotor currents is relatively high, the leakage reactance associated with the bars reduces the effective cross-sectional area of the bars, thus effectively increasing the rotor resistance. During normal operation at low values of slip, the effect is much less pronounced and the rotor resistance is essentially the dc resistance of the rotor bars.

EXAMPLE 3.3 From an applications viewpoint, induction machine characteristics as a function of machine loading are of interest. Determine the machine efficiency, real and reactive power requirements, power factor, and armature current magnitude as a function of machine loading if the machine load is varied from 0 to 125% of rated output.

Typical machine data for a three-phase 100-hp 460-V, Y-connected machine are

$$\text{voltage (phase): } \frac{460}{\sqrt{3}} = 266 \text{ V}$$

$$r_s = 0.09 \ \Omega$$

$$r_r' = 0.07 \ \Omega$$

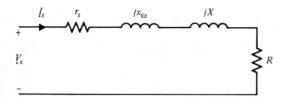

FIGURE 3.15

$$x_{\ell s} = 0.18 \ \Omega$$

$$x\ell_r = 0.15 \ \Omega$$

$$x_m = 6.0 \ \Omega$$

Over the range of operation of interest, friction and windage losses are constant at 3.0 kW. Core losses are negligible.

We can find the results desired by solving our equivalent circuit for a range of slip values that we estimate will correspond to the desired range of loadings. The armature voltage will be assumed constant at the value specified. Thus

$$V_s = 266$$

The circuit of Figure 3.5 can be redrawn as shown in Figure 3.15. In Figure 3.15,

$$R + jX = \frac{jx_m(jx'_{\ell r} + r'_r/s)}{r'_r/s + j(x_m + x'_{\ell r})}$$

By evaluating R and X at particular values of slip, we can easily find the total circuit impedance and solve for current I_s. The mechanical power converted will be

$$P_m = 3(1 - s)P_g$$

$$= 3(1 - s)I_s^2 R$$

The power out will be the mechanical power converted less friction and windage losses:

$$P_o = P_m - 3000$$

The total real power requirements at the armature terminals will be

$$P_{in} = 3I_s^2(r_s + R)$$

The efficiency will be

$$\eta = \frac{P_o}{P_{in}} \times 100\%$$

The total reactive power requirements at the armature terminals will be

$$Q_{in} = 3I_s^2(x_{\ell s} + X)$$

The power factor will be

$$\text{P.F.} = \frac{P_{in}}{3V_s I_s}$$

Results for various values of slip are shown in Table 3.1. The value of slip denoted $0^+ > 0$ is the value of slip at which $P_m = 3000$ W. It may be taken as zero in calculations involving the equivalent circuit. The rated machine output is 100 hp or 74.6 kW.

Table 3.1

Slip	I_s (A)	P_o (kW)	P_{in} (kW)	η	Q_{in} (kvar)	Power Factor
0^+	43.0	0.0	3.5	0.00	34.3	0.102
0.005	46.7	11.1	14.7	0.76	34.3	0.394
0.01	56.5	24.5	28.7	0.85	34.8	0.637
0.02	84.6	49.9	56.0	0.89	37.7	0.829
0.03	116.1	73.1	82.1	0.89	42.9	0.886
0.04	147.7	93.8	106.7	0.88	50.1	0.905

Thus the rated slip is approximately 0.03. Selected data from Table 3.1 are plotted in Figure 3.16 against percent machine loading. The curves in the figure reflect typical induction machine behavior. The efficiency is relatively flat over a wide load range, whereas the power factor falls off sharply at light load. Reactive power requirements are of interest in that the curve is relatively flat. Capacitors in parallel with the motor (i.e., shunt capacitors) connected at the motor terminals and switched with the motor can supply motor reactive power requirements over a wide range of load conditions. This permits correction to near-unity power factor for a wide range of load conditions.[†]

3.4

INDUCTION MACHINE DYNAMIC ANALYSIS

To analyze induction machine dynamic behavior, we must examine the equation of motion for the rotor of the machine:

$$\frac{J \, d\omega_r}{dt} = T_e - T_L \tag{3.73}$$

[†]A phenomenon known as *self-excitation* limits the amount of reactive compensation that can be applied in practice. We will examine this phenomenon in Chapter 8.

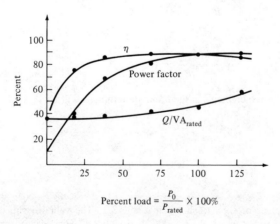

$$\text{Percent load} = \frac{P_0}{P_{\text{rated}}} \times 100\%$$

FIGURE 3.16

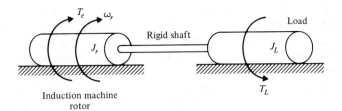

FIGURE 3.17

where J is the moment of inertia, T_e the developed electrical torque, and T_L the total load torque, including torque due to friction and windage. In reality the determination of the total load torque T_L may not be simple. We assume here that the induction machine rotor and its connected mechanical load can be viewed as a single rotating mass having total moment of inertia J. That is,

$$J = J_r + J_L \qquad (3.74)$$

where J_r is the moment of inertia of the induction machine rotor and J_L is the moment of inertia of the load. In essence we are neglecting shaft dynamics and dealing with a single rigid rotating mass. This is illustrated conceptually in Figure 3.17.

If steady-state load torque can be described as a function of speed, then $T_L = T_L(\omega_r)$ and we can superimpose the load torque–speed characteristic on our induction machine torque–speed characteristic to determine a steady-state operating point. This is illustrated in Figure 3.18. At the operating point ω_{ro}, load torque and developed electrical torque are equal, and [from (3.73)] acceleration $d\omega_r/dt = 0$. The torque T_e as a function of ω_r (or slip s) can be determined from (3.68). The load torque $T_L(\omega_r)$ depends on the load characteristics. Fans and centrifugal pumps exhibit torque–speed characteristics where steady-state torque $T_L(\omega_r)$ varies approximately as the square of speed ω_r. Conveyor systems typically have flat torque–speed characteristics; that is, steady-state load torque is independent of speed. Loads characterized by friction are often represented with $T_L(\omega_r)$ varying linearly with speed.

Strictly speaking, our curves are valid only for steady-state behavior and are

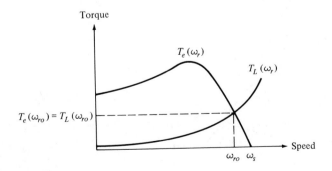

FIGURE 3.18

not suitable for dynamic analysis involving changing speed and perhaps electrical switching transients. Following a switching operation such as what we would have for an across-the-line start, we will have decaying transient stator and rotor currents, which cause a transient torque. These transient currents and the transient torque are present in addition to the currents and torque we would determine using our steady-state analysis techniques. They are often of relatively short duration compared to the time required for the rotor to accelerate to a steady-state operating speed. Moreover, the transient component of the torque waveform is typically a decaying sinusoid at a frequency of 60 Hz or higher. It does not appreciably affect the speed of the rotor for typical values of moments of inertia. Some interesting transient torque waveforms are presented and compared to those computed on the basis of steady-state theory in reference 1. We will examine transient currents and torque in greater detail in Chapter 6.

In essence, if we can neglect electrical transients associated with switching and assume that speed changes slowly so that our steady-state model can be used to determine a sequence in time of "steady-state" currents, torques, and so on, we can analyze dynamic behavior by considering torque characteristics as depicted in Figure 3.19. In the figure $T_a(\omega_r)$ is the accelerating torque:

$$T_a(\omega_r) = T_e(\omega_r) - T_L(\omega_r) \tag{3.75}$$

Substituting in (3.73) gives us

$$\frac{J \, d\omega_r}{dt} = T_e(\omega_r) - T_L(\omega_r) = T_a(\omega_r) \tag{3.76}$$

Since $T_a(\omega_r)$ is expressed as a function of the speed ω_r, we can separate variables,

$$\frac{J}{T_a(\omega_r)} \, d\omega_r = dt \tag{3.77}$$

and integrate over corresponding limits:

$$J \int_0^{\omega'_r} \frac{1}{T_a(\omega_r)} \, d\omega_r = \int_0^{t'} dt = t' \tag{3.78}$$

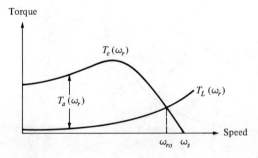

FIGURE 3.19

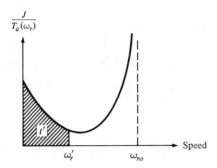

FIGURE 3.20

The graphical interpretation of (3.78) is shown in Figure 3.20. The time required to accelerate the rotor to a speed ω_r' is t. Plotting a curve such as the one shown in Figure 3.20 permits an approximate determination of time t' by graphical methods. A further simplification is sometimes helpful. It can be shown that the induction machine torque–slip characteristic given by (3.68) is reasonably approximated by (see Problem 3.13)

$$ T_e = \frac{2T_{em}}{s/s_m + s_m/s} \tag{3.79} $$

Equation (3.79) is sometimes useful in the evaluation of time t' in (3.78) provided that the load torque T_L can be expressed as a simple function of slip or rotor speed.

EXAMPLE 3.4 A 100-hp induction machine is supplied by a primary distribution feeder with a step-down transformer at the motor location. The voltage at the source end of the feeder may be regarded as constant in magnitude, frequency, and phase angle (i.e., an *infinite bus*). The motor parameters are the same as for the machine of Example 3.3. Assume that the machine is Y-connected and wound for two poles.

$$ \text{voltage (phase): } \frac{460}{\sqrt{3}} = 266 \text{ V} $$

$$ r_s = 0.09 \ \Omega $$

$$ r_r' = 0.07 \ \Omega $$

$$ x_{\ell s} = 0.18 \ \Omega $$

$$ x_{\ell r}' = 0.15 \ \Omega $$

$$ x_m = 6.0 \ \Omega $$

The combined moment of inertia for the motor and its connected load is

$$ J_r + J_L = 6.0 \text{ kg} \cdot \text{m}^2 $$

The combined feeder and transformer reactance referred to the motor circuit is

$$ x_i = x_{\text{feeder}} + x_{\text{transformer}} = 0.01 + 0.04 = 0.05 \ \Omega $$

The feeder and transformer resistance are negligible. Determine the time required for the motor to reach a rated slip of 0.03 if the motor is started from rest with an across-the-line start. Assume that the load torque T_L is zero during rotor acceleration. Determine

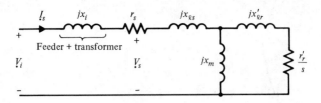

FIGURE 3.21

the rms motor current and armature voltage as a function of time during the speed buildup. Figure 3.21 illustrates the problem.

We start by determining the voltage V_i. We will assume that this voltage is adjusted to a value that will permit the motor to operate at rated voltage at rated slip. From Table 3.1 and Example 3.3, we determine the rated motor conditions.

$$V_s = 266 \text{ V}$$

$$\text{power factor} = 0.89, \quad \theta = \cos^{-1} 0.89 = 27.1°$$

$$I_s = 116.1 \text{ A}$$

$$I_s = 116.1 \underline{/-27.1°} \text{ A}$$

Therefore,

$$V_i = V_s + jx_i I_s$$

$$= 266 + (j.05) \, 116.1 \underline{/-27.1°}$$

$$= 269 \underline{/1.1°}$$

Now take the voltage V_i as a reference (i.e., rotate the phasor diagram):

$$V_i = 269 \underline{/0°}$$

We will use the approximate torque expression given by (3.79). The parameters are computed from (3.71) and (3.72), where x_i is included as a reactance in series with $x_{\ell s}$. Thus, from (3.65),

$$Z_{eq} = \frac{j6.0[0.09 + j(0.18 + 0.05)]}{0.09 + j(0.18 + 0.05 + 6.0)}$$

$$= 0.084 + j.223$$

$$= r_{eq} + jx_{eq}$$

From (3.71)

$$S_m = \frac{0.07}{\sqrt{(0.084)^2 + (0.223 + 0.15)^2}} = 0.183$$

From (3.64)

$$V_{eq} = (269) \frac{j6.0}{0.09 + j(0.05 + 0.18 + 6.0)} = 259 \underline{/0.83°}$$

From (3.72)

$$T_{em} = \frac{3}{2(377)} \frac{259^2}{0.084 + \sqrt{0.084^2 + (0.223 + 0.15)^2}}$$

$$= 572.4 \text{ N} \cdot \text{m}$$

THE INDUCTION MACHINE

From (3.76), where $T_L(\omega_r) = 0$,

$$\frac{J \, d\omega_r}{dt} = T_e(\omega_r)$$

Expressing this relationship in terms of slip ($\omega_r = (1 - s)\omega_s$), we have

$$-J\omega_s \frac{ds}{dt} = T_e(s) = \frac{2T_{em}}{s/s_m + s_m/s}$$

Separating variables and integrating yields

$$-\frac{J\omega_s}{2T_{em}} \int_1^{s'} \left(\frac{s}{s_m} + \frac{s_m}{s} \right) ds = \int_0^{t'} dt = t'$$

Thus to reach any value of slip s' the time t' required will be

$$t' = -\frac{J\omega_s}{2T_{em}} \left(\frac{s'^2 - 1}{2s_m} + s_m \ln s' \right)$$

We can address all of the problem requirements by determining the data shown in Table 3.2. Table 3.2 reveals that the total time required for the machine to reach the rated slip is 6.66 s. The last two columns of Table 3.2 were determined by solving the equivalent circuit shown in Figure 3.21 for the various values of slip. The table shows that inrush current remains relatively high until the slip reaches s_m. It then falls off to rated current in less than 1 s. The last two columns of the table can be used to determine time-delay settings for overcurrent and undervoltage devices that might be used to protect the motor circuit. The column of voltage ratios illustrates the phenomenon of *voltage dip*, which is a major area of concern in system design. Feeders are generally sized to ensure that V_s/V_{rated} remains ≥ 0.80 for industrial installations. If frequent motor starts are anticipated, requirements are generally more stringent.

Dynamic behavior of the machine in the vicinity of a steady-state operating point is often of interest. Small changes in machine load, or perhaps fluctuations in armature voltage, can occur which result in machine speed displacing slightly from an operating point. Another example is the abrupt displacement of rotor speed from a steady-state operating point due to an impact load. If the impact load is cyclic in nature, such as what we might have if an induction motor is driving a punch press, the time required for the rotor to return to its operating point is of interest in determining an upper limit on the frequency of the impact loading.

Table 3.2

Slip, s'	Time, t' (s)	I_s/I_{srated}	V_s/V_{srated}
1.0	0.0	5.66	0.898
0.75	2.46	5.54	0.904
0.50	4.30	5.26	0.914
0.25	5.56	4.39	0.943
0.183 (s_m)	5.83	3.84	0.958
0.125	6.07	3.09	0.975
0.0625	6.38	1.85	0.994
0.030 (s_{rated})	6.66	1.00	1.00

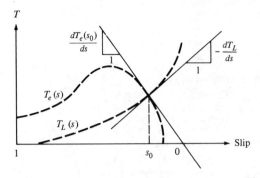

FIGURE 3.22

EXAMPLE 3.5 Find the linear differential equation required to describe small displacements in speed from a steady-state operating point for an induction machine driving a load that can be characterized by a load torque–speed curve.

The equation of motion of the rotor in terms of slip is

$$\frac{J \, d\omega_r}{dt} = -J\omega_s \frac{ds}{dt} = T_e(s) - T_L(s)$$

Now let the slip $s = s_0 + \Delta s$.

$$-J\omega_s \frac{ds}{dt} = -J\omega_s \frac{d \, \Delta s}{dt}$$

If Δs is assumed to be small,

$$T_e(s) \approx T_e(s_0) + \frac{dT_e(s_0)}{ds} \Delta s$$

$$T_L(s) \approx T_L(s_0) + \frac{dT_L(s_0)}{ds} \Delta s$$

Substituting in the equation of motion for the rotor, we have

$$-J\omega_s \frac{d \, \Delta s}{dt} = T_e(s_0) - T_L(s_0) + \left[\frac{dT_e(s_0)}{ds} - \frac{dT_L(s_0)}{ds} \right] \Delta s$$

$$= \left[\frac{dT_e(s_0)}{ds} - \frac{dT_L(s_0)}{ds} \right] \Delta s$$

$$= \frac{dT_a(s_0)}{ds} \Delta s$$

where $T_e(s_0) = T_L(s_0)$ at a steady-state operating point and $T_a(s)$ is accelerating torque. Figure 3.22 illustrates the concept. Our equation is in the form

$$\frac{d \, \Delta s}{dt} + \frac{1}{\tau} \Delta s = 0$$

which is a first-order homogeneous linear differential equation, where

$$\frac{1}{\tau} = \frac{1}{J\omega_s} \left[\frac{dT_e(s_0)}{ds} - \frac{dT_L(s_0)}{ds} \right]$$

$$= \frac{1}{J\omega_s} \frac{dT_a(s_0)}{ds}$$

The solution is a decaying exponential as illustrated in Figure 3.23.

THE INDUCTION MACHINE

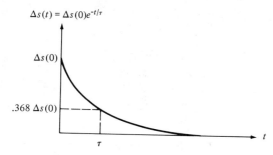

$$\Delta s(t) = \Delta s(0)e^{-t/\tau}$$

$\Delta s(0)$

$.368\,\Delta s(0)$

τ

t

FIGURE 3.23

EXAMPLE 3.6 Suppose that the machine of Example 3.4 is subjected to a sequence of impact loads that abruptly (i.e., in negligible time) cause a 1.5% reduction in machine speed. Determine an approximate time required for the machine to return to a steady-state operating speed following each impact loading.

If we assume that the machine is otherwise unloaded, its steady-state operating point will be at no-load speed or at a slip $s_0 \approx 0$. If we use the approximate expression for torque given by (3.79),

$$T_e = \frac{2T_{em}}{s/s_m + s_m/s}$$

we see that the second term in the denominator will dominate for slip near zero. Thus

$$T_e \simeq \frac{2T_{em}}{s_m}\,s = \frac{2T_{em}}{s_m}\,\Delta s$$

where $s = s_0 + \Delta s$ and $s_0 = 0$. Our equation of motion in terms of slip will be

$$-\omega_s J\frac{ds}{dt} = -\omega_s J\frac{d\,\Delta s}{dt} = \frac{2T_{em}}{s_m}\,\Delta s$$

or

$$\frac{d\,\Delta s}{dt} = -\frac{2T_{em}}{s_m \omega_s J}\,\Delta s = -\frac{1}{\tau}\,\Delta s$$

Thus our time constant τ will be

$$\tau = \frac{s_m \omega_s J}{2T_{em}}$$

Substituting data from Example 3.4 (assuming a two-pole machine), we have

$$\tau = \frac{(0.183)(377)(6)}{2(572.4)} = 0.362 \text{ s}$$

Machine speed approaches the operating point (no-load speed) asymptotically following each impact. We will have recovered in excess of 99% of the drop in speed after five time constants have elapsed. Thus an approximate recovery time would be $5(0.362) = 1.81$ s.

Equation (3.63) can be used to establish a relationship between the rotor energy loss during each recovery period and the kinetic energy of the rotor at synchronous speed.

$$T_e = \frac{3I_r'^2 r_r'}{s\omega_s}$$

$$-J\omega_s \frac{ds}{dt} = \frac{3I_r'^2 r_r'}{s\omega_s}$$

$$-J\omega_s^2 s \, ds = 3I_r'^2 r_r' \, dt$$

$$-J\omega_s^2 \int_{\Delta s(0)}^{s_0} s \, ds = \int_0^{t_f} (3I_r'^2 r_r') \, dt$$

In this case $\Delta s(0)$ is the change in slip corresponding to a 1.5% speed reduction; thus $\Delta s(0) = 0.015$. The integral on the right is energy loss in the rotor during each recovery period. Our steady-state operating point s_0 is approximately zero. Time t_f is the time required for slip to return to s_0. Thus during each recovery period our rotor losses will be

$$\tfrac{1}{2}J\omega_s^2 \, \Delta s(0)^2 = \tfrac{1}{2}(6)(377)^2(0.015)^2$$

$$= 95.9 \text{ J}$$

If the impact load were to occur every 2 s, we would dissipate on the average $95.9/2 = 47.95$ W in rotor losses. In this case the rotor loss would not be excessive since a 100-hp machine with a rated slip of 0.03 would typically have rotor losses on a continuous basis of approximately 3% of rated output or $(0.03)(100)(746) = 2238$ W. Larger speed excursions and/or higher moments of inertia may cause rotor heating to be a limiting factor. Other considerations, such as system voltage fluctuations, become important for large speed excursions.

3.5

UNBALANCED OPERATION OF INDUCTION MACHINES

Unbalanced operation of polyphase induction machines occurs whenever armature phase currents do not constitute a balanced polyphase set. The situation is typically an abnormal one caused by some undesirable system condition. Improper distribution of large single-phase loads in an industrial plant may cause various three-phase circuits to operate unbalanced. A blown fuse in one leg of the three-phase supply to a machine is another example.

In Chapter 2 we showed that balanced two-phase currents flowing in windings appropriately displaced in space could establish a sinusoidally distributed magnetic field intensity wave rotating at a speed determined by the frequency of the currents. We now repeat the development considering an unbalanced set of currents. Consider the two-phase machine shown conceptually in Figure 3.24. If we assume a sinusoidal distribution of the radial component of the air-gap magnetic field intensity for each winding, we have

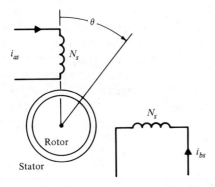

FIGURE 3.24

$$H_{as}(\theta) = H_{asmax} \cos \theta \tag{3.80}$$

$$H_{bs}(\theta) = H_{bsmax} \cos \left(\theta - \frac{\pi}{2} \right) \tag{3.81}$$

where $H_{as}(\theta)$ and $H_{bs}(\theta)$ are the magnetic field intensities due to windings a and b, respectively. From Ampère's law (as in Sections 2.5 and 2.6),

$$H_{asmax} = \frac{N_s i_{as}}{2g} \tag{3.82}$$

$$H_{bsmax} = \frac{N_s i_{bs}}{2g} \tag{3.83}$$

The resultant magnetic field intensity in the gap due to phases a and b is

$$H_s(\theta) = H_{as}(\theta) + H_{bs}(\theta) \tag{3.84}$$

$$= \frac{N_s}{2g} \left[i_{as} \cos \theta + i_{bs} \cos \left(\theta - \frac{\pi}{2} \right) \right]$$

Now suppose that the armature currents are given by

$$i_{as} = \sqrt{2}\, I_a \cos \omega_s t \tag{3.85}$$

$$i_{bs} = \sqrt{2}\, I_b \cos (\omega_s t + \varphi_b) \tag{3.86}$$

Substituting in (3.84), we obtain

$H_s(\theta, t)$

$$= \frac{\sqrt{2}\, N_s}{2g} \left[I_a \cos \omega_s t \cos \theta + I_b \cos (\omega_s t + \varphi_b) \cos \left(\theta - \frac{\pi}{2} \right) \right]$$

$$= \frac{\sqrt{2}\, N_s}{2g} \left\{ \frac{I_a}{2} [\cos (\omega_s t + \theta) + \cos (\omega_s t - \theta)] \right.$$

$$\left. + \frac{I_b}{2} \left[\cos \left(\omega_s t + \theta + \varphi_b - \frac{\pi}{2} \right) + \cos \left(\omega_s t - \theta + \varphi_b + \frac{\pi}{2} \right) \right] \right\}$$

$$\tag{3.87}$$

Note that if $I_b = I_a = I_s$ and $\varphi_b = -\pi/2$, the expression reduces to our previous result given as (2.74); that is, the equation of a single sinusoidally distributed magnetic field intensity wave rotating at speed ω_s. Examination of the four terms in (3.87) reveals that two yield sinusoidally distributed magnetic field intensity waves that rotate in a clockwise (forward) direction, and the remaining two yield sinusoidally distributed magnetic field intensity waves that rotate in a counterclockwise (backward) direction. These can be combined to yield single fields rotating in the forward and in the backward directions. The resultant can thus be expressed as

$$H_s(\theta, t) = H_{sf}(\theta, t) + H_{sb}(\theta, t) \tag{3.88}$$

where the forward-rotating field is given by

$$H_{sf}(\theta, t) = \frac{\sqrt{2}\,N_s}{2g}\left[\frac{I_a}{2}\cos(\omega_s t - \theta) + \frac{I_b}{2}\cos\left(\omega_s t - \theta + \varphi_b + \frac{\pi}{2}\right)\right] \tag{3.89}$$

and the backward-rotating field is given by

$$H_{sb}(\theta, t) = \frac{\sqrt{2}\,N_s}{2g}\left[\frac{I_a}{2}\cos(\omega_s t + \theta) + \frac{I_b}{2}\cos\left(\omega_s t + \theta + \varphi_b - \frac{\pi}{2}\right)\right] \tag{3.90}$$

It is easy to show that the forward-rotating field could have been established by the following armature currents [substitute currents in (3.84) to see this; note that the first term of (3.91) and the first term of (3.92) combine to yield the first term in (3.89), etc.]:

$$i_{a1} = \sqrt{2}\,\frac{I_a}{2}\cos\omega_s t + \sqrt{2}\,\frac{I_b}{2}\cos\left(\omega_s t + \varphi_b + \frac{\pi}{2}\right) \tag{3.91}$$

$$i_{b1} = \sqrt{2}\,\frac{I_a}{2}\cos\left(\omega_s t - \frac{\pi}{2}\right) + \sqrt{2}\,\frac{I_b}{2}\cos(\omega_s t + \varphi_b) \tag{3.92}$$

The phasors representing i_{as} and i_{bs} as given by (3.85) and (3.86) may be taken as

$$i_{as} = \sqrt{2}\,I_a\cos\omega_s t \leftrightarrow \underline{I}_a = I_a\underline{/0} \tag{3.93}$$

$$i_{bs} = \sqrt{2}\,I_b\cos(\omega_s t + \varphi_b) \leftrightarrow \underline{I}_b = I_b\underline{/\varphi_b} \tag{3.94}$$

The corresponding phasor relationships for (3.91) and (3.92) will be

$$\underline{I}_{a1} = \frac{\underline{I}_a}{2} + j\frac{\underline{I}_b}{2} \tag{3.95}$$

$$\underline{I}_{b1} = -j\frac{\underline{I}_a}{2} + \frac{\underline{I}_b}{2} = -j\underline{I}_{a1} \tag{3.96}$$

In a similar way we can show that the backward-rotating field could have been established by the following armature currents:

$$i_{a2} = \sqrt{2}\,\frac{I_a}{2}\cos\omega_s t + \sqrt{2}\,\frac{I_b}{2}\cos\left(\omega_s t + \varphi_b - \frac{\pi}{2}\right) \qquad (3.97)$$

$$i_{b2} = \sqrt{2}\,\frac{I_a}{2}\cos\left(\omega_s t + \frac{\pi}{2}\right) + \sqrt{2}\,\frac{I_b}{2}\cos\left(\omega_s t + \varphi_b\right) \qquad (3.98)$$

The corresponding phasor relationships for (3.97) and (3.98) will be

$$I_{a2} = \frac{I_a}{2} - j\frac{I_b}{2} \qquad (3.99)$$

$$I_{b2} = j\frac{I_a}{2} + \frac{I_b}{2} = + jI_{a2} \qquad (3.100)$$

Examination of (3.95) and (3.96) shows that the forward-rotating wave is established by a balanced set of currents with phase b current lagging phase a by $\pi/2$. This sequence will be referred to as a *positive sequence*. Examination of (3.99) and (3.100) shows that the backward-rotating wave is established by a balanced set of currents of opposite, or *negative, sequence*. The combined effect of positive- and negative-sequence currents is the same as the effect of our original unbalanced set of currents given by (3.85) and (3.86). From (3.95) and (3.99) we have

$$I_a = I_{a1} + I_{a2} \qquad (3.101)$$

From (3.96) and (3.100) we have

$$I_b = I_{b1} + I_{b2} \qquad (3.102)$$
$$= -jI_{a1} + jI_{a2}$$

In matrix form

$$\begin{bmatrix} I_a \\ I_b \end{bmatrix} = \begin{bmatrix} 1 & 1 \\ -j & j \end{bmatrix}\begin{bmatrix} I_{a1} \\ I_{a2} \end{bmatrix} = \overline{A}\begin{bmatrix} I_{a1} \\ I_{a2} \end{bmatrix} \qquad (3.103)$$

and from (3.95) and (3.99) we have

$$\begin{bmatrix} I_{a1} \\ I_{a2} \end{bmatrix} = \frac{1}{2}\begin{bmatrix} 1 & j \\ 1 & -j \end{bmatrix}\begin{bmatrix} I_a \\ I_b \end{bmatrix} = \overline{A}^{-1}\begin{bmatrix} I_a \\ I_b \end{bmatrix} \qquad (3.104)$$

Equations (3.103) and (3.104) define the classical symmetrical component transformation for two-phase systems. Essentially, any set of unbalanced phase currents at a particular frequency can be resolved into sets of balanced phase currents at the same frequency where one set is of positive phase sequence and the other is of negative phase sequence. From a physical viewpoint we see that if we have unbalanced phase currents in the armature of a symmetrical induction machine, we will establish a forward-rotating field and a backward-rotating field. The backward-rotating field typically establishes a braking torque and increases rotor winding losses. It also causes a double-frequency component to appear in the instantaneous torque waveform.

Our steady-state equivalent circuit for the polyphase induction machine was developed under the assumption of balanced stator and rotor phase variables. Since the circuit is valid for any constant speed (slip), and since it is a linear circuit, we can use the superposition principle to analyze unbalanced operation. Starting with a given set of unbalanced armature currents represented by phasors I_a and I_b, we can apply (3.104) to determine positive- and negative-sequence currents I_{a1} and I_{a2}. Positive-sequence components I_{a1} and $I_{b1} = jI_{a1}$ give rise to a forward-rotating wave. The rotor slip s with respect to this field is, from (3.37),

$$s = \frac{\omega_s - \omega_r}{\omega_s} \tag{3.105}$$

Negative-sequence components I_{a2} and $I_{b2} = +jI_{a2}$ give rise to a backward-rotating field. The rotor slip s' with respect to this field is, from (3.37),

$$s' = \frac{-\omega_s - \omega_r}{-\omega_s} = 1 + \frac{\omega_r}{\omega_s} = 2 - \frac{\omega_s - \omega_r}{\omega_s} \tag{3.106}$$

$$= 2 - s$$

when expressed in terms of slip s with respect to the forward field.

Currents I_{a1} and I_{a2} are applied to the appropriate equivalent circuits as shown in Figure 3.25. If we solve for V_{a1} and V_{a2} we can apply (3.103) (for voltages) to determine V_a and V_b. We could just as easily have started with unbalanced voltages V_a and V_b and found currents I_a and I_b.

The procedure is conceptually simple, but conclusions based on power dissipated in the various resistances of the equivalent circuits must be drawn with care. In a polyphase system operating under balanced conditions, total instantaneous electrical power converted is a constant. Torque developed is also a constant. If we operate an induction machine under unbalanced conditions, the counter-rotating fields established give rise to double-frequency torque components. Although these double-frequency torque components produce no time-average torque, they may cause a vibration problem that would not exist if phase variables were balanced.

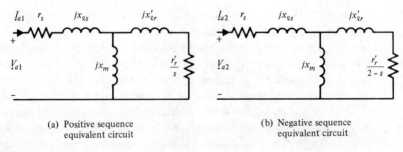

(a) Positive sequence equivalent circuit

(b) Negative sequence equivalent circuit

FIGURE 3.25

THE INDUCTION MACHINE

We can draw conclusions about *average* power and torque by examining the total average power input to the machine at the armature terminals.

$$P = \text{Re}\{V_a I_a^* + V_b I_b^*\}$$

$$= \text{Re}\{\bar{V}^T \bar{I}^*\} \tag{3.107}$$

where

$$V = \begin{bmatrix} V_a \\ V_b \end{bmatrix} \tag{3.108}$$

$$\bar{I}^* = \begin{bmatrix} I_a^* \\ I_b^* \end{bmatrix} \tag{3.109}$$

The asterisk denotes the complex conjugate of the variable. Now applying the transformation relationships given by (3.103) and (3.104), we have

$$\bar{V} = \begin{bmatrix} V_a \\ V_b \end{bmatrix} = \begin{bmatrix} 1 & 1 \\ -j & j \end{bmatrix} \begin{bmatrix} V_{a1} \\ V_{a2} \end{bmatrix} \tag{3.110}$$

$$\bar{V}^T = [V_a \quad V_b] = [V_{a1} \quad V_{a2}] \begin{bmatrix} 1 & -j \\ 1 & j \end{bmatrix} \tag{3.111}$$

$$\bar{I} = \begin{bmatrix} I_a \\ I_b \end{bmatrix} = \begin{bmatrix} 1 & 1 \\ -j & j \end{bmatrix} \begin{bmatrix} I_{a1} \\ I_{a2} \end{bmatrix} \tag{3.112}$$

$$\bar{I}^* = \begin{bmatrix} I_a^* \\ I_b^* \end{bmatrix} = \begin{bmatrix} 1 & 1 \\ j & -j \end{bmatrix} \begin{bmatrix} I_{a1}^* \\ I_{a2}^* \end{bmatrix}$$

Thus

$$\bar{V}^T \bar{I}^* = [V_{a1} \quad V_{a2}] \begin{bmatrix} 1 & -j \\ 1 & j \end{bmatrix} \begin{bmatrix} 1 & 1 \\ j & -j \end{bmatrix} \begin{bmatrix} I_{a1}^* \\ I_{a2}^* \end{bmatrix}$$

$$= [V_{a1} \quad V_{a2}] \begin{bmatrix} 2 & 0 \\ 0 & 2 \end{bmatrix} \begin{bmatrix} I_{a1}^* \\ I_{a2}^* \end{bmatrix} \tag{3.114}$$

$$= 2(V_{a1} I_{a1}^* + V_{a2} I_{a2}^*)$$

Thus, from (3.107),

$$P = 2 \,\text{Re}\{V_{a1} I_{a1}^* + V_{a2} I_{a2}^*\} \tag{3.115}$$

Equation (3.115) shows that we can compute average power P by simply computing twice the average power input to the sequence equivalent circuits of Figure 3.25(a) and (b) and summing the result. The factor of 2 appears because we have two phases. Thus average power may be computed by considering first positive-sequence effects, then negative-sequence effects. Similarly, we can determine the average mechanical power converted, average torque, average power converted to rotor winding losses, and so on, by considering first positive-

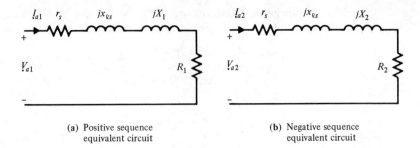

(a) Positive sequence
equivalent circuit

(b) Negative sequence
equivalent circuit

FIGURE 3.26

sequence effects, then negative-sequence effects. Our equivalent circuits can be simplified as shown in Figure 3.26, where

$$R_1 + jX_1 = \frac{jx_m\left(\dfrac{r_r'}{s} + jx_{\ell r}'\right)}{\dfrac{r_r'}{s} + j(x_m + x_{\ell r}')} \tag{3.116}$$

$$R_2 + jX_2 = \frac{jx_m\left(\dfrac{r_r'}{2-s} + jx_{\ell r}'\right)}{\dfrac{r_r'}{2-s} + j(x_m + x_{\ell r}')} \tag{3.117}$$

For the positive-sequence circuit, the power across the gap is

$$P_{g1} = I_{a1}^2 R_1 \tag{3.118}$$

The mechanical power converted is

$$P_{m1} = (1-s)P_{g1} \tag{3.119}$$

The rotor losses are

$$P_{r1} = sP_{g1} \tag{3.120}$$

For the negative-sequence circuit, the power across the gap is

$$P_{g2} = I_{a2}^2 R_2 \tag{3.121}$$

The mechanical power converted is

$$P_{m2} = (1-s')P_{g2} \tag{3.122}$$
$$= -(1-s)P_{g2}$$

The rotor losses are

$$P_{r2} = s'P_{g2} \tag{3.123}$$
$$= (2-s)P_{g2}$$

Thus the total quantities will be

$$P_g = 2(P_{g1} + P_{g2}) \tag{3.124}$$

$$P_m = 2(P_{m1} + P_{m2})$$

$$= 2(1 - s)(P_{g1} - P_{g2}) \tag{3.125}$$

$$P_r = 2(P_{r1} + P_{r2})$$

$$= 2(sP_{g1} + (2 - s)P_{g2})$$

$$= 2(s(P_{g1} - P_{g2}) + 2P_{g2}) \tag{3.126}$$

The torque can be found as

$$T_e = \frac{P_m}{\omega_r} = \frac{P_m}{(1 - s)\omega_s}$$

$$= \frac{2}{\omega_s}(P_{g1} - P_{g2}) \tag{3.127}$$

$$= T_{e1} - T_{e2}$$

In many cases $P_{g1} \gg P_{g2}$. Thus, from (3.125) and (3.127), we see that in such cases the mechanical power converted (and the total torque developed) will not be reduced significantly. The rotor and stator winding loss increases due to negative-sequence currents, however, may be significant. The increased i^2R heating raises the operating temperature, which in turn can seriously reduce insulation life.

EXAMPLE 3.7 A two-phase two-pole induction machine operates at rated slip when a fuse in phase b interrupts phase b current. Determine the machine mechanical power converted, electrical torque developed, and winding losses during normal operation and following the interruption of current in phase b assuming that phase a current is not interrupted. The machine parameters are

$$\text{rated voltage} = 115 \text{ V}$$

$$r_s = 0.35 \text{ }\Omega$$

$$r_r' = 0.25 \text{ }\Omega$$

$$x_{\ell s} = 1.50 \text{ }\Omega$$

$$x_{\ell r}' = 1.00 \text{ }\Omega$$

$$x_m = 50.0 \text{ }\Omega$$

$$\text{rated slip} = 0.035$$

Prior to the occurrence of the interruption, we have balanced two-phase operation. Armature voltages and currents constitute balanced two-phase sets of variables and thus contain only positive-sequence components. We thus analyze behavior using only the positive-sequence equivalent circuit. Immediately following the circuit interruption, we will have unbalanced operation, since currents no longer constitute a balanced two-phase set of variables. We thus require positive- and negative-sequence equivalent circuits.

Since we are operating at rated slip, the machine is under load and developing its rated output power. We anticipate that the machine will slow down after the circuit interruption because of the reduction in net average electrical torque developed. If we assume that the machine speed does not depart appreciably from rated speed, we can use the machine rated slip in all of our calculations. The validity of this assumption can be checked following the calculation. We will assume that phase a voltage can be described as

$$v_a(t) = \sqrt{2}\,(115)\cos \omega_s t \leftrightarrow V_a = 115 \underline{/0^\circ}\ \text{V}$$

before and after the circuit interruption; that is, the machine is connected to an infinite bus. Switching transients will be neglected.

We start by determining positive- and negative-sequence equivalent circuits for our machine at rated slip. We can simplify the circuits of Figure 3.25 to those shown in Figure 3.26 by applying (3.116) and (3.117):

$$R_1 + jX_1 = \frac{j50.0[(0.25/0.035) + j1.0]}{(0.25/0.035) + j(50.0 + 1.0)} = 6.86 + j1.98\ \Omega$$

As explained in Section 3.3, the effective rotor resistance at standstill (60-Hz rotor currents) may be several times higher than what we have at rated slip. For negative-sequence currents, we have induced rotor currents at roughly twice line frequency and the increase in effective rotor resistance is more pronounced. We will assume that rotor resistance appropriate for the negative-sequence equivalent circuit in this case is five times that for the positive-sequence circuit. Reference 2 presents some practical ramifications of overlooking this effect, and reference 3 presents an analytical development for more rigorously choosing an appropriate multiplier.

Thus, from (3.117),

$$R_2 + jX_2 = \frac{j50.0\{[(5)(0.25)/(2 - 0.035)] + j1.0\}}{[5(0.25)/(2 - 0.035)] + j(50.0 + 1.0)} = 0.611 + j.988\ \Omega$$

Our equivalent circuits will be as shown in Figure 3.27.
Before the fuse interruption in phase b we have

$$V_{a1} = V_a = 115 \underline{/0^\circ}\ \text{V}$$

$$V_{a2} = 0\ \text{V}$$

Solving for I_{a1} using the equivalent circuit of Figure 3.27(a) gives us

$$I_{a1} = \frac{115 \underline{/0^\circ}}{(0.35 + 6.86) + j(1.5 + 1.98)} = 14.4 \underline{/-25.8^\circ}\ \text{A}$$

Since $V_{a2} = 0$, all negative-sequence quantities will be zero. From (3.118)

$$P_{g1} = (14.4)^2(6.86) = 1422.5\ \text{W}$$

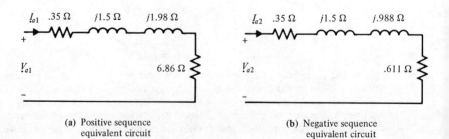

(a) Positive sequence
equivalent circuit

(b) Negative sequence
equivalent circuit

FIGURE 3.27 Equivalent Circuits at s = 0.035

From (3.125)

$$P_m = 2(1 - 0.035)(1,422.5) = 2745 \text{ W}$$
$$= 3.68 \text{ hp}$$

(Subtracting friction and windage losses from this value would yield available shaft output power. Thus the machine rating might be $3\frac{1}{2}$ hp.) From (3.27)

$$T_e = \frac{2745}{(1 - 0.035)377} = 7.55 \text{ N} \cdot \text{m}$$

From (3.126) the rotor winding losses are

$$P_r = 2(0.035)(1422.5) = 99.6 \text{ W}$$

The stator winding losses are

$$P_s = 2(14.4)^2(0.35) = 145 \text{ W}$$

The combined winding losses are

$$P_s + P_r = 245 \text{ W}$$

After the fuse interruption, the current in phase b is zero, and the voltage in phase b is unknown. From (3.104) we have

$$\begin{bmatrix} I_{a1} \\ I_{a2} \end{bmatrix} = \frac{1}{2} \begin{bmatrix} 1 & j \\ 1 & -j \end{bmatrix} \begin{bmatrix} I_a \\ 0 \end{bmatrix}$$

Thus

$$I_{a1} = I_{a2} = \frac{I_a}{2} \tag{1}$$

From (3.103) (for voltages), we have

$$\begin{bmatrix} V_a \\ V_b \end{bmatrix} = \begin{bmatrix} 1 & 1 \\ -j & j \end{bmatrix} \begin{bmatrix} V_{a1} \\ V_{a2} \end{bmatrix}$$

Thus

$$V_{a1} + V_{a2} = V_a \tag{2}$$

We can satisfy (1) and (2) by combining the equivalent circuits of Figure 3.27 in series as shown in Figure 3.28.

Solving for currents in the equivalent circuit of Figure 3.28 gives us

$$I_{a1} = I_{a2} = \frac{115 \underline{/0^\circ}}{(0.35 + 6.86 + 0.35 + 0.611) + j(1.5 + 1.98 + 1.5 + 0.988)}$$
$$= 11.4 \underline{/-36.1^\circ} \text{ A}$$

Note, from (1), that phase a current will be

$$I_a = 2I_{a1} = 22.8 \underline{/-36.1^\circ} \text{ A}$$

Thus phase a current increases by roughly 58%. From (3.118) and (3.121)

$$P_{g1} = (11.4)^2(6.86) = 893.1 \text{ W}$$

$$P_{g2} = (11.4)^2(0.611) = 79.4 \text{ W}$$

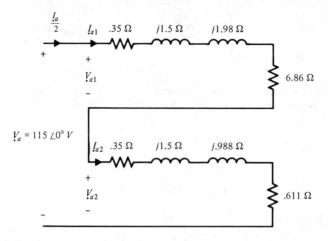

FIGURE 3.28

From (3.125)

$$P_m = 2(1 - 0.035)(893.1 - 79.4) = 1570.4 \text{ W}$$

$$= 2.11 \text{ hp}$$

From (3.27)

$$T_e = \frac{1570.4}{(1 - 0.035)(377)} = 4.32 \text{ N} \cdot \text{m}$$

From (3.126) the rotor winding losses are

$$P_r = 2[0.035(893.1 - 79.4) + 2(79.4)] = 374.6 \text{ W}$$

The stator winding losses are

$$P_s = 2[(11.4)^2(0.35) + (11.4)^2(0.35)] = 181.9 \text{ W}$$

The combined winding losses are

$$P_r + P_s = 556.5 \text{ W}$$

Since the average torque following the interruption of current in phase b is significantly less than the torque at rated conditions, an assumption of constant speed would not be valid. The actual speed reduction will depend on the load characteristics. Our solution is, accordingly, only appropriate immediately following the circuit interruption, before the speed changes appreciably. We see that total winding losses have increased significantly, and we would expect the machine to overheat. A significant speed reduction would compound the problem. Operation on a prolonged basis under these conditions would seriously reduce insulation life.

3.6

SINGLE-PHASE INDUCTION MACHINES

A single-phase induction machine is illustrated conceptually in Figure 3.29, and a photograph of such a machine appears in Figure 3.30. The auxiliary winding denoted by the subscript a is used for starting purposes and is typically switched out when the machine is at roughly 70% of rated speed. A centrifugal switch

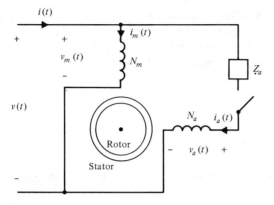

FIGURE 3.29

can be used to cause switching to occur at the proper machine speed. The machine then runs on only the main winding denoted by the subscript m. This winding must be designed for continuous operation under rated conditions and it typically differs from the auxiliary winding.

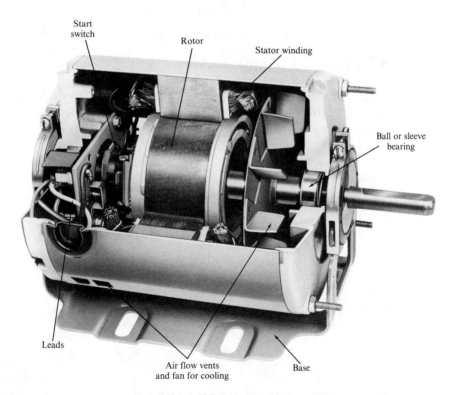

FIGURE 3.30 Typical Single-Phase Fractional-Horsepower Motor. (Courtesy of Marathon Electric)

When the auxiliary winding is switched out, the machine can be regarded as a symmetrical two-phase machine running on one phase. The situation is conceptually similar to what we encountered in Example 3.7. Under steady-state conditions with the auxiliary winding switched out, we have

$$I_m = I$$

$$I_a = 0$$

$$V_m = V$$

where phasors are used to represent the appropriate time-varying quantities. Applying (3.104) yields

$$\begin{bmatrix} I_{m1} \\ I_{m2} \end{bmatrix} = \frac{1}{2} \begin{bmatrix} 1 & 1 \\ 1 & -j \end{bmatrix} \begin{bmatrix} I_m \\ 0 \end{bmatrix} \tag{3.128}$$

$$I_{m1} = I_{m2} = \frac{I_m}{2} = \frac{I}{2}$$

Applying (3.103) gives us

$$\begin{bmatrix} V_m \\ V_a \end{bmatrix} = \begin{bmatrix} 1 & 1 \\ -j & j \end{bmatrix} \begin{bmatrix} V_{m1} \\ V_{m2} \end{bmatrix} \tag{3.129}$$

$$V = V_m = V_{m1} + V_{m2}$$

We can satisfy (3.128) and (3.129) by connecting the equivalent circuits of Figure 3.25 in series as shown in Figure 3.31.

The circuit can be further simplified by recognizing that if we halve all impedances, the current for a voltage V will be doubled. We thus arrive at the circuit shown in Figure 3.32. We can determine the average mechanical power converted, average torque, rotor losses, and so on, from the equivalent circuit of

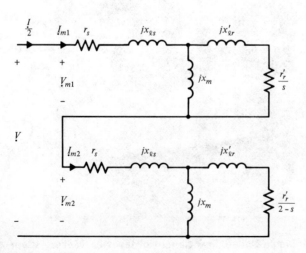

FIGURE 3.31

THE INDUCTION MACHINE

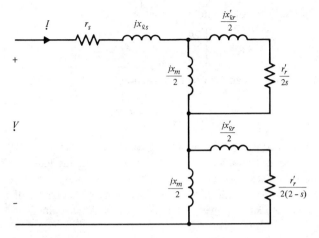

FIGURE 3.32 Equivalent Circuit for a Single-Phase Induction Machine Running on Its Main Winding Only

Figure 3.32. The circuit can easily be simplified to the circuit shown in Figure 3.33. The parameters R_1, X_1, R_2, and X_2 are determined by (3.116) and (3.117). The total average power across the gap will be

$$P_g = I^2 \left(\frac{R_1}{2} + \frac{R_2}{2} \right) \tag{3.130}$$

The average mechanical power converted will be

$$P_m = I^2 \left[(1 - s) \frac{R_1}{2} + (1 - s') \frac{R_2}{2} \right] \tag{3.131}$$

$$= (1 - s)I^2 \left(\frac{R_1}{2} - \frac{R_2}{2} \right)$$

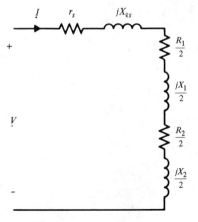

FIGURE 3.33 Equivalent Circuit for a Single-Phase Induction Machine at a Particular Value of Slip

where $s' = 2 - s$. The rotor winding losses will be

$$P_r = I^2 \left[\frac{sR_1}{2} + (2 - s) \frac{R_2}{2} \right] \tag{3.132}$$

The rotor winding losses can also be expressed as (see Problem 3.21)

$$P_r = \frac{s}{1 - s} P_m + I^2 R_2 \tag{3.133}$$

Note that in a balanced polyphase induction machine $P_r = [s/(1 - s)]P_m$. Thus for the single-phase machine, we have increased rotor winding losses due to the $I^2 R_2$ term. These increased losses reduce efficiency and cause the motor to run at a higher temperature. Single-phase machines must generally be of a larger frame size to have the same power rating as a polyphase machine.

The torque can be found as

$$T_e = \frac{P_m}{\omega_r} = \frac{P_m}{(1 - s)\omega_s} \tag{3.134}$$

$$= \frac{I^2}{\omega_s} \left(\frac{R_1}{2} - \frac{R_2}{2} \right)$$

Examination of Figures 3.32 and 3.33 shows that at slip $s = 1$, we will have $R_1 = R_2$. Thus the torque is zero at zero speed. The reason for the auxiliary winding is now evident. The impedance Z_a shown in Figure 3.29 may be chosen so that I_m and I_a differ in phase by approximately 90° for values of slip near unity. The machine then develops nonzero starting torque in the manner of a polyphase machine.

By expressing s in terms of rotor speed in the equivalent circuit of Figure 3.32, we have

$$s = \frac{\omega_s - \omega_r}{\omega_s} = 1 - \frac{\omega_r}{\omega_s}$$

$$2 - s = 1 + \frac{\omega_r}{\omega_s}$$

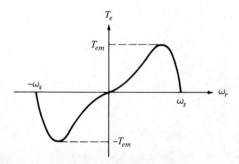

FIGURE 3.34 Average Torque Versus Speed for a Single-Phase Induction Machine Running on Its Main Winding Only

FIGURE 3.35 Stator for a Four-Pole Split-Phase Motor Showing Main Winding and Auxiliary Winding. (Courtesy of Bodine Electric Company)

and examining the torque expression above, we see that the torque will be an odd function of the speed ω_r:

$$T_e(\omega_r) = - T_e(-\omega_r)$$

The torque versus speed curve for a single-phase induction machine is illustrated by the somewhat typical shape shown in Figure 3.34. The reader should justify this shape on the basis of rotating magnetic field intensity waves, recalling that $I_{a1} = I_{a2}$ for a machine running on a single winding.

Often, the auxiliary winding consists of finer wire and has fewer turns than the main winding (Figure 3.35). If when the rotor is at standstill, the auxiliary winding is predominately resistive and the main winding is predominately reactive, the current in the auxiliary winding will lead the current in the main winding since both windings have the same source voltage across their terminals. The auxiliary winding is not designed for continuous duty and must be switched out of the circuit within several seconds to keep the circuit from overheating. A

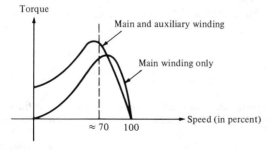

FIGURE 3.36

**FIGURE 3.37 Centrifugal Switch Cutout Mechanism for a Split-Phase Motor.
(Courtesy of Bodine Electric Company)**

capacitor in series with the starting winding can provide improved starting torque
characteristics and can reduce starting inrush current. Such motors are referred
to as *capacitor-start motors*. Motors that rely on a series impedance and/or
differences between the main- and auxiliary-winding characteristics to produce
an appropriate phase difference in winding currents at starting are often referred
to as *split-phase motors*. Typical torque–speed curves for a split-phase motor
are illustrated in Figure 3.36, and a centrifugal switch-cutout mechanism is
shown in Figure 3.37. Reference 4 summarizes characteristics for the various
types of split-phase motor configurations.

REFERENCES

[1] P. C. KRAUSE and C. H. THOMAS, "Simulation of Symmetrical Induction Machinery," *IEEE Transactions on Power Apparatus and Systems,* Vol. PAS-84, pp.
1038–1053, November 1965.

[2] M. SHAN GRIFFITH, "A Penetrating Gage at One Open Phase: Analyzing the Polyphase Induction Motor Dilemma," *IEEE Transactions on Industry Applications,* Vol.
IA-13, No. 6, pp. 504–516, November/December 1977.

[3] B. ADKINS and R. G. HARLEY, *The General Theory of Alternating Current Machines.*
London: Chapman & Hall Ltd., 1975.

[4] *Small Motor, Gearmotor, and Control Handbook,* 4th ed. Chicago: Bodine Electric
Company, 1978.

PROBLEMS

3.1   A four-pole three-phase 60-Hz induction machine has a rated slip of $s = 0.035$. Find:

  a. The speed (in rpm) of the rotating flux wave due to armature currents with respect
to the stator.

  b. The speed (in rpm) of the rotating flux wave due to rotor currents with respect to the
stator.

c. The speed (in rpm) of the rotating resultant flux wave due to armature currents and rotor currents with respect to the stator.

d. The speed (in rpm) of the rotor with respect to the stator.

e. The frequency (in hertz) of the armature currents.

f. The frequency (in hertz) of the rotor currents.

3.2 A general purpose four-pole 50 horsepower induction machine develops its rated power output at a slip of 3%. Find the speed of the machine at which rated power is developed and the torque developed by the machine at this speed.

3.3 A three-phase induction machine has a rated speed of 1120 rpm.

a. How many poles is the stator of the machine wound for?

b. What is the rated slip of the machine?

3.4 Find the impedance $Z_s = V_s/I_s$ for an induction machine represented by the equivalent circuit shown in Figure 3.4 if the rotor is blocked (i.e., if $\omega_r = 0$). Recalculate the impedance if the machine operates at a slip $s = 0.02$. The parameter values are

$$r_s = 0.30 \ \Omega$$

$$r_r' = 0.20 \ \Omega$$

$$x_m = 15.0 \ \Omega$$

$$x_{\ell s} = 0.45 \ \Omega$$

$$x_{\ell_r}' = 0.30 \ \Omega$$

3.5 The equivalent circuit for an induction machine is similar to that of a power transformer. If measurements are made at the terminals of the machine corresponding to the measurements made on a power transformer for open-circuit and short-circuit tests, approximate parameters for the machine can be determined. Assume that $r_s = r_r'$ and that $x_{\ell s} = x_{\ell r}'$ and find the parameters for a machine having the following test data:

Machine: three phase, "Y" connected

No-load test: line-to-line voltage	460 V
line current	2 A
three-phase power	negligible

Blocked-rotor test: line-to-line voltage	450 V
line current	44 A
three-phase power	4800 W

3.6 A three-phase Y-connected six-pole induction machine designed for operation at 220 V line to line has the following parameters:

$$r_s = 0.25 \ \Omega$$

$$r_r' = 0.16 \ \Omega$$

$$x_{\ell s} = 0.47 \ \Omega$$

$$x_{\ell_r}' = 0.28 \ \Omega$$

$$x_m = 12 \ \Omega$$

$$\text{phase voltage} = \frac{220}{\sqrt{3}} = 127 \ V$$

Find the starting torque for an across-the-line start, the running torque for operation at a rated slip $s = 0.025$, and the mechanical power output at a rated slip if the mechanical losses (friction and windage) are 350 W.

3.7 A three-phase, six-pole induction machine is loaded in such a way that speed is constant at 400 rpm. The stator winding and rotor winding resistance (referred to the stator winding) are equal, that is, $r_r' = r_s$. The magnetizing reactance x_m is relatively high and may be treated as an infinite reactance. Neglect the friction and windage losses and find the efficiency of the motor at this speed. What general conclusion can be drawn about the operation of induction machines based on the result?

3.8   A three-phase induction machine operates at a slip of 0.04. The machine data are

$$r_s = 0.10 \ \Omega \qquad\qquad x_{\ell s} = .20 \ \Omega$$

$$r_r' = 0.08 \ \Omega \qquad\qquad x_{\ell r}' = .15 \ \Omega$$

$$\text{voltage (phase)} = \frac{460}{\sqrt{3}} = 266 \ \text{V} \qquad x_m = 9.0 \ \Omega$$

The friction and windage losses at this value of slip are 2000 W. For operation at this value of slip, determine:
  a. Armature winding losses.
  b. Rotor winding losses.
  c. Available output shaft power in horsepower.
  d. Efficiency.

3.9   A 100-hp rated (1 hp = 746 W) three-phase induction machine has a rated slip of 0.04. The friction and windage losses for speed in the vicinity of the rated speed are known to be 3000 W. Find the rotor winding losses for operation at rated conditions.

3.10 A three-phase induction machine operates at its rated slip of 3%. If the phase sequence is reversed (i.e., if two of the stator leads are suddenly interchanged), at what value of slip does the machine operate before any appreciable change in rotor speed occurs? (Neglect the effects of transients due to the switching.)

3.11 For the machine described in Example 3.3, determine the slip at which maximum torque occurs, and find the output power if the machine actually operates at this value of slip.

3.12 For the machine described in Example 3.3, determine the angle between the resultant field due to stator and rotor currents and the field due to rotor currents for each of the slip values shown in Table 3.1, the value of slip at which torque is a maximum and at a slip of 1.0.

3.13 Show that if armature resistance r_s is neglected, the induction machine torque–slip characteristic given by (3.68) can be expressed as

$$T_e = \frac{2T_{em}}{s/s_m + s_m/s}$$

3.14 Examine the torque–slip characteristics for an induction machine for relatively high values of slip and for relatively low values of slip. Sketch curves for these two regions relating torque and slip. It may be helpful to consider the simplified expression for torque given by (3.79).

3.15 A certain induction machine has a slip at its maximum developed torque of $s_m = 0.20$. The rated slip for the machine is 0.06. Assume that the machine develops 100% torque at the rated slip. Use the approximate torque-slip expression given by (3.79) and determine:
  a. Maximum torque T_{em} in percent.
  b. Blocked rotor torque in percent.
  c. The value of braking torque in percent immediately after the phase sequence is reversed if the motor is initially operating at the rated slip.

3.16 Determine the rotor energy losses during starting for the motor of Example 3.4. Assume that the machine rotor is initially at rest and accelerates to its no-load speed with the load torque $T_L = 0$.

3.17 The induction machine of Example 3.4 drives a load having a flat torque–speed curve; that is, T_L is a constant value equal in magnitude to machine torque at rated slip. Determine the time constant τ for the first-order linear differential equation used to describe small displacements of machine slip from rated slip.

3.18 A symmetrical two-phase induction machine is operated in such a way that phase voltages are given by

$$V_a = 115 \underline{/0°}$$
$$V_b = 105 \underline{/-80°}$$

Find the values of voltages V_{a1} and V_{a2} that would be appropriate for use with the equivalent circuits of Figure 3.25.

3.19 A symmetrical two-phase induction machine has stator currents of

$$i_{as} = \sqrt{2}\,(10)\,\cos\omega_s t \quad \text{A}$$

$$i_{bs} = \sqrt{2}\,(15)\,\cos\left(\omega_s t - \frac{\pi}{4}\right) \quad \text{A}$$

Based on the accompanying figure, sketch the rotating space vectors representing the forward-rotating field and the backward-rotating field at time $t = 0$. The sketch should correctly reflect angles and relative magnitudes. [NOTE: Assume that a positive current

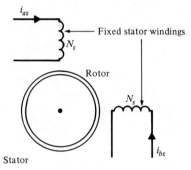

i_{as} with $i_{bs} = 0$ results in a vertically aligned space vector positive in the upward direction. A positive current i_{bs} with $i_{as} = 0$ results in a horizontally aligned space vector positive to the right.]

3.20 A two-pole two-phase induction machine running unloaded undergoes a fault (short circuit) on phase b such that the phase b armature voltage is reduced to zero. The phase a voltage is given by

$$v_a(t) = \sqrt{2}\,(100)\,\cos 377t \quad \text{V}$$

The machine parameters are

$$r_s = 0.5\ \Omega \qquad x_{\ell s} = 3.0\ \Omega$$
$$r'_r = 0.75\ \Omega \qquad x'_{\ell r} = 3.0\ \Omega$$
$$x_m = 97\ \Omega$$

Find the machine currents $i_a(t)$ and $i_b(t)$ after transients due to the fault have decayed to zero but before the machine speed changes appreciably.

3.21 Show that for a single-phase induction machine running on its main winding alone, the rotor winding losses can be expressed as

$$P_r = \frac{s}{1-s} P_m + I^2 R_2$$

where R_2 is defined in the equivalent circuit of Figure 3.33.

3.22 A $\frac{1}{4}$-hp, 115-V, four-pole single-phase motor has the following data:

$$x_{\ell s} = 2.8\ \Omega \qquad r_s = 1.8\ \Omega$$

$$x'_{\ell r} = 2.0\ \Omega \qquad r'_r = 3.6\ \Omega$$

$$x_m = 75\ \Omega$$

The friction and windage losses are 30 W.

Find the mechanical output power, torque, and motor efficiency for a slip of 0.05. Assume that the motor is running at a steady-state speed on its main winding only.

CHAPTER 4

The Synchronous Machine

4.1

INTRODUCTION

Nearly all three-phase power is generated by three-phase synchronous machines operated as generators. Bulk power generation is the most common application for large rotating synchronous machinery. An elementary two-pole three-phase generator is illustrated conceptually in Figure 4.1. The field circuit of the machine is on the rotor and it is excited by direct current. During normal operation the rotor is driven by a prime mover at constant speed and its magnetic field induces sinusoidal voltages in the windings of the armature, which are located on the stator of the machine. The frequency of the induced armature voltages depends on the speed at which the rotor turns and on the number of poles for which the machine is wound. The frequency of the armature voltages is given by

$$f_e = \frac{P}{2} f_m \qquad (4.1)$$

where f_e is the electrical frequency and f_m is the mechanical frequency of rotation. Induced voltages in each of the phases of the machine will be phase

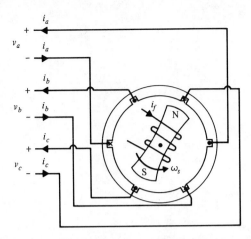

FIGURE 4.1 Elementary Two-Pole Three-Phase Synchronous Generator

displaced depending on the space displacement of the armature phase windings. The phase displacement is always $2\pi/3$ radians in a three-phase system. The machine normally operates with its armature windings connected to a three-phase power system. During normal steady-state operation the rotor must turn at a constant speed, which is dictated by the desired system frequency (60 Hz in the United States). This speed is referred to as the *synchronous speed,* and during normal operation the machine is said to operate *in synchronism* with the system to which it is connected.

Two fundamentally different types of large synchronous generators are commonly used: steam-turbine-driven generators and hydrogenerators. Roughly 85% of the power generation in the United States is by steam-turbine-driven generators. These generators are driven at relatively high speeds, such as 3600 or 1800 rpm (two- and four-pole machines respectively). The rotors of these generators are cylindrical, or round, in configuration and have a relatively large axial length and small diameter. A ratio of 10:1 is typical. The small diameter limits centrifugal forces at the normal operating speeds of these machines. The rotor is sometimes made from a single steel forging. Such machines are termed *solid-iron-rotor turbogenerators.* A cylindrical, or round, rotor configuration is shown conceptually in Figure 4.2. A typical, large turbine-driven generator rating might be 1100 MVA at a rated power factor of 0.90 lagging with a rated line-to-line armature voltage of 24 kV. Present-day large steam-turbine-driven generators range from about 200 to 1500 MVA.

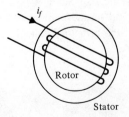

FIGURE 4.2 Two-Pole Cylindrical Rotor for a Synchronous Generator

THE SYNCHRONOUS MACHINE

The remaining generation (roughly 15%) is by water power. Hydraulic turbines generally operate at much lower speeds, say on the order of several hundred rpm. To generate power at 60 Hz, a relatively large number of poles are required. A typical hydrogenerator rating might be 50 MVA at a rated power factor of 0.80 lagging with a rated line-to-line armature voltage of 13.8 kV. The rated rpm might be 180 (40 poles). The relatively large number of poles required dictates a salient-pole structure for the rotor. This structure typically has a relatively short axial length and relatively large diameter. A ratio of 1:5 is typical. A salient-pole rotor is illustrated conceptually in Figure 4.3.

Although salient-pole machines differ in detail from cylindrical-rotor machines and have somewhat different operating characteristics, operation of the machines is similar in many ways. The armature windings of the machine are always on the stator and must be designed to develop the same number of magnetic poles as the field winding that is on the rotor. Almost all machines are wound with three phases, and (in the United States) are designed to develop an electrical frequency of 60 Hz. The field circuit is excited by direct current, necessitating a source of dc power. This can be accomplished by mounting a dc generator (generally referred to as an *exciter,* since it provides the field excitation) on the same shaft as the rotor of the synchronous machine. A dc voltage developed by the exciter can then be fed through slip rings to the rotating field winding of the synchronous machine. Advances in solid-state technology have permitted the development of efficient static excitation systems which can be used in lieu of dc-generator-type excitation systems. We will see that variation of the dc excitation voltage can be used to control the power factor at which the machine operates. This effectively controls the amount of reactive power the machine delivers (or receives) and to a somewhat lesser degree the rms armature terminal voltage of the machine. This important characteristic of the synchronous machine is a major reason for its widespread use in power generation. It is important to recognize that the dc power requirements of the field are generally much less than the synchronous machine rating. For example, a 790-MVA 0.90-power factor turbogenerator would deliver 790 × 0.90 = 711 MW of output power at its armature terminals under rated conditions. The dc excitation requirements to supply the field might typically be 2.0 MW or 0.28% of the machine output power. Indeed, were it not for the resistance of the field circuit the steady-state excitation power requirements would be zero. Thus the exciter always has a much lower rating than the synchronous machine itself.

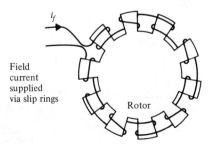

Field
current
supplied
via slip rings

FIGURE 4.3 Salient-Pole Rotor Wound for 10 Poles (the Stator Must Also Be Wound for 10 Poles)

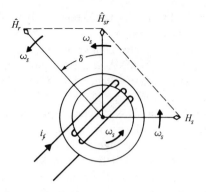

FIGURE 4.4 Typical Field Alignment for a Two-Pole Cylindrical-Rotor Synchronous Generator

The most common mode of operation for a synchronous generator is operation under essentially steady-state conditions and in synchronism with a power grid to which its armature terminals have been connected. Balanced three-phase currents in the armature of the machine establish a rotating magnetic field in which the speed of the rotating field is determined by the system frequency and the number of poles for which the machine is wound. A steady direct current in the field winding on the rotor establishes a magnetic field that rotates in synchronism with the rotating field due to armature currents, provided that the rotor itself turns at synchronous speed. As explained in Chapter 2, the only way that we can develop a constant torque on the rotor under such conditions is if the speed of the rotor itself is constant at precisely synchronous speed. A typical synchronous machine field alignment for operation as a generator is shown in Figure 4.4 using space vectors to represent the various fields.

It may be helpful for the reader to review the definition of the term *space vector,* which was introduced in Chapter 2. Referring to Figure 4.4, $\hat{H}_s$ is the rotating space vector which represents the rotating sinusoidally distributed air-gap magnetic field intensity wave due to armature currents. Its magnitude is directly proportional to rms value of the armature currents. The space vector $\hat{H}_r$ represents the magnetic field intensity wave due to the dc field current. Its magnitude is directly proportional to the field current. Note that its alignment is dictated by the physical position of the rotor and its winding. The space vector $\hat{H}_{sr}$ represents the resultant magnetic field intensity wave due to the combined effects of armature and field currents. Note the vector nature of the addition of $\hat{H}_s$ and $\hat{H}_r$, as shown in Figure 4.4. We could easily repeat this description in terms of the space vectors representing the respective magnetic flux density waves (i.e., $\hat{B}_s$, $\hat{B}_r$, and $\hat{B}_{sr}$, where $\hat{B}_s = \mu_0 \hat{H}_s$, etc.). If the leakage flux and armature resistance are neglected, the armature terminal voltage can be determined directly from $\hat{B}_{sr}$ using Faraday's law. Similarly, given the armature terminal voltage we can determine the magnitude and position of $\hat{B}_{sr}$ using Faraday's law. In essence, specification of a particular armature voltage condition fixes the resultant flux density wave $\hat{B}_{sr}$. Generally, the machine is connected to a power system that has a large number of other rotating machines connected as well. The armature terminal voltage magnitude and phase angle is dictated primarily by the combined effects of the large system to which any single machine is connected. In examining the behavior of any single machine, it is convenient to assume that the system voltage at the point of connection is absolutely constant in magnitude, phase angle, and frequency. Such a point in a power system is referred to as an *infinite bus*. It is "infinite" in the sense that voltage is unaffected by any machine that might be connected, regardless of any changes that might be made in the way the machine operates. We will make this simplifying assumption in order to explain the characteristic behavior of the synchronous machine. Referring again to Figure 4.4, we would say that if a machine were connected to an infinite bus and if the effects of armature leakage reactance and resistance were neglected, $\hat{B}_{sr}$ and hence $\hat{H}_{sr}$ would be fixed in magnitude, frequency of rotation, and alignment. The component magnetic field intensity waves $\hat{H}_s$ and $\hat{H}_r$, must adjust in such a way that for any given operating condition, they combine vectorially to produce the required resultant $\hat{H}_{sr}$.

To illustrate this concept, we consider an example. A simplified steam-turbine-driven generator and its prime mover are illustrated conceptually in Figure 4.5. Figure 4.5 shows turbines, generator, and exciter all on a common shaft. Speed governor control components and field excitation system control components that would be present in any modern-day system are not shown in the figure. Valves controlling steam flow to the high-pressure turbine and a field rheostat controlling field excitation are the only controllable items that we will discuss for the moment. If we assume that the generator is connected to an infinite bus and operating under normal steady-state conditions, the shaft must turn at synchronous speed and we will have fixed angles between $\hat{H}_{sr}$, $\hat{H}_s$, and $\hat{H}_r$ as shown in Figure 4.4. An electrical torque will be developed that tends to

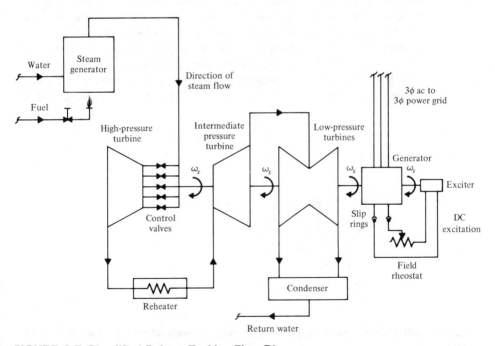

FIGURE 4.5 Simplified Reheat Turbine Flow Diagram

align the field of the rotor $\hat{H}_r$ with the resultant field $\hat{H}_{sr}$. This electrical torque acting on the rotor will be proportional to the product of the magnitudes of $\hat{H}_r$ and $\hat{H}_{sr}$ and the sine of the angle between the two:

$$T_e \propto |\hat{H}_r| \, |\hat{H}_{sr}| \sin \delta \qquad (4.2)$$

Referring to Figure 4.4, this torque will be in the clockwise direction. If we neglect any damping (friction and windage), the combined prime-mover torque T_s due to the four turbines must be equal in magnitude to the electrical torque T_e but in the opposite direction of rotation so that the net torque acting on the rotor will be zero for constant-speed operation at synchronous speed. That is, we must have

$$J \frac{d\omega_r}{dt} = T_s - T_e = 0 \qquad (4.3)$$

$$\omega_r = \omega_s \qquad (4.4)$$

where J is the combined moment of inertia of the shaft and the rotating masses associated with the turbines, generator rotor, and exciter rotor; ω_r the speed of the rotor; and ω_s the synchronous speed. If we neglect all mechanical and electrical losses, we must also have

$$P = T_s \omega_r = T_s \omega_s = \sqrt{3} \, VI \cos \theta \qquad (4.5)$$

where P is the total power converted, V the rms line-to-line armature voltage, I the rms armature line current, and θ the power-factor angle by which the armature phase voltage leads the armature phase current. If it is desired to increase the power converted, we must increase the prime-mover torque T_s. Typically, control valve openings must be increased, increasing the steam flow and fuel consumption requirements. If this is done while field excitation is held constant, $|\hat{H}_r|$ will remain constant and the angle δ between $\hat{H}_{sr}$ and $\hat{H}_r$ will increase. Since $\hat{H}_{sr}$ is fixed in position and magnitude by our infinite-bus assumption, the rotor must momentarily speed up, then return to operation at synchronous speed at a new angle $\delta' > \delta$. The change with respect to the

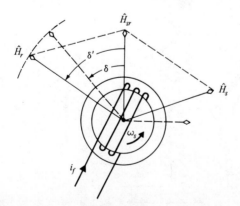

FIGURE 4.6 Field Alignments for an Increase in Prime-Mover Torque at Constant Field Excitation

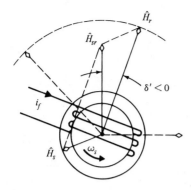

FIGURE 4.7 Field Alignments with the Synchronous Machine Operating as a Motor

conditions illustrated in Figure 4.4 is illustrated in Figure 4.6. Electrical torque T_e will have increased in accordance with (4.2), so that the required condition of (4.3) can be satisfied. Consideration of (4.2) shows that there will be an upper limit on prime-mover torque T_s that we might apply and still be able to develop a counter torque T_e of equal magnitude. This upper limit, which occurs at $\delta = 90°$, is sometimes referred to as the *steady-state stability limit,* and when expressed in terms of power converted it is the maximum power we might hope to convert at a particular field excitation setting.

In a similar manner, if prime-mover torque T_s is decreased while holding the field excitation constant, the angle δ will decrease. The rotor will momentarily slow down, then return to operation at synchronous speed at a new angle $\delta' < \delta$. If T_s is reduced to the point where it is insufficient to overcome friction and windage losses, the generator power flow must reverse in order to establish a new steady-state condition. The synchronous machine will "motor" slightly and δ will become negative. This is illustrated in Figure 4.7. Note that the electrical torque T_e undergoes a change in sign and it is developed in the direction of rotation in this case.

Increases in the field excitation at constant prime-mover torque result in changes in the magnitude of $\hat{H}_r$ and in changes in the angle δ such that new steady-state conditions satisfy (4.2) through (4.5). An increase in field excitation at constant prime-mover torque might result in the alignments illustrated in Figure 4.8, where changes are again shown with respect to the condition illustrated in Figure 4.4. The field $\hat{H}_s$ due to armature currents is of interest in this case. Since the prime-mover torque T_s remained constant, the power converted $P = T_s\omega_s$ is the same at the new operating point. It is clear from Figure 4.8 that changes in $\hat{H}_r$ at constant power will result in changes in the magnitude and angle of $\hat{H}_s$. This essentially requires changes in the magnitude and phase angle of armature current with respect to armature voltage. The result is a change in the reactive power output of the generator while the real power remains constant. A decrease in the field excitation at constant prime-mover torque is illustrated in Figure 4.9, where changes are again shown with respect to the conditions illustrated in Figure 4.4. Figure 4.9 suggests that we can cause armature current to lead or lag armature voltage at a particular real power output simply by adjusting the field excitation. This is an important capability that synchronous machines have,

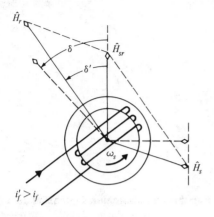

FIGURE 4.8 Field Alignments for an Increase in Field Excitation at Constant Prime-Mover Torque

whether operated as generators or motors. In essence, the machine reactive power output can be varied simply by making adjustments to the field excitation.

To quantify the effects we have discussed, we will establish a steady-state equivalent circuit for the synchronous machine which will permit us to demonstrate operating characteristics in terms of armature circuit quantities. We will develop the important steady-state characteristics of the synchronous machine from our equivalent circuit, then extend the concept to examine dynamic behavior.

Our introduction would not be complete without a remark or two on synchronous machines designed for dedicated operation as motors. Obviously, a synchronous motor is a fixed-speed motor. It is not self-starting and thus requires a field-application system. Typically, the field winding is short circuited and, on starting, armature voltage is applied and the machine runs as an induction machine up to a speed near synchronous. The field-application system is de-

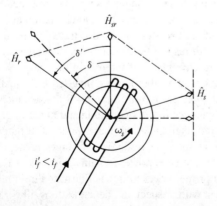

FIGURE 4.9 Field Alignments for a Decrease in Field Excitation at Constant Prime-Mover Torque

signed to sense this near-synchronous-speed condition and then apply a dc voltage to the field winding. The machine then pulls into synchronism with the system to which it is connected. A squirrel-cage induction machine provides a simpler alternative for most fixed-speed applications. The synchronous motor may be economically attractive in situations requiring a low-rpm fixed-speed drive where power-factor correction is required. Finally, the synchronous machine offers some advantages over the induction machine when fed from a solid-state variable-speed drive. Large variable-speed drive systems (i.e., in excess of 1500 hp) typically employ a synchronous rather than an induction machine.

4.2

THE STEADY-STATE EQUIVALENT CIRCUIT FOR A CYLINDRICAL-ROTOR SYNCHRONOUS MACHINE

The cylindrical-rotor synchronous machine may be regarded as a smooth-air-gap machine as defined in Chapter 2, and we can begin our treatment in a manner similar to the development presented in Chapter 3 for the induction machine. Although machines are generally wound for three phases on the stator, a properly designed two-phase armature winding can be used to establish the same rotating field due to armature currents as would be established by a three-phase winding. Thus, without loss of generality, we can treat all cases of interest by considering an equivalent two-pole two-phase cylindrical-rotor machine having two armature windings on the stator and a single field winding on the rotor. (We will examine the effects of additional rotor windings, called damper or amortisseur windings, later. They have no effect on machine steady-state performance and thus need not concern us at the moment.) We thus start with the elementary synchronous machine configuration shown conceptually in Figure 4.10.

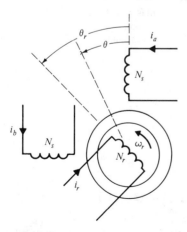

FIGURE 4.10 Elementary Two-Pole Two-Phase Synchronous Machine

The angle θ is an arbitrary angle measured from the axis of the phase a stator winding. The angle θ_r is the angle between the axis of the phase a stator winding and the axis of the field winding on the rotor of the machine. Our configuration constitutes a set of coupled coils. Thus

$$\bar{\lambda} = \bar{L}(\theta_r)\bar{i} \tag{4.6}$$

where

$$\bar{\lambda} = \begin{bmatrix} \lambda_a \\ \lambda_b \\ \lambda_r \end{bmatrix} \tag{4.7}$$

$$\bar{i} = \begin{bmatrix} i_a \\ i_b \\ i_r \end{bmatrix} \tag{4.8}$$

$$L(\theta_r) = \begin{bmatrix} L_{aa}(\theta_r) & L_{ab}(\theta_r) & L_{ar}(\theta_r) \\ & L_{bb}(\theta_r) & L_{br}(\theta_r) \\ & & L_{rr}(\theta_r) \end{bmatrix} \tag{4.9}$$

Consideration of Figure 4.10 suggests that some of the inductances in (4.9) will be independent of rotor angle θ_r and that some will be zero. We will examine these inductances in greater detail. We assume that:

1. All windings establish magnetic fields whose radial components of air-gap magnetic flux density and magnetic field intensity are sinusoidal functions of the angle θ for any value of θ_r.
2. Stator and rotor iron is highly permeable such that stored energy is associated predominantly with the air gap. Saturation effects are negligible (i.e., the overall magnetic circuit may be characterized as linear).
3. Tangential components of air-gap magnetic flux density and field intensity are negligible compared to radial components.
4. The air gap g is small compared to the radius r of the rotor so that magnetic flux density and field intensity are essentially invariant with radius in the gap.
5. The axial length ℓ of the rotor is sufficiently long that end effects are negligible.
6. The effects of slots are negligible.

Proceeding as in Sections 2.5 and 2.6, the radial components of the air-gap magnetic field intensity for each of the windings will be

$$H_a(\theta) = H_{amax} \cos \theta \tag{4.10}$$

$$H_b(\theta) = H_{bmax} \cos \left(\theta - \frac{\pi}{2} \right) \tag{4.11}$$

$$H_r(\theta) = H_{rmax} \cos (\theta - \theta_r) \tag{4.12}$$

The total resultant magnetic field intensity in the air gap at angle θ will be·

$$H(\theta = H_a(\theta) + H_b(\theta) + H_r(\theta) \tag{4.13}$$

The coenergy density in the gap will be

$$w_c(\theta) = \frac{\mu_0 H^2(\theta)}{2} \tag{4.14}$$

Substitute in (4.14) using (4.10) through (4.13) and simplify to obtain

$$
\begin{aligned}
w_c(\theta) = \frac{\mu_0}{2} \Bigg\{ &\frac{H_{a\max}^2}{2}(1 + \cos 2\theta) + \frac{H_{b\max}^2}{2}(1 - \cos 2\theta) \\
&+ \frac{H_{r\max}^2}{2}[1 + \cos 2(\theta - \theta_r)] + H_{a\max}H_{b\max}\sin 2\theta \\
&+ H_{a\max}H_{r\max}\cos(2\theta - \theta_r) + H_{a\max}H_{r\max}\cos\theta_r \\
&+ H_{b\max}H_{r\max}\sin(2\theta - \theta_r) + H_{b\max}H_{r\max}\sin\theta_r \Bigg\}
\end{aligned}
\tag{4.15}
$$

The total coenergy will be

$$W_c = \int_0^{2\pi} w_c(\theta) g\ell r \, d\theta \tag{4.16}$$

where g is the gap width, ℓ the rotor axial length, and r may be taken as the radius of the rotor since the gap is assumed to be small. Substitute in (4.16) using (4.15) and recognize that the integral over 2π will be zero for all sinusoidal terms in θ or 2θ.

$$
\begin{aligned}
W_c = \pi\mu_0\ell rg \Bigg(&\frac{H_{a\max}^2}{2} + \frac{H_{b\max}^2}{2} + \frac{H_{r\max}^2}{2} \\
&+ H_{a\max}H_{r\max}\cos\theta_r + H_{b\max}H_{r\max}\sin\theta_r \Bigg) \tag{4.17}
\end{aligned}
$$

The maximum field intensities are related to currents by applying Ampère's law [see Figure 2.26 and (2.63) and (2.64)]:

$$H_{a\max} = \frac{N_s i_a}{2g} \tag{4.18}$$

$$H_{b\max} = \frac{N_s i_b}{2g} \tag{4.19}$$

$$H_{r\max} = \frac{N_r i_r}{2g} \tag{4.20}$$

Substitute in (4.17) using (4.18) through (4.20):

$$W_c = \frac{\pi\mu_0\ell r}{4g}\left(\frac{N_s^2 i_a^2}{2} + \frac{N_s^2 i_b^2}{2} + \frac{N_r^2 i_r^2}{2} + N_s N_r i_a i_r \cos\theta_r + N_s N_r i_b i_r \sin\theta_r\right) \tag{4.21}$$

Now express the coenergy in terms of inductances:

$$
\begin{aligned}
W_c &= \tfrac{1}{2}\,\bar{i}^T \overline{L}(\theta_r)\bar{i} \\
&= \tfrac{1}{2}L_{aa}i_a^2 + \tfrac{1}{2}L_{bb}i_b^2 + \tfrac{1}{2}L_{rr}i_r^2 + L_{ab}i_a i_b + L_{ar}i_a i_r + L_{br}i_b i_r
\end{aligned}
\tag{4.22}
$$

4.2 THE STEADY-STATE EQUIVALENT CIRCUIT FOR A CYLINDRICAL-ROTOR SYNCHRONOUS MACHINE

Now equate coefficients by comparing (4.21) and (4.22):

$$L_{aa} = L_{bb} = N_s^2 \frac{\pi \mu_0 \ell r}{4g} \tag{4.23}$$

$$L_{rr} = N_r^2 \frac{\pi \mu_0 \ell r}{4g} \tag{4.24}$$

$$L_{ab} = L_{ba} = 0 \tag{4.25}$$

$$L_{ar} = L_{ra} = N_s N_r \frac{\pi \mu_0 \ell r}{4g} \cos \theta_r \tag{4.26}$$

$$L_{br} = L_{rb} = N_s N_r \frac{\pi \mu_0 \ell_r}{4g} \sin \theta_r \tag{4.27}$$

We have neglected stator and rotor leakage inductances in our development. We will assume that stator and rotor flux linkages due to leakage flux are independent of rotor position and linearly related to currents. Thus, for either phase on the armature, we have flux linkages due to leakage:

$$\lambda_{s\ell} = L_{\ell s} i_s \tag{4.28}$$

and for the rotor winding,

$$\lambda_{r\ell} = L_{\ell r} i_r \tag{4.29}$$

If we define M as

$$M = N_s^2 \frac{\pi \mu_0 \ell r}{4g} \tag{4.30}$$

and incorporate the leakage terms, we can summarize our results in matrix form as

$$\begin{bmatrix} \lambda_a \\ \lambda_b \\ \lambda_r \end{bmatrix} = \begin{bmatrix} L_{\ell s} + M & 0 & \frac{N_r}{N_s} M \cos \theta_r \\ & L_{\ell s} + M & \frac{N_r}{N_s} M \sin \theta_r \\ & & L_{\ell r} + \left(\frac{N_r}{N_s}\right)^2 M \end{bmatrix} \begin{bmatrix} i_a \\ i_b \\ i_r \end{bmatrix} \tag{4.31}$$

We ultimately seek an equivalent circuit referred to the armature. Thus, in the manner of referring variables from one winding of a transformer to the other, we refer rotor variables to the stator in the following way:

$$i_r' = \frac{N_r}{N_s} i_r \tag{4.32}$$

$$\lambda_r' = \frac{N_s}{N_r} \lambda_r \tag{4.33}$$

The rotor leakage inductance referred to the stator will be

$$L'_{\ell r} = \left(\frac{N_s}{N_r}\right)^2 L_{\ell r} \tag{4.34}$$

Substitution of (4.32) through (4.34) into (4.31) yields

$$\begin{bmatrix} \lambda_a \\ \lambda_b \\ \lambda'_r \end{bmatrix} = \begin{bmatrix} L_{\ell s} + M & 0 & M\cos\theta_r \\ & L_{\ell s} + M & M\sin\theta_r \\ \text{(symmetric matrix)} & L'_{\ell r} + M \end{bmatrix} \begin{bmatrix} i_a \\ i_b \\ i'_r \end{bmatrix} \tag{4.35}$$

Equation (4.35) is valid in general within the limits of our assumptions. Balanced operation under steady-state conditions is the important special case we now consider. For balanced operation, armature currents must constitute a balanced two-phase set of currents. Thus armature currents and their phasor representations can be chosen as

$$i_a = \sqrt{2}\,I_s \cos(\omega_s t + \varphi_s) \leftrightarrow \underline{I}_s = I_s\,\underline{/\varphi_s} \tag{4.36}$$

$$i_b = \sqrt{2}\,I_s \cos\left(\omega_s t + \varphi_s - \frac{\pi}{2}\right) \leftrightarrow -j\underline{I}_s = I_s\,\underline{/\varphi_s - \pi/2} \tag{4.37}$$

Under steady-state conditions, the rotor current i_r will be constant. Let

$$i'_r = \frac{N_r}{N_s} i_r = I_f \tag{4.38}$$

Under steady-state conditions, the rotor turns at synchronous speed ω_s. Thus, under steady-state conditions, the rotor angle θ_r can be expressed as

$$\theta_r = \omega_r t + \theta_r(0)$$

$$= \omega_s t + \theta_r(0) \tag{4.39}$$

Let the stator (armature) resistance be r_s. For the armature windings we will have

$$v_a = -i_a r_s + \frac{d\lambda_a}{dt} \tag{4.40}$$

$$v_b = -i_b r_s + \frac{d\lambda_b}{dt} \tag{4.41}$$

Equations (4.40) and (4.41) are written with armature currents assigned for operation of the machine as a generator. Our final equivalent circuit can be easily modified for operation of the machine as a motor. Substituting in (4.40) and (4.41) using (4.35), (4.38), and (4.39), we obtain

$$v_a = -i_a r_s + \frac{d}{dt}\{-(L_{\ell s} + M)i_a + MI_f\cos[\omega_s t + \theta_r(0)]\}$$

$$= -i_a r_s - \frac{d}{dt}(L_{\ell s} + M)i_a - \omega_s MI_f\sin[\omega_s t + \theta_r(0)] \tag{4.42}$$

Similarly,

$$v_b = -i_b r_s - \frac{d}{dt}(L_{\ell s} + M)i_b + \omega_s MI_f \cos[\omega_s t + \theta_r(0)] \quad (4.43)$$

The last term in (4.42) and in (4.43) is the field-induced armature voltage in the respective phase a and phase b armature windings. Under steady-state conditions the magnitude of this voltage is directly proportional to the field current. It will be convenient to define the field-induced voltage in the following way. Let

$$e_f(t) = -\omega_s MI_f \sin[\omega_s t + \theta_r(0)] \quad (4.44)$$

The phasor representation for this voltage will be

$$e_f(t) \leftrightarrow E_f = \frac{\omega_s MI_f}{\sqrt{2}} \underline{/\theta_r(0) + \pi/2} = \frac{+j\omega_s MI_f}{\sqrt{2}} \underline{/\theta_r(0)} \quad (4.45)$$

Our assumption of balanced two-phase armature currents [Eqs. (4.36) and (4.37)] and our definition of field-induced voltage [Eqs. (4.44) and (4.45)] permit us to write corresponding phasor equations for (4.42) and (4.43):

$$V_a = -[r_s + j\omega_s(L_{\ell s} + M)]I_s + E_f \quad (4.46)$$

$$V_b = -[r_s + j\omega_s(L_{\ell s} + M)](-jI_s) - jE_f \quad (4.47)$$

It is clear from (4.46) and (4.47) that $V_b = -jV_a$ and that all variables in the phase b armature winding are the same as those in the phase a winding but shifted in phase by $-\pi/2$ radians. Thus, for any analysis under the assumption of steady-state conditions, we need consider only one phase. We can easily synthesize an equivalent circuit consistent with (4.46) and (4.47) as shown in Figure 4.11, where we define $V_s = V_a = jV_b$. We see that armature terminal voltage V_s is essentially the field-induced voltage E_f less the voltage drop developed across an equivalent impedance due to armature winding resistance, leakage inductance, and the mutual inductance M. It is generally simpler to deal with reactances. The magnetizing reactance, sometimes referred to as the *reactance of armature reaction*, is

$$x_m = \omega_s M \quad (4.48)$$

The armature leakage reactance is

$$x_{\ell s} = \omega_s L_{\ell s} \quad (4.49)$$

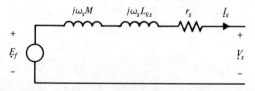

FIGURE 4.11 Steady-State Equivalent Circuit for a Cylindrical-Rotor Synchronous Machine

THE SYNCHRONOUS MACHINE

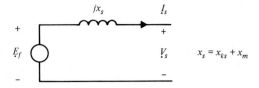

FIGURE 4.12

The sum of magnetizing reactance and armature leakage reactance is the synchronous reactance,

$$x_s = x_{\ell s} + x_m \tag{4.50}$$

The total armature impedance due to combined effects is defined as the *synchronous impedance*,

$$z_s = r_s + jx_s \tag{4.51}$$

The armature resistance r_s is often much less than the synchronous reactance x_s and can be neglected in such cases. We thus arrive at the simplified equivalent circuit shown in Figure 4.12.

To complete the development of a steady-state model for the cylindrical-rotor synchronous machine, we consider the voltage equation for the rotor (field) circuit. Since in normal steady-state operation the rotor turns in synchronism with the rotating magnetic field established by armature currents, and since the rotor current is a constant current, we will have no time rate of change of flux linking turns in the field winding. The field voltage (supplied by the excitation system) will thus be a constant voltage required to overcome any resistance associated with the field-winding circuit. We can demonstrate this more rigorously in the following way. Let v_r be the voltage supplied to the field winding by the exciter, and let r_r be resistance of the field winding. These quantities may be referred to the armature winding:

$$v_r' = \frac{N_s}{N_r} v_r \tag{4.52}$$

$$r_r' = \left(\frac{N_s}{N_r}\right)^2 r_r \tag{4.53}$$

The field current, field flux linkages, and field leakage inductance referred to the armature winding are given by (4.32), (4.33), and (4.34). The voltage equation for the field winding expressed in terms of variables referred to the armature winding will be

$$v_r' = r_r' i_r' + \frac{d\lambda_r'}{dt} \tag{4.54}$$

The field flux linkages λ_r' from (4.35) will be

$$\lambda_r' = -Mi_a \cos \theta_r - Mi_b \sin \theta_r + (L_{\ell r}' + M)i_r' \tag{4.55}$$

For an assumed steady-state condition we may substitute in (4.55) using (4.36) through (4.39) to obtain

$$\lambda'_r = -\sqrt{2}\, I_s M \cos[\varphi_s - \theta_r(0)] + (L'_{\ell r} + M)I_f \qquad (4.56)$$

Thus λ'_r will be a constant in the steady state $(d\lambda'_r/dt = 0)$ and from (4.54) we obtain

$$v'_r = r'_r i'_r = r'_r I_f = V_f \qquad (4.57)$$

where V_f is the constant dc value of required exciter voltage referred to the armature. We can summarize relationships for the field-induced armature voltage using (4.45) and (4.57):

$$E_f = |E_f| = \frac{\omega_s M I_f}{\sqrt{2}} = \frac{\omega_s M V_f}{\sqrt{2}\, r'_r} \qquad (4.58)$$

Equation (4.58) shows that the magnitude of E_f in our equivalent circuit [and, of course, in the relationships from which the equivalent circuit was synthesized, i.e., (4.46) and (4.47)] is directly proportional to the dc field current and to the dc field voltage in the steady state.

EXAMPLE 4.1   A three-phase two-pole 800-MVA cylindrical-rotor synchronous generator has a rated armature voltage (line-to-line) of 24 kV. The machine has a synchronous reactance of 1.20 Ω, an armature leakage reactance of 0.20 Ω, and an armature resistance of 0.002 Ω. The machine is Y-connected. Find the field-induced voltage E_f if the machine delivers rated MVA at rated voltage and at a power factor of 0.85 lagging.

The situation described is typical. A hypothetical operating condition at the terminals of the machine fixes the field excitation voltage E_f. The field excitation must be adjusted accordingly. The armature phase voltage is

$$V_s = \frac{24}{\sqrt{3}} \text{ kV} = 13.9 \text{ kV}$$

We may choose this voltage as a reference; thus

$$\underline{V}_s = V_s \underline{/0} = 13.9 \underline{/0} \text{ kV}$$

If the machine delivers the rated MVA, we must have

$$3V_s I_s = 800 \text{ MVA}$$

$$I_s = 19.2 \text{ kA}$$

At a power factor of 0.85 lagging we will have

$$\underline{I}_s = 19.2 \underline{/-31.8°} \text{ kA}$$

If armature resistance is neglected, the equivalent circuit of Figure 4.12 requires that

$$\begin{aligned} \underline{E}_f &= \underline{V}_s + jx_s \underline{I}_s \\ &= 13.9 + j(1.2)(19.2 \underline{/-31.8°}) \\ &= 32.6 \underline{/36.9°} \text{ kV} \end{aligned}$$

A phasor diagram showing the circuit quantities is shown in Figure 4.13. The angle between E_f and V_s, δ in Figure 4.13, is sometimes referred to as the *power* or *torque angle*, for reasons that will become apparent when we investigate steady-state characteristics in the next section.

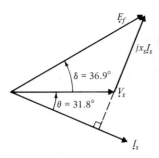

$\delta = 36.9°$

$\theta = 31.8°$

E_f

$jx_s I_s$

V_s

I_s

FIGURE 4.13

Consideration of flux linkages in the armature windings permits us to relate phasors representing the circuit quantities shown in Figure 4.13 to the rotating space vectors described in Section 4.1. For an assumed steady-state condition, the flux linkages in phase a will be [Eqs. (4.35), (4.38), and (4.39)]

$$\lambda_a = -L_{\ell s} i_a - M i_a + M i_f \cos [\omega_s t + \theta_r(0)]$$

or

$$\lambda_a = \lambda_{\ell s} + \lambda_m + \lambda_f$$

where $\lambda_{\ell s}$ represents the flux linkages in the armature winding due to the leakage flux caused by the armature current, λ_m represents the flux linkages in the armature winding due to the air-gap flux caused by the armature current, and λ_f represents the flux linkages in the armature winding due to the air-gap flux established by the field current. The flux linkages are scalar quantities which under steady-state conditions vary sinusoidally with time. We can thus represent flux linkages by phasors. Choose phasor representations for the flux linkages in the following way:

$$-\lambda_a \leftrightarrow \psi_{res} \quad \text{(resultant flux linkages)}$$

$$-\lambda_{\ell s} \leftrightarrow \psi_{\ell s}$$

$$-\lambda_m \leftrightarrow \psi_m$$

$$-\lambda_f \leftrightarrow \psi_f$$

The minus sign is introduced arbitrarily here. It permits somewhat simpler phasor diagrams. As a result of this choice,

$$\psi_{\ell s} = L_{\ell s} I_s$$

$$\psi_m = M I_s$$

From Faraday's law

$$v_a = \frac{+d\lambda_a}{dt}$$

Thus

$$V_s = -j\omega_s \psi_{res}$$

$$e_f = \frac{+d\lambda_f}{dt}$$

Thus

$$E_f = -j\omega_s \psi_f$$

From the expression for resultant flux linkages, we have

$$\lambda_a = \lambda_{\ell s} + \lambda_m + \lambda_f$$

$$-\psi_{res} = -\psi_{\ell s} - \psi_m - \psi_f$$

Now multiply through by $+j\omega_s$:

$$-j\omega_s \psi_{res} = -j\omega_s \psi_{\ell s} - j\omega_s \psi_m - j\omega_s \psi_f$$

Thus

$$V_s = -jx_{\ell s} I_s - jx_m I_s + E_f$$

The last expression is our steady-state circuit equation derived in a slightly different way. The last three relationships suggest the symmetry we must have between circuit quantities V_s, I_s, and E_f and flux linkages. This is illustrated in Figure 4.14. The reader should examine this figure carefully and recognize the necessary phase relationships that we must have between various quantities. The time-varying quantity that a phasor represents is formally obtained by multiplying by $e^{j\omega_s t}$ and taking the real part of the product. For example,

$$e_f(t) = \mathrm{Re}\,\{e^{j\omega_s t} E_f\}$$

It is thus reasonable to view the phasors as rotating at a speed ω_s but retaining their angular relationships with respect to each other. If we superimpose a phasor diagram of rotating phasors on a diagram of rotating space vectors, we can establish the "link" between the two concepts. The angles between phasors correspond to phase differences in time. These angles must be related to the space angles between particular rotating space vectors. Similarly, the magnitudes

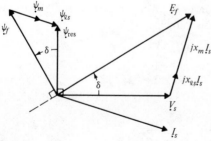

FIGURE 4.14

THE SYNCHRONOUS MACHINE

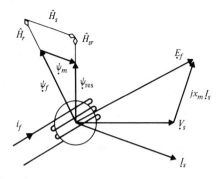

FIGURE 4.15

of phasors must be related to the magnitudes of particular rotating space vectors. These relationships are shown in Figure 4.15, where the effects of armature leakage flux are neglected to simplify the figure. Figure 4.15 illustrates the geometry we must have if we superimpose diagrams aligning ψ_{res} and $\hat{H}_{sr}$. The diagram is useful in understanding synchronous machine behavior in the steady state. Perhaps the most important point to recognize is that we need deal only with circuit quantities V_s, I_s, and E_f in order to characterize operation of the machine for a particular operating condition. Flux linkages and/or space-vector quantities, if desired, can be determined from circuit quantities. Steady-state characteristics can thus be completely described with our equivalent-circuit concept, as will be demonstrated in the next section.

4.3

SYNCHRONOUS MACHINE STEADY-STATE CHARACTERISTICS

We can determine steady-state characteristics for the synchronous machine by analyzing the equivalent circuit developed in Section 4.2. Armature resistance is typically quite low in a well-designed machine, and we can neglect it without introducing significant error. The equivalent circuit with armature resistance neglected for a cylindrical rotor synchronous machine operated as a generator is shown in Figure 4.16. The terminal voltage relationship and a phasor diagram showing a typical generator operating condition are also shown in the figure.

To analyze operation of the machine as a motor it is convenient to redefine the armature current as shown in Figure 4.17. The equivalent circuit with armature resistance neglected for a cylindrical-rotor synchronous machine operated as a motor is shown in Figure 4.17. The terminal voltage relationship and a phasor diagram showing a typical motor operating condition are also shown in the figure. Note the phase relationship between the field-induced voltage E_f and the terminal voltage V_s for generator versus motor action. If a synchronous machine is generating, E_f will lead V_s. If a synchronous machine is "motoring," V_s will lead E_f. The physical significance of this is best illustrated by examining the appropriate phase relationships between the various phasors representing

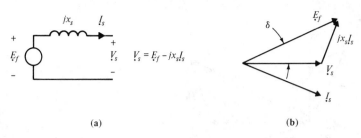

(a)   (b)

FIGURE 4.16 Synchronous Machine Relationships with the Machine Operated as a Generator

flux linkages in the generating versus motoring modes of operation (see Problem 4.6). In essence, the rotating magnetic field established by the field current leads the rotating resultant magnetic field if the machine is operating as a generator. If the machine operates as a motor, the rotating field established by the field current lags the rotating resultant magnetic field.

EXAMPLE 4.2 Two identical three-phase synchronous machines are connected such that one operates as a generator and one operates as a motor. Each Y-connected machine is rated at 1000 kVA, unity power factor, and 4160 V line-to-line. The synchronous reactance is 18.0 Ω and the armature resistance is negligible. The feeder connecting the two machines has a reactance of 3.0 Ω and negligible resistance. The situation is depicted in Figure 4.18. Suppose that the machines are operated in such a way that the generator delivers its rated kVA at a power factor of 0.80 lagging with its terminal voltage at rated value. Determine the field-induced voltages E_{fg} and E_{fm} and the voltage at the terminals of the machine, which is operated as a motor. Determine the reactive power requirements for the various devices in the system.

The example will illustrate the important capability that synchronous machines have insofar as reactive power is concerned. In any discussion of reactive power, certain conventions are necessary and it is worthwhile to reemphasize the ones used in this text before proceeding with the example. Consider Figure 4.19, which shows a source connected to a load by a circuit having zero impedance. The active power or average power P is determined by $P = VI \cos \theta$, where θ may be taken as $\alpha - \beta$ or $\beta - \alpha$, in that the cosine function is an even function [i.e., $\cos \theta = \cos(-\theta)$]. The source is said to "supply," "generate," or "deliver" power to the load, which is said to "consume," "receive," or "ab-

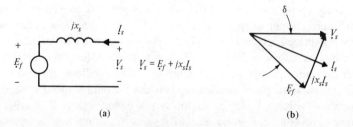

(a)   (b)

FIGURE 4.17 Synchronous Machine Relationships with the Machine Operated as a Motor

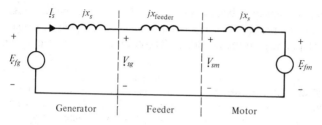

FIGURE 4.18

sorb'' it. If, upon evaluating the expression for P, we find that $P < 0$, the source could be regarded as supplying negative power (consuming positive power) and the load could be regarded as consuming negative power (supplying positive power). Such an occurrence would imply that we had failed to identify the source and load properly to start with. Reversing the reference direction I and redefining the devices would be a reasonable course of action, although not absolutely necessary. It is important to recognize that when we apply the fundamental relationship for active power P, it yields power supplied if the variables V and I associated with the device conform to the *active* sign convention. It yields power consumed if the variables V and I associated with the device conform to the *passive* sign convention. Note that, in Figure 4.19, the assignment of V and I conforms to the active sign convention for the source and to the passive sign convention for the load.

The reactive power Q may be treated in much the same way as the active power P provided that a troublesome ambiguity concerning its algebraic sign is resolved. The reactive power is defined as $Q = VI \sin \theta$ and it is necessary to define θ carefully in that the sine function is an odd function [i.e., $\sin(-\theta) = -\sin\theta$]. It can be shown that the choice of θ as $\alpha - \beta$ or $\beta - \alpha$ is an arbitrary one. In this text, θ is defined as the angle by which the voltage "leads" the current, specifically, $\theta = \alpha - \beta$ for the variables depicted in Figure 4.19. The angle θ is referred to as the *power-factor angle* and $\cos \theta$ is referred to as the *power factor*. The terms *"lagging"* power factor or *"lagging"* power-factor angle refer to the case where the current lags the voltage (i.e., $\theta = \alpha - \beta > 0$). The terms *"leading"* power factor or *"leading"* power factor angle refer to the case where the current leads the voltage (i.e., $\theta = \alpha - \beta < 0$). Based on these conventions, it follows that a load which is predominantly inductive will have a lagging power factor and will consume positive reactive power. A load that is predominantly capacitive will have a leading power factor and will consume negative reactive power. Because capacitive loads tend to compensate

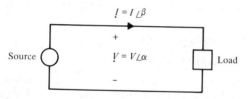

FIGURE 4.19

inductive loads, capacitor banks are regarded by power system engineers as sources of reactive power. Synchronous machines can supply or consume reactive power depending on their field excitation levels, as will be shown in this example.

EXAMPLE 4.2 (continued) We have specified the terminal conditions of the generator completely and can thus effectively "solve" the equivalent circuit shown in Figure 4.18. This permits determination of all quantities of interest. Choose V_{sg} as the reference.

$$V_{sg} = \frac{4160}{\sqrt{3}} \underline{/0°} = 2400 \underline{/0°} \text{ V}$$

$$I_s = \frac{VA}{V_{sg}} = \frac{1000(10^3)/3}{2400} = 139 \text{ A}$$

Since the generator is operating at a lagging power factor, the generator current lags the generator voltage by $\cos^{-1} 0.8 = 36.9°$. Thus

$$I_s = 139 \underline{/-36.9°} \text{ A}$$

Solving for E_{fg} yields

$$E_{fg} = V_{sg} + jx_sI_s$$

$$= 2400 \underline{/0°} + j(18.0)(139 \underline{/-36.9°})$$

$$= 3900 + j2000 = 4380 \underline{/27.1°}$$

It is of interest to note that if the generator were operated at unity power factor at rated voltage and current, the field-induced voltage would be

$$E'_{fg} = 2400 \underline{/0°} + (j18.0) \, 139 \underline{/0°}$$

$$= 3470 \underline{/46.2°} \text{ V}$$

For our lagging power factor condition, we have $E_{fg} > E'_{fg}$. The field excitation requirements are thus increased over what we would have at unity power factor operation and the generator is said to be operating *overexcited*. Furthermore, it supplies reactive power. That is,

$$Q_g = 3V_{sg}I_s \sin \theta$$

$$= 3(2400)(139) \sin 36.9°$$

$$= 601 \text{ kvar}$$

The field-induced voltage for the motor will be

$$E_{fm} = V_{sg} - jx_{\text{feeder}}I_s - jx_sI_s$$

$$= 2400 \underline{/0°} - j(3.0 + 18.0)(139 \underline{/-36.9°})$$

$$= 647 - j2340 = 2430 \underline{/-74.5°}$$

We note that if the motor were operated at unity power factor at rated voltage and current, the field-induced voltage would be

$$E'_{fm} = 2400 \underline{/0°} - j(18.0)139 \underline{/0°}$$

$$= 3470 \underline{/-46.2°}$$

Thus, for our specified condition, $E_{fm} < E'_{fm}$. The field excitation requirements are thus decreased from what we would have at unity power factor operation, and the motor is said to be operating *underexcited*. In finding the remaining information required, we will see that the motor is consuming reactive power. The terminal voltage at the motor will be

$$V_{sm} = V_{sg} - jx_{\text{feeder}}I_s$$

$$= 2400 \underline{/0°} - (j3.0)(139) \underline{/-36.9°}$$

$$= 2150 - j333 = 2180 \underline{/-8.8°} \text{ V}$$

Thus, at the terminals of the motor, we have

$$3V_{sm}I_s = 3(2180)(139) = 909 \text{ kVA}$$

The power-factor angle for the motor is the angle by which motor phase voltage leads motor phase current. Thus

$$\theta_m = -8.8° - (-36.9°) = 28.1°$$

and the motor power factor is

$$\cos \theta_m = 0.88 \text{ lagging}$$

We further note that the reactive power consumed by the motor is

$$Q_m = 3V_{sm}I_s \sin \theta_m$$

$$= 3(2180)(139) \sin 28.1°$$

$$= 428 \text{ kvar}$$

The difference in the reactive power supplied by the generator and the reactive power consumed by the motor must be the reactive power consumed by the feeder. Thus, as a check,

$$Q_g - Q_m = 601 - 428 = 173 \text{ kvar}$$

$$3I_s^2 x_{\text{feeder}} = 3(139)^2(3.0) = 174 \text{ kvar}$$

which differs only slightly due to round-off error.

A more general result, which we will examine later in greater detail, has been illustrated in this example. Synchronous machines operating overexcited supply reactive power. Synchronous machines operating underexcited consume reactive power.

We now examine in detail steady-state characteristics for the synchronous machine. We will consider operation of the machine as a generator, but it will be apparent as we do this that corresponding characteristics apply as well to a synchronous machine operated as a motor.

Our development of the steady-state characteristics of the machine will be based on the equivalent-circuit and steady-state relationships developed in Section 4.2. We will assume that armature resistance is small enough to be neglected without appreciably altering machine behavior.

Consideration of steady-state operation of the machine under open-circuit and short-circuit conditions is of interest in that the synchronous reactance x_s can be approximately determined in a simple way from open- and short-circuit test data. If the machine is driven at constant speed by its prime mover and the armature terminals are left on open circuit, $I_s = 0$, and we must have $V_s = E_f$. The steady-state relationship between E_f and the field circuit current I_f or the field circuit voltage V_f is given by (4.58) and stated again here for convenience.

$$E_f = |E_f| = \frac{\omega_s M I_f}{\sqrt{2}} = \frac{\omega_s M V_f}{\sqrt{2} \, r_r'} \qquad (4.58)$$

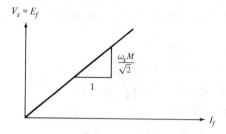

FIGURE 4.20 Open-Circuit Armature Voltage Versus Field Current

If we vary I_f (by varying V_f) and measure V_s, we should produce the straight-line relationship shown in Figure 4.20 under open-circuit conditions. In reality, saturation, which we have neglected in our development, will cause us to determine experimentally a different curve. Saturation limits and distorts the rotating magnetic flux density wave established by the field current, causing its space distribution to become nonsinusoidal. Induced armature open-circuit voltage, in turn, will be reduced and nonsinusoidal in time. If we neglect harmonics in this voltage waveform and measure the rms value of the fundamental, we might obtain the more or less typical nonlinear curve shown in Figure 4.21, where our previous curve is also shown for comparison. The dashed line shown in Figure 4.21 is a straight line drawn from the origin through a point on the nonlinear curve that corresponds to rated machine armature voltage. It illustrates an approach to incorporating the effects of saturation in an approximate way while retaining a linear model. Other more elaborate methods for dealing with saturation are possible, but for our purposes, a linear representation will suffice. The original straight-line representation shown in Figure 4.20 and repeated in Figure 4.21 is sometimes referred to as the *air-gap line,* so named because it is effectively determined by the reluctance of the air gap. To see this it may be helpful to reexamine (4.30), which gives the mutual inductance constant M in terms of the physical dimensions of the machine.

If armature terminals are short circuited, $V_s = 0$, and we must have $I_s = E_f/jx_s$. Thus $I_s = \omega_s M I_f/\sqrt{2}\, x_s$, and if the rotor is driven at constant speed by a prime mover with the armature windings short circuited, we should have a

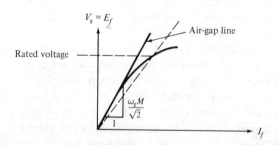

FIGURE 4.21 Open-Circuit Armature Voltage Versus Field Current Showing Effects of Saturation

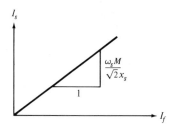

FIGURE 4.22 Short-Circuit Armature Current Versus Field Current

linear relationship between the rms armature current and the field current I_f. This is illustrated in Figure 4.22. If we can determine experimentally the air-gap line shown in Figure 4.20 (also in Figure 4.21) and the short-circuit-current line shown in Figure 4.22, we can find the synchronous reactance x_s by simply taking the ratio of the slope of the air-gap line to the slope of the short-circuit-current line. (We could also replace the air-gap line with the dashed line shown in Figure 4.21 and determine, in a similar way, a "saturated" value of synchronous reactance.) Fortunately, saturation does not affect the short-circuit-current line. This can be explained by referring to Figure 4.23. In the figure we see that flux linkages due to armature current are phase displaced by 180° from flux linkages due to field current. In essence, flux due to armature currents subtracts from that due to field current. The resultant flux is low and saturation does not occur. Ideally, we would have $\psi_s = -\psi_f$, so that resultant flux linkages $\psi_{res} = \psi_s + \psi_f$ and terminal voltage $V_s = -j\omega_s\psi_{res}$ would be zero. This is not the case in practice since armature resistance, which is small but not zero, will alter the diagram somewhat.

We examine real power P per phase and reactive power Q per phase at the terminals of the machine by considering complex power $S = P + jQ$. Since $P = V_s I_s \cos\theta$ and $Q = V_s I_s \sin\theta$, where θ is the power-factor angle (the angle by which voltage V_s leads current I_s), it is easy to show that

$$S = P + jQ = V_s I_s^* \tag{4.59}$$

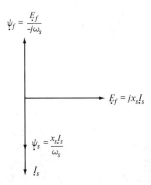

FIGURE 4.23 Phasor Diagram for a Synchronous Machine with Its Armature Short Circuited (Armature Resistance Neglected)

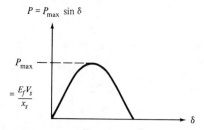

FIGURE 4.24

where the asterisk denotes the complex conjugate of current I_s. For any operating condition we must have

$$I_s = \frac{E_f - V_s}{jx_s} \qquad (4.60)$$

Thus

$$I_s^* = \frac{E_f^* - V_s^*}{-jx_s} \qquad (4.61)$$

$$S = V_s I_s^* = \frac{V_s E_s^* - V_s^2}{-jx_s} \qquad (4.62)$$

$$= \frac{jV_s E_f^*}{x_s} - \frac{jV_s^2}{x_s}$$

Let $V_s = V_s \underline{/0°}$ and $E_f = E_f \underline{/\delta}$. Substituting in (4.62), we have

$$S = \frac{V_s E_f}{x_s} \sin \delta + j\left(\frac{V_s E_f \cos \delta - V_s^2}{x_s}\right) \qquad (4.63)$$

Taking the real part on both sides of (4.63), we have

$$P = \frac{V_s E_f}{x_s} \sin \delta \qquad (4.64)$$

Taking the imaginary part on both sides of (4.63), we have

$$Q = \frac{V_s E_f \cos \delta - V_s^2}{x_s} \qquad (4.65)$$

Equations (4.64) and (4.65) yield P and Q in terms of the armature terminal voltage magnitude V_s, the field-induced voltage magnitude E_f, the synchronous reactance x_s, and the power angle δ. Equation (4.64) yields the power-angle characteristic shown in Figure 4.24. This result is of fundamental importance. The angle δ between E_f and V_s is the same as the angle between ψ_{res} and ψ_f. This is illustrated in Figure 4.25. These angular relationships are a direct consequence of Faraday's law. The angle δ can also be described as the angle between the axis of the field winding and a rotating reference axis whose speed

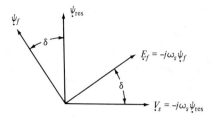

FIGURE 4.25

of rotation and position are determined by the armature terminal voltage frequency and phase angle. If armature leakage reactance is zero, the latter axis coincides with the axis of the resultant air-gap magnetic field intensity wave $\hat{H}_{sr}$ (see Figures 4.14 and 4.15).

Our development in Chapter 2 showed that torque is typically proportional to the sine of the angle between magnetic fields. Thus the relationship for power given by (4.64) and illustrated in Figure 4.24 is not surprising. We see that P_{max} depends on E_f, which in turn depends on the field current I_f, as shown in (4.58). For a particular value of the field current and rms armature terminal voltage, we have an upper limit P_{max} on the real power that the machine can deliver. P_{max} is the steady-state stability limit referred to earlier in Section 4.1.

If we plot P as given by (4.64) and Q as given by (4.65) in the manner illustrated in Figure 4.26, some further relationships among quantities become evident. If a circular arc of radius $V_s E_f / x_s$ is described with its center at the point $-V_s^2/x_s$ on the Q axis, points along the arc determine P and Q for particular values of V_s, E_f, and δ. The armature terminal voltage V_s will typically be at or near rated value. The field-induced voltage E_f will be limited to a maximum value based on the maximum field current, which is limited by heating due to ohmic losses in the field circuit. The VA rating of the machine can be used to

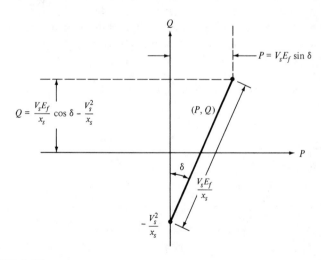

FIGURE 4.26

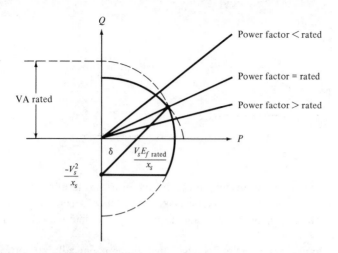

FIGURE 4.27 Reactive Power Capability Curve

determine a second circular arc within which points (P, Q) must lie in order not to exceed armature circuit heating constraints. That is, we must have

$$P^2 + Q^2 \leq \mathrm{VA}_{\mathrm{rated}}^2 \tag{4.66}$$

For stable operation with a fixed value of E_f, we must have $\delta < 90°$, as we will see when we investigate the dynamic behavior of the machine. These three constraints can be used to establish the reactive power capability curve illustrated in Figure 4.27. The reader should be able to identify regions in the P–Q plane where armature circuit heating, field circuit heating, and stability considerations would be limiting. It should be noted that improved machine cooling methods can extend the limits based on heating effects.

If the machine prime-mover torque (and hence the power input to the machine running at synchronous speed) is held constant and the field current is varied, we vary the reactive power Q, the power factor, and the armature current magnitude while the machine output power remains constant. Figure 4.28 illustrates the concept. Since $P = V_s I_s \cos \theta$, we see that for constant power operation at a fixed armature voltage V_s, we must have $I_s \cos \theta$ a constant. Thus the tips of the armature current phasors must fall on a vertical line as the power factor is

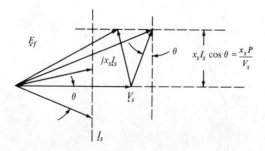

FIGURE 4.28 Variation of Field Current at Constant Power

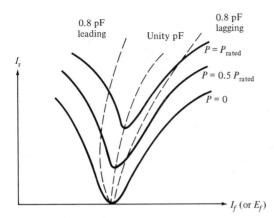

FIGURE 4.29 Armature Current Variation with Field Current (Generator Operation)

varied by varying the field current, as shown in the figure. The armature current will be at a minimum at unity power factor. We obtain characteristic V-shaped curves if we plot the armature current magnitude against the field current for various values of power. This is illustrated in Figure 4.29. The curves shown in Figure 4.29 can be obtained graphically by recognizing, as shown in Figure 4.28, that the tips of the phasors representing the field-induced voltage E_f must fall on a horizontal line whose vertical distance from the reference axis depends on the specified value of power.

4.4

EFFECTS OF SALIENT POLES

A salient-pole machine can be analyzed by modifying the steady-state relationships for a cylindrical-rotor machine. In a cylindrical-rotor machine the rotating air-gap flux wave established by armature currents is essentially independent of rotor position, and the various armature winding flux linkages due to armature current are obtained by simply multiplying armature current by an appropriate inductance. As discussed in Section 4.2, the phasor relationships for these quantities are

$$\psi_{\ell s} = L_{\ell s}I_s \tag{4.67}$$

$$\psi_m = MI_s \tag{4.68}$$

$$\psi_s = \psi_{\ell s} + \psi_m = (L_{\ell s} + M)I_s = L_s I_s \tag{4.69}$$

In the expressions above, $\psi_{\ell s}$ is the phasor representation for the flux linkages in the armature winding due to the leakage flux caused by the armature current, ψ_m is the phasor representation for flux linkages in the armature winding due to the air-gap flux caused by the armature current, and ψ_s is simply the sum of the two. In a salient-pole machine, the air-gap flux due to the armature current is

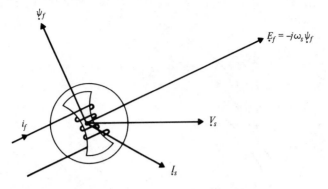

FIGURE 4.30

not independent of rotor position, and the simple relationships between flux linkages and I_s given above are not appropriate. This is illustrated in Figure 4.30, which shows a two-pole salient-pole machine operated as a generator. (Two-pole salient-pole machines are atypical in that salient-pole machines generally have a relatively large number of poles. We use a two-pole machine here for purposes of illustration.) In the manner discussed in Section 4.2, we have superimposed diagrams in Figure 4.30, aligning the axis of the field winding with the phasor ψ_f. It is clear from the figure that ψ_m will depend not only on I_s but also on its angle with respect to the axis of the field winding. If I_s is aligned with the axis of the field winding, we would expect $\psi_m = |\psi_m|$ to be relatively large, since the air gap along this axis is relatively small. (We have relatively low reluctance.) If I_s is in quadrature with the axis of the field winding (i.e., aligned with phasor E_f), we would expect $\psi_m = |\psi_m|$ to be relatively small, since the air gap along this axis is relatively large. (We have relatively high reluctance.) In reality the situation is complicated considerably by having an unsymmetrical rotor. For example, in the latter case it can be shown that when

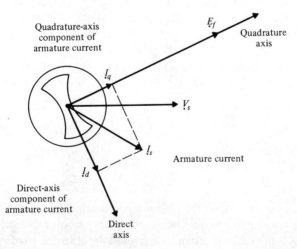

FIGURE 4.31 Resolution of Armature Current into Direct- and Quadrature-Axis Components

I_s and ψ_m are in quadrature with the field winding axis, a significant third harmonic appears in the spatial air-gap flux density distribution due to armature current. A less pronounced distortion in the flux density distribution occurs when I_s and ψ_m are aligned with the field winding axis. The appearance of harmonics in these spatial distributions would, in general, cause harmonics in time-varying quantities and preclude the use of phasors if we require that phasors represent sinusoidal time functions. If we neglect harmonics and consider only the fundamental component of all relevant flux density distributions, our phasor techniques apply, and we can resolve the armature current into two components: one component aligned with and one component in quadrature with the field-winding axis. These components are referred to as *direct-* and *quadrature-axis* components, as illustrated in Figure 4.31. As shown in Figure 4.31, the direct axis is aligned with the axis of the field winding. If diagrams are superimposed in the manner of Figures 4.30 and 4.31, the quadrature axis will always be aligned with the phasor E_f. Direct- and quadrature-axis components of the armature current I_s cause flux linkages along the respective axes:

$$\psi_{md} = L_{md}I_d \qquad (4.70)$$

$$\psi_{mq} = L_{mq}I_q \qquad (4.71)$$

where L_{mq} is always less than L_{md}. (L_{mq} might typically be 0.6 to 0.7 L_{md}.) Incorporating the effects of flux linkages due to the armature leakage flux, we have

$$\psi_d = L_{\ell s}I_d + \psi_{md} = (L_{\ell s} + L_{md})I_d = L_dI_d \qquad (4.72)$$

$$\psi_q = L_{\ell s}I_q + \psi_{mq} = (L_{\ell s} + L_{mq})I_q = L_qI_q \qquad (4.73)$$

The total resultant flux linkages ψ_{res} will be

$$\psi_{\text{res}} = \psi_d + \psi_q + \psi_f \qquad (4.74)$$

Multiply through by $-j\omega_s$:

$$-j\omega_s\psi_{\text{res}} = -j\omega_s\psi_d - j\omega_s\psi_q - j\omega_s\psi_f \qquad (4.75)$$

Now substitute using (4.72) and (4.73) and apply Faraday's law:

$$V_s = -j\omega_sL_dI_d - j\omega_sL_qI_q + E_f \qquad (4.76)$$

By defining direct- and quadrature-axis synchronous reactances as $x_d = \omega_sL_d$ and $x_q = \omega_sL_q$, (4.76) can be expressed as

$$V_s = -jx_dI_d - jx_qI_q + E_f \qquad (4.77)$$

Equation (4.77) is the steady-state equation for a salient-pole synchronous machine (armature resistance neglected) operated as a generator. A phasor diagram showing a typical operating condition is shown in Figure 4.32.

EXAMPLE 4.3 A three-phase 40-pole 100-MVA salient-pole generator has a rated armature voltage (line-to-line) of 13.8 kV. The machine has a direct-axis synchronous reactance of 2.75 Ω and a quadrature-axis synchronous reactance of 2.0 Ω. The armature

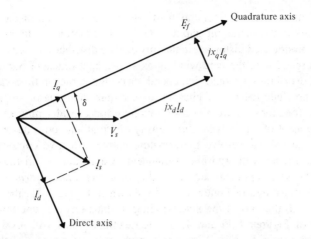

FIGURE 4.32 Phasor Diagram for a Salient-Pole Synchronous Machine Operated as a Generator

resistance is negligible. The machine is Y-connected. Find the field-induced voltage E_f if the machine delivers rated MVA at rated voltage and at a power factor of 0.85 lagging.

The situation described is similar to the one described in Example 4.1 but involves a salient-pole machine rather than a cylindrical-rotor machine. The solution is similar but must take into account the modification required to deal with saliency. It may be helpful to compare the two situations in the manner suggested by Figure 4.33. Note that whereas

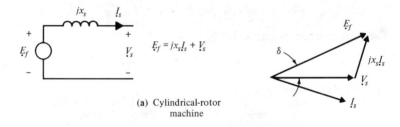

$E_f = jx_s I_s + V_s$

(a) Cylindrical-rotor
machine

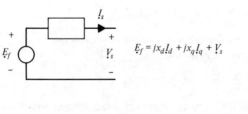

$E_f = jx_d I_d + jx_q I_q + V_s$

(b) Salient-pole
machine

FIGURE 4.33

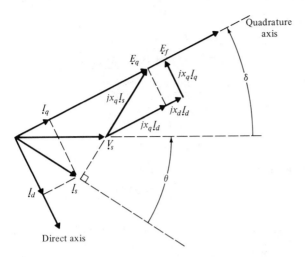

FIGURE 4.34

for the cylindrical-rotor machine a single inductive reactance is appropriate for an equiv-alent circuit, in the case of the salient-pole machine we cannot synthesize an equivalent circuit for (4.77) using simple circuit elements. Furthermore, whereas the solution for E_f in the case of the cylindrical-rotor machine requires only specification of V_s and I_s, the solution in the case of the salient-pole machine requires additionally that we resolve I_s into its components I_d and I_q. This would require knowing beforehand the position of the quadrature axis, which requires knowing the angle of E_f itself. Fortunately, the geometry of the phasor diagram for a salient-pole machine yields a simple way out of this dilemma. The armature current in terms of direct- and quadrature-axis components is

$$I_s = I_d + I_q$$

Multiplying through by jx_q, we obtain

$$jx_q I_s = jx_q I_d + jx_q I_q$$

The quantity $jx_q I_s$ can be used to determine a fictitious voltage E_q that must lie along the quadrature axis. This is illustrated in Figure 4.34.

The solution for our Y-connected machine can proceed as follows:

$$V_s = \frac{13.8}{\sqrt{3}} \text{ kV} = 7.97 \text{ kV}$$

Choose the terminal voltage as reference:

$$V_s = V_s \underline{/0°} = 7.97 \underline{/0°} \text{ kV}$$

If the machine delivers rated MVA, we must have

$$3V_s I_s = 100 \text{ MVA}$$

Thus

$$I_s = 4.18 \text{ kA}$$

At a power factor of 0.85 lagging we will have

$$I_s = 4.18 \underline{/-31.8°} \text{ kA}$$

From Figure 4.34 we see that

$$E_q = V_s + jx_qI_s$$

$$= 7.97 + j(2.0)(4.18 \underline{/-31.8°})$$

$$= 14.3 \underline{/29.9°} \text{ kV}$$

Again referring to Figure 4.34, we see that

$$E_f = E_q + (x_d - x_q)I_d$$

$$I_d = I_s \sin(\delta + \theta)$$

The angle δ is 29.9° and θ is the power-factor angle:

$$\theta = \cos^{-1}(0.8) = 31.8°$$

Thus

$$I_d = 4.18 \sin(29.9° + 31.8°) = 3.68 \text{ kA}$$

$$E_f = E_q + (x_d - x_q)I_d$$

$$= 14.3 + (2.75 - 2.0)(3.68) = 17.1 \text{ kV}$$

Thus

$$E_f = E_f \underline{/\delta} = 17.1 \underline{/29.9°}$$

The power-angle curve for a salient-pole machine differs from that of a cylindrical-rotor machine due to the presence of reluctance torque. Even if field current is reduced to zero, the rotor will tend to align itself with the field established by armature currents.

An expression for power per phase P in terms of V_s, E_f, and the angle δ can be obtained by examining the phasor diagram for a salient-pole machine. Power can be obtained by multiplying the "in-phase" components of voltage and current:

$$P = V_dI_d + V_qI_q \tag{4.78}$$

$$= (V_s \sin \delta)I_d + (V_s \cos \delta)I_q$$

From the geometry of the diagram (see Figure 4.32),

$$x_dI_d = E_f - V_s \cos \delta \tag{4.79}$$

$$x_qI_q = V_s \sin \delta \tag{4.80}$$

Substituting in (4.78) using (4.79) and (4.80) gives us

$$P = V_s \sin \delta \frac{E_f - V_s \cos \delta}{x_d} + \frac{V_s^2 \cos \delta \sin \delta}{x_q}$$

$$= \frac{V_sE_f}{x_d} \sin \delta + V_s^2 \cos \delta \sin \delta \left(\frac{1}{x_q} - \frac{1}{x_d}\right) \tag{4.81}$$

$$= \frac{V_sE_f}{x_d} \sin \delta + V_s^2 \frac{x_d - x_q}{2x_dx_q} \sin 2\delta$$

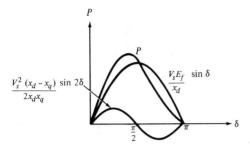

FIGURE 4.35

The second term in (4.81) is the contribution due to reluctance torque. The first term is the contribution due to an excited field. The effects of both terms are illustrated in Figure 4.35. It should be noted that if $x_d = x_q = x_s$, the reluctance torque term is zero and we obtain the power expression for a cylindrical-rotor machine, that is, (4.64).

EXAMPLE 4.4   An open-circuit slip test can be conducted on a salient-pole synchronous machine to determine direct- and quadrature-axis synchronous reactances x_d and x_q. The test is conducted with the field winding open circuited. The machine is connected to an infinite bus and the rotor is turned at a constant speed differing from synchronous speed by a small amount. The armature current waveform is observed and recorded. If the armature current waveform from the test is as shown in Figure 4.36, determine x_d and x_q. The machine is rated 200 MVA with a rated armature terminal voltage of 24 kV line to line. The machine is Y-connected.

If the field is left an open circuit, the resultant flux required for the terminal voltage (recall that $\psi_{res} = V_s/-j\omega_s$) must be established entirely by the armature current. If we assume that the machine is connected to an infinite bus, V_s, and hence ψ_{res}, will be constant in magnitude and angle. If the machine is driven at a constant speed slightly above the synchronous speed, the angle δ between V_s and the quadrature axis will increase linearly with time. If δ increases slowly, we can apply steady-state analysis techniques, essentially treating a sequence in time of quasi-steady-state operating points.

The minimum and maximum points on the armature current waveform will occur when the rotor is aligned as shown in Figure 4.37(a) and (b). Figure 4.37 illustrates that a relatively low value of armature current is required to establish the resultant flux with

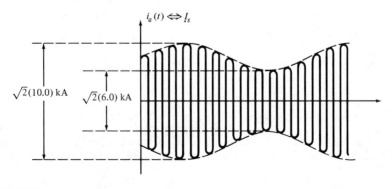

FIGURE 4.36

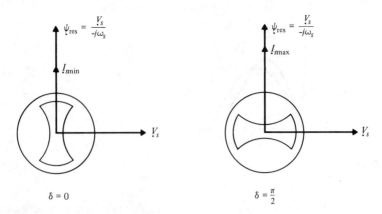

$\psi_{res} = \dfrac{V_s}{-j\omega_s}$

$\psi_{res} = \dfrac{V_s}{-j\omega_s}$

$\delta = 0$

$\delta = \dfrac{\pi}{2}$

FIGURE 4.37

the rotor positioned as shown in (a) and a relatively high value is required with the rotor positioned as shown in (b). With the rotor winding an open circuit, the field-induced voltage E_f is zero and from (4.77) we must have

$$V_s = -jx_d I_d - jx_q I_q$$

With the rotor aligned as shown in (a), $I_q = 0$ and we have

$$x_d = \frac{V_s}{I_d} = \frac{V_s}{I_{smin}} = \frac{24.0}{\sqrt{3}\,(3.0)} = 4.62 \ \Omega$$

With the rotor aligned as shown in (b), $I_d = 0$ and we have

$$x_q = \frac{V_s}{I_q} = \frac{V_s}{I_{smax}} = \frac{24.0}{\sqrt{3}\,(5.0)} = 2.77 \ \Omega$$

Further details on this particular test can be found in reference 1.

EXAMPLE 4.5 Two identical 2500-hp synchronous motors are used to drive a grinding mill. The motors are mechanically connected to the mill in the dual-drive scheme illustrated conceptually in Figure 4.38. Each motor has the following manufacturer-furnished data:

Number of poles	10
Armature winding	Y-connected
Armature terminal voltage	4160 V line to line
Rated power factor	0.9 leading
Direct-axis synchronous reactance	10.0 Ω
Quadrature-axis synchronous reactance	6.0 Ω
Armature resistance	0.02 Ω
Efficiency at rated conditions	0.96

Because of tolerances (small but not zero) in the gear trains of each of the drives and perhaps because of minor installation discrepancies, we anticipate a certain amount of misalignment between the rotors of the two motors. Specifically, it is estimated that if the rotor of one machine is at an angle θ_m with respect to a horizontal axis, the rotor of the other machine will be at an angle $\theta_m \pm 0.8°$ when the slack in both gear trains has been taken up. Determine the imbalance in machine loading if the combined power output of both motors is 5000 hp. If the grinding mill itself turns at 15 rpm, find the torques

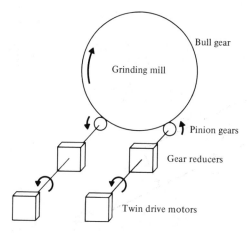

FIGURE 4.38

applied by each of the drives at the bull gear of the grinding mill if gear train losses are neglected.

Under ideal conditions, misalignment would be zero and both machines would deliver $(5000/2)$ hp = 2500 hp and operate under essentially rated conditions. We can use rated conditions to establish the field-induced voltage for each motor, which will remain constant if field excitation is constant. For a Y-connected machine,

$$V_s = \frac{4160}{\sqrt{3}} = 2402 \ \Omega$$

Taking the armature voltage as a reference gives us

$$V_s = 2402 \ \underline{/0°} \ \text{V}$$

$$P_{out} = (3)\eta V_s I_s \cos \theta$$

Thus

$$I_s = \frac{(2500 \ \text{hp})(746 \ \text{W/hp})}{3(0.96)(0.9)(2402 \ \text{V})} = 299.6 \ \text{A}$$

For a leading power factor of 0.9,

$$\theta = -\cos^{-1}(0.9) = -25.8°$$

where θ is the power-factor angle (i.e., the angle by which V_s leads I_s). Thus, if $V_s = V_s \ \underline{/0°}$, we must have for I_s,

$$I_s = 299.6 \ \underline{/+25.8} \ \text{A}$$

For a motor (neglecting armature resistance),

$$E_f = V_s - jx_d I_d - jx_q I_q$$

An appropriate phasor diagram for our condition is shown in Figure 4.39. The voltage E_q locates the quadrature axis:

$$E_q = V_s - jx_q I_s$$

$$= 2402 \ \underline{/0°} - j(6.0)(299.6 \ \underline{/25.8°})$$

$$= 3572 \ \underline{/-26.9°}$$

4.4 EFFECTS OF SALIENT POLES

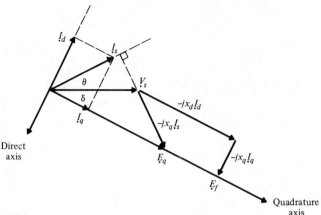

FIGURE 4.39

From Figure 4.39 we see that

$$E_f = E_q + (x_d - x_q)I_d$$

where

$$I_d = I_s \sin (\theta + \delta)$$

Thus

$$E_f = 3572 + (10.0 - 6.0)(299.6) \sin (25.8° + 26.9°)$$

$$= 4525$$

$$E_f = 4525 \underline{/-26.9°}$$

The power per phase for a salient-pole machine is given by (4.81)

$$P = \frac{V_s E_f}{x_d} \sin \delta + \frac{V_s^2(x_d - x_q)}{2x_d x_q} \sin 2\delta$$

$$= P_f \sin \delta + P_r \sin 2\delta$$

where for our example

$$P_f = \frac{(2402)(4525)}{10.0} = 1087 \text{ kW}$$

$$P_r = \frac{(2402)^2(10.0 - 6.0)}{2(10.0)(6.0)} = 192.3 \text{ kW}$$

We must convert our mechanical angle discrepancy to electrical degrees:

$$\Delta\theta_e = \frac{p}{2} \Delta\theta_m$$

$$= \frac{10}{2} (0.8°) = 4° \text{ (degrees electrical)}$$

For an operating point taking into account the discrepancy $\Delta\theta_e$, one machine will operate at an angle δ' and the other will operate at δ' at $\Delta\theta_e$. Assuming that the combined power output is 5000 hp, we must have (on a per phase basis)

$$P_1 + P_2 = \frac{(5000 \text{ hp})(746 \text{ W/hp})}{(3)(0.96)}$$

$$= 1295 \text{ kW}$$

and

$$P_1 = P_f \sin \delta' + P_r \sin 2\delta'$$

$$P_2 = P_f \sin (\delta' + \Delta\theta_e) + P_r \sin (2\delta' + 2 \Delta\theta_e)$$

Combining the relationships above gives us

$$P_1 + P_2 = P_f[\sin \delta' + \sin (\delta' + \Delta\theta_e)]$$

$$+ P_r[\sin 2\delta' + \sin (2\delta' + 2 \Delta\theta_e)]$$

The relationship above can be solved for the single unknown quantity δ' by trial and error. An approximate solution is

$$\delta' = 24.96° \text{ electrical}$$

This yields

$$P_1 = 605.8 \text{ kW per phase}$$

$$P_2 = 689.3 \text{ kW per phase}$$

Thus we see that one machine supplies roughly 14% more power than the other. Torques applied at the bull gear will be in the same ratio as P_1 and P_2. Thus if the total power converted is 5000 hp, the total torque at the bull gear at 15 rpm is

$$T = \frac{(5000 \text{ hp})(746 \text{ W/hp})}{(2\pi)(15/60)}$$

$$= 2.37(10^6) \text{ N} \cdot \text{m}$$

$$T_1 = \frac{P_1}{P_1 + P_2} T = 1.11(10^6) \text{ N} \cdot \text{m}$$

$$T_2 = \frac{P_2}{P_1 + P_2} T = 1.26(10^6) \text{ N} \cdot \text{m}$$

The percentage differences in torques will occur at any pair of corresponding points in the two drive trains. In practice the problem is more complicated than what we have described thus far. Because the bull gear itself is generally not perfectly circular in shape, the angular discrepancy $\Delta\theta_e$ will vary slowly as a periodic function of time, where the frequency is the frequency of rotation of the grinding mill itself. In this case the mill turns at 15 rpm; thus the frequency would be $15/60 = \frac{1}{4}$ Hz. Were it not for the time-varying nature of the angular discrepancy, a one-time corrective action might solve the load-sharing problem. For example, we might make an adjustment to a variable coupling in one of the drive trains or even rotate slightly the stator of one of the machines to achieve equal power conversion by each. The time-varying nature of $\Delta\theta_e$ compounds the problem, resulting in significant shifts back and forth between the amount of load carried by each of the two synchronous machines. This can significantly reduce gear life. Induction machines may be a preferable alternative. Because the angle $\Delta\theta_e$ varies slowly with time, changes in slip would be small. The torque in an induction machine is approximately linearly related to the slip at values near the rated slip. If changes in the slip are small, changes in the torque will be small. Thus induction machines tend to share loads better in such situations. The synchronous machine's somewhat improved efficiency and capability to correct the power factor have encouraged development of schemes to adjust field excitation automatically to improve load sharing. Reference 2 describes such a scheme and lists some of the trade-offs between synchronous and induction machines in dual-drive situations.

4.5

SYNCHRONOUS MACHINE DYNAMIC ANALYSIS

To analyze synchronous-machine dynamic behavior, we must examine the equation of motion for the rotor of the machine. For a generator

$$J \frac{d\omega_{rm}}{dt} = T_{pm} - T_e \tag{4.82}$$

In (4.82) J is combined moment of inertia of the rotating components of the prime-mover system, generator, and exciter if a rotating exciter is used. T_{pm} is the total prime-mover torque, T_e is the electrical torque developed by the generator, and ω_{rm} is the rotor speed in mechanical radians per second. It will be convenient to express all angles in electrical degrees or electrical radians and all speeds of rotation in electrical degrees per second or electrical radians per second. If the subscript m denotes variables expressed in mechanical degrees or mechanical radians, the conversion to electrical degrees or electrical radians is given by the following relationships:

$$\theta = \frac{p}{2}\,\theta_m \tag{4.83}$$

$$\omega_r = \frac{p}{2}\,\omega_{rm} \tag{4.84}$$

$$\omega_s = \frac{p}{2}\,\omega_{sm} \tag{4.85}$$

where p is the number of poles for which the machine is wound. Note that ω_s is synchronous speed and for 60-Hz systems we will always have $\omega_s = 2\pi f = 2\pi(60) \simeq 377$ rad/s.

In reality, the prime-mover system typically has a number of rotating masses with prime-mover torque applied at various points. A steam-turbine-driven generator is an example. Figure 4.40 illustrates conceptually the various stages of a steam-turbine-driven generator. In investigations where shaft dynamics are of interest, it is necessary to write an equation of motion for each of the rotating masses depicted in Figure 4.40, taking into account the stiffness of the shafts between the various stages. In this section we neglect shaft dynamics and treat

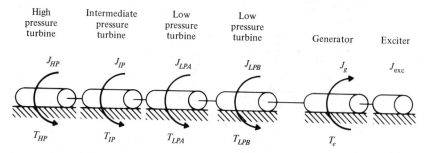

FIGURE 4.40 Stages of a Steam-Turbine-Driven Generator System

the system as a single rigid rotating mass where T_{pm} is the combined prime-mover torque. This implies that

$$J = J_{HP} + J_{IP} + J_{LPA} + J_{LPB} + J_g + J_{exc} \qquad (4.86)$$

$$T_{pm} = T_{HP} + T_{IP} + T_{LPA} + T_{LPB} \qquad (4.87)$$

In (4.86) the moments of inertia in the summation must also include the moments of inertia of the various shaft sections. Damping due to friction and windage is often negligible, and we will omit it in our discussion.

To better understand dynamic behavior, we will start by examining a typical steady-state operating condition for a generator. Substituting in (4.82) using (4.84), our general equation of motion becomes

$$\frac{2}{p} J \frac{d\omega_r}{dt} = T_{pm} - T_e \qquad (4.88)$$

At a steady-state operating point we must have

$$\frac{2}{p} J \frac{d\omega_r}{dt} = T_{pm} - T_e = 0 \qquad (4.89)$$

$$\omega_r = \omega_s \qquad (4.90)$$

where ω_s is synchronous speed. A typical steady-state operating point for a cylindrical-rotor machine is illustrated in Figure 4.41.

For a three-phase cylindrical-rotor machine, the steady-state torque T_e is related to the power per phase P by

$$T_e = \frac{3P}{\omega_{sm}} = \frac{3}{\omega_{sm}} \frac{V_s E_f}{x_s} \sin \delta \qquad (4.91)$$

Thus the maximum torque T_{emax} (the steady-state stabililty limit) is

$$T_{emax} = \frac{3}{\omega_{sm}} \frac{V_s E_f}{x_s} \qquad (4.92)$$

Referring to Figure 4.41, we see that a second possible operating point exists at $\delta_0' = \pi - \delta_0$. This point is an unstable point for constant field excitation. To see this, consider the diagram shown in Figure 4.42, which illustrates a

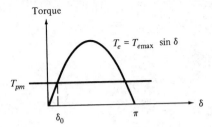

$$T_e = T_{emax} \sin \delta$$

FIGURE 4.41

typical generator operating condition. Figure 4.42 shows a rotating phasor diagram superimposed on the rotor of the machine with the axis of the field winding aligned with the phasor ψ_f. The machine is assumed to be connected to an infinite bus such that V_s and ψ_{res} are fixed in magnitude and rotate at speed ω_s. During normal steady-state operation, $\omega_r = \omega_s$ and δ will be a constant. If we consider operation at $\delta_0 < \pi/2$ as illustrated in Figure 4.41, we have a stable operating point. For example, if some disturbance should cause the angle δ to increase slightly from $\delta = \delta_0$ to $\delta = \delta_0^+ > \delta_0$, we would have (see Figure 4.41) $T_e(\delta_0^+) > T_{pm}$, assuming that T_{pm} remained constant. We see from (4.88) that this would result in a deceleration of the rotor. If the rotor speed ω_r becomes less than the synchronous speed ω_s (see Figure 4.42), the angle δ will decrease thus moving generally back toward the operating point δ_0. It is easy to see that for the operating point δ_0, disturbances resulting in small displacements to either side of the operating point result in accelerating or decelerating torque, tending to restore operation to δ_0. At the operating point $\pi - \delta_0$, the opposite is true, and the point is said to be an *unstable* operating point.

To quantify dynamic behavior, we must examine the equation of motion of the rotor [Eq. (4.88)] in greater detail. In a dynamic situation the angle δ can be taken as the angle between the axis of the field winding, which rotates at speed ω_r, and a reference axis, which rotates at the synchronous speed ω_s. If

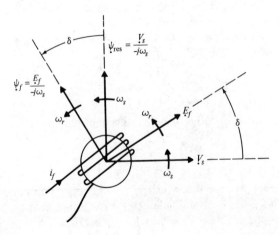

FIGURE 4.42

THE SYNCHRONOUS MACHINE

the machine is connected to an infinite bus, it is convenient to choose the synchronously rotating reference axis as aligned with phasor ψ_{res}, as shown in Figure 4.42. The angle so chosen is then the same as the steady-state power angle, that is, the angle between E_f and V_s (see Figure 4.42). In general, we have

$$\delta(t) = \int_0^t (\omega_r(\tau) - \omega_s) \, d\tau + \delta(0) \tag{4.93}$$

In (4.93) we see that $\delta(t)$ is obtained by simply integrating the rotor speed with respect to a synchronously rotating reference frame, taking into account the initial angle. Equivalently, $\delta(t)$ is seen to be proportional to the integral of the rotor slip with respect to a synchronously rotating reference frame. Differentiating (4.93), we have

$$\frac{d\delta(t)}{dt} = \omega_r(t) - \omega_s = -s\omega_s \tag{4.94}$$

where slip is defined as $s = (\omega_s - \omega_r)/\omega_s$. Differentiating (4.94), we have

$$\frac{d^2\delta}{dt^2} = \frac{d\omega_r(t)}{dt} \tag{4.95}$$

Equations (4.93) through (4.95) establish the relationships between $\delta(t)$, the rotor speed ω_r, the rotor slip s, and the rotor angular acceleration $d\omega_r/dt$.

The relationship for the electrical torque T_e illustrated in Figure 4.41 and given by (4.91) and (4.92) is a steady-state torque relationship. In general, it is not valid for dynamic analysis. If we restrict cases we consider to situations where the angle δ varies slowly such that $d\delta(t)/dt \approx 0$ (but not exactly zero), we can analyze dynamic behavior as a sequence in time of "steady-state" currents, voltages, and torques and thus apply steady-state concepts. Examination of Figure 4.42 illustrates the problem. If ω_r differs significantly from ω_s, the resultant flux linkages ψ_{res} would cause a significant time rate of change of the flux linking the field circuit. The resulting induced rotor currents would violate the assumption of constant field current. A rigorous general treatment of such behavior requires consideration of coupled coils under more general time-varying conditions. We will defer such treatment to Chapter 6 and for the moment augment in a simple way the steady-state torque expression by accounting for rotor damping. Damper windings on the rotor of a synchronous machine are often used to damp out rotor oscillations. Such windings are effectively short-circuited windings generally realized by embedding copper bars in the pole faces of the rotor and connecting them at the ends of the rotor. The configuration is somewhat similar to that used for the rotor of a squirrel-cage induction machine. During normal steady-state operation of the machine with the rotor turning at synchronous speed, the damper windings have no effect. If the machine develops a nonzero slip, a torque is developed in a manner similar to that of an induction machine. This torque, due to damper windings, tends to decelerate a rotor running above synchronous speed and to accelerate a rotor running below synchronous speed. The torque due to damper windings for small values of slip is approximately linearly related to the slip. Thus the torque in the direction of rotation due to damping can be expressed as

$$T_{\text{damping}} = ks \tag{4.96}$$

$$= \frac{-k}{\omega_s} \frac{d\delta}{dt}$$

Thus the total electrical torque T_e for a cylindrical-rotor machine, incorporating the damping torque, will be

$$T_e = T_{e\max} \sin \delta - T_{\text{damping}} \tag{4.97}$$

$$= T_{e\max} \sin \delta + \frac{k}{\omega_s} \frac{d\delta}{dt}$$

The equation of motion with these modifications becomes [substitute in (4.88) using (4.95) and (4.97)]

$$\frac{2}{p} J \frac{d^2\delta}{dt^2} + \frac{k}{\omega_s} \frac{d\delta}{dt} + T_{e\max} \sin \delta = T_{pm} \tag{4.98}$$

Equation (4.98) is a second-order ordinary differential equation that is nonlinear due to the sin δ term. In many cases the angle δ is small, and we can make the further assumption sin $\delta \simeq \delta$, which when substituted into (4.98), yields a second-order linear differential equation that can be used for approximate dynamic analysis. We have already implied that the equation is reasonable only if $d\delta/dt \approx 0$ (i.e., if $\omega_r \simeq \omega_s$, which implies that $\omega_{rm} \approx \omega_{sm}$). We thus incur little error by multiplying both sides of (4.98) by ω_{sm}, converting an equation expressed in torque to one expressed in power:

$$\frac{2}{p} J\omega_{sm} \frac{d^2\delta}{dt^2} + \frac{k\omega_{sm}}{\omega_s} \frac{d\delta}{dt} + P_{\max} \sin \delta = P_{pm} \tag{4.99}$$

where $P_{\max} = 3V_sE_f/x_s$, which is the steady-state stability limit expressed in total three-phase power and P_{pm} is the prime-mover input power. Further simplifications are helpful in ascertaining reasonable orders of magnitude for the various constants. The inertia constant H is defined as the ratio of the kinetic energy of the machine rotor at synchronous speed to the machine volt-ampere rating:

$$H = \frac{\frac{1}{2} J\omega_{sm}^2}{VA_{\text{rating}}} \quad s \tag{4.100}$$

This ratio falls within a relatively narrow range for most machines (3 to 9 s for synchronous generators). If we divide both sides of (4.99) by the volt-ampere rating of the machine and substitute using (4.100), we obtain

$$\frac{2H}{\omega_s} \frac{d^2\delta}{dt^2} + P_d \frac{d\delta}{dt} + P_{\max} \sin \delta = P_{pm} \tag{4.101}$$

where in (4.101) $P_{\max}$ and P_{pm} are normalized quantities, sometimes referred to as *per unit quantities* on the machine volt-ampere rating as base. Generally, P_{pm} so expressed will be on the order of 1.0. Typical steady-state operating angles δ_0 are roughly 20 to 30°; thus we surmise that $P_{\max}$ will be approximately 2.0

to 3.0. The constant P_d is typically 0.025 to 0.09 per unit power per electrical radian per second. This can be surmised by considering approximate analytical expressions for average torque developed by a synchronous machine running at a constant asynchronous ($\omega_r \neq \omega_s$) speed. References 3 and 4 provide the details.

A philosophical note is in order. Equation (4.101) is a second-order differential equation, which results from a number of simplifying assumptions found to be reasonable for many practical application problems. It is easily "linearized" to a second-order linear differential equation, which is the minimum level of mathematical complexity required to describe what is often observed about machine behavior, specifically an underdamped oscillatory response to changes in or disturbances to a steady-state operating condition. A caution is also in order. If we stage a test on an actual machine in an attempt to determine appropriate constants for (4.101), the resulting model will generally be valid only for applications where the assumptions leading up to (4.101) are reasonable.

EXAMPLE 4.6   A three-phase steam-turbine-driven synchronous generator delivers its rated real power at rated conditions. A three-phase fault (short circuit) occurs at the terminals of the machine. A circuit breaker opens, clearing the fault in eight cycles. Determine the percent change in machine speed at the time the fault is cleared.

We will neglect switching transients, which do not appreciably affect machine speed, and use (4.101) to obtain an approximate answer. We will assume that P_{pm} remains constant over the eight cycle time period. (Typical prime-mover speed governor time constants are on the order of 0.5 s; thus governor action would be minimal.) An appropriate choice for P_{pm} would be 1.0. Examining the various terms in (4.101), we see that, on a steady-state basis, P_{max} must be zero during the fault because $V_s = 0$. In essence, if terminals of the machine are short-circuited, there can be no power output, and the prime-mover power will be converted in machine losses and in accelerating (increasing the kinetic energy of) the rotor. The damping constant P_d in (4.101) will be significantly reduced by the reduction of machine terminal voltage to zero (recall that induction machine torque is essentially proportional to terminal voltage squared). We can thus obtain a conservative (pessimistic) estimate of the departure from the synchronous speed by solving the differential equation

$$\frac{2H}{\omega_s} \frac{d^2\delta}{dt^2} = P_{pm} = 1.0$$

Integrating the relationship above gives us

$$\frac{d\delta}{dt} = \frac{\omega_s}{2H} t + \frac{d\delta}{dt}(0)$$

From (4.94) we see that $d\delta(0)/dt = \omega_r(0) - \omega_s = 0$, since the rotor speed cannot depart instantaneously from the synchronous speed. Thus

$$\frac{d\delta}{dt} = \omega_r - \omega_s = \frac{\omega_s t}{2H}$$

Taking a typical value of H, say $H = 3$ s, and evaluating at $t = 8T = 8(1/f) = 8/60$ s, we have, on a percentage basis, at the end of eight cycles:

$$\frac{\omega_r - \omega_s}{\omega_s} \times 100 = \frac{8/60}{2(3)} \times 100$$

$$= 2.22\%$$

The percentage increase in speed calculated above will be somewhat high in that all of the assumptions made tend to result in more pronounced speed changes.

EXAMPLE 4.7 A three-phase cylindrical-rotor synchronous generator is synchronized to a power grid with its terminal voltage out of phase by 15°. Specifically, the machine terminal voltage leads the system voltage by 15° at the instant the circuit breaker is closed. The machine data are

Machine rating	100 MVA
Rated terminal voltage	24 kV line to line
Armature winding	Y-connected
Synchronous reactance	9.0 Ω
Inertia constant (H)	5 s

If the circuit breaker is closed at time $t = 0$, find $\delta(t)$ for $t \geq 0$.

When a machine is synchronized to a power grid, it is first brought up to speed by its prime mover while disconnected from the system, that is, with its armature windings open circuited. Field excitation is then adjusted so that machine open-circuit voltage is equal in magnitude to the voltage of the system to which it is to be connected. Ideally, at the instant of closing the circuit breaker which will connect the machine to the system, machine voltage and system voltage will be equal in magnitude, phase angle, and frequency. The speed governor system on the prime mover causes the prime mover to turn the machine rotor at a speed near the synchronous speed it must have when connected to the system. Typically, the phase-angle difference between the machine terminal voltage and the system voltage varies slowly prior to synchronization, since the machine speed is not exactly correct. The circuit breaker should be closed when the two voltages are in phase, and a device known as a synchroscope can be used to determine the correct point in time to close the breaker. In this example we have an out-of-phase synchronization due perhaps to device malfunction or operator error. The problem is illustrated conceptually on a single-phase basis in Figure 4.43.

We will assume that the system to which the machine is to be synchronized can be represented as an infinite bus. We will neglect switching transients and solve the problem using (4.101):

$$\frac{2H}{\omega_s}\frac{d^2\delta}{dt^2} + P_d\frac{d\delta}{dt} + P_{max}\sin\delta = P_{pm}$$

The prime-mover power P_{pm} can be taken as zero prior to, during, and immediately following the circuit breaker closing. The initial angle of 15° is sufficiently small that if δ is expressed in radians, we can assume that $\sin\delta \approx \delta$. The synchronizing power P_{max} is given by

$$P_{max} = \frac{3V_sE_f}{x_s}$$

$$= \frac{3V_s^2}{x_s}$$

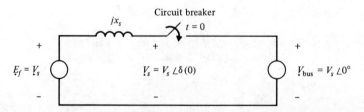

FIGURE 4.43 Situation Prior to Circuit Breaker Closing on Out-of-Phase Synchronization

where

$$V_s = \frac{24}{\sqrt{3}} = 13.86 \text{ kV}$$

Thus, when expressed on a normalized (per unit) basis with respect to machine volt-ampere rating, we have

$$P_{max} = \frac{3[13.86(10^3)]^2}{(9.0)[100(10^6)]} = 0.64$$

We will assume a typical value for P_d:

$$P_d = 0.05 \text{ per unit per rad/s}$$

The inertia constant H is given as 5 s; thus we must solve

$$\frac{2(5)}{377} \frac{d^2\delta}{dt^2} + 0.05 \frac{d\delta}{dt} + 0.64 \delta = 0$$

with

$$\delta(0) = 15° = 0.262 \text{ rad}$$

$$\frac{d\delta(0)}{dt} = \omega_r(0) - \omega_s \simeq 0$$

assuming that the machine is running sufficiently close to synchronous speed to neglect the initial speed discrepancy. The differential equation can be rewritten as

$$\frac{d^2\delta}{dt^2} + 1.885 \frac{d\delta}{dt} + 24.13 \delta = 0$$

The roots of the characteristic equation are given by

$$r = \frac{-1.885 \pm \sqrt{1.885^2 - (4)(24.13)}}{2}$$

$$= -0.943 \pm j4.82$$

Thus the solution will be of the form

$$\delta(t) = Ae^{-0.943t} \cos(4.82t + \varphi)$$

where the constants A and φ are determined from the initial conditions

$$\delta(0) = 0.262 = A \cos \varphi$$

$$\frac{d\delta(0)}{dt} = 0 = -4.82A \sin \varphi - 0.943A \cos \varphi$$

We find that

$$\varphi = -11.07° = -0.193 \text{ rad}$$

$$A = 0.267$$

Finally,

$$\delta(t) = 0.267e^{-.943t} \cos(4.82t - 0.193) \text{ rad}$$

which is a damped sinusoid of frequency $4.82/2\pi = 0.77$ Hz decaying to zero in approximately $5\tau = 5/0.943 = 5.3$ s.

EXAMPLE 4.8 A synchronous motor drives a conveyor belt that is loaded at equally spaced intervals in time by a feed system whose rate of feed can be varied. The repetitive loadings cause the load torque to vary from the rated load torque T_{L0} such that the total load torque as a function of time can be described as

$$T_L(t) = T_{L0} + f(t)$$

where $f(t)$ is periodic, having a time period determined by the loading frequency. Find an expression for the magnitude of the displacement of the angle δ from its operating point δ_0 due to the fundamental component of $f(t)$. Express the result as a function of the frequency and magnitude of the fundamental component of $f(t)$. Evaluate the expression if the fundamental component of $f(t)$ is at the undamped natural frequency of the system and if the fundamental component has a magnitude of 5% of rated load torque. Assume the following data:

$$H = 8 \text{ s}$$

$$P_{max} = 2.0 \text{ per unit}$$

$$P_d = 0.04 \text{ per unit per rad/s}$$

$$T_{L0} = 1.0 \text{ per unit}$$

Equation (4.101) was developed for a generator but applies as well for a motor if the angle δ is redefined as the angle by which the synchronously rotating reference axis leads the field winding axis and if P_{pm} is replaced by P_L, which is the load power. Let $\delta = \delta_0 + \Delta\delta$ and $P_L = P_{L0} + \Delta P \cos \omega_1 t$, where ω_1 is the radian frequency of the fundamental component of $f(t)$. Substituting in (4.101) gives us

$$\frac{2H}{\omega_s} \frac{d^2 \, \Delta\delta}{dt^2} + P_d \frac{d \, \Delta\delta}{dt} + P_{max} \sin (\delta_0 + \Delta\delta) = P_{L0} + \Delta P \cos \omega_1 t$$

where $P_L(t) \approx T_L(t)$ on a normalized basis assuming that machine speed does not depart significantly from the synchronous speed. If $\Delta\delta$ is assumed small,

$$P_{max} \sin (\delta_0 + \Delta\delta) \approx P_{max} \sin \delta_0 + (P_{max} \cos \delta_0) \, \Delta\delta$$

But

$$P_{max} \sin \delta_0 = P_{L0}$$

Thus the differential equation describing $\Delta\delta$ can be written as

$$\frac{2H}{\omega_s} \frac{d^2 \, \Delta\delta}{dt^2} + P_d \frac{d \, \Delta\delta}{dt} + (P_{max} \cos \delta_0) \, \Delta\delta = \Delta P \cos \omega_1 t$$

The equation above is a second-order linear differential equation with a sinusoidal forcing function, and phasor techniques can be used to find the steady-state response to the forcing function. Choose phasor representations as

$$\Delta P \cos \omega_1 t \leftrightarrow \underset{\cdot}{\Delta P} = \Delta P \underline{/0°}$$

$$\Delta\delta \leftrightarrow \underset{\cdot}{\Delta}$$

The phasor relationship for the differential equation in $\Delta\delta(t)$ will be

$$\underset{\cdot}{\Delta} \left[(j\omega_1)^2 \frac{2H}{\omega_s} + j\omega_1 P_d + P_{max} \cos \delta_0 \right] = \underset{\cdot}{\Delta P}$$

Thus

$$\frac{\underset{\cdot}{\Delta}}{\underset{\cdot}{\Delta P}} = \frac{\omega_s/2H}{(\omega_s/2H) P_{max} \cos \delta_0 - \omega_1^2 + j\omega_1 (\omega_s/2H) P_d}$$

The magnitude of this transfer relationship will be

$$\left| \frac{\underset{\cdot}{\Delta}}{\underset{\cdot}{\Delta P}} \right| = \frac{\omega_s/2H}{\sqrt{[(\omega_s/2H) P_{max} \cos \delta_0 - \omega_1^2]^2 + \omega_1^2 (\omega_s/2H)^2 P_d^2}}$$

which is the desired expression. This expression will now be evaluated for the given conditions.

P_{L0} must be 1.0 in that T_{L0} is 1.0. This requires that $\delta_0 = 30°$ if $P_{max} = 2.0$ (i.e., $P_{max} \sin \delta_0 = P_{L0}$). The undamped natural frequency ω_n is obtained by setting the denominator of the transfer relationship equal to zero with $P_d = 0$. Thus if the fundamental frequency ω_1 of $f(t)$ is at the undamped natural frequency ω_n,

$$\omega_1 = \omega_n = \sqrt{\frac{\omega_s}{2H} P_{max} \cos \delta_0}$$

$$= 6.39 \text{ rad/s}$$

and at this frequency

$$\left| \frac{\Delta}{\Delta P} \right| = \frac{1}{\omega_1 P_d} = \frac{1}{(6.39)(0.04)} = 3.91$$

Thus if $\Delta P = 0.05$, we would have a steady-state sinusoidal displacement of the angle δ from its operating point δ_0 of amplitude:

$$|\Delta| = (0.05)(3.91) = 0.1955 \text{ rad}$$

$$= 11.2°$$

The result suggests that if the feed system is operated such as to produce a fundamental frequency near 6.39 rad/s, we would have a significant oscillation in the angle δ. Consideration of $P_{max} \sin \delta = 2.0 \sin (\delta_0 + \Delta\delta)$ suggests that input power to the machine would vary between about 64 and 132% of its rated value. Such power fluctuations would be intolerable, and sustained operation under such conditions should be avoided. The role of the damping term P_d in reducing the oscillation magnitude is obvious. Increasing the moment of inertia through use of a flywheel is a possible remedial course of action if load torque oscillations at frequencies near the undamped natural frequency (sometimes referred to as the *hunting* frequency) of the machine cannot be avoided. In general, synchronous machines must be applied with caution whenever a possibility exists for sustained power fluctuations at a frequency near their hunting frequency.

EXAMPLE 4.9
A cylindrical-rotor synchronous generator is known to deliver its rated MVA at rated terminal voltage and at a power factor of 0.80 lagging at a power angle δ of 30°. The machine is connected to a large system, the terminals of which may be regarded as an infinite bus. The field excitation is held constant and the prime-mover power is very gradually reduced by 75% and held at this reduced value for an extended period of time. If the prime-mover power is suddenly increased to its original value, determine the maximum power-angle excursion on the first rotor swing following the sudden increase in the prime-mover power.

At the initial operating point, we have on a normalized basis with variables expressed as a fraction of the machine MVA rating:

$$P = 0.8 = P_{max} \sin 30°$$

Thus $P_{max} = 1.6$. If prime-mover power is gradually reduced by 75% without changing the field excitation, a new steady-state operating point will be established:

$$(0.25)(0.8) = 0.2 = P_{max} \sin \delta_0'$$

where $P_{max} = 1.6$. Thus $\delta_0' = 7.18°$. If the prime-mover power is suddenly increased to its original value, we anticipate that the rotor angle will initially increase, moving past the original steady-state angle of 30°. Accelerating power will become negative and the rotor will decelerate. Assuming that it decelerates to a speed below synchronous, the rotor angle will decrease. If the angle decreases below 30°, accelerating power will become positive and the rotor will accelerate. When the rotor speed exceeds the synchronous speed, the angle will again increase. If events occur as we have described them,

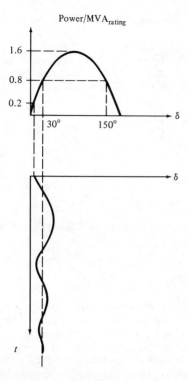

FIGURE 4.44 Example of Rotor Dynamics Following a Sudden Increase in Prime Mover Power from 0.2 to 0.8

damping will cause the rotor ultimately to settle out at the original steady-state angle of 30°. The situation is illustrated in Figure 4.44. It is clear that there will be some limit on the maximum amount of suddenly applied prime-mover power the machine can sustain such that the rotor will, in fact, decelerate to a speed below the synchronous speed on the first swing. The maximum power-angle excursion on the first rotor swing is of interest in that if it exceeds (in this case) an angle of $180° - 30° = 150°$ (see Figure 4.44), the accelerating power would become positive and the rotor would accelerate, resulting in a further increase in angle.

If we neglect damping in (4.101), we can determine a pessimistic estimate of the maximum angle swing. Equation (4.101) with damping neglected can be written as

$$\frac{2H}{\omega_s} \frac{d^2\delta}{dt^2} = P_{pm} - P_{max} \sin \delta$$

$$= P_{acc}$$

where P_{acc} is the accelerating power. In the absence of damping, nonzero accelerating power is converted in accelerating or decelerating the rotor, depending on whether it is positive or negative. It thus increases or decreases the rotor kinetic energy. Since P_{acc} is proportional to the accelerating power T_{acc} (recall that we originally converted an equation in torque to one in power by multiplying through by the rotor speed, assumed to be approximately synchronous) and since $\int T_{acc} \, d\delta$ is work, the change in rotor kinetic energy in moving between two angles δ_1 and δ_2 will be proportional to $\int_{\delta_1}^{\delta_2} P_{acc} \, d\delta$. Figure 4.45 illustrates the conditions of the example. In Figure 4.45 area A is proportional to the increase in rotor kinetic energy as the rotor moves from an angle of 7.18° to an angle

THE SYNCHRONOUS MACHINE

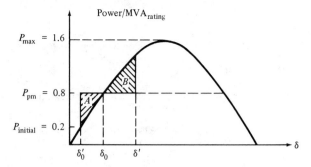

FIGURE 4.45

of 30°. Area B is proportional to the decrease in rotor kinetic energy as the rotor moves from 30° to the angle δ'. By setting area B equal to area A, we can determine the angle δ_{max} where kinetic energy will have returned to the value of kinetic energy corresponding to the synchronous speed. At this point, since the accelerating power is negative and the rotor speed is at the synchronous speed, the angle δ must begin to decrease. Thus, for the example ($\delta'_0 = 7.18° = 0.125$ rad; $\delta_0 = 30° = \pi/6$ rad $= 0.524$ rad), we must have

$$\int_{\delta_0}^{\delta_{max}} (P_{max} \sin \delta - P_{pm}) \, d\delta = \int_{\delta'_0}^{\delta_0} (P_{pm} - P_{max} \sin \delta) \, d\delta$$

which yields

$$P_{pm} \delta_{max} + P_{max} \cos \delta_{max} = P_{pm} \delta'_0 + P_{max} \cos \delta'_0$$

The equation above can be solved for the single unknown δ_{max} by trial and error. We find

$$\delta_{max} \approx 0.9563 \text{ rad} = 54.8°$$

The criterion we have used to determine δ_{max} is sometimes referred to as the *equal-area criterion*. It is useful in understanding the basic phenomenon of machine action in such situations but limited in its application to systems of, at most, two machines. Where it is applied, improved results may be obtained by using a "transient" power-angle curve rather than a steady-state power-angle curve. Reference 4 illustrates some interesting steady-state versus transient power-angle curves.

REFERENCES

[1] Test Procedures for Synchronous Machines, IEEE Standard 115.

[2] R. D. Valentine, J. G. Trasky, and R. D. Rippin, "Load Sharing of Dual Grinding Mills," *IEEE Transactions on Industry Applications,* Vol. IA-13, No. 2, pp. 161–168, March/April 1977.

[3] B. Adkins and R. G. Harley, *The General Theory of Alternating Current Machines.* London: Chapman & Hall Ltd., 1975.

[4] Charles Concordia, *Synchronous Machines Theory and Performance.* Schenectady, N.Y.: The General Electric Company, 1951.

PROBLEMS

4.1 Figure 4.1 illustrates a two-pole three-phase synchronous machine. Sketch the configuration we must have for:
a. A two-pole two-phase machine.
b. A four-pole single-phase machine.
At what speed in rpm must the rotor turn in each of these configurations to develop 60-Hz armature voltages?

4.2 Figure 4.1 illustrates a two-pole three-phase synchronous machine. Suppose that we consider operation with the armature windings open circuited such that $i_a = i_b = i_c = 0$. Field current i_f is set to a value of 10 A dc. A prime mover turns the rotor at a constant speed ω_s. The voltage on the phase a winding under these conditions is found to be $v_a = \sqrt{2} \, (100) \cos \omega_s t$. Find expressions for v_a, v_b, and v_c:
a. For the conditions described above.
b. If the field current is reduced to 5 A.
c. If the field current is reduced to 5 A and the rotor is turned at a constant speed of $0.8 \, \omega_s$.
d. If the field current is set at 10 A, and the rotor is turned at a constant speed ω_s but in the opposite direction.

4.3 A diesel-engine-driven synchronous generator is to be used to supply a 50-Hz system. The synchronous generator is wound for six poles. At what speed in revolutions per minute (rpm) must the shaft of the diesel engine turn, assuming that it is directly coupled to the shaft of the generator?

4.4 Consider the machine described in Example 4.1. For the operating condition described, find the armature I^2r losses. Express the answer as a percentage of the real power (megawatt) output of the machine at this particular operating condition. If the field current is 7500 A dc and the field voltage is 600 V dc at this operating point, find the field I^2r losses expressed as a percentage of synchronous machine real power output.

4.5 Consider the machine described in Example 4.1. Find the field-induced voltage E_f if the machine delivers rated MVA at rated voltage and at unity power factor. Repeat for a power factor of 0.90 leading.

4.6 Figures 4.16 and 4.17 show phasor diagrams illustrating synchronous-machine operation as a generator and as a motor, respectively. Extend the phasor diagrams in each case to show the phasor representing field flux linkages ψ_f, the phasor representing resultant flux linkages ψ_{res}, and the phasor representing flux linkages established by armature current $\psi_s = \psi_{\ell s} + \psi_m$. (See Figure 4.14 for generator operation.) Now assume that the armature leakage reactance is zero such that $\psi_{\ell s} = 0$ and $\psi_s = \psi_m$, and extend the diagram to show space vectors $\hat{H}_s$, $\hat{H}_{sr}$, and $\hat{H}_r$ and the orientation of the field winding axis. (See Figure 4.15 for generator operation.)

4.7 Suppose in Example 4.2 that operation is such that the generator delivers its rated kVA at *unity power factor* with its terminal voltage at rated value. Find the motor field-induced voltage E_{fm}, its terminal voltage V_{sm}, and its power factor. Determine the reactive power for the generator, feeder, and motor, stating in each case whether it is supplied by or consumed by the respective device. Is the motor operating underexcited or overexcited in this situation?

4.8 A three-phase Y-connected synchronous generator rated 1000 kVA and having a rated line-to-line armature terminal voltage of 4.16 kV has a synchronous reactance of 25 Ω. The armature resistance is negligible. The machine delivers 60% of its rated kVA at rated terminal voltage and at a *leading* power factor of 0.80. Find the line-to-line armature terminal voltage if the machine is disconnected from the system such that the armature windings are open circuited. Assume that the field excitation remains constant and that the speed regulator on the prime mover maintains the rotor speed constant at the synchronous speed.

4.9 A three-phase Y-connected synchronous generator rated 1000 kVA and having a rated line-to-line voltage of 4.16 kV has a synchronous reactance of 25 Ω. The armature resistance is negligible. The machine delivers 60% of its rated kVA at rated terminal voltage and at a *leading* power factor of 0.80. (NOTE: This is the same operating condition initially described for Problem 4.8.) What is the maximum possible total three-phase real power that the generator can deliver if the field excitation is held constant and the prime-mover torque is very gradually increased?

4.10 A three-phase, Y-connected, round-rotor synchronous motor is connected to an infinite bus. Thus the armature winding terminal voltages are constant in magnitude, phase angle, and frequency. The shaft load of the motor is removed and the field excitation of the motor is adjusted so that the motor is equivalent at its terminals to a three-phase bank of capacitors drawing currents equal in magnitude to the rated current of the machine. Find the field induced voltage E_f. The synchronous reactance is 7.0 Ω, armature resistance is negligible, line-to-line voltage is 220 V, and rated machine current is 25 A.

4.11 A three-phase, Y-connected, 100-kVA, 460-V cylindrical-rotor synchronous generator has a synchronous reactance of 3.5 Ω and negligible armature resistance. The machine is connected to an infinite bus so that the terminal voltage is constant at rated voltage magnitude. The machine initially is operated so that it supplies its rated kVA at unity power factor. The field current is then very gradually reduced until the machine is just beginning to pull out of synchronism with the system to which it is connected. Find the field induced voltage E_f and armature current I_s just prior to loss of synchronism. Is the machine supplying or consuming reactive power for this condition?

4.12 A three-phase, Y-connected, 1500-kVA, 13.8-kV cylindrical-rotor synchronous generator has a synchronous reactance of 190 Ω and negligible armature resistance. The machine supplies its rated kVA at unity power factor. If a three-phase short circuit occurs at the machine armature winding terminals, find the resulting steady-state armature current. Assume that the prime mover is regulated to turn the shaft of the machine at constant speed regardless of system conditions.

4.13 A three-phase Y-connected cylindrical-rotor synchronous machine is to be synchronized to a power system. The voltage of the power system at the point of connection may be regarded as constant in magnitude, phase angle, and frequency, that is, an infinite bus. Prior to synchronization, the field excitation of the machine is adjusted so that the machine terminal voltage is equal in magnitude to that of the system. If the system voltage is 13.8 kV line-to-line and the machine synchronous reactance is 80 Ω, what is the maximum possible power that the machine can deliver to the system on a steady-state basis following synchronization if the field current remains set at its presynchronization level?

4.14 Equation (4.77) is the steady-state equation for a salient-pole synchronous machine operated as a generator. Figure 4.32 shows a phasor diagram illustrating a typical generator operating condition. Redefine the armature current for motor operation and write the steady-state equation for a salient-pole synchronous machine operated as a motor. Sketch a phasor diagram illustrating a typical motor operating condition if the power factor is lagging. Indicate on the diagram the voltage E_q defined in Example 4.3.

4.15 A 1000-hp Y-connected salient-pole synchronous *motor* has a rated armature terminal voltage of 4160 V. The direct-axis synchronous reactance is 24 Ω, the quadrature-axis synchronous reactance is 15 Ω, and the armature resistance is negligible. The motor efficiency is 92%. If the motor develops its rated shaft power output of 1000 hp at unity power factor, what is the field-induced voltage $E_f = E_f \underline{/\delta}$?

4.16 The general power relationship for a salient-pole synchronous machine with negligible armature resistance is

$$P = \frac{V_s E_f}{x_d} \sin \delta + \frac{V_s^2 (x_d - x_q)}{2 x_d x_q} \sin 2\delta \quad \text{(per phase)}$$

Suppose that a three-phase Y-connected salient-pole synchronous motor has the following parameters:

Armature voltage	4160 V (line-to-line)
Power rating	1000 hp at unity power factor
Direct-axis synchronous reactance	$x_d = 25 \ \Omega$
Quadrature-axis synchronous reactance	$x_q = 15 \ \Omega$
Armature resistance	negligible
Losses	negligible

Find the maximum possible steady-state real power that the machine can convert at rated armature terminal voltage if the field current is reduced to zero.

4.17 Example 4.4 describes the open-circuit slip test, which can be used to determine values of the direct-axis synchronous reactance x_d and quadrature-axis synchronous reactance x_q for a salient-pole machine. Suppose that, in Example 4.4, the prime mover drives the rotor at a speed of $\omega_r = 1.01\omega_s$, where ω_s is the synchronous speed. Find an expression for torque (as a function of time) that the prime mover must develop in order to run the test. Sketch the torque waveform for the data of Example 4.4, indicating critical numerical values if the machine is a 24-pole machine.

4.18 The general real power relationship for a salient-pole synchronous machine with negligible armature resistance is

$$P = \frac{V_s E_f}{x_d} \sin \delta + \frac{V_s^2 (x_d - x_q)}{2 x_d x_q} \sin 2\delta$$

Find an expression for the reactive power Q.

4.19 A six-pole cylindrical-rotor synchronous motor is operated in such a way that its rotor makes an angle with respect to a fixed reference that can be described as

$$\theta_m(t) = \omega_m t + \Delta\theta_m \sin \frac{2\pi}{8} t \quad \text{rad}$$

where ω_m is the synchronous speed in mechanical radians per second and $\Delta\theta_m$ is the amplitude of a $\frac{1}{8}$-cycle per second variation in the mechanical angle. In the absence of any angle variation (i.e., if $\Delta\theta_m = 0$), the motor is known to develop its rated power at the rated field excitation at a power angle $\delta = 30°$ (electrical degrees). Determine the maximum and minimum values of field excitation voltage (expressed as a percentage of the rated value) which would be required if the excitation voltage were to be varied automatically by a control scheme that kept the real power constant when $\Delta\theta_m = 0.0087$ rad (0.5 degree).

4.20 A four-pole 100-MVA synchronous generator has an inertia constant H of 6 s. What is the moment of inertia J in kg · m²?

4.21 A four-pole, 100-kVA synchronous machine has a moment of inertia $J = 30$ kg · m². What is the inertia constant H for the machine?

4.22 Extend Example 4.6 to determine the change in the angle δ over the fault duration. That is, if $\delta(0) = \delta_0$, what is $\Delta\delta = \delta(t') - \delta_0$, where $t' = 8/60$ s? Assume the same data as those used in Example 4.6.

CHAPTER 5

The DC Machine

INTRODUCTION

The versatility of the direct-current machine is its greatest asset. The conventional dc machine configuration, introduced in Chapter 2, can be employed in a variety of ways either as a motor or as a generator. Examples of dc machines are shown in Figures 5.1 and 5.2. Depending on the manner in which the field and armature circuits are supplied, motor torque–speed characteristics or generator voltage–current characteristics vary considerably and in general can be tailored for a variety of applications. The dc motor features efficient operation over a wide speed range and is suitable for applications requiring precise speed or position control. Machines range in size from miniature permanent-magnet motors to machines rated for continuous operation at several thousand horsepower.

The advantages of the dc machine must be weighed against its relatively high initial cost and the maintenance problems associated with its brush-commutator system. Obviously, for a dc motor, a source of dc power is required. The advantages of ac over dc for power transmission have resulted in the widespread use and availability of ac power. This must be converted in order to supply a

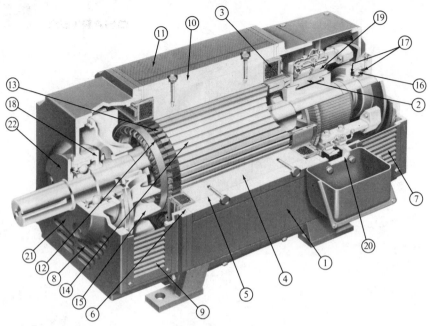

1. Frame laminations	10. Frame yoke	16. Ball bearings
2. Commutator assembly	11. Welded seam	17. Grease reservoirs
3. Field coils	12. Armature coils	18. Grease cap
4. Field poles	13. Slot wedges	19. Brush holders
5. Field pole seats	14. Rotor laminations	20. Commutator bracket
6. End brackets	15. Cooling fan	21. Grease plug
7-9. Ventilation louvers		22. Bearing cap inspection plate

FIGURE 5.1 Large Integral Horsepower DC Machine: The Frame Is Laminated to Reduce Eddy Current Losses Resulting from Use of a Static Power Supply (Courtesy of Reliance Electric)

FIGURE 5.2 Small DC Torque Motor, Measuring 1 Inch on a Side and Weighing 2.93 Ounces, That Could Be Used as a Servomotor in Instrumentation Devices (Courtesy of Inland Motors Specialty Products Division, Kollmorgen Corporation.)

dc motor. A motor–generator set is one option for performing this task. An ac-powered induction or synchronous machine can be used to turn the shaft of a dc machine operated as a generator. The dc generator can supply both the field and armature circuits of a dc motor. Alternatively, solid-state devices can be used in rectifier systems to provide sources of dc power. We examine such systems in greater detail in this chapter.

FUNDAMENTAL DC MACHINE RELATIONSHIPS

The conventional dc machine was introduced in Chapter 2 in Example 2.3. A four-pole machine was illustrated conceptually in Figure 2.6. For a sufficient number of uniformly distributed armature conductors, the commutating action of the machine results in a stationary pattern of armature currents which is independent of rotor position. This stationary distribution of armature currents results in a stationary magnetic field whose main axis is perpendicular to the main axis of the magnetic field due to field current. This is illustrated in Figure 5.3 (and in Figure 2.35).

The fixed pattern of armature currents is maintained by proper commutator action. A single-coil elementary commutator system is shown in Figure 5.4. The winding starts on one commutator segment and goes down the rotor (into the page) on one side (a) and comes back (out of the page) on the other side (a'). Several turns in the same set of slots might be made (the end turns are not shown) before the winding is finally terminated on the other commutator segment. It is easy to see that if the current i_a is constant and positive, the current in a slot positioned to the right of the vertical reference line will always be directed into the page except perhaps for a small angle where the coil is undergoing commutation. Similarly, the current in a slot positioned to the left of the

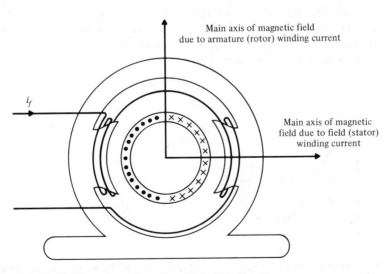

FIGURE 5.3 Field Alignments in a Two-Pole DC Machine

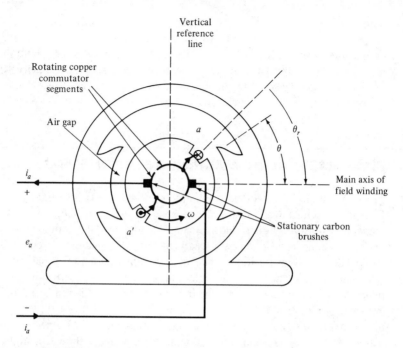

FIGURE 5.4 A Single-Coil Elementary Commutator System

reference line will be directed out of the page. A problem associated with commutation is evident here in that the brushes short the two commutator segments when the coil is undergoing commutation. Sparking generally occurs. Minimization of sparking at the brushes is a major objective in dc machine design.

Ignoring for the moment problems associated with commutation, we can determine the open-circuit voltage at the brushes using the results of Example 2.4. To illustrate, suppose that the radial component of magnetic flux density due to field current is sinusoidally distributed as shown in Figure 5.5. (A more realistic distribution is shown in Figure 2.7. A sinusoidal distribution is simpler to deal with for the points about to be made.)

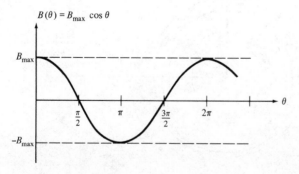

FIGURE 5.5 Assumed Magnetic Flux Density due to Field Current for the Simplified Machine of Figure 5.4

THE DC MACHINE

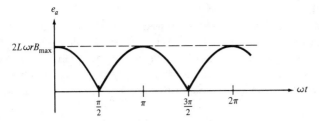

FIGURE 5.6 Open-Circuit Voltage at the Brushes of the Simplified Machine of Figure 5.4

As shown in Example 2.4, the voltage induced in a single-turn winding is given by

$$e = 2L\omega rB(\theta_r) \qquad (5.1)$$

$$= 2L\omega rB(\omega t + \theta_0)$$

where L is the axial length, ω the speed of rotation, r the rotor radius, and θ_0 the initial rotor angle. If we assume constant speed, the waveform of the open-circuit voltage across the commutator segments would have the same shape as the assumed flux density distribution shown in Figure 5.5. The voltage across the brushes would be rectified and for a single-turn winding would be as shown in Figure 5.6, where the initial angle θ_0 is arbitrarily taken as zero.

The average value of the open-circuit voltage can be increased and the "ripple" reduced by adding more windings to the rotor and connecting them in series. Figure 5.7 illustrates one technique for a multiple-coil winding. Referring to Figure 5.7, consider current from the right side brush fed into the bottom winding of slot 1. This winding goes down (into the page) slot 1, returns in the

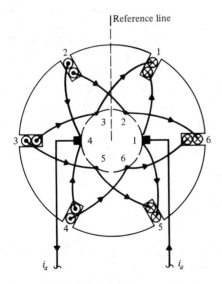

FIGURE 5.7 Multiple-Coil Commutator System

top of slot 4, and terminates on commutator segment 6. It is in series with a similar winding which goes down the bottom of slot 6, returns in the top of slot 3, and terminates on commutator segment 5. A similar winding in slots 5 and 2 is in series, finally terminating on commutator segment 4, which is in contact with the left-side brush. Note that a second circuit exists which is similar to and in parallel with the circuit just described. Figure 5.7 illustrates only one technique for realizing an armature current distribution which for a sufficient number of slots approaches the ideal distribution shown in Figure 5.3. Other possibilities are described in references 1 and 2.

EXAMPLE 5.1   Determine the open-circuit armature voltage waveform for the multiple-coil commutator system illustrated in Figure 5.7. Assume single-turn armature windings and a sinusoidal field flux density distribution.

Consider the windings in the bottom of slots 1, 5, and 6 (top of slots 2, 3, and 4). Each of these windings will have an open-circuit voltage waveform similar to that illustrated in Figure 5.6 but phase displaced based on the angular displacement between them ($60°$ or $\pi/3$ radians in this case). Note that the remaining windings yield a circuit effectively in parallel which need not be considered in determining the open-circuit voltage. Thus the voltages of three windings must be added in series to obtain the open-circuit armature voltage at the brushes. The addition is graphically illustrated in Figure 5.8. The example shows that the "ripple" can be significantly reduced with as few as three windings in series. Most dc machines have many armature slots and develop nearly ripple-free voltage waveforms. In small machines where other considerations limit the number of slots, the ripple in voltage waveforms (and in torque waveforms) can be significant.

For a sufficiently large number of armature slots distributed around the periphery of the rotor, we approach the uniform armature current distribution suggested in Figure 5.3. The armature winding and field-winding-induced fields are illustrated in the developed drawing of a generalized dc machine shown in Figure 5.9. The mmf wave due to a uniform distribution of armature currents is triangular. The flux density wave due to armature current would have the same shape as this mmf wave were it not for the increased air gap (and hence reluctance) in the interpolar regions, which tends to reduce it.

To determine the machine voltage (or torque) when the armature current is nonzero (i.e., for other than open-circuit conditions), a resultant flux density wave should be considered. This resultant flux density wave should include the effects of both the field and armature currents. As shown in Example 2.3, the torque depends on the flux per pole Φ (or equivalently on the average flux density). We will show shortly that the induced armature voltage also depends on the flux per pole. If saturation is neglected, the flux per pole obtained by integrating the flux density over an appropriate surface will be that due to the field current alone. For example, the flux per pole for a P-pole machine can be expressed as

$$\Phi = \int_{-\pi/P}^{\pi/P} B(\theta)Lr \, d\theta \tag{5.2}$$

If $B(\theta)$ is the resultant flux density wave due to the field and armature effects and saturation is neglected, we have

$$B(\theta) = B_a(\theta) + B_f(\theta) \tag{5.3}$$

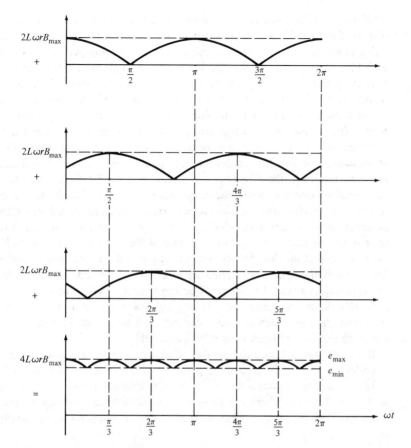

FIGURE 5.8 Open-Circuit Voltage at the Brushes of the Machine of Figure 5.7.
Peak voltage $e_{max} = 4L\omega r B_{max} = \dfrac{2}{\sqrt{3}} \, e_{min} \approx 1.15 e_{min}$

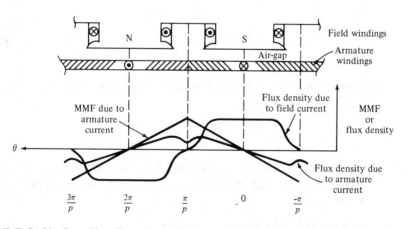

FIGURE 5.9 Air-Gap Flux Densities due to Field and Armature Circuits (the Armature Winding Is Uniformly Distributed Resulting in a Triangular MMF Wave)

where $B_a(\theta)$ is the flux density due to armature current and $B_f(\theta)$ is the flux density due to field current. If (5.3) is substituted in (5.2) and the integral is written as a sum of two integrals, one of which is the flux per pole due to $B_a(\theta)$, and the other of which is the flux per pole due to $B_f(\theta)$, the flux per pole due to $B_a(\theta)$ will be zero. This can be surmised by consideration of Figure 5.9. In reality, saturation effects require modification of (5.3). Because of saturation, the increase in the flux per pole due to $B_a(\theta)$ is less than the decrease due to $B_a(\theta)$. This causes the resultant flux per pole to be reduced somewhat by the armature current. This effect is sometimes referred to as the *demagnetizing* effect of armature reaction. It degrades machine performance somewhat and steps are taken to minimize the effect in a well-designed machine. In large machines additional windings are sometimes used to compensate for the effects of armature reaction. Typically, these additional windings are placed on the stator and are in series with the armature winding. They are positioned in such a way that their mmf distribution tends to cancel that of the armature winding itself. The net result is that the flux density is essentially that due to the field circuit alone. The demagnetizing effects of armature reaction and sparking at the brushes can be significantly reduced by properly designed compensating windings. We will generally assume that machines are properly compensated in what follows and will later discuss techniques that can be used to account for the effects of armature reaction where it cannot be neglected.

If we assume that the total number of axially aligned armature conductors Z is uniformly distributed around the periphery of the rotor as suggested in Figure 5.3, the number of conductors in series in a differential element of arc length will be $(Z/2\pi a)\,d\theta$, where a is the number of parallel paths through the armature winding. The voltage induced in an axially aligned conductor (see Example 2.4) is

$$e_s = B(\theta_r)L\omega_r \tag{5.4}$$

The voltage induced in a differential element of arc length $d\theta_r$ is obtained by multiplying e_s by the number of conductors in series in the arc length:

$$de = \frac{Z}{2\pi a} e_s\, d\theta_r$$

$$= \frac{Z}{2\pi a} B(\theta_r)L\omega r\, d\theta_r \tag{5.5}$$

The voltage induced per pole is obtained by integrating (5.5) over an arc length corresponding to one pole:

$$e_p = \int_{-\pi/P}^{+\pi/P} de = \frac{Z}{2\pi a}\omega\Phi \tag{5.6}$$

where Φ is the flux per pole as given by (5.2). The total voltage at the brushes, neglecting resistance and brush drops, is obtained by multiplying by the number of poles:

$$e_a = \frac{PZ}{2\pi a}\Phi\omega \tag{5.7}$$

This expression should be compared to the expression for torque similarly derived in Example 2.3:

$$T = \frac{PZ}{2\pi a} \Phi i_a \tag{5.8}$$

Equations (5.7) and (5.8) are fundamental to dc machinery. Note that electrical power at the armature terminals is the same as the mechanical power converted:

$$p_e = e_a i_a = \frac{PZ}{2\pi a} \Phi \omega i_a = T\omega = p_m \tag{5.9}$$

This is to be expected in that all cases of losses have been neglected in the development thus far.

EXAMPLE 5.2 A certain manufacturer of small permanent-magnet servomotors states that one of its products will develop 71.0 oz · in of torque per ampere of armature current. Find a suitable constant (sometimes referred to as the *back emf constant*) that could be used to relate induced armature voltage to speed of rotation.

Equations (5.7) and (5.8) provide the answer. In this case the flux Φ is established by permanent magnets and is thus a constant. Thus

$$T = \frac{PZ}{2\pi a} \Phi i_a = K_m i_a$$

Similarly,

$$e_a = \frac{PZ}{2\pi a} \Phi \omega = K_m \omega$$

The desired constant is K_m, which from the torque relationship is simply torque per ampere. If we choose to express e_a in volts and ω in rad/s, we must express torque in newton-meters (N · m).

$$1 \text{ oz} = 0.27801 \text{ N}$$

$$1 \text{ in} = 0.0254 \text{ m}$$

Thus

$$K_m = 71.0(0.27801)(0.0254)$$

$$= 0.501 \text{ V per rad/s}$$

We now complete development of a suitable analytical model for the dc machine. The armature circuit will have resistance associated with the conductors themselves, the brushes, contact surfaces between brushes and commutator segments, and leads out to the motor terminals. In many applications it is sufficient to combine these effects into a single equivalent resistance in series with the armature voltage. It is important to recognize that this is a simplified way of representing all armature circuit losses. The armature self-inductance also appears in series with the armature voltage. The field circuit may be treated as an inductance in series with a resistance. Figure 5.10 illustrates equivalent circuits for the armature and field of a dc machine. In Figure 5.10, the armature current has been assigned in such a way that it will be positive for operation of the machine as a motor. For analysis of the machine as a generator it is convenient to assign current in the opposite direction.

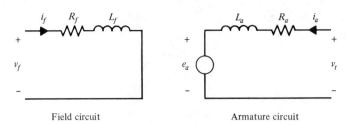

Field circuit   Armature circuit

FIGURE 5.10 Equivalent Circuits for the Armature and Field of a DC Machine

The flux per pole Φ depends on the field current i_f. If saturation is neglected, it is linearly related to the field current by the parameters of the magnetic circuit associated with the field:

$$\Phi = \frac{N_f i_f}{\mathcal{R}_f} \tag{5.10}$$

In (5.10) N_f is the number of turns on the field winding, i_f the field current, and $\mathcal{R}_f$ the magnetic circuit reluctance. The relationship between flux per pole and field current can be determined indirectly by turning the machine shaft at a constant speed and plotting the open-circuit armature voltage as a function of field current. A typical curve is illustrated in Figure 5.11. Note that the curve essentially relates the flux Φ to the field current i_f in that for the condition described, v_t and Φ are related by a constant. The dashed line is the air-gap line, determined primarily by the reluctance of the air gap [see (5.10)]. The nonzero value of v_t at $i_f = 0$ is due to residual magnetism. Saturation limits the curve at relatively high values of i_f.

The equation of motion for a motor can be written as

$$J \frac{d\omega}{dt} = T - T_L - T_D \tag{5.11}$$

where J is the moment of inertia, T the developed motor torque given by (5.8), T_L the load torque, and T_D the torque due to damping. For a generator, the equation of motion can be written as

$$J \frac{d\omega}{dt} = T_{pm} - T - T_D \tag{5.12}$$

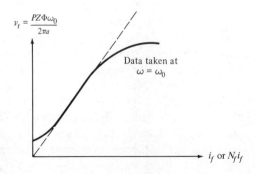

FIGURE 5.11 Typical DC Machine Magnetization Curve

where T_{pm} is the prime-mover torque and T is the opposing torque developed by the machine and given by (5.8).

The equivalent circuits of Figure 5.10, voltage and torque relationships (5.7) and (5.8), a suitable relationship between the flux per pole Φ and the field current i_f, and an equation of motion for the rotor are the analytical tools required for dc machine analysis.

DC MACHINE STEADY-STATE CHARACTERISTICS

If all voltages and currents are treated as constants for a dc machine operating in a steady-state condition, voltages developed across armature and field inductances shown in the equivalent circuits of Figure 5.10 will be zero. Rotor acceleration is zero, which implies that the summation of torques on the right-hand side of (5.11) [or (5.12)] must be zero. Thus for steady-state analysis,

$$v_f = i_f R_f \qquad (5.13a)$$

$$v_t = +i_a R_a + e_a \qquad \text{(for a motor)} \qquad (5.13b)$$

$$v_t = -i_a R_a + e_a \qquad \text{(for a generator)} \qquad (5.13c)$$

$$T = T_L + T_D \qquad \text{(for a motor)} \qquad (5.13d)$$

$$T = T_{pm} - T_D \qquad \text{(for a generator)} \qquad (5.13e)$$

The voltage and torque relationships given by (5.7) and (5.8) and a relationship between the flux per pole Φ and the field current i_f are required to complete the model for steady-state analysis.

EXAMPLE 5.3   A dc generator is operated with a separately excited field winding. The machine is rated 1200 kW, 500 V at 600 rpm. The armature and field circuit resistances are

$$R_a = 0.10 \ \Omega$$

$$R_f = 20.0 \ \Omega$$

The magnetization curve for the machine at the rated rpm is given in Figure 5.12. The field winding of the machine has 200 turns per pole. The prime mover for the generator is a 12-pole induction machine of the same power rating with a rated slip of 0.05.

The terminal voltage of the dc generator can be maintained constant as the load current (armature current) varies by properly adjusting the terminal voltage of the field winding circuit. Determine the required field voltage at no load and at full load if the generator terminal voltage is to be the rated voltage for each of the two conditions.

The machine terminal voltage differs from the generated voltage e_a only by the armature resistance drop. At no load, the armature current is zero and $e_a = v_t = 500$ V. The induction machine used as a prime mover must overcome only the combined no-load losses of itself and the generator. It would thus turn at nearly synchronous speed $n = (2/p)3600 = 600$ rpm, assuming it is supplied from a 60-Hz source. From the curve of Figure 5.12, we see that an mmf of 5000 A · t per pole is required. The required field current is thus

$$i_f = \frac{N_f i_f}{N_f} = \frac{5000 \ \text{A} \cdot \text{t}}{200 \ \text{t}} = 25 \ \text{A}$$

Thus the required field voltage at no load is

$$v_f = i_f R_f$$

$$= 25(20) = 500 \text{ V}$$

At full load, the armature current is

$$\frac{1200 \text{ kW}}{500 \text{ V}} = 2400 \text{ A}$$

The induced voltage must be

$$e_a = v_t + i_a R_a$$

$$= 500 + (2400)(0.010) = 524 \text{ V}$$

The prime mover under rated load will turn at a speed close to the rated speed:

$$n = (1 - s)n_s$$

$$= (1 - 0.05)(600) = 570 \text{ rpm}$$

We thus have a requirement for an induced voltage of 524 V at 570 rpm. The curve of Figure 5.12 is based on a speed of 600 rpm. At 600 rpm the induced voltage would be

$$e_a' = \frac{n'}{n} e_a$$

$$= \frac{600}{570}(524) = 552$$

From the curve of Figure 5.12, we see that an mmf of roughly 6000 A · t per pole is required. The required field current is thus

$$i_f = \frac{6000 \text{ A} \cdot \text{t}}{200 \text{ t}} = 30 \text{ A}$$

and the required field voltage at full load is

$$v_f = i_f R_f$$

$$= 30(20) = 600 \text{ V}$$

The example illustrates how adjustments to the field circuit of the machine can compensate for armature resistance voltage drops and voltage drop due to

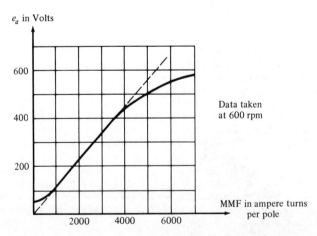

FIGURE 5.12 Magnetization Curve for the 1200-kW 500-V Machine of Example 5.3

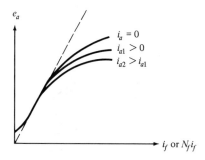

FIGURE 5.13 Magnetization Curves for a DC Machine with Significant Demagnetization due to Armature Reaction Effects.

prime-mover speed reduction. Often the adjustment to the field is accomplished using a field rheostat in series with the field. Because the field circuit is a relatively low power circuit, the field rheostat control does not appreciably increase overall losses. The field circuit can be supplied by the armature circuit if desired by simply connecting it across the terminals of the armature. The machine is then said to be self-excited. Such a configuration is referred to as a *shunt machine* in that the field circuit is in shunt with (parallel with) the armature. We will examine this and other configurations shortly.

If the machine is not compensated for armature reaction effects, it may be necessary to use a family of magnetization curves with armature current as a parameter. This is illustrated in Figure 5.13. Such curves can be determined indirectly by making measurements of current and voltage at the terminals of the armature winding under various loading conditions. A calculated value for the voltage drop across the armature winding resistance is added to the armature terminal voltage to obtain the voltage e_a assuming that the machine is operated as a generator during the tests. As stated earlier, a properly compensated machine will exhibit negligible armature reaction effects.

Examination of generator terminal characteristics for steady-state operation yields additional insights. To simplify the notation, let $K = PZ/2\pi a$. For the separately excited machine, terminal characteristics are as illustrated in Figure 5.14. The intercept e_a on the voltage axis can be varied by changing the prime-

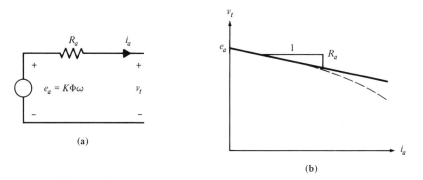

FIGURE 5.14 Separately Excited Generator Characteristics at Constant Speed

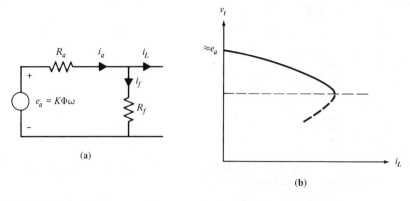

(a)

(b)

FIGURE 5.15 Shunt Generator Characteristics at Constant Speed

mover speed or by varying the machine field current. The voltage essentially falls off due to the armature resistance voltage drop. The dashed line shows the additional voltage drop that might occur due to the demagnetizing effects of armature reaction.

Terminal characteristics for a dc shunt machine are illustrated in Figure 5.15. The initial slope of the curve is somewhat steeper than the armature-resistance-determined slope of the separately excited machine. In this case, the terminal voltage is reduced by the armature resistance drop and by the reduction in e_a due to reduction in the field flux. The curve is concave downward and a multivalued function of the line current. This shape can be explained by considering the magnetization curve for the machine with the field resistance line superimposed as shown in Figure 5.16, where the machine is assumed to be fully compensated such that the demagnetizing effects of armature reaction are negligible.

A field winding designed for shunt connection will have a relatively large number of turns and consist of conductors of relatively small cross-sectional area. Typically, the current i_f in such a field winding will be much less than the

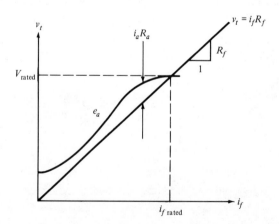

FIGURE 5.16 Magnetization Curve with Field Resistance Line Superimposed

THE DC MACHINE

rated armature current (perhaps 1 to 2% of the rated armature current). The no load ($i_L = 0$) operating point for the shunt generator is the point where the two curves of Figure 5.16 intersect if we neglect the small armature resistance drop due to the rated field current through the armature winding. As the armature current i_a increases, the operating point moves to the left, with the difference in the two curves being equal to the armature resistance voltage drop. The terminal voltage and field current decrease. The line current $i_L = i_a - i_f$ is approximately equal to the armature current for appreciable values of armature current. Note that terminal voltage always decreases as the operating point moves to the left, whereas the armature current increases to a maximum, then decreases, yielding the curve illustrated in Figure 5.15). The machine would typically be operated on the upper portion of the curve, where the voltage falls off with increasing line current.

The curves of Figure 5.16 can be used to explain the open-circuit ($i_L = 0$) voltage *buildup* of the machine. If the rotor of the machine is turned at rated speed by a prime mover when the field current is initially zero, the induced voltage e_a will be nonzero due to residual magnetism. The difference between e_a and the shunt field resistance drop i_fR_f, which is illustrated graphically in Figure 5.16, must be

$$e_a - i_fR_f = (L_a + L_f)\frac{di_f}{dt} + R_ai_f \approx L_f\frac{di_f}{dt} \tag{5.14}$$

neglecting the small voltage drop due to field current through the armature winding resistance and assuming that L_f is much greater than L_a. This voltage difference and hence time rate of change of the field current i_f will be positive for all points to the left of the point where the curves intersect. The field current, field flux, and terminal voltage increase to rated values at the point where the curves intersect. If the field resistance is too high, the point of intersection may be at a value of the terminal voltage considerably less than rated. If the flux due to the field current does not add to the flux due to residual magnetism, the buildup will not occur. This problem can be corrected by reversing the terminals of the shunt field winding.

Terminal characteristics for a dc series machine are illustrated in Figure 5.17. The field winding for a series machine will differ considerably from that of a

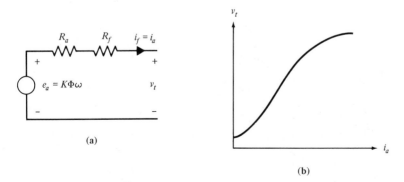

(a)

(b)

FIGURE 5.17 Series Generator Characteristics at Constant Speed

shunt field winding. The series winding must carry rated armature current. Conductors must be of relatively large cross-sectional area and fewer turns will be required to establish rated field mmf. Assuming that armature and field winding resistances are relatively small, the terminal characteristics will be similar to the magnetization curve of the generator. Stability problems can occur if the generator is operated where the slope of the terminal characteristic is positive. A *drooping* voltage characteristic is generally preferable and series generators are not frequently used.

Series field windings are often used to provide compensating effects in separately excited or shunt-field machines. In such cases the field mmf consists of an mmf due to field current and an mmf due to armature current flowing in a series winding wrapped around the pole pieces. It is possible to obtain very flat terminal voltage characteristics with machines having both shunt and series fields. Such machines are referred to as *compound machines*. Generally speaking, their characteristics lie somewhere between those of a shunt machine and those of a series machine.

Speed versus torque characteristics are of interest for dc machines operated as motors. Consider a separately excited motor. The armature terminal voltage equation is

$$v_t = i_a R_a + e_a \tag{5.15}$$

Substituting for e_a and i_a using (5.7) and (5.8) with $PZ/2\pi a = K$ gives us

$$v_t = \frac{R_a}{K\Phi} T + K\Phi\omega \tag{5.16}$$

Now solve for ω:

$$\omega = -\frac{R_a}{(K\Phi)^2} T + \frac{v_t}{K\Phi} \tag{5.17}$$

Equation (5.17) yields a speed–torque characteristic where armature terminal voltage v_t and field flux per pole Φ are parameters. This is illustrated in Figure 5.18.

If the terminal voltage is increased, the intercept $v_t/K\Phi$ on the speed axis increases and the slope of the line remains the same. A family of parallel lines

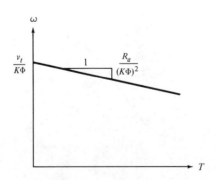

FIGURE 5.18 Speed–Torque Curve for a Separately Excited Motor

THE DC MACHINE

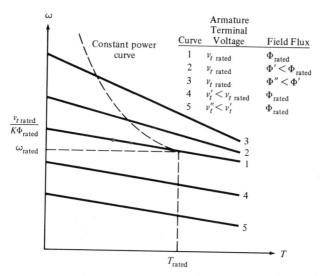

FIGURE 5.19 Speed–Torque Curves for a Separately Excited Motor Illustrating Effects of Varying Armature Voltage and Field Flux

is obtained by varying the armature terminal voltage. If the field flux is decreased by reducing the field voltage or increasing the field rheostat resistance in the separately excited field, the intercept on the speed axis increases and the magnitude of the slope of the line increases. These changes are illustrated in Figure 5.19. In a well-designed machine, the drop in speed in going from no load ($T = 0$) to full load ($T = T_{rated}$) along the rated-voltage rated-flux curve is typically small, say 2 to 5%. For speed changes below the rated speed, armature voltage control is used. For speed changes above the base speed, *field weakening* is used. Maximum speeds are typically limited to four times rated speed. Minimum speeds are generally at least 10% of the rated speed. In the region where armature voltage control is used, the field flux remains constant and armature current restrictions imply a torque limit for continuous operation. That is, we must have $T \leq T_{rated} = K\Phi_{rated}i_{arated}$. In the region where field weakening is used, the terminal voltage remains constant and armature current restrictions imply a power limit for continuous operation. That is, we must have $P < P_{rated} = v_{trated}i_{arated}$.

The actual behavior of the machine for either armature voltage or field circuit control will depend on the speed–torque characteristics of the load. Superimposing load characteristics on motor characteristics permits a fuller appreciation of the effects of various adjustments that might be made. When the machine is lightly loaded, care must be taken not to reduce the field too much since overspeeds can occur. The rotor should not be permitted to reach a speed close to its *bursting speed,* where centrifugal forces may cause it to come apart.

A typical dc drive system is illustrated in Figure 5.20. The system illustrated in Figure 5.20 is based on the versatility achieved with a separately excited motor. The rectifier/controller can be designed to supply separate field and armature circuits in order to realize the two modes of speed control. Rectifier/controller packages often use solid-state technology, where thyristors are

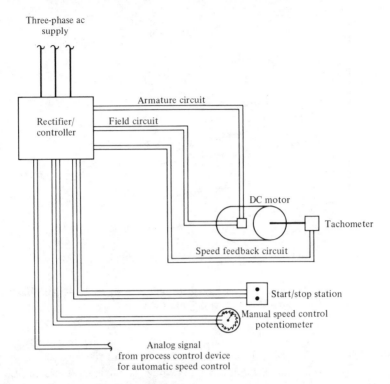

FIGURE 5.20 Typical DC Drive System

used for rectification and control. A typical system is illustrated in Figure 5.21. Drive systems can be adapted to provide a significant range of protective and control features. The armature current is generally monitored and limited. A short time limit of 200% of the rated armature current is typical. This prevents excessive armature current and torque during startup when the field-induced voltage e_a is low. This can be achieved by varying the firing angle on thyristors in such a way as to reduce the average armature terminal voltage during startup. Earlier, less sophisticated systems inserted resistance into the armature circuit to limit the armature current during startup. The additional resistance was then switched out sometimes in several steps as the machine came up to speed. The armature current at sustained speeds should be maintained at the rated armature current or less and drive systems typically have overload protection to prevent motor overheating. Loss-of-field protection should be provided. If the machine is lightly loaded and the field circuit is interrupted, the field flux is reduced to residual magnetism levels and potentially damaging overspeeds can occur. If the field current falls below a predetermined level, the armature circuit should be interrupted. Automatic changeover from armature voltage control to field circuit control should be provided. When precise speed control is required, a tachometer is used to furnish the necessary feedback signal. In some cases it may be desirable to control speed on the basis of one or more process variables. An analog signal from a control device is sometimes used in automatic-speed-control schemes.

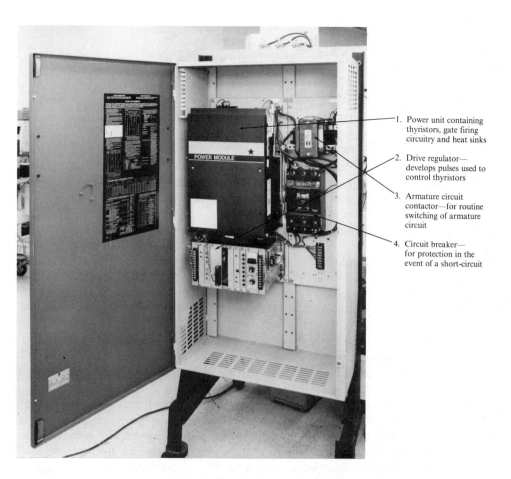

1. Power unit containing thyristors, gate firing circuitry and heat sinks

2. Drive regulator—develops pulses used to control thyristors

3. Armature circuit contactor—for routine switching of armature circuit

4. Circuit breaker—for protection in the event of a short-circuit

FIGURE 5.21 100-HP Solid-State DC Drive Controller and DC Motor (Courtesy of Reliance Electric)

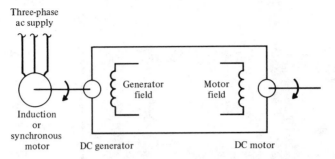

Three-phase
ac supply

Induction
or
synchronous
motor

DC generator

Generator
field

Motor
field

DC motor

FIGURE 5.22 Speed Control of a DC Motor Using a Motor–Generator Set

Motor–generator sets can be used to supply a dc motor as illustrated in Figure 5.22. Armature terminal voltage control of the dc motor can be achieved by varying the field of the dc generator. For speed changes above the motor base speed, adjustments can be made to the field of the dc motor. Solid-state drive systems are often more economical, but motor–generator sets provide an extremely simple and flexible system for realizing the two modes of speed control. Furthermore, the stored inertial energy in the rotating masses of a motor–generator set (which can be increased using a flywheel) tends to level out load-power fluctuations. That is, ac input power to the system can be relatively constant, with fluctuations about average load power supplied by increases and decreases of stored inertial energy.

Dc series motors have a significantly different speed–torque characteristic, as illustrated in Figure 5.23. As in the case of the series generator, the field is in series with the armature and must be capable of carrying the rated armature current. At low values of torque, the armature current and field flux are low. The speed must be relatively high so that the armature-induced voltage e_a matches terminal voltage at low values of field flux. As torque increases, the armature current and field flux increase, requiring a drop in machine speed to maintain the armature-induced voltage. Speed falls off quickly with increases in torque, and speed regulation is poor. The high starting torque makes the machine attractive for applications such as cranes, hoists, and car dumpers. Series motors are often used as traction motors. Automobile starter motors are generally series-wound motors.

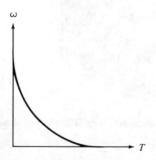

FIGURE 5.23 Speed Versus Torque for a DC Series Motor

THE DC MACHINE

A series-wound motor can be designed to operate on either ac or dc. Such motors are referred to as *universal series motors* and are designed to have similar characteristics whether operated on ac or dc. Fractional-horsepower universal motors can operate at relatively high speed, say 10,000 rpm, whereas induction machines are limited to 3600 rpm with a 60-Hz supply. Because of this, they tend to have much higher power ratings for total weight (and cost) than those of induction machines of comparable rating.

As in the case of generators, the dc motor characteristics can be altered significantly by combining the effects of shunt and series fields. The characteristics of compound-wound motors lie somewhere between those of a separately excited motor and those of a series motor. This versatility of the dc machine is one of its greatest attributes.

5.4

DC MACHINE DYNAMIC ANALYSIS

Dynamic analysis is required when system behavior as a function of time is desired. Generally, we seek the response of a system to a change in operating conditions. For example, changes in supply voltage or load torque cause a dc motor to respond in a dynamic way. In some cases we are simply moving from one steady-state operating point to another, and a dynamic analysis reveals how the change occurs.

EXAMPLE 5.4   A 500-hp 250-V separately excited dc motor draws an armature current of 1400 A when driving a particular load. Determine the motor speed variation as a function of time if the armature voltage is abruptly reduced by 20%. Assume that the load torque remains constant. The motor data are

$$R_a = 0.0035 \ \Omega$$

$$L_a = 0.000175 \ \text{H}$$

$$K_m = K\Phi = 2.65 \ \text{V per rad/s}$$

$$J = 74.6 \ \text{kg} \cdot \text{m}^2$$

rotational losses: 5.4 kW at 900 rpm

If the load torque remains constant, the load torque–speed curve will be a vertical line on a set of characteristics such as those illustrated in Figure 5.19. A 20% drop in voltage will result in a new steady-state speed that is roughly 20% below the initial speed. The damping torque T_D due to rotational losses will affect the final steady-state speed only slightly, assuming that it is significantly less than the load torque. We will assume that the damping torque increases linearly with speed:

$$T_D = B\omega$$

The rotational losses would then be given by

$$P_{\text{rot}} = B\omega^2 = 5.4 \ \text{kW at 900 rpm}$$

The constant B must then be

$$B = \frac{5.4(10^3)}{[2\pi(900/60)]^2} = 0.608 \ \text{N} \cdot \text{m per rad/s}$$

The armature terminal voltage equation for dynamic analysis is

$$v_t = R_a i_a + L_a \frac{di_a}{dt} + e_a$$

where $e_a = K\Phi\omega = K_m\omega$. The equation of motion for the rotor is

$$J \frac{d\omega}{dt} = T - T_L - T_D$$

where $T = K\Phi i_a = K_m i_a$ and $T_D = B\omega$. After substituting for e_a, T, and T_D, the two differential equations can be rearranged and expressed as

$$\frac{di_a}{dt} = -\frac{R_a}{L_a} i_a - \frac{K_m}{L_a} \omega + \frac{v_t}{L_a} \tag{1}$$

$$\frac{d\omega}{dt} = \frac{K_m}{J} i_a - \frac{B}{J} \omega - \frac{T_L}{J} \tag{2}$$

Equations (1) and (2) are expressed in state-variable form, where i_a and ω are state variables and v_t and T_L are forcing functions. This form is particularly suitable for computer-aided solutions when the order of the system (number of the first-order differential equations when expressed in state-variable form) is high or when non-linearities exist. In this case the equations describe a second-order linear system and an analytical solution for $\omega(t)$ is not difficult. We start by determining the initial and final steady-state values of ω. The initial value of ω is obtained by solving for ω in (1) with $i_a = 1400$ A, $v_t = 250$ V, and $di_a/dt = 0$:

$$\omega = \frac{v_t - R_a i_a}{K_m}$$

$$= \frac{250 - (0.0035)(1400)}{2.65} = 92.5 \text{ rad/s}$$

The value of constant load torque T_L is obtained by solving (2) with $i_a = 1400$ A, $\omega = 92.5$ rad/s, and $d\omega/dt = 0$:

$$T_L = K_m i_a - B\omega$$

$$= (2.65)(1400) - (0.608)(92.5) = 3653.8 \text{ N} \cdot \text{m}$$

The final steady-state value of ω is obtained by solving simultaneously (1) and (2) with derivatives set to zero, T_L held constant at 3653.8 N · m, and v_t reduced to $(0.8)(250) = 200$ V. The desired final value of ω is found to be

$$\omega_f = 73.6 \text{ rad/s}$$

This value is roughly 20% less than the initial value of ω, as discussed earlier. Now recognize that the solution for $\omega(t)$ consists of a natural response and forced response:

$$\omega(t) = \omega_{nr}(t) + \omega_{fr}(t)$$

In this case the forced response is the final value of ω. The form of the natural response can be determined by finding the eigenvalues of the $\bar{A}$ matrix associated with (1) and (2)[†]

$$\bar{A} = \begin{bmatrix} \dfrac{-R_a}{L_a} & \dfrac{-K_m}{L_a} \\ \dfrac{K_m}{J} & \dfrac{-B}{J} \end{bmatrix} = \begin{bmatrix} -\dfrac{1}{\tau_a} & \dfrac{-K_m}{L_a} \\ \dfrac{K_m}{J} & -\dfrac{1}{\tau_L} \end{bmatrix}$$

where $\tau_a = L_a/R_a$ is the armature time constant and $\tau_L = J/B$ is the load time constant.

[†]See Appendix G for further details on the solution of differential equations in state-variable form.

The eigenvalues are obtained by setting $\det(\bar{A} - \lambda\bar{I}) = 0$, where $\bar{I}$ is the identity matrix and λ is a scalar. The resulting second-order equation in λ (sometimes referred to as the *characteristic equation*) is given by

$$\det(\bar{A} - \lambda\bar{I}) = \left(\frac{1}{\tau_a} + \lambda\right)\left(\frac{1}{\tau_L} + \lambda\right) + \frac{K_m^2}{JL_a}$$

$$= \lambda^2 + \left(\frac{1}{\tau_a} + \frac{1}{\tau_L}\right)\lambda + \frac{K_m^2}{JL_a} + \frac{1}{\tau_a\tau_L} = 0$$

Upon substituting numerical data, the eigenvalues are found to be

$$\lambda_1 = -10.0 + j20.9 \text{ s}^{-1}$$

$$\lambda_2 = -10.0 - j20.9 \text{ s}^{-1}$$

The response is underdamped and the total solution can be expressed as

$$\omega(t) = Me^{-10t}\cos(20.9t + \theta) + 73.6 \text{ rad/s}$$

where constants M and θ must be evaluated based on the initial conditions

$$\omega(0) = 92.5 \text{ rad/s}$$

$$\frac{d\omega}{dt}(0) = 0$$

The second initial condition is obtained by recognizing that state variables ω and i_a cannot change instantaneously [see (2)]. Upon evaluation of the constants the complete solution is given by

$$\omega(t) = 20.95e^{-10t}\cos(20.9t - 25.6°) + 73.6 \text{ rad/s}$$

This is illustrated in Figure 5.24. The effects of rotational losses are negligible in the final result. The damping torque is small compared to the constant load torque for all speeds of interest and thus has little or no effect on steady-state values. Similarly, $\tau_L = J/B \gg \tau_a = L_a/R_a$ and eigenvalues do not differ significantly if the constant B is set to zero in the initial formulation of equations. In many cases damping due to rotational losses can be neglected.

The approach illustrated in Example 5.4 is sufficiently general for analysis of the separately excited dc motor to warrant some additional observations. Equations (1) and (2) of the example are restated below as (5.18) and (5.19):

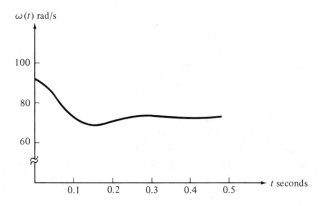

FIGURE 5.24 Speed Versus Time for the Machine of Example 5.4

$$\frac{di_a}{dt} = -\frac{1}{\tau_a} i_a - \frac{K_m}{L_a} \omega + \frac{v_t}{L_a} \tag{5.18}$$

$$\frac{d\omega}{dt} = \frac{K_m}{J} i_a - \frac{1}{\tau_L} \omega - \frac{T_L}{J} \tag{5.19}$$

where τ_a and τ_L are time constants defined previously. The characteristic equation, as developed in Example 5.4, is given by

$$\lambda^2 + \left(\frac{1}{\tau_a} + \frac{1}{\tau_L}\right)\lambda + \frac{K_m^2}{JL_a} + \frac{1}{\tau_a \tau_L} = 0 \tag{5.20}$$

If mechanical damping is neglected, $\tau_L \to \infty$ and (5.20) becomes

$$\lambda^2 + \frac{1}{\tau_a} \lambda + \frac{K_m^2}{JL_a} = 0 \tag{5.21}$$

If (5.21) is multiplied through by τ_a, we have

$$\tau_a \lambda^2 + \lambda + \frac{1}{\tau_m} = 0 \tag{5.22}$$

where

$$\tau_m = \frac{JR_a}{K_m^2} \tag{5.23}$$

The time constant τ_m is referred to as the *inertial time constant* or *mechanical time constant*. It is the single time constant describing behavior if mechanical damping and armature inductance are neglected [i.e., if in (5.22) $\tau_a = L_a/R_a = 0$]. Although mechanical damping can be neglected in most cases, relative magnitudes of τ_a and τ_m should be examined to determine whether further simplifying assumptions can be made. The roots of (5.22) are given by

$$\lambda = -\frac{1}{2\tau_a} \pm j \sqrt{\frac{1}{\tau_a \tau_m} - \left(\frac{1}{2\tau_a}\right)^2} \tag{5.24}$$

$$= -\frac{1}{2\tau_a} \pm j \frac{1}{\tau_a} \sqrt{\frac{\tau_a}{\tau_m} - \frac{1}{4}}$$

For values of $\tau_a/\tau_m > \frac{1}{4}$, the roots occur in complex-conjugate pairs and the response is said to be underdamped (undamped if armature resistance R_a is neglected). At $\tau_a/\tau_m = \frac{1}{4}$, the roots are real and equal and the response is said to be critically damped. For $\tau_a/\tau_m < \frac{1}{4}$ the roots are real and unequal and the response is said to be overdamped. As τ_a/τ_m approaches zero, one of these real roots approaches $-1/\tau_a$ and the other approaches $-1/\tau_m$. The shorter time constant τ_a can be neglected with little resulting error when $\tau_a/\tau_m \le \frac{1}{10}$, in which case behavior is determined entirely by the mechanical time constant τ_m.

Further insights are obtained by considering frequency-domain-based techniques. If the Laplace transforms of all terms in (5.18) and (5.19) are taken assuming zero initial conditions, we obtain

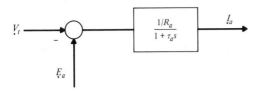

FIGURE 5.25 Block Diagram Showing I_a as a Function of V_t and E_a

$$sI_a = -\frac{1}{\tau_a} I_a - \frac{K_m}{L_a} \Omega + \frac{V_t}{L_a} \tag{5.25}$$

$$s\Omega = \frac{K_m}{J} I_a - \frac{1}{\tau_L} \Omega - \frac{T_L}{J} \tag{5.26}$$

The variables with dots below are used to identify the transformed variables [i.e., $i_a(t) \leftrightarrow I_a$, etc., where dependence on the complex variable s is omitted to simplify notation]. Solving for I_a in (5.25) where $\tau_a = L_a/R_a$ gives us

$$\begin{aligned} I_a &= \frac{1/R_a}{1 + \tau_a s} (V_t - K_m \Omega) \\ &= \frac{1/R_a}{1 + \tau_a s} (V_t - E_a) \end{aligned} \tag{5.27}$$

Variables are related as shown in the block diagram of Figure 5.25. Solving for Ω in (5.26), where $\tau_L = J/B$, we have

$$\begin{aligned} \Omega &= \frac{1/J}{s + 1/\tau_L} (K_m I_a - T_L) \\ &= \frac{1/J}{s + 1/\tau_L} (T - T_L) \end{aligned} \tag{5.28}$$

Variables are related as shown in the block diagram of Figure 5.26. The block diagrams of Figures 5.25 and 5.26 can be combined to yield the block diagram shown in Figure 5.27 for a separately excited motor. The block diagram of Figure 5.27 shows the relationships between variables I_a and Ω and system inputs V_t and T_L. The diagram can be simplified by block diagram reduction techniques. Where speed or position control is required, the diagram can be extended to show additional feedback loops. The subject is well covered in introductory texts on automatic control.

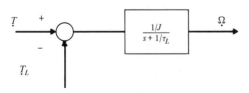

FIGURE 5.26 Block Diagram Showing Ω as a Function of T and T_L

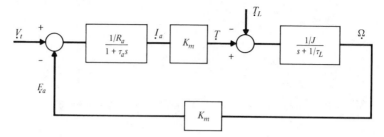

FIGURE 5.27 Block Diagram for a Separately Excited Motor

EXAMPLE 5.5 A small permanent-magnet motor initially at rest has a step voltage of 100 V applied to its armature. The motor drives a load whose torque increases linearly with machine speed. Determine the motor speed buildup as a function of time. The motor data are

$$K_m = 0.25 \text{ V per rad/s}$$

$$R_a = 2.2 \ \Omega$$

$$L_a = 3.0 \text{ mH}$$

$$J = 3.0(10^{-4}) \text{ kg} \cdot \text{m}^2$$

mechanical losses are negligible

load torque $= 5 \cdot \text{m}$ at $\omega = 200 \text{ rad/s}$

The speed buildup can be found by solving (5.18) and (5.19) or by consideration of the block diagram shown in Figure 5.27. A block-diagram-based approach will be followed here. If mechanical losses are negligible, $\tau_L \to \infty$ and the portion of the block diagram relating torques and speed will be as shown in Figure 5.28. The constant B' characterizes the linear buildup of load torque with speed. The block diagram of Figure 5.28 is of the general form shown in Figure 5.29, which yields the familiar transfer relationship between input and output variables also shown in the figure. The diagram of Figure 5.28 thus yields a simple transfer relationship between the developed motor torque T and the speed Ω:

$$\frac{\Omega}{T} = \frac{G}{1 + GH} = \frac{1/Js}{1 + B'/Js} = \frac{1/J}{s + 1/\tau_L'}$$

where $\tau_L' = J/B'$ is a load time constant which takes into account the assumed load torque characteristic. Primed parameters are used here to distinguish quantities based on load characteristics from those based on mechanical losses. The block diagram of Figure 5.27 thus reduces to the diagram shown in Figure 5.30. The diagram yields a transfer relationship between terminal voltage V_t and speed Ω:

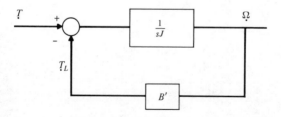

FIGURE 5.28 Block Diagram Relating Torques and Speed

THE DC MACHINE

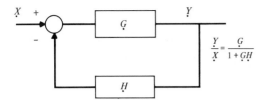

$$\frac{Y}{X} = \frac{G}{1 + GH}$$

FIGURE 5.29 Generalized Block Diagram for Systems with Feedback

$$\frac{\dot{\Omega}}{V_t} = \frac{G}{1 + GH} = \frac{\dfrac{K_m}{R_a J} \dfrac{1}{(1 + \tau_a s)(s + 1/\tau_L')}}{1 + \dfrac{K_m^2}{R_a J} \dfrac{1}{(1 + \tau_a s)(s + 1/\tau_L')}}$$

$$= \frac{1/K_m}{\tau_m(1 + \tau_a s)(s + 1/\tau_L') + 1}$$

where τ_m is the previously defined mechanical time constant, $\tau_m = JR_a/K_m^2$. This transfer relationship can be rearranged and expressed as

$$\frac{\dot{\Omega}}{V_t} = \frac{1/\tau_m \tau_a K_m}{s^2 + \left(\dfrac{1}{\tau_a} + \dfrac{1}{\tau_L'}\right)s + \dfrac{1}{\tau_a \tau_L'} + \dfrac{1}{\tau_a \tau_m}}$$

For the given data, the time constants are

$$\tau_m = \frac{JR_a}{K_m^2} = \frac{3.0(10^{-4})(2.2)}{(0.25)^2} = 10.6 \text{ ms}$$

$$\tau_a = \frac{L_a}{R_a} = \frac{3.0(10^{-3})}{2.2} = 1.36 \text{ ms}$$

$$\tau_L' = \frac{J}{B'} = \frac{3.0(10^{-4})}{5/200} = 12.0 \text{ ms}$$

Substituting numerical data and factoring the denominator polynomial, we find

$$\frac{\dot{\Omega}}{V_t} = \frac{2.77(10^5)}{(s + 601.5)(s + 217.1)}$$

For a step input of 100 V, $V_t = 100/s$. Substituting and expanding by partial fractions, speed can be expressed as

$$\dot{\Omega} = \frac{212.1}{s} + \frac{119.8}{s + 601.5} - \frac{331.9}{s + 217.1}$$

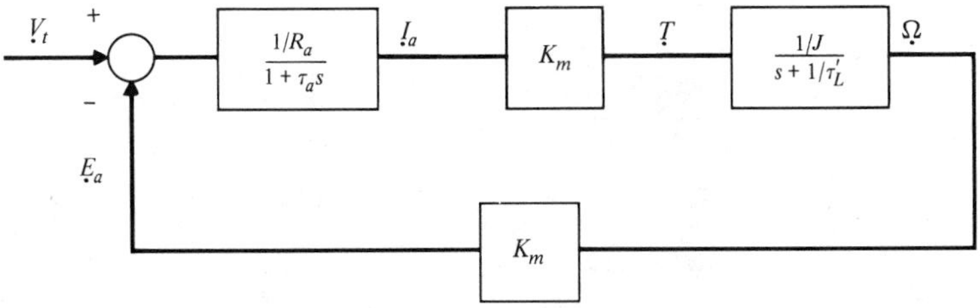

FIGURE 5.30 Block Diagram for the Separately Excited Motor of Example 5.5

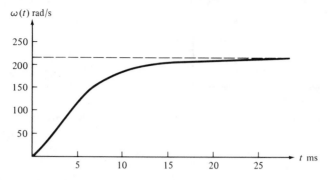

FIGURE 5.31 Speed Buildup for the Machine of Example 5.5

The required inverse transform is

$$\underset{\cdot}{\Omega} \leftrightarrow \omega(t) = 212.1 + 119.8e^{-t/\tau_1} - 331.9e^{-t/\tau_2} \text{ rad/s}$$

where $\tau_1 = 1/601.5 = 1.66$ ms and $\tau_2 = 1/217.1 = 4.06$ ms. Note that the expression yields $\omega(0) = 0$, as required for an initial rest condition. The speed buildup is illustrated in Figure 5.31.

EXAMPLE 5.6 A small permanent-magnet servomotor is used in a position control scheme to drive an inertial load. Feedback is obtained by employing a position-sensing device that supplies a feedback voltage proportional to shaft position. The block diagram for the system is illustrated in Figure 5.32. The motor data are

Mechanical time constant	τ_m =	12.0 ms
Back emf constant	K_m =	1.5 V per rad/s
Armature time constant		negligible
Load inertia		same as for the motor rotor
Mechanical losses		negligible

Investigate system performance as a function of the gain K_f of the feedback system.

The transfer relation for motor speed Ω and terminal voltage V_t for negligible mechanical damping, zero load torque, and negligible armature time constant is given by (see Problem 5.14)

$$\frac{\underset{\cdot}{\Omega}}{\underset{\cdot}{V_t}} = \frac{1/K_m}{\tau_m s + 1}$$

The motor shaft position θ is obtained by integrating the shaft speed ω, which corresponds

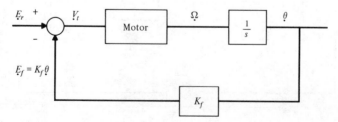

FIGURE 5.32 Block Diagram for an Elementary Position Control Device

THE DC MACHINE

to dividing Ω by s, as shown in Figure 5.32. The overall transfer relation for the system is

$$\frac{\theta}{E_r} = \frac{G}{1 + GH} = \frac{(1/K_m)/s(\tau_m s + 1)}{1 + \dfrac{K_f/K_m}{s(\tau_m s + 1)}} = \frac{1/\tau_m K_m}{s^2 + (1/\tau_m)s + K_f/\tau_m K_m}$$

The nature of system damping is determined by examining the roots of the denominator polynomial

$$s^2 + \frac{1}{\tau_m} s + \frac{K_f}{\tau_m K_m} = 0$$

This polynomial is of the form

$$s^2 + 2\omega_n \zeta s + \omega_n^2 = 0$$

where ζ is the damping factor and ω_n is the undamped natural frequency. In this case

$$\omega_n = \sqrt{\frac{K_f}{\tau_m K_m}} \tag{1}$$

$$\zeta = \frac{1}{2} \sqrt{\frac{K_m}{\tau_m K_f}} \tag{2}$$

The roots of the characteristic equation in terms of ζ and ω_n are

$$-\zeta\omega_n \pm j\omega_n \sqrt{1 - \zeta^2}$$

If K_f is too high, damping is reduced and overshoot tends to be excessive. If K_f is too low, the response is sluggish. A value of ζ of 0.5 is a typical compromise choice. Specification of ζ permits us to determine a value of K_f for a given K_m and τ_m. The mechanical time constant $\tau_m = JR_a/K_m^2$ must be adjusted to accommodate the load moment of inertia. If the load moment of inertia is equal to the motor inertia, the motor mechanical time constant should be increased by a factor of 2. Solving for K_f in (2) yields

$$K_f = \frac{K_m}{\tau_m} \left(\frac{1}{2\zeta}\right)^2$$

$$= \frac{1.5}{24.0(10^{-3})} \left[\frac{1}{2(0.5)}\right]^2 = 62.5 \text{ V per rad}$$

The response to a step input of the reference voltage E_r is of interest. The ratio of the final value of θ to the magnitude of the step is obtained by setting s to zero in the transfer relation, which yields

$$\frac{\theta_f}{E_r} = \frac{1}{K_f} \text{ rad/V}$$

The undamped natural frequency is, from (1),

$$\omega_n = \sqrt{\frac{62.5}{24(10^{-3})(1.5)}} = 41.67 \text{ rad/s}$$

Thus roots of the characteristic polynomial will be

$$-\zeta\omega_n \pm j\omega_n \sqrt{1 - \zeta^2} = -20.8 \pm j36.1 \text{ s}^{-1}$$

The form of the solution is thus

$$\theta(t) = \frac{E_r}{K_f} + Me^{-20.8t} \sin(36.1t + \varphi)$$

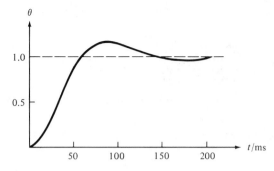

FIGURE 5.33 θ_n **Versus Time for the System of Example 5.6**

where M and φ are constants that can be evaluated from the initial conditions

$$\theta(0) = 0$$

$$\frac{d\theta}{dt}(0) = \omega(0) = 0$$

The solution on a normalized basis (i.e., where the final value is taken as 1.0) is

$$\theta_n(t) = \frac{\theta(t)}{E_r/K_f} = 1.0 - 1.154e^{-t/\tau} \sin (36.1t + 60.05°)$$

where $\tau = 1/20.8 = 48.1$ ms. This result is illustrated graphically in Figure 5.33. The overshoot is roughly 16% with the response remaining within 2% of its final value for all time $t > 200$ ms.

REFERENCES

[1] S. A. NASAR and L. E. UNNEWEHR, *Electromechanics and Electric Machines.* New York: John Wiley & Sons, Inc., 1983.

[2] DONALD G. FINK and H. WAYNE BEATY, *Standard Handbook for Electrical Engineers,* 11th ed. New York: McGraw-Hill Book Company, 1978.

PROBLEMS

5.1 Consider the machine illustrated in Figure 5.3. If field and armature currents flow in the directions shown in the figure, in what direction is torque developed on the rotor of the machine (i.e., clockwise or counterclockwise)? If the rotor turns at a constant speed in the counterclockwise direction, is the machine operating as a motor or as a generator?

5.2 In Example 5.1, the open-circuit armature voltage waveform is developed for the multiple-coil commutator system illustrated in Figure 5.7. This waveform can be expressed in the following way:

$$e_a(t) = 4I\omega r B_{max} \cos \omega t \qquad \frac{-\pi}{6} \leq \omega t < \frac{+\pi}{6} \text{ (periodic)}$$

Find an expression for the time-average open-circuit armature voltage.

5.3 Evaluate (5.7) for the multiple-coil commutator system illustrated in Figure 5.7. Compare your result to the result obtained in Problem 5.2. What conclusion can be drawn concerning the validity of (5.7) if the number of slots is not large?

5.4 Dc machine capacity is limited by certain fundamental considerations. The maximum flux density B_{max} due to the field is limited by saturation. Centrifugal forces on the armature windings and commutation considerations limit rotor peripheral speed $S = \omega r$. The armature current is limited by conductor heating and perhaps armature reaction flux considerations. This limit is often expressed in terms of the maximum allowable ampere conductors per unit of circumferential length $q = Zi_a/2\pi ra$. Assume a rectangular flux density wave due to the field as shown in the figure and find expressions for torque and

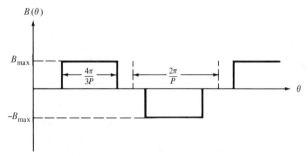

power in terms of B_{max}, S, q, axial length L, and rotor radius r. [HINT: Apply (5.2) and (5.8).]

5.5 Use the expressions developed in Problems 5.4 to estimate the rotor radius and length for a 1500-kW 720-rpm dc machine. Assume that $B_{max} = 1.0$ T, $S_{max} = 40$ m/s, and $q_{max} = 60,000$ A $\cdot$ conductors/m.

5.6 A certain dc machine has a rotor radius of 0.6 m and a rotor effective length of 0.35 m. The machine is wound for six poles. Assume that the maximum flux density B_{max} is 1.0 T and find flux per pole Φ if the flux density distribution is as shown for Problem 5.4.

5.7 A six-pole dc machine has a rated speed of 500 rpm and a rated armature terminal voltage of 700 V. The flux per pole Φ for the machine is 0.100 Wb per pole. The armature is wound so as to produce two parallel paths between the brushes. Determine the approximate number of total conductors Z for the armature.

5.8 Consider the motor–generator set of Example 5.3. Suppose that the field voltage is adjusted so that the generator terminal voltage at full load is 500 V (as in the example). If the field voltage remains the same and the load is disconnected, find the steady-state no-load voltage across the armature terminals.

5.9 Consider the motor–generator set of Example 5.3. What torque must be developed by the induction machine that serves as the prime mover when the generator is at the full-load condition described in the example?

5.10 A small permanent-magnet servomotor has a back emf constant of 0.4 V per rad/s. The armature resistance is 3.0 Ω. Determine the torque developed by the motor if an applied terminal voltage of 60 V results in a steady-state speed of 1200 rpm.

5.11 A separately excited dc motor has a rated armature voltage of 300 V. The armature resistance is 0.06 Ω. The no-load speed of the motor at the rated armature voltage is 1000 rpm. What armature current does the motor draw if a load is applied to the shaft resulting in a machine steady-state speed of 900 rpm at the rated armature voltage?

5.12 A 250-V, separately excited dc motor has a no-load speed of 1200 rpm. The motor is rated at 40 hp and has an efficiency of 94% at its rated output. The armature resistance is 0.06 Ω. Find the machine full-load speed, that is, the speed at the rated output.

5.13 The magnetization curve for a dc machine is shown in the figure. The machine was run at 900 rpm as a separately excited generator in order to develop the data. The number of turns on the field winding is 500. The effects of armature reaction are negligible.

What open-circuit voltage will be developed by a machine identical in all respects but having 250 turns on its field winding if its field current is adjusted to 8 A and its speed of rotation is 600 rpm?

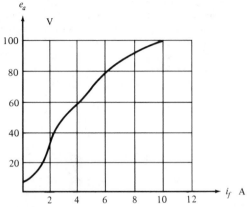

5.14 A 250-V 100-hp dc shunt motor has an adjustable rheostat in series with its field winding. The armature resistance is 0.05 Ω. If the total field resistance is 25 Ω, the machine runs at a speed of 500 rpm with no load on the shaft. Find the unloaded speed of the machine if the total field resistance is set at 35 Ω. (Assume a linear magnetization curve.)

5.15 A series motor develops a blocked-rotor torque of 1000 N · m when the motor line current is 20 A. The combined series winding and armature winding resistance is 0.10 Ω. Determine the required terminal voltage if the motor is to develop a torque of 100 N · m at a speed of 100 rpm. Assume a linear magnetization curve.

5.16 A dc shunt motor runs at the rated speed when the line voltage is 250 V and the line current is 150 A. The field resistance is 35 Ω and the armature resistance including brush drop is 0.10 Ω. If no-load rotational losses are 1000 W, what is the available shaft output power? What is the overall efficiency of the motor?

5.17 A separately excited 150-hp 250-V dc motor has a rated armature current of 450 A. The armature resistance is 0.04 Ω. The armature inductance is negligible. The field current is held constant and the motor is started by applying rated voltage to the armature circuit.
  a. Find the resistance required in series with the armature to limit the starting current to 150% of the rated current.
  b. The field is set for a no-load speed of 1000 rpm at the rated voltage. The machine is started as described above, with the value of the resistance determined in part (a) in series with the armature. What value of speed does the machine reach when the armature current has decreased to its rated value?

5.18 In Example 5.5 the speed buildup as a function of time was determined for a separately excited motor under load. Find the speed buildup as a function of time if the machine is unloaded (i.e., if $T_L = 0$). Use the same numerical data as those specified for Example 5.5.

5.19 A block diagram for the separately excited motor is shown in Figure 5.27. Find a simplified block diagram that could be used to represent the motor if the load torque is zero, mechanical damping is negligible, and $\tau_a \ll \tau_m = JR_a/K_m^2$.

5.20 A permanent-magnet dc motor with no shaft load has a rated armature voltage applied at $t = 0$. The rotor of the machine is initially at rest. Mechanical damping is negligible and the armature circuit inductance may be neglected (i.e., the time constant associated with the armature circuit is negligibly small). Find an expression for the developed torque as a function of time.

CHAPTER 6

Analysis of AC Machine Transients

6.1

INTRODUCTION

The analysis of ac machinery under the assumption of balanced steady-state operating conditions is readily accomplished using the techniques described in Chapters 3 and 4. It is often possible to extend steady-state analysis techniques to investigate a sequence in time of relatively slowly changing "steady-state" operating points. Rotor speed or rotor angle variation with time can then be approximated as shown in the examples presented in Chapters 3 and 4. In the earlier examples it was necessary to neglect transients due to switching. Occasionally, the complete instantaneous waveform of certain machine variables is of interest. For example, if a short circuit occurs, armature currents in the various phases of an ac machine vary with time in a rather complicated way. In general, they consist of sinusoidally varying components with decaying magnitudes, decaying unidirectional components that give rise to what is sometimes referred to as *dc offset*, and sinusoidal terms with constant amplitude. Figure 6.1 illustrates typical waveforms for a synchronous machine.

The magnitude of each of the phase current waveforms decays to a value that could be determined on the basis of steady-state theory (i.e., $I_s = E_f/x_s$). This

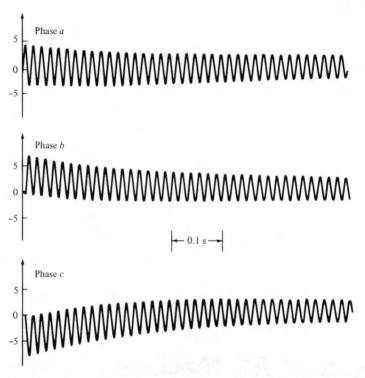

FIGURE 6.1 Typical Armature Current Waveforms Resulting from a Short Circuit at the Terminals of an Initially Open-Circuited Synchronous Machine. (Currents Are Normalized Such That 1.0 Corresponds to Rated Armature Current)

asymptotically approached value might be roughly 60% of the rated armature current for typical values of machine parameters. The initial value of the various currents can be considerably higher, as shown in Figure 6.1, where peak values roughly 8.0 times the rated armature current occur. Peak values are of interest in that machine windings must be braced to withstand the forces that occur at any time following a short circuit. Such information is also of interest with regard to other components that might be connected to a machine subjected to a short circuit. For example, circuit breakers used to clear short circuits normally interrupt current in three to eight cycles (0.05 to 0.133 s). They must therefore be designed to interrupt the initial currents that flow immediately following a short circuit.

The electromagnetic torque of a machine following a disturbance can exhibit behavior somewhat similar to the time variation of currents shown in Figure 6.1. Such torque transients interact with machine mechanical systems and can produce excessive stresses on gear trains, shafts, and other components.

The phenomena described above are examples of ac machine transients. Analysis of such transients requires a general method of attack which is specifically not restricted to an underlying assumption of steady-state behavior. It is important to recognize, however, that the steady-state analysis techniques of Chapters 3 and 4 are particularly appropriate for determining initial operating conditions prior to a disturbance and, in many cases, conditions that must exist when transients have decayed to negligible levels. The ability to determine such "before" and "after" conditions is often of considerable assistance in general analyses dealing with transients.

GENERALIZED ANALYSIS OF AC MACHINES

Relationships required for generalized analysis will be illustrated by considering an elementary two-pole two-phase synchronous machine. In Section 4.2 we examined the configuration shown in Figure 6.2. By making suitable simplifying assumptions, we were able to determine a matrix relationship between flux linkages and currents:

$$\bar{\lambda} = \bar{L}(\theta_r)\bar{i} \tag{6.1}$$

where

$$\bar{\lambda} = \begin{bmatrix} \lambda_a \\ \lambda_b \\ \lambda'_r \end{bmatrix} \tag{6.2}$$

$$\bar{i} = \begin{bmatrix} i_a \\ i_b \\ i'_r \end{bmatrix} \tag{6.3}$$

$$\bar{L}(\theta_r) = \begin{bmatrix} L_{\ell s} + M & 0 & M\cos\theta_r \\ & L_{\ell s} + M & M\sin\theta_r \\ \text{(symmetric matrix)} & & L'_{\ell r} + M \end{bmatrix} \tag{6.4}$$

The constant M depends on machine dimensions and the number of turns on the

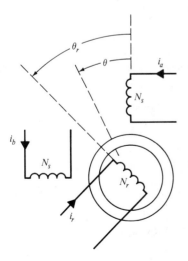

FIGURE 6.2 Elementary Two-Pole Two-Phase Round-Rotor Synchronous Machine

armature windings [see (4.30)]. The leakage inductance terms $L_{\ell s}$ and $L'_{\ell r}$ similarly depend on machine parameters with slot dimensions playing a major role. Primed quantities are quantities associated with the field circuit (on the rotor) but referred to the armature windings (on the stator) by appropriate turns ratios:

$$i'_r = \frac{N_r}{N_s} i_r \tag{6.5}$$

$$\lambda'_r = \frac{N_s}{N_r} \lambda_r \tag{6.6}$$

$$L'_{\ell r} = \left(\frac{N_s}{N_r}\right)^2 L_{\ell r} \tag{6.7}$$

Voltage equations for the armature circuits can be written as (with variables assigned positive for motor action)

$$v_a = i_a r_s + \frac{d\lambda_a}{dt} \tag{6.8a}$$

$$v_b = i_b r_s + \frac{d\lambda_b}{dt} \tag{6.8b}$$

The voltage equation for the field circuit is given by

$$v'_r = r'_r i'_r + \frac{d\lambda'_r}{dt} \tag{6.9}$$

where rotor resistance is referred to the armature:

$$r'_r = \left(\frac{N_s}{N_r}\right)^2 r_r \tag{6.10}$$

The equation of motion for the rotor can be written (for operation as a motor)

$$\frac{J \, d\omega_r}{dt} = T - T_L - T_D \tag{6.11}$$

where J is the combined moment of inertia of the rotor and load, T_L the load torque, and T_D the torque due to mechanical damping. The developed electrical torque for a linear system is, from (2.42) and (2.53),

$$
\begin{aligned}
T &= \frac{\partial W_c}{\partial \theta_r} = \frac{1}{2} \frac{\partial}{\partial \theta_r} \left[\bar{i}^T \bar{L}(\theta_r) \bar{i} \right] \\[2mm]
&= \frac{1}{2} \bar{i}^T \left[\frac{d}{d\theta_r} \bar{L}(\theta_r) \right] \bar{i} \\[2mm]
&= \frac{1}{2} [i_a, \, i_b, \, i'_r] \begin{bmatrix} 0 & 0 & -M \sin \theta_r \\ & 0 & M \cos \theta_r \\ \text{(symmetric} & & 0 \\ \text{matrix)} & & \end{bmatrix} \begin{bmatrix} i_a \\ i_b \\ i'_r \end{bmatrix} \\[2mm]
&= M i'_r (\cos \theta_r i_b - \sin \theta_r i_a)
\end{aligned} \tag{6.12}
$$

Note that the differentiation is with respect to θ_r with currents held constant. The angle θ_r is related to the machine speed ω_r by

$$\omega_r = \frac{d\theta_r}{dt} \tag{6.13}$$

The equations above must be satisfied simultaneously in any investigation of machine transients. Analytical solutions are difficult to obtain in most practical cases of interest. Computer-assisted numerical solution of the equations is an option. We might choose currents, speed ω_r, and angle θ_r as state variables with voltages v_a, v_b, and v'_r and load torque T_L as independent variables or "inputs" to a system of equations expressed in state-variable form. Damping torque could be eliminated by expressing it as a suitable function of speed. The algebraic manipulations required to do this are formidable even for a simplified two-phase machine. Additional rotor windings, referred to as *damper* or *amortisseur windings*, will increase the complexity of the final system of equations. Nevertheless, the approach outlined above summarizes the fundamental relationships that must be addressed. It is not difficult to develop in a similar way appropriate relationships for synchronous or induction machines of any number of phases. In essence, we typically must determine and satisfy:

1. Relationships between machine flux linkages and machine currents such as the ones given by (6.1) through (6.4).
2. Voltage relationships for each winding of the machine such as the ones given by (6.8) and (6.9).
3. A relationship for developed machine torque such as the one given by (6.12).
4. Relationships relevant to the mechanical equation of motion such as the ones given by (6.11) and (6.13).

6.3

THE USE OF TRANSFORMATIONS IN AC MACHINE ANALYSIS

Transformations can be used to simplify machine equations. The transformations typically used in ac machine analysis also provide additional insights into machine behavior. We will illustrate the application of such transformations by considering a two-phase round-rotor synchronous machine wound for an arbitrary number of poles. Modifications required for other machines will then be described.

6.3.1 FLUX LINKAGE RELATIONSHIPS

The variation of machine inductances with rotor position adds considerably to the difficulty in manipulating equations formulated in the manner of Section 6.2. If the stationary armature windings of a machine are replaced with a fictitious set of windings that rotate with the rotor of a machine, the fictitious windings will have a fixed angular relationship with respect to the axis of the rotor. The inductances required to describe the resulting system of windings will be independent of rotor angle and can be treated as constants. The technique is sometimes described as one of referring stator variables to a frame of reference fixed in the rotor of the machine. A simple mathematical transformation that accomplishes this for a two-phase machine is given by

$$\bar{f}^{ab} = \bar{T}(\varphi)\bar{f}^{qd} \tag{6.14}$$

where

$$\bar{f}^{ab} = \begin{bmatrix} f_a \\ f_b \end{bmatrix} \tag{6.15}$$

$$\bar{f}^{qd} = \begin{bmatrix} f_q \\ f_d \end{bmatrix} \tag{6.16}$$

$$\bar{T}(\varphi) = \begin{bmatrix} \cos\varphi & \sin\varphi \\ \sin\varphi & -\cos\varphi \end{bmatrix} \tag{6.17}$$

In (6.17), φ is the rotor angle with respect to some fixed reference. It must be expressed in electrical radians (or degrees). The inverse transformation is obtained by premultiplying both sides of (6.14) by the inverse of $\bar{T}(\varphi)$:

$$\bar{f}^{qd} = \bar{T}^{-1}(\varphi)\bar{f}^{ab} \tag{6.18}$$

where the inverse of $\bar{T}(\varphi)$ is

$$\bar{T}^{-1}(\varphi) = \begin{bmatrix} \cos\varphi & \sin\varphi \\ \sin\varphi & -\cos\varphi \end{bmatrix}$$
$$= T(\varphi) \tag{6.19}$$

In the transformation relationships above, the variables f_a and f_b are used to denote phase a and phase b variables, respectively. The general designation f is used in that the variables may be currents, voltages, or flux linkages, that is, any of the time-varying scalar quantities associated with the armature windings. The variables f_d and f_q are corresponding variables but referred to the fictitious windings. The d and q subscripts refer to the direct and quadrature axes, respectively, of a frame of reference fixed in the rotor of the machine. The transformation relationships can be visualized (and verified) by referring to Figure 6.3. Figure 6.3 should be compared with Figure 6.2. The angle φ in the transformation relationships differs from our earlier defined rotor angle θ_r by $\pi/2$ radians; $\varphi = \theta_r + \pi/2$. Note that our designation of direct and quadrature axes is similar to the designation in Chapter 4 for steady-state analysis of the salient-pole synchronous machine. The angle φ is related to machine speed by

$$\varphi = \int_0^t \omega_r(\tau)\, d\tau + \varphi(0) \tag{6.20}$$

or

$$\frac{d\varphi}{dt} = \omega_r(t) \tag{6.21}$$

EXAMPLE 6.1 Show that for the elementary synchronous machine of Figure 6.2, the transformation given by (6.14) through (6.21) results in relationships between flux linkages and currents which are independent of rotor position.

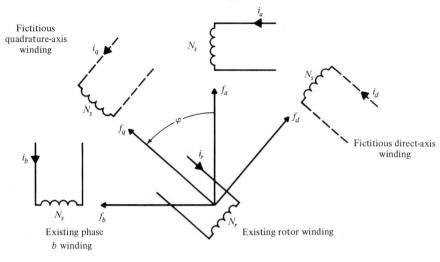

FIGURE 6.3 Graphical Depiction of the Transformation Relationships

The relationship between flux linkages and currents for the machine of Figure 6.2 is given by (6.1) through (6.4):

$$\bar{\lambda}^{abr'} = \bar{L}^{abr'}(\theta)\bar{i}^{abr'}$$

or

$$\begin{bmatrix} \lambda_a \\ \lambda_b \\ \lambda'_r \end{bmatrix} = \begin{bmatrix} L_s & 0 & M\cos\theta_r \\ 0 & L_s & M\sin\theta_r \\ M\cos\theta_r & M\sin\theta_r & L'_f \end{bmatrix} \begin{bmatrix} i_a \\ i_b \\ i'_r \end{bmatrix}$$

where we have simplified notation somewhat by defining inductances:

$$L_s = L_{\ell s} + M \qquad \text{(synchronous inductance)}$$

$$L'_f = L'_{\ell r} + M \qquad \text{(a field winding inductance referred to the armature windings)}$$

In terms of the angle $\varphi = \theta_r + \pi/2$, the relationships between flux linkages and currents are given by

$$\begin{bmatrix} \lambda_a \\ \lambda_b \\ \lambda'_r \end{bmatrix} = \begin{bmatrix} L_s & 0 & M\sin\varphi \\ 0 & L_s & -M\cos\varphi \\ M\sin\varphi & -M\cos\varphi & L'_f \end{bmatrix} \begin{bmatrix} i_a \\ i_b \\ i'_r \end{bmatrix}$$

Now formulate a transformation that will refer phase variables to the direct and quadrature axes of a frame of reference fixed in the rotor of the machine while leaving rotor variables unchanged. An appropriate transformation is

$$\begin{bmatrix} f_a \\ f_b \\ f'_r \end{bmatrix} = \begin{bmatrix} \cos\varphi & \sin\varphi & 0 \\ \sin\varphi & -\cos\varphi & 0 \\ 0 & 0 & 1 \end{bmatrix} \begin{bmatrix} f_q \\ f_d \\ f'_r \end{bmatrix}$$

with inverse relationship given by

$$\begin{bmatrix} f_q \\ f_d \\ f'_r \end{bmatrix} = \begin{bmatrix} \cos\varphi & \sin\varphi & 0 \\ \sin\varphi & -\cos\varphi & 0 \\ 0 & 0 & 1 \end{bmatrix} \begin{bmatrix} f_a \\ f_b \\ f'_r \end{bmatrix}$$

Equivalently,

$$\bar{f}^{abr'} = \bar{A}(\varphi)\bar{f}^{qdr'}$$
$$\bar{f}^{qdr'} = \bar{A}^{-1}(\varphi)\bar{f}^{abr'}$$

where

$$\bar{A}(\varphi) = \begin{bmatrix} \begin{bmatrix} \bar{T} & (\varphi) \end{bmatrix} & \begin{matrix} 0 \\ 0 \end{matrix} \\ \begin{matrix} 0 & 0 \end{matrix} & 1 \end{bmatrix}$$

$$\bar{A}^{-1}(\varphi) = \begin{bmatrix} \begin{bmatrix} \bar{T}^{-1}(\varphi) \end{bmatrix} & \begin{matrix} 0 \\ 0 \end{matrix} \\ \begin{matrix} 0 & 0 \end{matrix} & 1 \end{bmatrix}$$

Note that $\bar{A}(\varphi) = \bar{A}^{-1}(\varphi)$ in that $\bar{T}(\varphi) = \bar{T}^{-1}(\varphi)$. Now substitute in the original matrix flux linkage relationship:

$$\bar{\lambda}^{abr'} = \bar{L}^{abr'}\bar{i}^{abr'}$$
$$\bar{A}\bar{\lambda}^{qdr'} = \bar{L}^{abr'}\bar{A}\bar{i}^{qdr'}$$

Premultiply both sides by $\bar{A}^{-1}$:

$$\bar{\lambda}^{qdr'} = \bar{A}^{-1}\bar{L}^{abr'}\bar{A}\bar{i}^{qdr'}$$
$$= \bar{L}^{qdr'}\bar{i}^{qdr'}$$

from which it must be true that

$$\bar{L}^{qdr'} = \bar{A}^{-1}\bar{L}^{abr'}\bar{A}$$

Thus

$\bar{L}^{qdr'}$

$$= \begin{bmatrix} \cos\varphi & \sin\varphi & 0 \\ \sin\varphi & -\cos\varphi & 0 \\ 0 & 0 & 1 \end{bmatrix} \begin{bmatrix} L_s & 0 & M\sin\varphi \\ 0 & L_s & -M\cos\varphi \\ M\sin\varphi & -M\cos\varphi & L'_f \end{bmatrix} \begin{bmatrix} \cos\varphi & \sin\varphi & 0 \\ \sin\varphi & -\cos\varphi & 0 \\ 0 & 0 & 1 \end{bmatrix}$$

$$= \begin{bmatrix} \cos\varphi & \sin\varphi & 0 \\ \sin\varphi & -\cos\varphi & 0 \\ 0 & 0 & 1 \end{bmatrix} \begin{bmatrix} L_s\cos\varphi & L_s\sin\varphi & M\sin\varphi \\ L_s\sin\varphi & -L_s\cos\varphi & -M\cos\varphi \\ 0 & M & L'_f \end{bmatrix}$$

$$= \begin{bmatrix} L_s & 0 & 0 \\ 0 & L_s & M \\ 0 & M & L'_f \end{bmatrix}$$

Thus $\bar{L}^{qdr'}$ is an inductance matrix that is independent of angle φ (or θ_r). The relationships between flux linkages λ_q, λ_d, and λ'_r and currents i_q, i_d, and i'_r are independent of rotor angle.

The final result of Example 6.1 suggests the conceptual viewpoint illustrated in Figure 6.4 if winding resistances are neglected. From the inductance matrix $L^{qdr'}$ of Example 6.1 we have $\lambda_d = L_s i_d + M i'_r$. Flux linkages in the fictitious direct-axis winding depend only on currents in windings whose axes are aligned with the direct axis. Similarly, $\lambda'_r = M i_d + L'_f i'_r$. Flux linkages in the field

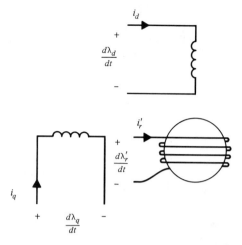

FIGURE 6.4 Fixed Angular Relationships Between Direct- and Quadrature-Axis Windings for a Machine with a Single Direct-Axis Rotor Winding

winding depend only on currents in windings whose axes are aligned with the direct axis. In this case there is only one winding whose axis is aligned with the quadrature axis, namely the fictitious quadrature-axis winding; thus $\lambda_q = L_s i_q$. The flux linkage relationships given mathematically by the matrix equation $\overline{\lambda}^{qdr'} = \overline{L}^{qdr'} \overline{i}^{qdr'}$ suggest the coupled-coil equivalent circuits shown in Figure 6.5.

6.3.2 VOLTAGE RELATIONSHIPS

Appropriate voltage relationships for the fictitious d- and q-axis windings are determined by starting with the original voltage relationships for a machine in terms of phase variables and simply applying the transformation. For a two-phase machine, the armature voltage relationships are given by (6.8):

$$v_a = i_a r_s + \frac{d\lambda_a}{dt}$$

$$v_b = i_b r_s + \frac{d\lambda_b}{dt}$$

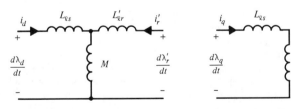

FIGURE 6.5 Coupled-Coil Equivalent Circuits for a Round-Rotor Synchronous Machine with a Single Direct-Axis Rotor Winding

Substitute in the first equation using the transformation relationships

$$\cos \varphi v_q + \sin \varphi v_d = r_s (\cos \varphi i_q + \sin \varphi i_d)$$

$$+ \frac{d}{dt} (\cos \varphi \lambda_q + \sin \varphi \lambda_d)$$

$$= \left(r_s i_d - \omega_r \lambda_q + \frac{d\lambda_d}{dt} \right) \sin \varphi$$

$$+ \left(r_s i_q + \omega_r \lambda_d + \frac{d\lambda_q}{dt} \right) \cos \varphi$$

Now equate coefficients of the $\cos \varphi$ and $\sin \varphi$ terms:

$$v_q = r_s i_q + \omega_r \lambda_d + \frac{d\lambda_q}{dt} \tag{6.22a}$$

$$v_d = r_s i_d - \omega_r \lambda_q + \frac{d\lambda_d}{dt} \tag{6.22b}$$

This same set of relationships results if appropriate substitutions are made in the phase b voltage relationship.

The field circuit voltage equation remains unchanged:

$$v_r' = r_r' i_r' + \frac{d\lambda_r'}{dt} \tag{6.23}$$

Armature voltage relationships for the fictitious windings given by (6.22) should be compared to the original armature voltage relationships in terms of phase variables. The sets of relationships differ in structure due to the product terms $+\omega_r \lambda_d$ and $-\omega_r \lambda_q$. These terms are sometimes referred to as speed–voltage terms. In general, they cause interaction (coupling) between direct- and quadrature-axis windings. If the machine is described by a set of equations in state-variable form, the speed–voltage terms will cause the system of equations to be characterized as nonlinear in that products of state variables occur.

Flux linkage and voltage relationships considered together suggest the equivalent circuits of Figure 6.6.

EXAMPLE 6.2   A two-phase two-pole synchronous generator is driven at constant speed by a prime mover while disconnected from any system (i.e., its armature windings are left as open circuits). Its field voltage is initially zero. Find an expression for the open-circuit armature voltage as a function of time if a constant field voltage V_{field} is suddenly applied at the terminals of the field winding.

The armature voltage is initially zero. Since the armature windings are open circuits, the armature phase currents will be zero for all time. The currents in the fictitious d- and q-axis windings must also be zero for all time [apply the inverse transformation (6.18) to see this]. Referring to the equivalent circuits of Figure 6.6, we see that field current i_r' will be described by a single first-order linear differential equation:

$$v_r' = r_r' i_r' + (L_{\ell r}' + M) \frac{di_r'}{dt}$$

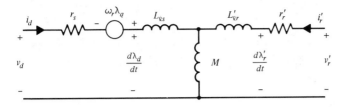

(a) Direct-axis equivalent circuit

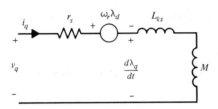

(b) Quadrature-axis equivalent circuit

FIGURE 6.6 Equivalent Circuits for a Round-Rotor Synchronous Machine with a Single Direct-Axis Rotor Winding

where v_r' is a step input:

$$v_r' = \begin{cases} 0 & t < 0 \\ \dfrac{N_s}{N_r} V_{\text{field}} = V_f & t \geq 0 \end{cases}$$

Note that V_f is simply excitation voltage referred to the armature. The solution for i_r' is easily obtained:

$$i_r' = \frac{V_f}{r_r'}(1 - e^{-t/\tau_{do}'})$$

where the time constant τ_{do}' is sometimes referred to as the machine direct-axis open-circuit time constant:

$$\tau_{do}' = \frac{L_{\ell r}' + M}{r_r'}$$

By referring to Figure 6.6, we can easily find appropriate relationships for armature terminal voltage. The quadrature-axis terminal voltage for zero armature currents is given by

$$v_q = \omega_r \lambda_d$$

$$= \omega_r M i_r'$$

$$= \omega_r M \frac{V_f}{r_r'}(1 - e^{-t/\tau_{do}'})$$

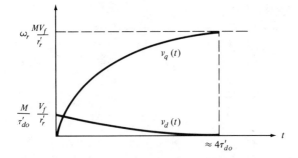

FIGURE 6.7 Direct- and Quadrature-Axis Terminal Voltage Variation with Time for Example 6.2

The direct-axis terminal voltage for zero armature currents is given by

$$v_d = \frac{d\lambda_d}{dt} = \frac{d}{dt} M i'_r$$

$$= \frac{M}{\tau'_{do}} \frac{V_f}{r'_r} e^{-t/\tau'_{do}}$$

Plots of the direct- and quadrature-axis terminal voltages are shown in Figure 6.7.

In large synchronous generators, the effect of the direct-axis voltage $v_d(t)$ will be negligible. This can be seen by comparing the initial (maximum) value of $v_d(t)$ with the final value of $v_q(t)$. The field circuit leakage inductance $L'_{\ell r}$ is typically small compared to the mutual term M. The direct-axis open-circuit time constant is thus given approximately by

$$\tau'_{do} = \frac{L'_{\ell r} + M}{r'_r} \approx \frac{M}{r'_r}$$

With this approximation, the final value of $v_q(t)$ is given by

$$\lim_{t\to\infty} v_q(t) = \omega_r M \frac{V_f}{r'_r} \approx \omega_r \tau'_{do} V_f$$

The initial value of $v_d(t)$ is given approximately by

$$v_d(0) = \frac{M}{\tau'_{do}} \frac{V_f}{r'_r} \approx V_f$$

A typical value of τ'_{do} for a synchronous generator is 5.0 s [1]. The speed ω_r for an electrical frequency of 60 Hz (corresponding to a prime mover turning the machine at rated speed) will be roughly 377 rad/s. Thus

$$\lim_{t\to\infty} \frac{v_d(0)}{v_q(t)} \approx \frac{1}{\omega_r \tau'_{do}} = 5(10^{-4})$$

which suggests that $v_d(t)$ could be neglected.

The armature terminal voltage in terms of phase variables is obtained by applying the transformation relationship (6.14). If $v_d(t)$ is neglected, we have

$$\bar{v}^{ab} = \begin{bmatrix} v_a(t) \\ v_b(t) \end{bmatrix} = \begin{bmatrix} \cos\varphi & \sin\varphi \\ \sin\varphi & -\cos\varphi \end{bmatrix} \begin{bmatrix} v_q(t) \\ 0 \end{bmatrix}$$

$$= \begin{bmatrix} v_q(t) \cos\varphi \\ v_q(t) \sin\varphi \end{bmatrix}$$

For a constant rotor speed, $\omega_r = \omega_s$, the angle φ is, from (6.20),

$$\varphi = \omega_s t + \varphi(0)$$

Substituting for $v_q(t)$, we have

$$v_a(t) = \omega_s M \frac{V_f}{r'_r} (1 - e^{-t/\tau'_{do}}) \cos [\omega_s t + \varphi(0)]$$

$$v_b(t) = \omega_s M \frac{V_f}{r'_r} (1 - e^{-t/\tau'_{do}}) \sin [\omega_s t + \varphi(0)]$$

A plot of the phase a voltage is shown in Figure 6.8.

Several observations are in order. Armature voltage approaches a steady-state value which is consistent with what we would determine on the basis of steady-state theory. In particular, see (4.58) and Figure 4.20. Armature voltage in terms of actual quantities is formally obtained by applying the appropriate transformation relationship. By comparing Figures 6.7 and 6.8, an intuitive appreciation is obtained for the relationship between actual phase variables and fictitious winding variables. Note that as $v_q(t)$ approaches a final constant value, $v_a(t)$ approaches a sinusoidal waveform. In general, steady-state conditions imply sinusoidal time variation of phase variables while qd variables must be constants. The initial rotor angle $\theta_r(0)$ implies a phase angle that is arbitrary in this case. It may often be taken as zero without loss of generality. The reader might well question the general applicability of the equivalent circuits of Figure 6.6, which were obtained by consideration of an elementary two-phase machine. It will be shown that they apply as well for a three-phase machine, assuming that it is of the round-rotor configuration with a single direct-axis winding (the field winding). We will address shortly the formal relationship between phase variables and qd variables when a three-phase machine is involved. As familiarity with transformation techniques increases, the ability to draw meaningful conclusions without formally transforming qd variables back to phase variables is enhanced. The distinction between two-phase and three-phase machines accordingly becomes less important. In any event, it should not be regarded as an obstacle.

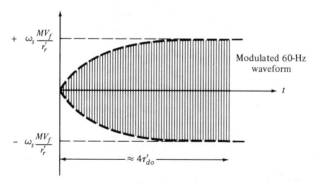

FIGURE 6.8 Buildup of Armature Voltage for the Machine of Example 6.2

6.3 THE USE OF TRANSFORMATIONS IN AC MACHINE ANALYSIS

EXAMPLE 6.3 A synchronous generator connected to a power system operates under steady-state conditions. A disturbance occurs at a time that may be taken as $t = 0$. The disturbance causes the machine to depart from its original steady-state operating point. The d- and q-axis fictitious winding currents before and after the disturbance are given by

$$i_d = 2.0 \qquad\qquad t < 0$$

$$i_q = 1.0 \qquad\qquad t < 0$$

$$i_d(t) = 2.0 - 0.5(1 - e^{-2t}) \qquad t \geq 0$$

$$i_q(t) = 1.0 + 2.0(1 - e^{-2t}) \qquad t \geq 0$$

Find expressions for the armature currents $i_a(t)$ and $i_b(t)$ before and after the disturbance occurs. Assume that the machine speed remains constant at $\omega_r = \omega_s$ (i.e., at the synchronous speed).

From the transformation relationship (6.14) we have

$$i_a(t) = i_q(t) \cos \varphi + i_d(t) \sin \varphi$$

$$i_b(t) = i_q(t) \sin \varphi - i_d(t) \cos \varphi$$

The angle φ is from (6.20), where $\omega_r = \omega_s$,

$$\varphi = \omega_s t + \varphi(0)$$

By applying appropriate trigonometric identities the phase currents can be expressed in general as

$$i_a(t) = \sqrt{i_d^2 + i_q^2} \, \sin \left[\omega_s t + \varphi(0) + \tan^{-1} \frac{i_q}{i_d} \right]$$

$$i_b(t) = \sqrt{i_d^2 + i_q^2} \, \sin \left[\omega_s t + \varphi(0) - \tan^{-1} \frac{i_d}{i_q} \right]$$

Note that phase b current is the same as phase a current, but phase displaced by $\pi/2$ radians. This can be seen by comparing the phase angles of the two currents. Let

$$\psi_a(t) = \varphi(0) + \tan^{-1} \frac{i_q}{i_d}$$

$$\psi_b(t) = \varphi(0) - \tan^{-1} \frac{i_d}{i_q}$$

For any value of the ratio of direct- and quadrature-axis currents,

$$\psi_a(t) - \psi_b(t) = \frac{\pi}{2}$$

(Construction of a right triangle with sides i_d, i_q and hypotenuse $\sqrt{i_d^2 + i_q^2}$ may be helpful here.)

We proceed, considering further only the phase a current. For time $t < 0$ we have

$$i_a = \sqrt{2^2 + 1^2} \, \sin \left[\omega_s t + \varphi(0) + \tan^{-1} \frac{1}{2} \right]$$

$$= 2.24 \sin [\omega_s t + \varphi(0) + 26.6°]$$

For time $t \geq 0$ we have, after substitution and simplification,

$$i_a = \sqrt{11.25 - 10.5e^{-2t} + 4.25e^{-4t}} \, \sin [\omega_s t + \varphi(0) + \psi_a(t)]$$

where

$$\psi_a(t) = \tan^{-1} \frac{3.0 - 2.0e^{-2t}}{1.5 + 0.5e^{-2t}}$$

The result shows that when i_d and i_q are constants, we have simple sinusoidal time variation of the phase currents that is characteristic of steady-state behavior. When i_d and i_q vary as functions of time, the phase-variable time variation can be characterized as a sinusoidal function with time-varying amplitude and time-varying angle. Problem 6.5 affords some additional insights.

EXAMPLE 6.4 A two-phase 2 kVA round-rotor synchronous generator has a rated armature voltage of 120 V. The machine has a synchronous reactance of 12.5 Ω and a negligible armature resistance. The armature leakage reactance is 20% of synchronous reactance. The machine delivers its rated kVA at rated voltage and at a power factor of 0.85 lagging. Find values of i_d, i_q, i'_r, and $\varphi(0)$ to be used as initial values in an analysis of machine transients for a disturbance assumed to occur at time $t = 0$.

The problem is typical in that the steady-state operating condition prior to the disturbance is often at or near the machine's rated operating point. The disturbance might be a sudden change in machine terminal voltage, field excitation, or prime-mover torque. An analysis of transients following the disturbance generally will require a set of initial values for state variables. These initial values are obtained by determining values of qd variables, rotor variables, and machine angle for the assumed steady-state operating condition prior to the disturbance.

We start by establishing the relationship between qd variables and the phasors customarily used for steady-state analysis. As shown in earlier examples, the qd variables are constant when the machine operates in a steady-state condition with balanced phase variables. The machine speed is constant at the synchronous speed. Thus, in general for all qd variables,

$$f_q = F_q$$
$$f_d = F_d$$

where capital letters are used to denote constants. The machine speed is constant at the synchronous speed. Thus

$$\omega_r = \omega_s$$
$$\varphi = \omega_s t + \varphi(0)$$

From the transformation relationship (6.14),

$$f_a(t) = f_q \cos \varphi + f_d \sin \varphi$$
$$= F_q \cos [\omega_s t + \varphi(0)] + F_d \sin [\omega_s t + \varphi(0)]$$

An appropriate phasor relationship would be

$$f_a(t) \leftrightarrow \underline{F} = Fe^{j\theta} = \frac{F_q}{\sqrt{2}} \underline{/\varphi(0)} - j\frac{F_d}{\sqrt{2}} \underline{/\varphi(0)}$$

$$= \frac{e^{j\varphi(0)}}{\sqrt{2}} (F_q - jF_d)$$

(1)

Note that the angle θ introduced here is simply the angle associated with the phasor $\underline{F}$. Next, we establish a relationship between the initial rotor angle $\varphi(0)$ and the torque angle δ. The torque angle δ as defined in Chapter 4 is illustrated in Figure 6.9. In the figure the angle δ is the angle between the machine terminal voltage phasor $\underline{V}_s$ and the phasor

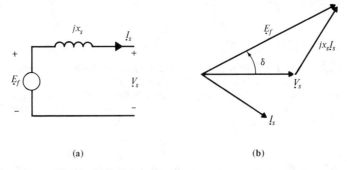

(a)   (b)

FIGURE 6.9 Steady-State Equivalent Circuit for a Round-Rotor Synchronous Machine with a Negligible Armature Resistance and a Phasor Diagram for a Typical Operating Point

representing field-induced voltage E_f. As shown in Chapter 4, the voltage E_f is given by [see (4.45)]

$$E_f = \frac{j\omega_s MI_f}{\sqrt{2}} \underline{/\theta_r(0)}$$

$$= \frac{\omega_s MI_f}{\sqrt{2}} \underline{/\theta_r(0) + \pi/2}$$

$$= \frac{\omega_s MI_f}{\sqrt{2}} \underline{/\varphi(0)}$$

where I_f is the assumed constant field winding current referred to the armature windings and in general $\varphi = \theta_r + \pi/2$ (see Figure 6.3). In many cases it is convenient to take the angle of V_s as zero, which implies that $\delta = \varphi(0)$. In essence, the assumption of a particular steady-state operating condition together with the assumption $V_s = V_s \underline{/0°}$ effectively fixes $\theta_r(0)$ [and $\varphi(0)$, which differs by $\pi/2$ radians].

Substituting in the relationship above between F and $F_q - jF_d$, that is, (1), we find

$$F_q - jF_d = \sqrt{2}\, Fe^{j\theta}e^{-j\varphi(0)}$$

$$= \sqrt{2}\, Fe^{j\theta}e^{-j\delta}$$

$$= \sqrt{2}\, Fe^{-j(\delta - \theta)}$$

$$= \sqrt{2}\, F[\cos(\delta - \theta) - j\sin(\delta - \theta)]$$

from which

$$F_q = \sqrt{2}\, F \cos(\delta - \theta) \tag{2}$$

$$F_d = \sqrt{2}\, F \sin(\delta - \theta) \tag{3}$$

Equations (2) and (3) suggest the geometric relationships illustrated in Figure 6.10. This figure should be compared to Figures 4.31, 4.32, and 4.34, where direct and quadrature axes were first introduced. It should be apparent that direct- and quadrature-axis variables for a particular steady-state condition can be obtained by multiplying the appropriate phasor by $\sqrt{2}$ and then resolving it into components along the direct and quadrature axes. That is, $F_d = |F_d'|$ and $F_q = |F_q'|$. The $\sqrt{2}$ factor can be eliminated through the use of normalized (per unit) quantities with properly chosen base values, as will be shown in Section 6.3.9.

In this particular case we first find the angle δ as in Example 4.1. The voltage is arbitrarily assigned the zero reference angle:

$$V_s = 120 \underline{/0°}$$

Armature current is determined based on the given steady-state operating condition:

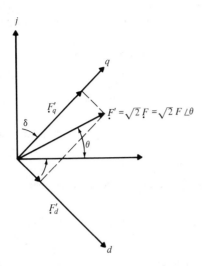

FIGURE 6.10 Steady-State Relationships Between Phasors and *dq* Variables

$$2V_s I_s = 2 \text{ kVA}$$

$$I_s = \frac{2000}{2(120)} = 8.33 \text{ A}$$

At a lagging power factor of 0.85, we have

$$I_s = 8.33 \underline{/-\cos^{-1}(0.85)}$$

$$= 8.33 \underline{/-31.8°}$$

The field-induced voltage $E_f = E_f \underline{/\delta}$ is given by

$$E_f = jx_s I_s + V_s$$

$$= j(12.5)(8.33 \underline{/-31.8°}) + 120 \underline{/0°}$$

$$= 174.9 + j88.5$$

$$= 196.0 \underline{/26.84°}$$

Thus the angle δ is 26.84°. Figure 6.11 illustrates the results.

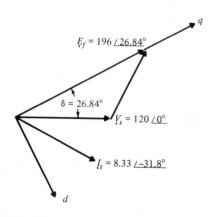

FIGURE 6.11

Resolving appropriate phasors into components along the direct and quadrature axes, we find

$$V_q = \sqrt{2}\,(120)\cos 26.84° = 151.4 \text{ V}$$

$$V_d = \sqrt{2}\,(120)\sin 26.84° = 76.62 \text{ V}$$

$$I_q = \sqrt{2}\,(8.33)\cos(26.84° + 31.8°) = 6.133 \text{ A}$$

$$I_d = \sqrt{2}\,(8.33)\sin(26.84° + 31.8°) = 10.06 \text{ A}$$

The angle $\varphi(0)$ is the angle δ:

$$\varphi(0) = \delta = 26.84° \text{ or } 0.468 \text{ rad}$$

The remaining variables can be found by considering the armature voltage equations given previously as (6.22):

$$v_d = -r_s i_d - \omega_r \lambda_q + \frac{d\lambda_d}{dt}$$

$$v_q = -r_s i_q + \omega_r \lambda_d + \frac{d\lambda_q}{dt}$$

For an assumed steady-state operating condition and negligible armature resistance, these relationships can be expressed as

$$V_d = -\omega_s \lambda_q$$

$$V_q = +\omega_s \lambda_d$$

where the constant flux linkages are related to the constant currents by (see Figure 6.5 or 6.6)

$$\lambda_d = -L_s I_d + M I_f$$

$$\lambda_q = -L_s I_q$$

Substituting in the relationship for V_q yields

$$V_q = \omega_s(-L_s I_d + M I_f)$$

$$= -x_s I_d + \omega_s M I_f$$

Thus

$$i_r' = I_f = \frac{V_q + x_s I_d}{\omega_s M} = \frac{V_q + x_s I_d}{x_m}$$

$$= \frac{151.4 + (12.5)(10.06)}{(0.8)(12.5)}$$

$$= 27.72 \text{ A}$$

where the magnetizing reactance $x_m = \omega_s M$ is 80% of the synchronous reactance. As an alternative (or check) we can find $i_r' = I_f$ from (4.45).

$$E_f = |E_f| = \frac{\omega_s M I_f}{\sqrt{2}}$$

Thus

$$I_f = \frac{\sqrt{2}\,E_f}{x_m}$$

$$= \frac{\sqrt{2}\,(196.0)}{(0.8)(12.5)} = 27.72 \text{ A}$$

The example shows the relationship between qd variables and the phasors customarily used in steady-state analysis. Had we chosen flux linkages as state variables rather than currents, initial values of flux linkages would be required, and they can similarly be determined following the procedure described above.

6.3.3 TORQUE RELATIONSHIPS

Referring to the equivalent circuits of Figure 6.6, we can formulate appropriate relationships for mechanical power converted and developed electrical torque. The resistances r_s and r_r' in the equivalent circuit give rise to i^2R losses. The inductances $L_{\ell s}$, $L_{\ell r}'$, and M determine stored magnetic energy in the machine. The dependent voltage sources determine the mechanical power converted:

$$p_m = p_{mq} + p_{md} \tag{6.24}$$

$$p_m = \omega_r \lambda_d i_q - \omega_r \lambda_q i_d$$

The developed electrical torque is given by

$$T = \frac{p_m}{\omega_{rm}} \tag{6.25}$$

where ω_{rm} is the rotor speed of rotation in mechanical radians per second. Recall that $\omega_{rm} = (2/P)\,\omega_r$, where P is the number of poles. Thus the torque can be expressed as

$$T = \frac{P}{2}(\lambda_d i_q - \lambda_q i_d) \tag{6.26}$$

Care must be exercised in determining the torque in fundamental units using (6.26). Depending on the version of the transformation used, it may be necessary to multiply the result by a factor of $3/2$ for three-phase machines. If the transformation given by (6.17) and (6.19) is used, the factor is 1 and (6.26) yields the correct result. Troublesome factors associated with other transformations can generally be avoided by expressing quantities on a normalized or per unit basis, as will be discussed in Section 6.3.9.

6.3.4 THE MECHANICAL EQUATION OF MOTION

If the rotor of a machine operated as a motor and its connected load are treated as a single rigid, rotating mass with combined moment of inertia J, the equation of motion for the rotor can be expressed as

$$\frac{J\,d\omega_{rm}}{dt} = T - T_L - T_D \tag{6.27}$$

where ω_{rm} is the speed of rotation in mechanical radians per second. The developed torque T can be obtained from (6.26). T_L is the load torque. T_D is the torque due to mechanical damping. It is often expressed as a function of speed of rotation. For a generator, the equation of motion for the rotor with similar assumptions can be expressed as

$$\frac{J\,d\omega_{rm}}{dt} = T_P - T - T_D \tag{6.28}$$

The developed torque T in (6.28) is also given by (6.26) provided that currents i_d and i_q are assigned positive for generator action (i.e., with the active sign convention rather than the passive sign convention shown in Figure 6.6). T_p is the prime-mover torque. T_D is the torque due to mechanical damping. It is often neglected in the analysis of synchronous generators.

The machine speed and rotor angle are related by

$$\varphi = \int_0^t \omega_r(\tau) \, d\tau + \varphi(0) \tag{6.29}$$

or

$$\frac{d\varphi}{dt} = \omega_r(t) \tag{6.30}$$

The angle $\varphi(t)$ used in the transformation relationships must be expressed in electrical degrees or radians. Thus, in (6.29), we must have

$$\omega_r(t) = \frac{P}{2} \omega_{rm}(t) \tag{6.31}$$

Equations (6.27) and (6.28) are of the general form

$$\frac{J \, d\omega_{rm}}{dt} = T'_{acc} \tag{6.32}$$

where T'_{acc} is the accelerating torque expressed in appropriate units (i.e., newton-meters in the SI system). The inertia constant H defined in Chapter 4 is given by

$$H = \frac{\frac{1}{2}J\omega_{sm}^2}{VA_{rating}} \qquad \text{s} \tag{6.33}$$

Thus

$$J = \frac{2H \, VA_{rating}}{\omega_{sm}^2} \qquad \text{kg} \cdot \text{m}^2$$

Substituting in (6.32), we have

$$\frac{2H}{\omega_{sm}} \frac{d\omega_{rm}}{dt} = \frac{T'_{acc}}{VA_{rating}/\omega_{sm}}$$

or $\qquad\qquad\qquad\qquad\qquad\qquad\qquad\qquad\qquad\qquad\qquad$ (6.34)

$$\frac{2H}{\omega_s} \frac{d\omega_r}{dt} = T_{acc}$$

where ω_r and ω_s are in electrical radians per second and T_{acc} is expressed as a normalized variable with a base of

$$T_{base} = \frac{VA_{rating}}{\omega_{sm}} \tag{6.35}$$

As discussed in Chapter 4, the constant H falls within a relatively narrow range for most machines. Typical values for synchronous machines are given in reference 1.

It is sometimes preferable to work with the angle $\delta(t)$ defined in Chapter 4.

$$\delta(t) = \int_0^t (\omega_r(\tau) - \omega_s)\, d\tau + \delta(0) \tag{6.36}$$

or

$$\frac{d\delta}{dt} = \omega_r(t) - \omega_s \tag{6.37}$$

The initial angle $\delta(0)$ is generally taken as the torque angle δ determined at a steady-state operating point prior to a disturbance. The angle $\delta(t)$ gives directly the manner in which the rotor moves with respect to a synchronously rotating reference frame following a disturbance.

6.3.5 MODIFICATIONS REQUIRED FOR SALIENT POLES

In Chapter 4 salient-pole effects for steady-state analysis were accounted for by using different magnetizing inductances (or reactances) for the direct and quadrature axes of a machine:

$$L_d = L_{\ell s} + L_{md} \tag{6.38a}$$

$$L_q = L_{\ell s} + L_{mq} \tag{6.38b}$$

If the qd axis representation for a round-rotor synchronous machine is similarly modified, we might conceptualize a salient pole machine as illustrated in Figure 6.12.

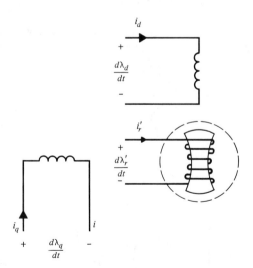

FIGURE 6.12 Fixed Angular Relationships Between Direct- and Quadrature-Axis Windings for a Salient-Pole Machine

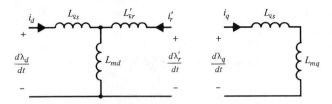

FIGURE 6.13 Coupled-Coil Equivalent Circuits for a Salient-Pole Synchronous Machine with a Single Direct-Axis Rotor Winding

The increased air gap along the quadrature axis implies higher reluctance and less magnetizing inductance along this axis, and we would expect L_{mq} to be less than L_{md}. An appropriate relationship between flux linkages and currents would be

$$\begin{bmatrix} \lambda_q \\ \lambda_d \\ \lambda_r' \end{bmatrix} = \begin{bmatrix} L_q & 0 & 0 \\ 0 & L_d & L_{md} \\ 0 & L_{md} & L_f' \end{bmatrix} \begin{bmatrix} i_q \\ i_d \\ i_r' \end{bmatrix} \tag{6.39}$$

where

$$L_f' = L_{\ell r}' + L_{md} \tag{6.40}$$

The flux linkage relationships given by (6.38), (6.39), and (6.40) suggest the coupled-coil equivalent circuits shown in Figure 6.13. The coupled-coil equivalent circuits shown in Figure 6.13 can be extended in the same manner as described in Section 6.3.2 to include the armature and field voltage relationships.

The foregoing assumptions imply a relationship between flux linkages and currents when armature circuit variables are expressed as phase variables. This relationship can be determined by following a procedure similar to that described in Example 6.1. Equation (6.39) is of the form

$$\overline{\lambda}^{qdr'} = \overline{L}^{qdr'}\overline{i}^{qdr'} \tag{6.41}$$

The transformation that refers phase variables to direct and quadrature axes of a frame of reference fixed in the rotor while leaving rotor variables unchanged was presented in Example 6.1:

$$\overline{f}^{abr'} = \overline{A}(\varphi)\overline{f}^{qdr'} \tag{6.42a}$$

$$\overline{f}^{qdr'} = \overline{A}^{-1}(\varphi)\overline{f}^{abc} \tag{6.42b}$$

where

$$\overline{A} = \begin{bmatrix} \begin{bmatrix} \overline{T}(\varphi) \end{bmatrix} & \begin{matrix} 0 \\ 0 \end{matrix} \\ 0 \ \ 0 & 1 \end{bmatrix} \tag{6.43a}$$

$$\overline{A}^{-1} = \begin{bmatrix} \begin{bmatrix} -1 \\ \overline{T}(\varphi) \end{bmatrix} & \begin{matrix} 0 \\ 0 \end{matrix} \\ 0 \ \ 0 & 1 \end{bmatrix} = \overline{A} \tag{6.43b}$$

Substituting in (6.41) yields

$$\overline{\lambda}^{qdr'} = \overline{L}^{qdr'}\overline{i}^{qdr'} \tag{6.44}$$

$$\overline{A}^{-1}\overline{\lambda}^{abr'} = \overline{L}^{qdr'}\overline{A}^{-1}\overline{i}^{abr'}$$

Premultiply both sides of (6.44) by $\overline{A}$:

$$\overline{\lambda}^{abr'} = \overline{A}\overline{L}^{qdr'}\overline{A}^{-1}\overline{i}^{abr} \tag{6.45a}$$

$$= \overline{L}^{abr'}\overline{i}^{abr}$$

It then follows that

$$\overline{L}^{abr'} = \overline{A}\overline{L}^{qdr'}\overline{A}^{-1} \tag{6.45b}$$

The evaluation of (6.45b) is somewhat tedious but yields in terms of $\theta_r = \varphi - \pi/2$,

$$\overline{L}^{abr'} = \begin{bmatrix} L_1 + L_2 \cos 2\theta_r & L_2 \sin 2\theta_r & L_{md} \cos \theta_r \\ & L_1 - L_2 \cos 2\theta_r & L_{md} \sin \theta_r \\ \text{(symmetric matrix)} & & L'_{\ell r} + L_{md} \end{bmatrix} \tag{6.45c}$$

where

$$L_1 = L_{\ell s} + \frac{L_{md} + L_{mq}}{2} \tag{6.46}$$

$$L_2 = \frac{L_{md} - L_{mq}}{2} \tag{6.47}$$

The matrix $\overline{L}^{abr'}$ should be compared with the matrix of (6.4) for a round-rotor machine. Note that if $L_{md} = L_{mq} = M$, the two matrices are the same as would be expected. The reader should attempt to justify the general nature of the variation of inductances in (6.45) with angle θ_r. A figure similar to Figure 6.2 but with a salient-pole rotor rather than a round rotor is helpful. The flux linkage relationships given by (6.45) are reasonable if harmonics in the spatial distributions of air-gap magnetic fields are neglected. The reader will recall that a similar assumption was necessary in Section 4.4 for steady-state analysis of the salient-pole machine.

6.3.6 MODIFICATIONS REQUIRED FOR ADDITIONAL ROTOR WINDINGS

Additional short-circuited rotor windings referred to as damper windings or amortisseur windings are sometimes used to improve rotor dynamics. Eddy currents induced in the iron of solid-iron rotor turbogenerators cause a similar effect. The effects of damper windings and eddy currents can be accounted for approximately by modifying the equivalent circuits of the machine as shown in Figure 6.14. Note that the additional rotor windings are similar to the winding used to represent the field circuit, but their terminal voltages are necessarily zero in that they are short circuited. Primed quantities are used to emphasize the fact that they are quantities referred to the armature by appropriate turns ratios or ratios squared.

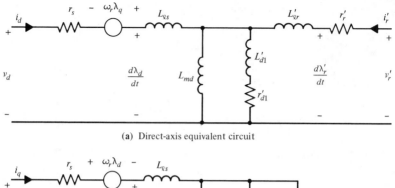

(a) Direct-axis equivalent circuit

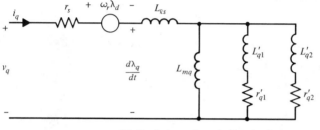

(b) Quadrature-axis equivalent circuit

FIGURE 6.14 Equivalent Circuits for a Synchronous Machine with Two Rotor Circuits Represented in Each Axis

Determination of an appropriate equivalent-circuit structure and suitable element values for a particular analysis is not always an easy task. References 2 and 3 address this issue. For transient stability studies relatively simple models are often adequate [4]. When machine transients are of interest, two rotor windings in each axis may be required to achieve reasonable results [5].

6.3.7 SYMMETRICAL MACHINES

The synchronous machine examples considered thus far treat a machine with an unsymmetrical rotor. The rotor is unsymmetrical in that the single field winding on the direct axis supplied by the field excitation system does not have a corresponding winding on the quadrature axis. A salient-pole synchronous machine has additional dissymmetry due to the configuration of the rotor itself.

It was shown in Chapter 2 that if windings displaced appropriately in space are excited by sinusoidal currents appropriately displaced in time, it is possible to establish a magnetic field that rotates with respect to the windings. A two-phase system is the simplest configuration capable of this. A more general statement can be made regarding magnetic fields established by symmetrical polyphase systems of windings. To see this, consider the configuration illustrated in Figure 6.15. The configuration is similar to that of Figure 2.28. Each of the two stator windings has N_s turns, and they are identical in all respects except for the fixed angle difference of $\pi/2$ radians between their axes. As shown in Chapter 2, the resultant magnetic field intensity in the air gap, assuming that each winding establishes a sinusoidally distributed magnetic field, can be expressed as

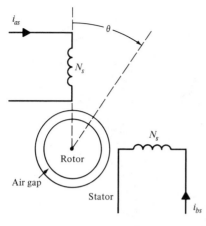

FIGURE 6.15 Machine with a Symmetrical Two-Phase Stator Winding

$$H_s(\theta) = H_{as}(\theta) + H_{bs}(\theta)$$

$$= \frac{N_s}{2g} \left[i_{as} \cos\theta + i_{bs} \cos\left(\theta - \frac{\pi}{2}\right) \right]$$

$$= \frac{N_s}{2g} (i_{as} \cos\theta + i_{bs} \sin\theta) \tag{6.48}$$

This can also be expressed as (using appropriate trigonometric identities)

$$H_s(\theta) = \frac{N_s}{2g} \sqrt{i_{as} + i_{bs}} \cos\left(\theta - \tan^{-1}\frac{i_{bs}}{i_{as}}\right) \tag{6.49}$$

Equation (6.49) shows that the *resultant* magnetic field intensity is also sinusoidally distributed with respect to the angle θ. The maximum value H_{smax} is given by

$$H_{smax} = \frac{N_s}{2g} \sqrt{i_{as}^2 + i_{bs}^2} \tag{6.50}$$

This maximum value occurs at an angle ψ with respect to the axis of the phase *a* winding, where

$$\psi = \tan^{-1}\frac{i_{bs}}{i_{as}} \tag{6.51}$$

It is easy to see that by appropriately choosing values for the currents i_{as} and i_{bs}, the magnitude H_{smax}, and the angle ψ of the resultant could be made to take on any values we might arbitrarily choose. Thus a symmetrical two-phase winding can establish *any* desired sinusoidally distributed field (including a rotating one) provided that currents are appropriately chosen. The same field could be established by a different symmetrical two-phase winding perhaps displaced in space from the first. Consider the system of windings illustrated in Figure 6.16.

6.3 THE USE OF TRANSFORMATIONS IN AC MACHINE ANALYSIS

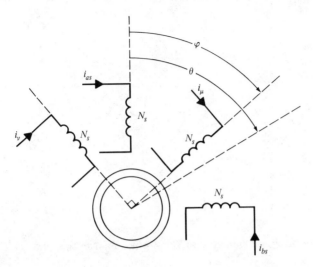

FIGURE 6.16 Machine with a Symmetrical Two-Phase Stator Winding and a Second Symmetrical Two-Phase Winding at an Arbitrary Angle θ

The configuration of Figure 6.16 is the same as that of Figure 6.15 but with a second symmetrical set of windings which is displaced by an arbitrary angle φ. This second set of windings carries currents i_u and i_v. The magnetic field in the gap due to the original set of windings is given by (6.48) [or (6.49)]. The magnetic field due to the second set of windings carrying currents i_u and i_v is given by

$$H(\theta) = \frac{N_s}{2g} \left[i_u \cos (\theta - \varphi) + i_v \cos \left(\theta - \varphi + \frac{\pi}{2} \right) \right] \qquad (6.52)$$

$$= \frac{N_s}{2g} [i_u \cos (\theta - \varphi) - i_v \sin (\theta - \varphi)]$$

Now apply the trigonometric identities:

$$\cos (\theta - \varphi) = \cos \theta \cos \varphi + \sin \theta \sin \varphi \qquad (6.53a)$$

$$\sin (\theta - \varphi) = \sin \theta \cos \varphi - \cos \theta \sin \varphi \qquad (6.53b)$$

and rewrite (6.52) as

$$H(\theta) = \frac{N_s}{2g} [(i_u \cos \varphi + i_v \sin \varphi) \cos \theta + (i_u \sin \varphi - i_v \cos \varphi) \sin \theta] \qquad (6.54)$$

If the fields due to the two sets of windings are to be the same, it must be true that $H(\theta) = H_s(\theta)$. This will be true if

$$i_{as} = i_u \cos \varphi + i_v \sin \varphi \qquad (6.55a)$$

$$i_{bs} = i_u \sin \varphi - i_v \cos \varphi \qquad (6.55b)$$

as can be seen by comparing (6.48) and (6.54). Equations (6.55) yield a transformation between ab variables and uv variables. The transformation is the same as the one given by (6.14) through (6.21) except that the angle φ is arbitrary rather than an angle associated with rotor position. This suggests that any symmetrical two-phase winding can be replaced with a fictitious symmetrical two-phase winding displaced by an arbitrary angle. The fictitious set of windings will establish the same sinusoidally distributed field as the original set of windings. The angle φ can be time varying or constant. In the synchronous machine examples considered thus far, the angle φ was chosen to be $\theta_r + \pi/2$. This resulted in a configuration of windings fixed with respect to the rotor. When both stator *and* rotor windings of a machine can be treated as symmetrical two-phase windings, other reference frames can be chosen. Krause and Thomas showed that if stator and rotor variables of a symmetrical induction machine are referred to any common reference frame, the variation of inductances with rotor angle will be eliminated [6, 7]. Consider the two-phase induction machine as an example. In Chapter 3 we considered the configuration illustrated in Figure 6.17. Note that both stator and rotor windings constitute symmetrical sets of windings. If variables associated with each set of windings are referred to a common reference frame (fixed or rotating), the inductances required to describe the resulting equivalent set of windings will be independent of the actual rotor

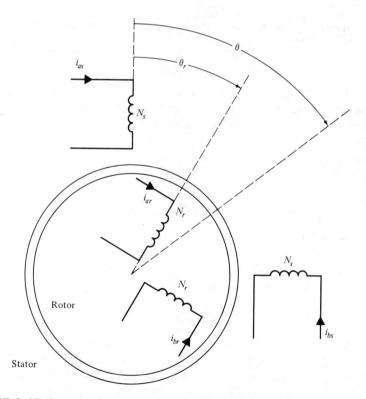

FIGURE 6.17 Symmetrical Two-Phase Induction Machine

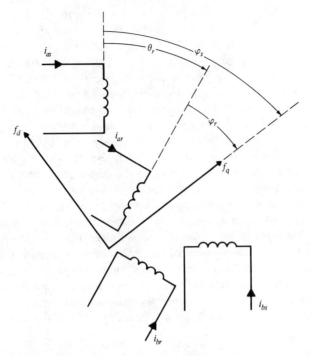

FIGURE 6.18 Referring Stator and Rotor Variables of an Induction Machine to a Common Reference Frame

angle θ_r. The common reference frame can be chosen as the orthogonal qd (or we might label them uv) set of axes shown in Figure 6.18.

An appropriate transformation relationship is

$$\begin{bmatrix} f_{as} \\ f_{bs} \\ f'_{ar} \\ f'_{br} \end{bmatrix} = \begin{bmatrix} \overline{T} \ (\varphi_s) & \begin{matrix} 0 & 0 \\ 0 & 0 \end{matrix} \\ \begin{matrix} 0 & 0 \\ 0 & 0 \end{matrix} & \overline{T} \ (\varphi_r) \end{bmatrix} \begin{bmatrix} f_{qs} \\ f_{ds} \\ f'_{qr} \\ f'_{dr} \end{bmatrix} \tag{6.56}$$

The inverse relationship is

$$\begin{bmatrix} f_{qs} \\ f_{ds} \\ f'_{qr} \\ f'_{dr} \end{bmatrix} = \begin{bmatrix} \overline{T}^{-1}(\varphi_s) & \begin{matrix} 0 & 0 \\ 0 & 0 \end{matrix} \\ \begin{matrix} 0 & 0 \\ 0 & 0 \end{matrix} & \overline{T}^{-1}(\varphi_r) \end{bmatrix} \begin{bmatrix} f_{as} \\ f_{bs} \\ f'_{ar} \\ f'_{br} \end{bmatrix} \tag{6.57}$$

In (6.56), the transformation matrix $\overline{T}$ is the same as that given by (6.17). The angles φ_s and φ_r are as shown in Figure 6.18. Note that $\varphi_r = \varphi_s - \theta_r$, where θ_r is the rotor angle. It is left as an exercise (Problem 6.6) to show that the flux linkage relationships given by (3.34) following the transformation can be expressed as

$$
\begin{bmatrix} \lambda_{qs} \\ \lambda_{ds} \\ \lambda'_{qr} \\ \lambda'_{dr} \end{bmatrix} = \begin{bmatrix} L_s & 0 & M & 0 \\ 0 & L_s & 0 & M \\ M & 0 & L'_r & 0 \\ 0 & M & 0 & L'_r \end{bmatrix} \begin{bmatrix} i_{qs} \\ i_{ds} \\ i'_{qr} \\ i'_{dr} \end{bmatrix}
\tag{6.58}
$$

where

$$L_s = L_{\ell s} + M$$

$$L'_r = L'_{\ell r} + M$$

and M is given by (3.27). The flux linkage relationships of (6.58) suggest the coupled-coil equivalent circuits shown in Figure 6.19.

The voltage equations for the machine of Figure 6.17 are for the armature windings

$$
v_{as} = i_{as} r_s + \frac{d\lambda_{as}}{dt}
\tag{6.59a}
$$

$$
v_{bs} = i_{bs} r_s + \frac{d\lambda_{bs}}{dt}
\tag{6.59b}
$$

and for the rotor windings

$$
v'_{ar} = i'_{ar} r'_r + \frac{d\lambda'_{ar}}{dt}
\tag{6.60a}
$$

$$
v'_{br} = i'_{br} r'_r + \frac{d\lambda'_{br}}{dt}
\tag{6.60b}
$$

These equations are transformed in the manner described in Section 6.3.2. For the armature voltage equations apply the transformation $\overline{T}(\varphi_s)$ and find

$$
v_{ds} = r_s i_{ds} - \frac{d\varphi_s}{dt} \lambda_{qs} + \frac{d\lambda_{ds}}{dt}
\tag{6.61a}
$$

$$
v_{qs} = r_s i_{qs} + \frac{d\varphi_s}{dt} \lambda_{ds} + \frac{d\lambda_{qs}}{dt}
\tag{6.61b}
$$

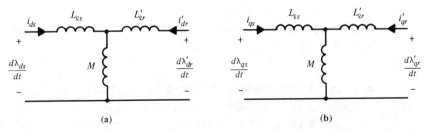

(a)　　　　　　　　　　　　(b)

FIGURE 6.19 Coupled-Coil Equivalent Circuits for a Symmetrical Two-Phase Induction Machine

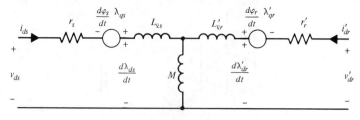

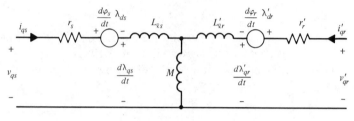

(b) Quadrature-axis equivalent circuit

FIGURE 6.20 Equivalent Circuits for a Symmetrical Two-Phase Induction Machine

For the rotor voltage equations apply the transformation $\overline{T}(\varphi_r)$ and find

$$v'_{dr} = r_s i'_{dr} - \frac{d\varphi_r}{dt} \lambda'_{qr} + \frac{d\lambda'_{dr}}{dt} \tag{6.62a}$$

$$v'_{qr} = r_s i'_{qr} + \frac{d\varphi_r}{dt} \lambda'_{dr} + \frac{d\lambda'_{qr}}{dt} \tag{6.62b}$$

Flux linkage and voltage relationships considered together suggest the equivalent circuits of Figure 6.20. Primed quantities are rotor quantities referred to the armature windings by appropriate turns ratios or ratios squared. If the induction machine operates with its rotor windings short circuited, voltages v'_{dr} and v'_{qr} would be set to zero. The speed voltage terms depend on the choice of the reference frame. For example, if the reference frame is fixed in the rotor with the q axis aligned with the axis of the phase a rotor winding, we would have $\varphi_r = 0$, $d\varphi_r/dt = 0$, $\varphi_s = \theta_r$, and $d\varphi_s/dt = d\theta_r/dt = \omega_r$.

Analysis of a symmetrical induction machine using the equivalent circuits of Figure 6.20 is similar in many ways to the analysis of a synchronous machine using the equivalent circuits of Figure 6.6.

6.3.8 THREE-PHASE VERSUS TWO-PHASE MACHINES

The behavior of three-phase machines can be explained in terms of equivalent two-phase machines. This technique has been used in Chapters 3 and 4 (as well as in this chapter) to reduce algebraic complexity. It is sometimes necessary in the analysis of three-phase machines to have a direct relationship between three-phase variables and qd variables. A transformation similar to the two-phase

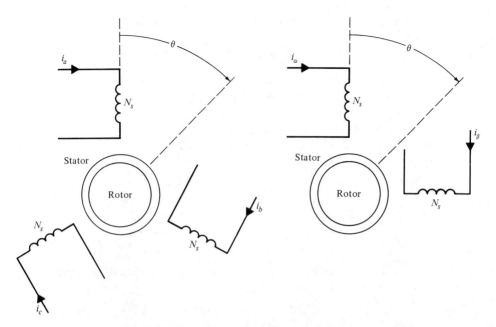

FIGURE 6.21 Symmetrical Three-Phase and Two-Phase Stator Windings

transformation can be used. Consider the winding configuration illustrated in Figure 6.21.

As shown in Chapter 2, the magnetic field intensity in the air gap of a symmetrical three-phase machine can be expressed as

$$H(\theta) = \frac{N_s}{2g} \left[i_a \cos\theta + i_b \cos\left(\theta - \frac{2\pi}{3}\right) + i_c \cos\left(\theta + \frac{2\pi}{3}\right) \right] \quad (6.63)$$

Now apply the trigonometric identity:

$$\cos(\theta + \varphi) = \cos\theta \cos\varphi - \sin\theta \sin\varphi \quad (6.64)$$

to the second and third terms of (6.63)

$$\cos\left(\theta - \frac{2\pi}{3}\right) = -\frac{1}{2}\cos\theta + \frac{\sqrt{3}}{2}\sin\theta \quad (6.65a)$$

$$\cos\left(\theta + \frac{2\pi}{3}\right) = -\frac{1}{2}\cos\theta - \frac{\sqrt{3}}{2}\sin\theta \quad (6.65b)$$

Substitute in (6.63) using (6.65) and rewrite (6.63) as

$$H(\theta) = \frac{N_s}{2g} \left[\left(i_a - \frac{1}{2}i_b - \frac{1}{2}i_c \right) \cos\theta + \left(+\frac{\sqrt{3}}{2}i_b - \frac{\sqrt{3}}{2}i_c \right) \sin\theta \right]$$

$$(6.66)$$

For the symmetrical two-phase configuration of Figure 6.21 we have

$$H'(\theta) = \frac{N_s}{2g} \left[i_\alpha \cos \theta + i_\beta \cos \left(\theta - \frac{\pi}{2} \right) \right]$$

(6.67)

$$= \frac{N_s}{2g} (i_\alpha \cos \theta + i_\beta \sin \theta)$$

The two-phase winding configuration establishes the same field intensity if $H'(\theta) = H(\theta)$. If we compare (6.66) and (6.67), it is evident that this can be accomplished by choosing i_α and i_β as

$$i_\alpha = i_a - \frac{1}{2} i_b - \frac{1}{2} i_c$$

(6.68a)

$$i_\beta = + \frac{\sqrt{3}}{2} i_b - \frac{\sqrt{3}}{2} i_c$$

(6.68b)

To obtain an inverse relationship, it is necessary to define a third variable. This is generally taken as

$$i_0 = \tfrac{1}{3} (i_a + i_b + i_c)$$

(6.68c)

This last variable is defined in a manner similar to the *zero-sequence* variable used in symmetrical components analysis and is so named. It will be zero for a Y-connected machine with no connection at the neutral point. Equations (6.68) yield a transformation between *abc* variables and $\alpha\beta0$ variables that can be used to obtain an equivalent two-phase machine (with an additional variable referred to as the zero-sequence variable) for any three-phase machine. It is easy to show by applying (6.68) that if the three-phase currents constitute a balanced three-phase set, that is, if $i_a = \cos \omega t$, $i_b = \cos (\omega t - 2\pi/3)$, and $i_c = \cos (\omega t + 2\pi/3)$, then $i_\alpha = \frac{3}{2} \cos \omega t$, $i_\beta = \frac{3}{2} \sin \omega t$, and $i_0 = 0$. Thus, if quantities are expressed on a normalized basis, balanced three-phase currents of unit magnitude would yield currents having a magnitude of $\frac{3}{2}$ in the equivalent two-phase machine. This can be avoided by introducing a $\frac{2}{3}$ factor in (6.68a) and (6.68b). With this modification, a transformation can be formulated as

$$\begin{bmatrix} f_\alpha \\ f_\beta \\ f_0 \end{bmatrix} = \frac{2}{3} \begin{bmatrix} 1 & -\frac{1}{2} & -\frac{1}{2} \\ 0 & +\frac{\sqrt{3}}{2} & -\frac{\sqrt{3}}{2} \\ \frac{1}{2} & \frac{1}{2} & \frac{1}{2} \end{bmatrix} \begin{bmatrix} f_a \\ f_b \\ f_c \end{bmatrix}$$

(6.69a)

The inverse relationship is given by

$$\begin{bmatrix} f_a \\ f_b \\ f_c \end{bmatrix} = \begin{bmatrix} 1 & 0 & 1 \\ -\frac{1}{2} & +\frac{\sqrt{3}}{2} & 1 \\ -\frac{1}{2} & -\frac{\sqrt{3}}{2} & 1 \end{bmatrix} \begin{bmatrix} f_\alpha \\ f_\beta \\ f_0 \end{bmatrix}$$

(6.69b)

The transformation given by (6.69) is one version of the transformation used in analysis by Clarke's components. Some authors describe use of the transformation as one of referring three-phase variables to a stationary reference frame and accordingly label variables $qd0$ rather than $\alpha\beta0$.

Analysis of a three-phase machine can be accomplished by using the transformation relationships of (6.69) to obtain an equivalent two-phase machine. The transformation given by (6.14) through (6.21) can be used to transform $\alpha\beta$ variables to qd variables in whatever reference frame (fixed or rotating) is desired. The zero-sequence quantities associated with $\alpha\beta0$ variables are used without modification. Thus, for a three-phase machine, typically we consider $qd0$ variables. Equivalently, the two transformations required can be expressed as a single transformation.

$$
\begin{bmatrix} f_q \\ f_d \\ f_0 \end{bmatrix} = \begin{bmatrix} \cos\varphi & \sin\varphi & 0 \\ \sin\varphi & -\cos\varphi & 0 \\ 0 & 0 & 1 \end{bmatrix} \begin{bmatrix} f_\alpha \\ f_\beta \\ f_0 \end{bmatrix}
$$

(6.70a)

$$
= \frac{2}{3} \begin{bmatrix} \cos\varphi & \sin\varphi & 0 \\ \sin\varphi & -\cos\varphi & 0 \\ 0 & 0 & 1 \end{bmatrix} \begin{bmatrix} 1 & -\frac{1}{2} & -\frac{1}{2} \\ 0 & +\frac{\sqrt{3}}{2} & -\frac{\sqrt{3}}{2} \\ \frac{1}{2} & \frac{1}{2} & \frac{1}{2} \end{bmatrix} \begin{bmatrix} f_a \\ f_b \\ f_c \end{bmatrix}
$$

$$
= \frac{2}{3} \begin{bmatrix} \cos\varphi & \cos\left(\varphi - \frac{2\pi}{3}\right) & \cos\left(\varphi + \frac{2\pi}{3}\right) \\ \sin\varphi & \sin\left(\varphi - \frac{2\pi}{3}\right) & \sin\left(\varphi + \frac{2\pi}{3}\right) \\ \frac{1}{2} & \frac{1}{2} & \frac{1}{2} \end{bmatrix} \begin{bmatrix} f_a \\ f_b \\ f_c \end{bmatrix}
$$

The inverse transformation is given by

$$
\begin{bmatrix} f_a \\ f_b \\ f_c \end{bmatrix} = \begin{bmatrix} \cos\varphi & \sin\varphi & 1 \\ \cos\left(\varphi - \frac{2\pi}{3}\right) & \sin\left(\varphi - \frac{2\pi}{3}\right) & 1 \\ \cos\left(\varphi + \frac{2\pi}{3}\right) & \sin\left(\varphi + \frac{2\pi}{3}\right) & 1 \end{bmatrix} \begin{bmatrix} f_q \\ f_d \\ f_0 \end{bmatrix}
$$

(6.70b)

Transformations used by various authors differ somewhat depending on the choice of reference angles, but the underlying theory is essentially the same. The choice of reference angle for the transformation given by (6.70) is as shown in Figure 6.22. The transformation given by (6.70) is similar to the transformation used by R. H. Park [8]. It differs only in the definition of the angle φ. Many authors loosely refer to all such transformations for three-phase machines as *Park's transformation*.

6.3 THE USE OF TRANSFORMATIONS IN AC MACHINE ANALYSIS

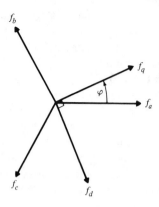

FIGURE 6.22 Conventions Assumed in the Transformation of Equation (6.70)

6.3.9 ANALYSIS IN TERMS OF NORMALIZED QUANTITIES

Significant advantages follow if quantities are expressed on a normalized or *per unit* basis. The per unit value of a quantity (variable or parameter) is obtained by dividing the quantity, expressed in fundamental units, by an appropriately chosen base value. For example, if machine terminal voltage is 13.0 kV and the chosen base is 13.8 kV, which is a standard rms line-to-line voltage, voltage on a per unit basis or per unit voltage is $13.0/13.8 = 0.942$. Conversely, a per unit value of 0.942 implies a line-to-line rms voltage of 13.0 kV, a line-to-neutral rms voltage of $13.0/\sqrt{3} = 7.51$ kV, and a line-to-neutral peak voltage of $\sqrt{2}\,(13.0)/\sqrt{3}$ kV $= 10.61$ kV. In essence, it is understood that base $kV_L = \sqrt{3}$ base kV_{LN} and that base $kV_{peak} = \sqrt{2}$ base kV, base V $= 1000$ base kV, and so on.

Base quantities must be chosen in a particular way if familiar relationships are to remain valid with all quantities expressed in per unit. For example, volt-amperes for a single-phase ac circuit is given by

$$VA = VI \tag{6.71}$$

If base quantities are chosen such that

$$base\ VA = (base\ V)(base\ I) \tag{6.72a}$$

then

$$\frac{VA}{base\ VA} = \frac{V}{base\ V} \cdot \frac{I}{base\ I} \tag{6.72b}$$

$$VA_{p.u.} = V_{p.u.}I_{p.u.}$$

The relationship between per unit quantities given by (6.72b) is the same as the relationship between variables expressed in fundamental units (6.71). The subscript "p.u." is used here only for purposes of illustration. It is normally omitted when use of the per unit system is understood. The base power factor is customarily chosen as 1.0 in that power factor is a dimensionless quantity. This implies that

$$\text{base } P = \text{base VA} = \text{base } V \cdot \text{base } I \tag{6.73}$$

We also must have

$$\text{base } Z = \frac{\text{base } V}{\text{base } I} \tag{6.74}$$

It is easy to show that

$$\text{base } Z = \frac{(\text{base kV}_{\text{LN}})^2 \times 1000}{\text{base kVA}_{1\varphi}} = \frac{(\text{base kV}_{\text{L}})^2 \times 1000}{\text{base kVA}_{3\varphi}} \tag{6.75}$$

where base $\text{kVA}_{1\varphi}$ is the kVA base on a single-phase basis and base $\text{kVA}_{3\varphi} = 3 \cdot \text{base kVA}_{1\varphi}$ is the kVA base on a three-phase basis. A more complete treatment of the per unit system is contained in reference 9. A major advantage of the per unit system is that impedances of various power system components tend to fall in a relatively narrow range when they are expressed in "per unit" on base values corresponding to rated values. For example, the synchronous reactance of a machine is typically 1.0 to 2.0 in per unit when the base VA is taken as the machine volt-ampere rating and base V is taken as the machine voltage rating.

The transformation of machine phase variables to qd or $qd0$ variables requires a few additional manipulations in order to realize the advantages of the system. The armature voltage equations (6.22) were developed in Section 6.3.2. They are restated here for convenience:

$$v_q = r_s i_q + \omega_r \lambda_d + \frac{d\lambda_q}{dt} \tag{6.76a}$$

$$v_d = r_s i_d - \omega_r \lambda_q + \frac{d\lambda_d}{dt} \tag{6.76b}$$

These equations can be expressed as

$$v_q = r_s i_q + \frac{\omega_r}{\omega_s} \psi_d + \frac{1}{\omega_s} \frac{d\psi_q}{dt} \tag{6.77a}$$

$$v_d = r_s i_d - \frac{\omega_r}{\omega_s} \psi_q + \frac{1}{\omega_s} \frac{d\psi_d}{dt} \tag{6.77b}$$

where $\psi_d = \omega_s \lambda_d$ and $\psi_q = \omega_s \lambda_q$. [The use of the symbol ψ differs here from the usage in Chapter 4, where it denoted the phasor representation for flux linkages varying sinusoidally with time in a steady-state condition. No such restriction is implied in (6.77).] Consideration of (6.77) shows that the ψ variables will be expressed in volts. (Alternatively, $\psi = \omega_s \lambda$ is in flux linkages per second, which is dimensionally the same as volts.) Furthermore, dimensionally we have $\psi = \omega_s \lambda = \omega_s L i = xi$, where x is reactance in ohms. It is common practice to replace inductances with reactances in all of our equivalent circuits developed earlier. When this option is chosen, it is important to remember that, in general, we have for voltage developed across a reactance,

$$v = L\frac{di}{dt} = \frac{x}{\omega_s}\frac{di}{dt} \tag{6.78}$$

where i is the current through the reactance.

Armed with these notions, the armature-voltage equations in per unit are the same as when expressed in fundamental units, that is, (6.77) provided that base $V_{qd0} =$ base $Z_{qd0} \cdot$ base I_{qd0}. This can be seen by simply dividing both sides of the equations by base V_{qd0}.

Factors of $\frac{3}{2}$ and $\sqrt{2}$ that occurred earlier can be eliminated by appropriately relating phase base variables to $qd0$ base variables. Consider the three-phase transformation given by (6.70). It is not difficult to show (see Problem 6.8) that total instantaneous power at the terminals of a three-phase machine is given by

$$p = v_a i_a + v_b i_b + v_c i_c = \tfrac{3}{2}(v_q i_q + v_d i_d + 2v_0 i_0) \tag{6.79}$$

If we choose

$$\text{base VA}_{3\varphi} = \tfrac{3}{2}\text{ base VA}_{qd0} \tag{6.80}$$

the $\frac{3}{2}$ factor in (6.79) will be eliminated when the relationship is expressed in per unit. Thus we have in per unit

$$v_a i_a + v_b i_b + v_c i_c = v_q i_q + v_d i_d + 2v_0 i_0 \tag{6.81}$$

It follows from (6.80) that

$$\text{base VA}_{1\varphi} = \tfrac{1}{2}\text{ base VA}_{qd0} \tag{6.82}$$

We may choose

$$\text{base V}_{\text{LN}} = \frac{\text{base }V_{qd0}}{\sqrt{2}} \tag{6.83a}$$

$$\text{base }I_{\text{L}} = \frac{\text{base }I_{qd0}}{\sqrt{2}} \tag{6.83b}$$

and satisfy (6.82). With these choices,

$$\text{base }Z_{\text{abc}} = \frac{\text{base V}_{\text{LN}}}{\text{base }I_{\text{L}}} = \frac{\text{base }V_{qd0}}{\text{base }I_{qd0}} = \text{base }Z_{qd0} \tag{6.84}$$

Furthermore, our relationships between phasors and constant qd variables for steady-state operation [developed in Example 6.4 and given as (2) and (3) of the example] can be written without the $\sqrt{2}$ factor if expressed in per unit.

$$F_q = F\cos(\delta - \theta) \tag{6.85a}$$

$$F_d = F\sin(\delta - \theta) \tag{6.85b}$$

(Recall that a phasor determined on the basis of steady-state analysis is expressed as $F = F\,\underline{/0}$. The angle δ is the angle between the quadrature axis of a synchronously rotating reference frame and the phase a winding axis at time $t = 0$. It is the steady-state power angle for a synchronous machine.)

The developed torque in fundamental units for a three-phase machine in terms of qd variables is

$$T = \frac{3}{2}\frac{P}{2}(\lambda_d i_q - \lambda_q i_d)$$

$$= \frac{3}{2}\frac{P}{2}\frac{1}{\omega_s}(\psi_d i_q - \psi_q i_d)$$

(6.86)

The equation of motion in per unit was developed in Section 6.3.4:

$$\frac{2H}{\omega_s}\frac{d\omega_r}{dt} = T_{acc}$$

(6.87)

where the base torque was

$$\text{base } T = \frac{VA_{rated}}{\omega_{sm}} = \frac{P}{2}\frac{VA_{rated}}{\omega_s}$$

(6.88)

If $VA_{3\varphi}$ is taken as VA_{rated},

$$\text{base } T = \frac{P}{2}\frac{VA_{3\varphi}}{\omega_s} = \frac{P}{2}\frac{3}{2}\frac{1}{\omega_s}\text{ base } V_{qd0}\text{ base } I_{qd0}$$

(6.89)

If (6.86) is divided by base T as given by (6.89), it is easy to see that the expression for developed torque with all variables expressed in per unit will be

$$T = \psi_d i_q - \psi_q i_d$$

(6.90)

As an example, consider a generator with mechanical damping neglected and the prime-mover torque T_P assumed to be expressed in per unit. The equation of motion could be written

$$\frac{2H}{\omega_s}\frac{d\omega_r}{dt} = T_P - T$$

$$= T_P - (\psi_d i_q - \psi_q i_d)$$

(6.91)

Equations (6.77), (6.85), (6.87), (6.90), and (6.91) apply as well for a two-phase machine when all variables are properly expressed in per unit. This can be seen by following a similar development starting with base $VA_{2\varphi}$ = base VA_{qd} and base $VA_{1\varphi}$ = base $VA_{qd}/2$, and so on.

EXAMPLE 6.5 A 100-MVA 13.8 kV Y-connected salient-pole generator without damper windings has the following parameters:

$$x_d = 1.25$$

$$x_q = 0.70$$

$$x_{\ell s} = 0.20$$

$$x_d' = 0.30$$

$$r_s = \text{negligible}$$

$$\tau_{d0}' = 5\text{ s}$$

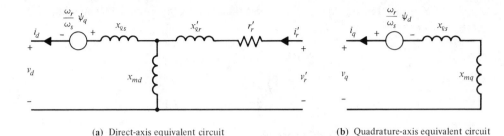

(a) Direct-axis equivalent circuit   (b) Quadrature-axis equivalent circuit

FIGURE 6.23 Direct- and Quadrature-Axis Equivalent Circuits for Example 6.5

Reactances are expressed in per unit with the machine rated MVA and kV as base values. If the machine delivers rated MVA at rated voltage and at a power factor of 0.90 lagging, find:

(a) Equivalent circuits for the direct and quadrature axes of the machine.
(b) ψ_d and ψ_q at the operating point described.

Equivalent circuits for a salient-pole machine with a single direct-axis rotor winding (the field winding) are appropriate. The circuits are as shown in Figure 6.23. The armature leakage reactance $x_{\ell s}$ is assumed to be the same in each axis and is given in the initial data as $x_{\ell s} = 0.20$. From our earlier definitions, $x_d = x_{md} + x_{\ell s}$ and $x_q = x_{mq} + x_{\ell s}$. Thus

$$x_{md} = x_d - x_{\ell s}$$
$$= 1.25 - 0.20 = 1.05$$
$$x_{mq} = x_q - x_{\ell s}$$
$$= 0.70 - 0.20 = 0.50$$

The reactance x_d' is the direct-axis transient reactance. It is by definition the limiting value (asymptote) for the direct-axis operational reactance at high frequency when there are no direct-axis damper windings. To clarify this, consider an alternative but equivalent description of the coupled-coil equivalent circuits for the synchronous machine. In terms of reactances, the coupled-coil equivalents for a machine with a single direct-axis rotor winding are as shown in Figure 6.24.

A generalized description of the equivalent circuits in terms of operational reactances and transfer relationships can be formulated as

$$\psi_d(s) = -X_d(s)I_d(s) + G(s)V_r'(s)$$
$$\psi_q(s) = -X_q(s)I_q(s)$$

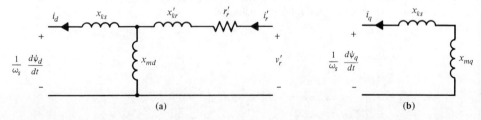

(a)   (b)

FIGURE 6.24 Coupled-Coil Equivalent Circuits for a Machine with a Single Direct-Axis Rotor Circuit

     ANALYSIS OF AC MACHINE TRANSIENTS

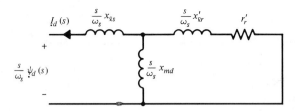

FIGURE 6.25 Equivalent Circuit for Determination of the Direct-Axis Operational Reactance for a Machine with a Single Direct-Axis Rotor Circuit

where Laplace transforms of variables are related by operational reactances $X_d(s)$ and $X_q(s)$ and the transfer relationship $G(s)$. These two relationships for ψ_d and ψ_q can be used to describe machines with any number of rotor circuits. Note that the direct-axis operational reactance $X_d(s)$ is given by

$$X_d(s) = \frac{\psi_d(s)}{-I_d(s)}\bigg|_{V_r'(s) = 0}$$

For the machine of this example an expression for $X_d(s)$ can be obtained by considering the circuit of Figure 6.25.

The operational reactance $X_d(s)$ is related to impedance $Z_d(s)$ at the terminals of the equivalent circuit by

$$Z_d(s) = \frac{s}{\omega_s}\frac{\psi_d(s)}{-I_d(s)} = \frac{s}{\omega_s}X_d(s)$$

The impedance $Z_d(s)$ can be obtained by making appropriate series–parallel combinations in the equivalent circuit of Figure 6.25. The high-frequency asymptote for $Z_d(s)$ is obtained by neglecting the resistance in the field circuit:

$$\lim_{s\to\infty} Z_d(s) = \frac{s}{\omega_s}\left(x_{\ell s} + \frac{x_{md}x_{\ell r}'}{x_{md} + x_{\ell r}'}\right)$$

It follows that

$$\lim_{s\to\infty} X_d(s) = x_d' = x_{\ell s} + \frac{x_{md}x_{\ell r}'}{x_{md} + x_{\ell r}'}$$

The transient reactance x_d' was specified as 0.30 in the given data. Solving for the remaining reactance $x_{\ell r}'$ with $x_{\ell s} = 0.20$ and $x_{md} = 1.05$, we find

$$x_{\ell r}' = 0.1105$$

The remaining circuit element r_r' is obtained using the given direct-axis open-circuit time constant introduced earlier in Example 6.2:

$$\tau_{do}' = 5 = \frac{1}{\omega_s}\frac{x_{md} + x_{\ell r}'}{r_r'}$$

We find

$$r_r' = 0.000616$$

Equivalent circuits with element values in per unit are shown in Figure 6.26.

In order to eliminate ψ_d and ψ_q at the given operating point, locate the quadrature axis of the machine using x_q. (It may be helpful to review the technique for doing this illustrated in Section 4.4.) Armature voltage can be taken as

$$V_s = 1\,\underline{/0°}$$

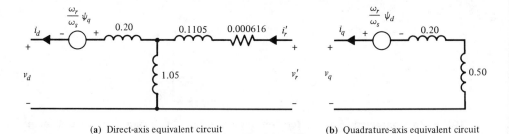

(a) Direct-axis equivalent circuit **(b)** Quadrature-axis equivalent circuit

FIGURE 6.26 Equivalent Circuits for the Machine of Example 6.5 with Element Values in Per Unit

At a power factor of 0.90 lagging, armature current will be

$$I_s = 1 \,\underline{/-\cos^{-1}(0.9)} = 1 \,\underline{/-25.84°}$$

The voltage E_q locates the quadrature axis:

$$E_q = V_s + jx_q I_s$$

$$= 1 + j(0.70)(1 \,\underline{/-25.84°})$$

$$= 1.305 + j0.630$$

$$= 1.449 \,\underline{/25.77°}$$

Thus the angle δ is 25.77 °. Direct- and quadrature-axis values of armature voltage and current can be found using (6.85) or equivalently by considering the geometry illustrated in Figure 6.27.

$$V_q = |V_q| = \cos \delta = 0.9005$$

$$V_d = |V_d| = \sin \delta = 0.4348$$

$$I_q = |I_q| = \cos (\delta + 25.84°) = 0.6210$$

$$I_d = |I_d| = \sin (\delta + 25.84°) = 0.7838$$

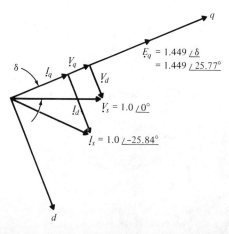

FIGURE 6.27 Diagram for Determining Direct- and Quadrature-Axis Components of Armature Voltage and Current for the Machine of Example 6.5

The flux linkages ψ_d and ψ_q are found by considering the armature voltage equations (6.77)

$$v_q = -r_s i_q + \frac{\omega_r}{\omega_q} \psi_d + \frac{1}{\omega_s} \frac{d\psi_q}{dt}$$

$$v_d = -r_s i_d - \frac{\omega_r}{\omega_s} \psi_q + \frac{1}{\omega_s} \frac{d\psi_d}{dt}$$

For a steady-state operating point with negligible armature resistance, we see that

$$V_q = \psi_d \qquad \text{and} \qquad V_d = -\psi_q$$

Thus

$$\psi_d = V_q = 0.9005$$

$$\psi_q = -V_d = -0.4348$$

As a check, the flux linkages can be found from

$$\psi_q = -x_q I_q$$

$$= -(0.70)(0.6210) = -0.4347$$

$$\psi_d = -x_d I_d + x_{md} I_f$$

The voltage $x_{md} I_f$ is equal in magnitude to the field-induced voltage E_f. Thus

$$x_{md} I_f = |E_f| = E_q + (x_d - x_q) I_d$$

$$= 1.449 + (1.25 - 0.70)(0.7838)$$

$$= 1.8801$$

$$\psi_d = -x_d I_d + x_{md} I_f$$

$$= -(1.25)(0.7838) + 1.8801 = 0.9004$$

which agrees with our previous result except for round-off error.

To illustrate a further advantage of the per unit system, consider the developed torque as given by (6.90):

$$T = \psi_d i_q - \psi_q i_d$$

At a steady-state operating point with negligible armature resistance, we have (substitute for ψ_d and ψ_q to see this)

$$T = V_q I_q + V_d I_d$$

Thus per unit torque is equal to per-unit power in the steady state if armature resistance losses are neglected. In this case

$$T = (0.9005)(0.6210) + (0.4348)(0.7378)$$

$$= 0.90$$

which is the per unit power (the power factor in this case) specified for the example.

SYNCHRONOUS MACHINE TRANSIENTS

A broad class of synchronous machine transients can be investigated assuming constant rotor speed. In reality, the machine speed does not change appreciably from the synchronous speed for the first few cycles following a disturbance. Thus initial current and torque waveforms following a disturbance can be obtained by analyzing the linear system of equations that results when speed is assumed constant.

To illustrate some of the analytical techniques that might be employed, consider a machine with a single direct-axis rotor winding. If the speed is assumed constant at the synchronous speed, we have $\omega_r = \omega_s$ and the equivalent circuits will be as shown in Figure 6.28. If the disturbance of interest can be described by changes in the armature voltages v_d and v_q or field voltage v_r', the equivalent circuits can be solved to obtain the resulting changes in the currents i_d, i_q, and i_r'. Changes in the currents can be added to the initial currents to obtain total currents. If desired, flux linkages can then be obtained as simple linear combinations of the currents. Torque can then be obtained by applying (6.90), that is, $T = \psi_d i_q - \psi_q i_d$. For example, suppose that voltages are given by

$$v_d = V_d + \Delta v_d(t) \tag{6.92a}$$

$$v_q = V_q + \Delta v_q(t) \tag{6.92b}$$

$$v_r' = V_f + \Delta v_r'(t) \tag{6.92c}$$

where V_d, V_q, and V_f are constant values corresponding to a steady-state condition in existence prior to a disturbance. The changes $\Delta v_d(t)$, $\Delta v_q(t)$, and $\Delta v_r'(t)$

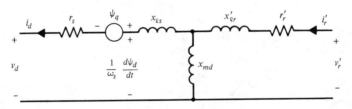

(a) Direct-axis equivalent circuit

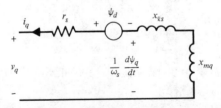

(b) Quadrature-axis equivalent circuit

FIGURE 6.28 Equivalent Circuits for a Synchronous Machine with a Single Direct-Axis Rotor Winding Assuming Constant Speed

would typically be zero for $t < 0$, assuming that the disturbance occurs at $t = 0$, and nonzero for $t \geq 0$. Currents would be given by

$$i_d = I_d + \Delta i_d(t) \tag{6.93a}$$

$$i_q = I_q + \Delta i_q(t) \tag{6.93b}$$

$$i_r' = I_f + \Delta i_r'(t) \tag{6.93c}$$

The equivalent circuits of Figure 6.28 are linear and the superposition principle can be applied. First find I_d, I_q, and I_f due to V_d, V_q, and V_f; then find Δi_d, Δi_q, and $\Delta i_r'$ due to Δv_d, Δv_q, and $\Delta v_r'$. In general, I_d, I_q, and I_f can be obtained by applying steady-state analysis techniques as in Examples 6.4 and 6.5. It then remains to find Δi_d, Δi_q, and $\Delta i_r'$. If the field voltage is assumed constant, $\Delta v_r'(t) = 0$ and appropriate equivalent circuits relating changes in variables would be as shown in Figure 6.29. If we refer to the figure, the following armature voltage relationships can be written:

$$\frac{1}{\omega_s}\frac{d}{dt}\Delta\psi_d = r_s\,\Delta i_d + \Delta\psi_q + \Delta v_d \tag{6.94a}$$

$$\frac{1}{\omega_s}\frac{d}{dt}\Delta\psi_q = r_s\,\Delta i_q - \Delta\psi_d + \Delta v_q \tag{6.94b}$$

The armature resistance in practical machines is generally small. If the armature resistance is neglected in (6.94), we have

$$\frac{1}{\omega_s}\frac{d}{dt}\Delta\psi_d = \Delta\psi_q + \Delta v_d \tag{6.95a}$$

$$\frac{1}{\omega_s}\frac{d}{dt}\Delta\psi_q = -\Delta\psi_d + \Delta v_q \tag{6.95b}$$

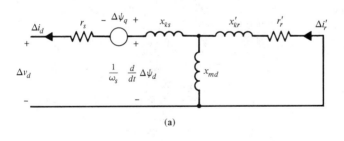

(a)

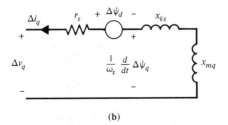

(b)

FIGURE 6.29 Equivalent Circuits Relating Changes in Variables Assuming Constant Field Voltage and Constant Speed

Note that for specified values of Δv_d and Δv_q, (6.95) can be solved directly for $\Delta \psi_d$ and $\Delta \psi_q$. Again, referring to Figure 6.29, we can easily see that

$$\Delta \psi_q = -(x_{\ell s} + x_{mq}) \Delta i_q \qquad (6.96)$$

$$= -x_q \Delta i_q$$

Equation (6.96) can be used to determine Δi_q once $\Delta \psi_q$ is known. The relationship for $\Delta \psi_d$ in terms of Δi_d is somewhat more complicated due to the presence of the field winding. Even so, for a given $\Delta \psi_d(t)$ it is not difficult to obtain $(1/\omega_s)(d/dt) \Delta \psi_d$ (a voltage), which could then be applied in the direct-axis equivalent circuit of Figure 6.29 to obtain both Δi_d and $\Delta i'_r$. If the field winding resistance is neglected, it is easy to see from Figure 6.29 that

$$\Delta \psi_d = -\left(x_{\ell s} + \frac{x_{md} x'_{\ell r}}{x_{md} + x'_{\ell r}} \right) \Delta i_d \qquad (6.97)$$

$$= -x'_d \Delta i_d$$

$$\Delta i'_r = \frac{x_{md}}{x'_{\ell r} + x_{md}} \Delta i_d \qquad (6.98)$$

Thus, if all resistances are neglected, (6.95) can be used to obtain $\Delta \psi_d$ and $\Delta \psi_q$. Then Δi_q can be obtained from (6.96), Δi_d from (6.97), and $\Delta i'_r$ from (6.98).

EXAMPLE 6.6 A synchronous machine is driven at constant synchronous speed by a prime mover. The armature windings are initially open circuited and field voltage is adjusted so that the armature terminal voltage is at rated value (i.e., 1.0 per unit). Find the armature current and machine torque immediately following a three-phase fault assumed to occur at time $t = 0$.

First find an appropriate set of initial values. If the angle of the armature terminal voltage is arbitrarily taken as zero,

$$V_s = 1.0 \, \underline{/0°}$$

Prior to the fault, the machine current I_s is zero. The quadrature axis is located by determining E_q:

$$E_q = V_s + jx_q I_s$$

$$= V_s$$

$$= 1.0 \, \underline{/0°}$$

Thus the angle δ is zero. The initial voltages V_q and V_d are

$$V_q = V_s \cos \delta = 1.0$$

$$V_d = V_s \sin \delta = 0.0$$

Following the three-phase fault, the armature winding terminal voltages will be zero. Thus we must have

$$\Delta v_q(t) = -1.0$$

$$\Delta v_d(t) = 0.0$$

Equations (6.95) can be written

$$\frac{d}{dt} \Delta\psi_d = \omega_s \, \Delta\psi_q$$

$$\frac{d}{dt} \Delta\psi_q = -\omega_s \, \Delta\psi_d - \omega_s$$

Solving for $\Delta\psi_d$ and $\Delta\psi_q$, where $\Delta\psi_d(0) = \Delta\psi_q(0) = 0$,

$$\Delta\psi_d = \cos \omega_s t - 1$$

$$\Delta\psi_q = -\sin \omega_s t$$

The currents Δi_d and Δi_q are obtained from (6.96) and (6.97):

$$\Delta i_d = \frac{-\Delta\psi_d}{x'_d} = \frac{1 - \cos \omega_s t}{x'_d}$$

$$\Delta i_q = \frac{-\Delta\psi_q}{x_q} = \frac{\sin \omega_s t}{x_q}$$

The changes in currents are the same as the total currents in that $I_d = I_q = 0$. Thus

$$i_d = \Delta i_d = \frac{1 - \cos \omega_s t}{x'_d}$$

$$i_q = \Delta i_q = \frac{\sin \omega_s t}{x_q}$$

Now apply the transformation (6.70) to find phase currents:

$$i_a = i_q \cos \varphi + i_d \sin \varphi + i_0$$

$$i_b = i_q \cos \left(\varphi - \frac{2\pi}{3} \right) + i_d \sin \left(\varphi - \frac{2\pi}{3} \right) + i_0$$

$$i_c = i_q \cos \left(\varphi + \frac{2\pi}{3} \right) + i_d \sin \left(\varphi + \frac{2\pi}{3} \right) + i_0$$

The angle φ for constant synchronous speed is given by

$$\varphi = \omega_s t + \varphi(0)$$

$$= \omega_s t + \delta$$

$$= \omega_s t$$

where the angle δ is zero. For a three-phase fault without connection to ground, or for a three-wire machine, the phase currents must sum to zero. In such cases, $i_0 = 0$ from (6.70a). We will make this assumption in what follows. Substituting, we find for the phase currents:

$$i_a(t) = \frac{1}{x'_d} \sin \omega_s t + \frac{1}{2} \left(\frac{1}{x_q} - \frac{1}{x'_d} \right) \sin 2\omega_s t$$

$$i_b(t) = \frac{1}{x'_d} \sin \left(\omega_s t - \frac{2\pi}{3} \right) + \frac{1}{2} \left(\frac{1}{x_q} - \frac{1}{x'_d} \right) \sin \left(2\omega_s t - \frac{2\pi}{3} \right) + \frac{1}{2} \left(\frac{1}{x_q} + \frac{1}{x'_d} \right) \sin \frac{2\pi}{3}$$

$$i_c(t) = \frac{1}{x'_d} \sin \left(\omega_s t + \frac{2\pi}{3} \right) + \frac{1}{2} \left(\frac{1}{x_q} - \frac{1}{x'_d} \right) \sin \left(2\omega_s t + \frac{2\pi}{3} \right) - \frac{1}{2} \left(\frac{1}{x_q} + \frac{1}{x'_d} \right) \sin \frac{2\pi}{3}$$

Examination of the expressions for the phase currents reveals that each consists of a fundamental component at ω_s, a second harmonic at $2\omega_s$, and a constant term (the constant term for phase a current is zero). Note that at $t = 0$, $i_a(0) = i_b(0) = i_c(0) = 0$, as would be expected in that currents which were zero prior to the fault cannot change instantaneously. If the effects of resistances in the armature and field-winding circuits are taken into account, the various terms will decay exponentially. When a final steady-state operating condition is reached, the amplitude of the fundamental term at ω_s must have decayed to a value of $E_f/x_d = 1/x_d$ as would be obtained from steady-state analysis. The second harmonic and dc offset terms would have decayed to zero. Thus the solution we have arrived at would apply only for the first few cycles following the short circuit. It will be necessary to include the effects of resistances in order to ascertain the length of time for which the solution is reasonable. It may be helpful at this point for the reader to reexamine Figure 6.1 in light of the expressions we have developed for short-circuit currents with resistances neglected. The initial amplitudes of the fundamental currents at ω_s are of interest for purposes of determining the short-circuit duty required of circuit breakers or fuses. Power-system analysts often represent a machine as a constant voltage "back of" transient reactance (subtransient reactance if damper winding effects are to be included; it is the high-frequency asymptote for reactance that is relevant) in order to calculate short-circuit currents. Multiplying factors are sometimes used to account for the effects of the initial dc offset.

The torque can be obtained as $T = \psi_d i_q - \psi_q i_d$. In this case $I_d = I_q = 0$ and consideration of the armature voltage equations for steady-state conditions shows that $\psi_{q0} = 0$ if $V_d = 0$ and $\psi_{d0} = 1.0$ if $V_q = 1.0$. Thus the torque can be expressed as

$$T = \psi_d i_q - \psi_q i_d$$

$$= (1 + \Delta\psi_d)\,\Delta i_q - \Delta\psi_q\,\Delta i_d$$

$$= \Delta i_q + \Delta\psi_d\,\Delta i_q - \Delta\psi_q\,\Delta i_d$$

Upon substituting, the torque is given by

$$T = \frac{1}{x_d'}\sin \omega_s t + \frac{1}{2}\left(\frac{1}{x_q} - \frac{1}{x_d'}\right)\sin 2\omega_s t$$

If winding resistances are considered, both terms in the expression for torque will decay and torque will be constant and approximately zero in the steady state. A small nonzero torque in the steady state is necessary to account for steady-state armature resistance losses.

It is necessary to consider resistances in order to determine decay rates of the various components of currents or torque. Furthermore, it is necessary to include both armature resistance and field-winding resistance. Consider (6.94) and (6.95), which give, respectively, the armature voltage relationships with and without armature resistance. It is easy to see that if armature resistance is neglected, solutions for $\Delta\psi_d$ and $\Delta\psi_q$ will contain fundamental frequency components that do not decay to zero. Thus armature resistance must be included if ψ_d and ψ_q are to become constant in the steady state as would be required if Δv_d and Δv_q are constants. It is necessary to consider field winding resistance in order to have the proper relationships between ψ_d, i_d, and i_r' as a final steady-state operating point is approached.

Consider a machine with a single direct-axis rotor winding. The equivalent circuits relating changes in variables are as shown in Figure 6.29 if speed and field voltage are assumed constant. Substitute in (6.94) using (6.96) and (6.97) to obtain

$$\frac{1}{\omega_s}\frac{d}{dt}\Delta\psi_d = \frac{-r_s}{x_d'}\Delta\psi_d + \Delta\psi_q + \Delta v_d \tag{6.99a}$$

$$\frac{1}{\omega_s}\frac{d}{dt}\Delta\psi_q = \frac{-r_s}{x_q}\Delta\psi_q - \Delta\psi_d + \Delta v_q \tag{6.99b}$$

The substitution in (6.99a) using (6.97) introduces an approximation in that it was necessary to neglect field resistance in order to obtain (6.97). This substitution is justifiable if armature resistance is sufficiently small to limit the error so introduced. The armature voltage equations given by (6.99) can be solved for $\Delta\psi_d$ and $\Delta\psi_q$ if Δv_d and Δv_q are known. Again, consider a three-phase fault on an initially unloaded machine. As shown in Example 6.6, appropriate choices for Δv_d and Δv_q are

$$\Delta v_q = -1.0 \tag{6.100a}$$

$$\Delta v_d = 0.0 \tag{6.100b}$$

Now substitute in (6.99) using (6.100) and take Laplace transforms in each of the equations to obtain

$$\frac{s}{\omega_s}\Delta\psi_d(s) = \frac{-r_s}{x_d'}\Delta\psi_d(s) + \Delta\psi_q(s) \tag{6.101a}$$

$$\frac{s}{\omega_s}\Delta\psi_q(s) = \frac{-r_s}{x_q}\Delta\psi_q(s) - \Delta\psi_d(s) - \frac{1}{s} \tag{6.101b}$$

Now solve for $\Delta\psi_d(s)$ and $\Delta\psi_q(s)$:

$$\Delta\psi_d(s) \simeq -\frac{\omega_s^2}{s(s^2 + 2s/\tau_a + \omega_s^2)} \tag{6.102a}$$

$$\Delta\psi_q(s) \simeq -\frac{\omega_s}{s^2 + 2s/\tau_a + \omega_s^2} \tag{6.102b}$$

In (6.102), terms in r_s^2 have been neglected in the denominator polynomials and a term in r_s has been neglected in the numerator of the expression for $\Delta\psi_q(s)$. These approximations are reasonable if armature resistance is small. The time constant τ_a is defined by

$$\frac{1}{\tau_a} = \frac{r_s\omega_s}{2}\left(\frac{1}{x_d'} + \frac{1}{x_q}\right) \tag{6.103}$$

The direct-axis current is obtained by considering Figure 6.29 in light of the solution for $\Delta\psi_d$ given by (6.102a). Figure 6.30 illustrates the circuit that must be solved for $\Delta I_d(s)$.

As shown in Example 6.5, an operational reactance $X_d(s)$ can be used to relate $\Delta\psi_d(s)$ and $\Delta I_d(s)$:

$$\frac{\Delta I_d(s)}{\Delta\psi_d(s)} = -\frac{1}{X_d(s)} \tag{6.104}$$

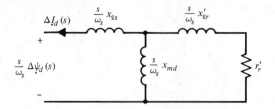

FIGURE 6.30 Equivalent Circuit Relating ΔI_d (s) and $\Delta \psi_d$ (s) for a Machine with a Single Direct-Axis Rotor Circuit and Constant Field Voltage

It is left as an exercise (Problem 6.9) to show that $X_d(s)$ can be expressed as

$$X_d(s) = x_d \frac{1 + \tau'_d s}{1 + \tau'_{d0} s} \qquad (6.105)$$

where τ'_{d0} is the direct-axis open-circuit time constant:

$$\tau'_{d0} = \frac{1}{\omega_s r'_r} (x'_{\ell r} + x_{md}) \qquad (6.106)$$

and τ'_d is the direct-axis short-circuit time constant:

$$\tau'_d = \frac{1}{\omega_s r'_r} \left(x'_{\ell r} + \frac{x_{\ell s} x_{md}}{x_{\ell s} + x_{md}} \right) \qquad (6.107)$$

It follows that

$$\frac{1}{X_d(s)} = \frac{1}{x_d} \frac{1 + \tau'_{d0} s}{1 + \tau'_d s} \qquad (6.108)$$

Now as $s \to \infty$, $1/X_d(s) \to 1/x'_d$. Thus

$$\frac{1}{x'_d} = \frac{1}{x_d} \frac{\tau'_{d0}}{\tau'_d} \qquad (6.109)$$

Equation (6.109) can be used to eliminate τ'_{d0} in (6.108):

$$\frac{1}{X_d(s)} = \frac{1}{x_d} \frac{1}{1 + \tau'_d s} + \frac{1}{x'_d} \frac{\tau'_d s}{1 + \tau'_d s}$$

$$= \frac{1}{x_d} + \left(\frac{1}{x'_d} - \frac{1}{x_d} \right) \frac{\tau'_d s}{1 + \tau'_d s} \qquad (6.110)$$

Now solve for $\Delta I_d(s)$ using (6.102a), (6.104), and (6.110):

$$\Delta I_d(s) = \frac{\omega_s^2}{s \left(s^2 + \dfrac{2}{\tau_a} s + \omega_s^2 \right)} \left[\frac{1}{x_d} + \left(\frac{1}{x'_d} - \frac{1}{x_d} \right) \frac{\tau'_d s}{1 + \tau'_d s} \right] \qquad (6.111)$$

Application of partial-fraction-expansion techniques to (6.111) yields for the first term,

$$\frac{1}{x_d}\frac{\omega_s^2}{s(s^2 + 2/\tau_a\, s + \omega_s^2)} = \frac{1}{x_d}\frac{1}{s} - \frac{1}{x_d}\frac{s + 2/\tau_a}{s^2 + 2/\tau_a\, s + \omega_s^2} \quad (6.112)$$

and for the second term,

$$\left(\frac{1}{x_d'} - \frac{1}{x_d}\right)\frac{\tau_d'\omega_s^2}{(1 + \tau_d's)(s^2 + 2/\tau_a\, s + \omega_s^2)} \approx$$

$$\left(\frac{1}{x_d'} - \frac{1}{x_d}\right)\left(\frac{1}{s + 1/\tau_d'} - \frac{s}{s^2 + 2/\tau_a\, s + \omega_s^2}\right) \quad (6.113)$$

The approximation in (6.113) is based on the assumptions that $1/\tau_d' \ll \omega_s$ and $1/\tau_a \ll \omega_s$. The inverse transform for (6.112) is for $t \geq 0$:

$$\Delta i_{d1}(t) = \frac{1}{x_d} - \frac{1}{x_d}e^{-t/\tau_a}\cos\omega_s t - \frac{e^{-t/\tau_a}}{\tau_a\omega_s x_d}\sin\omega_s t \quad (6.114)$$

where again it is assumed that $1/\tau_a \ll \omega_s$. Similarly, the inverse transform for (6.113) is for $t \geq 0$:

$$\Delta i_{d2}(t) = \left(\frac{1}{x_d'} - \frac{1}{x_d}\right)\left(e^{-t/\tau_d'} - e^{-t/\tau_a}\cos\omega_s t + \frac{e^{-t/\tau_a}}{\tau_a\omega_s}\sin\omega_s t\right) \quad (6.115)$$

The last terms in (6.114) and (6.115) are small and can be neglected and we finally obtain

$$i_d = \Delta i_d = \Delta i_{d1} + \Delta i_{d2} \quad (6.116)$$

$$= \frac{1}{x_d} + \left(\frac{1}{x_d'} - \frac{1}{x_d}\right)e^{-t/\tau_d'} - \frac{1}{x_d'}e^{-t/\tau_a}\cos\omega_s t$$

Quadrature-axis current is considerably simpler to obtain. Substitute in (6.102b) and (6.96) to obtain

$$\Delta I_q(s) = \frac{1}{x_q}\frac{\omega_s}{s^2 + (2/\tau_a)\, s + \omega_s^2} \quad (6.117)$$

Take the inverse transform of (6.117):

$$i_q = \Delta i_q \simeq \frac{1}{x_q}e^{-t/\tau_a}\sin\omega_s t \quad (6.118)$$

Now apply the transformation relationships given by (6.70) to (6.116) and (6.118) to find phase currents. Assuming no zero-sequence current, the phase currents are found to be, where $\varphi = \omega_s t$,

$$i_a = \left[\frac{1}{x_d} + \left(\frac{1}{x_d'} - \frac{1}{x_d}\right) e^{-t/\tau_d'}\right] \sin \omega_s t$$

$$+ \frac{e^{-t/\tau_a}}{2} \left(\frac{1}{x_q} - \frac{1}{x_d'}\right) \sin 2\omega_s t \qquad (6.119a)$$

$$i_b = \left[\frac{1}{x_d} + \left(\frac{1}{x_d'} - \frac{1}{x_d}\right) e^{-t/\tau_d'}\right] \sin\left(\omega_s t - \frac{2\pi}{3}\right)$$

$$+ \frac{e^{-t/\tau_a}}{2}\left[\left(\frac{1}{x_q} - \frac{1}{x_d'}\right) \sin\left(2\omega_s t - \frac{2\pi}{3}\right) + \left(\frac{1}{x_q} + \frac{1}{x_d'}\right) \sin\frac{2\pi}{3}\right]$$

$$(6.119b)$$

$$i_c = \left[\frac{1}{x_d} + \left(\frac{1}{x_d'} - \frac{1}{x_d}\right) e^{-t/\tau_d'}\right] \sin\left(\omega_s t + \frac{2\pi}{3}\right)$$

$$+ \frac{e^{-t/\tau_a}}{2}\left[\left(\frac{1}{x_q} - \frac{1}{x_d'}\right) \sin\left(2\omega_s t + \frac{2\pi}{3}\right) - \left(\frac{1}{x_q} + \frac{1}{x_d'}\right) \sin\frac{2\pi}{3}\right]$$

$$(6.119c)$$

Equation set (6.119) should be compared term by term to the expressions for phase currents developed in Example 6.6, where resistances were neglected. The amplitude of the fundamental decays from an initial value of $1/x_d'$ to a final value of $1/x_d$. The direct-axis short-circuit time constant determines the decay rate. The second-harmonic terms at $2\omega_s$ decay to zero, as do the dc offset terms with decay rate determined by the time constant τ_a.

The technique we have followed in developing equation set (6.119) necessarily relies on assumptions regarding relative magnitudes of machine time constants. Short-circuit tests [10] tend to validate the assumptions with time constants and reactances falling in relatively narrow ranges [1]. The technique becomes unwieldy when a number of rotor circuits are to be represented in each axis. The trend in modern practice is to represent machines with at least two rotor windings in each axis for studies involving transients. In such cases alternative analytical procedures may be preferable. The assumption of constant speed yields linear equations, as we have seen. Eigenvalue techniques can be applied to develop closed-form solutions for currents and torque. With computer assistance such techniques can be applied regardless of the number of rotor windings represented and simplifying assumptions resulting in approximate solutions are not required. These techniques are illustrated in Chapter 8. The type of approximations used in obtaining (6.119) can be used in many practical applications to gain insights on machine behavior.

INDUCTION MACHINE TRANSIENTS

Induction machine transients can be analyzed in a manner similar to that described in Section 6.4 for the synchronous machine. Equivalent circuits appropriate for a symmetrical induction machine were developed and presented in Section 6.3.7. To illustrate, an example will be considered.

EXAMPLE 6.7 A 200-hp 460-V three-phase Y-connected squirrel-cage induction machine has the following per unit data:

$$r_s = 0.020$$

$$r'_r = 0.035$$

$$x_{\ell s} = 0.085$$

$$x'_{\ell r} = 0.075$$

$$x_m = 3.0$$

$$H = 3.0 \text{ s}$$

The machine is started from rest with a full voltage start; that is, rated armature winding voltages are applied at time $t = 0$. Examine the initial current and torque transients.

The equivalent circuits for a symmetrical induction machine are given in Figure 6.20. It is a simple matter to adapt the circuits for analysis in terms of per unit quantities. The machine rotor is initially at rest. It will be assumed that the rotor remains at rest (i.e., $\omega_r = 0$). In practice this would occur only if the rotor were blocked. In most cases the rotor will accelerate and the current and torque transients calculated on the basis of zero rotor speed will be valid only over some short-time duration before rotor speed is appreciable. As discussed in Section 6.3.7, if the reference frame is fixed in the rotor with the q axis aligned with the axis of the equivalent phase a rotor winding, $\varphi_r = 0$, $d\varphi_r/dt = 0$, $\varphi_s = \theta_r$, and $d\varphi_s/dt = d\theta_r/dt = \omega_r$. Rotor windings are short circuited and the equivalent circuits will be as shown in Figure 6.31.

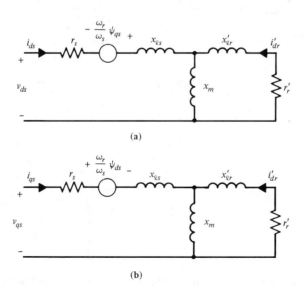

(a)

(b)

FIGURE 6.31 Equivalent Circuits for the Induction Machine of Example 6.7

Now if the speed ω_r is assumed constant and zero, the speed voltage terms in the equivalent circuits will be zero and the direct- and quadrature-axis equivalent circuits are uncoupled. Without loss of generality, the angle $\varphi_s = \theta_r$ can be taken as zero. Now, if the phase voltages applied at time $t = 0$ are

$$v_a = \cos \omega_s t$$

$$v_b = \cos \left(\omega_s t - \frac{2\pi}{3} \right)$$

$$v_c = \cos \left(\omega_s t + \frac{2\pi}{3} \right)$$

the corresponding qd voltages can be found by applying (6.70a). If we assume a three-wire machine, the zero-sequence variables will be zero and application of the transformation yields for $t \geq 0$:

$$v_{qs} = \cos \omega_s t$$

$$v_{ds} = -\sin \omega_s t$$

It will be sufficient to solve the quadrature-axis equivalent circuit. The solution for the direct-axis circuit will then be apparent. The circuit to be solved is as shown in Figure 6.32. The solution for current i_{qs} can be obtained by finding an operational admittance $Y(s) = 1/Z(s)$ at the armature terminals. The impedance $Z(s)$ is

$$Z(s) = r_s + \frac{sx_{\ell s}}{\omega_s} + \frac{(sx_m/\omega_s)(r_r' + sx_{\ell r}'/\omega_s)}{r_r' + (s/\omega_s)(x_m + x_{\ell r}')}$$

$$= \frac{(s/\omega_s)^2(x_{\ell s}x_m + x_{\ell s}x_{\ell r}' + x_m x_{\ell r}') + (s/\omega_s)(r_s x_m + r_s x_{\ell r}' + r_r' x_{\ell s} + r_r' x_m) + r_r' r_s}{r_r' + (s/\omega_s)(x_m + x_{\ell r}')}$$

Now divide by the coefficient of s^2 in the numerator and neglect products of small terms (i.e., resistances and leakage reactances) in the expressions in parentheses to obtain for $Y(s)$:

$$Y(s) = \frac{\dfrac{\omega_s^2}{x_m(x_{\ell s} + x_{\ell r}')} \left[r_r' + \dfrac{s}{\omega_s}(x_m + x_{\ell r}') \right]}{s^2 + \dfrac{\omega_s(r_s + r_r')s}{x_{\ell s} + x_{\ell r}'} + \dfrac{\omega_s^2 r_r' r_s}{x_m(x_{\ell s} + x_{\ell r}')}}$$

If the following time constants are defined and evaluated for the given data,

$$\frac{1}{\tau_1} = \frac{\omega_s(r_r' + r_s)}{x_{\ell r}' + x_{\ell s}} = 129.6$$

$$\frac{1}{\tau_2} = \frac{\omega_s r_r' r_s}{x_m(r_r' + r_s)} = 1.60$$

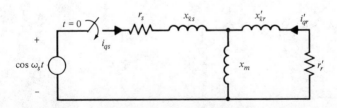

FIGURE 6.32 Equivalent Circuit for the Quadrature Axis of the Induction Machine of Example 6.7

ANALYSIS OF AC MACHINE TRANSIENTS

the denominator polynomial can be expressed as

$$D(s) = s^2 + \frac{1}{\tau_1} s + \frac{1}{\tau_1 \tau_2}$$

In general, $\tau_2 \gg \tau_1$ in that τ_2 is the ratio of magnetizing inductance to the parallel combination of r_r' and r_s, whereas τ_1 is the ratio of series leakage inductance to series resistance. Thus

$$D(s) = s^2 + \frac{1}{\tau_1} s + \frac{1}{\tau_1} \frac{1}{\tau_2} \approx s^2 + \left(\frac{1}{\tau_2} + \frac{1}{\tau_1} \right) s + \frac{1}{\tau_1 \tau_2} = \left(s + \frac{1}{\tau_1} \right) \left(s + \frac{1}{\tau_2} \right)$$

The numerator can be expressed as

$$N(s) = \frac{\omega_s}{x_m} \frac{x_m + x_{\ell r}'}{x_{\ell s} + x_{\ell r}'} \left(s + \frac{\omega_s r_r'}{x_m + x_{\ell r}'} \right)$$

$$= K \left(s + \frac{1}{\tau_3} \right)$$

where

$$K = \frac{\omega_s}{x_m} \frac{x_m + x_{\ell r}'}{x_{\ell s} + x_{\ell r}'} = 2415.2$$

$$\frac{1}{\tau_3} = \frac{\omega_s r_r'}{x_m + x_{\ell r}'} = 4.29$$

Note that $\tau_2 = 0.625$ s and $\tau_3 = 0.233$ s are both considerably larger than $\tau_1 = 0.0077$ s. This will be true for typical per unit values. The Laplace transform of v_{qs} is given by

$$V_{qs} = \frac{s}{s^2 + \omega_s^2}$$

and the Laplace transform of i_{qs} will be

$$I_{qs} = V_{qs} Y(s)$$

$$= K \frac{s}{s^2 + \omega_s^2} \frac{s + 1/\tau_3}{(s + 1/\tau_1)(s + 1/\tau_2)}$$

Upon expanding by partial fractions and evaluating for the given data, I_{qs} is given by

$$I_{qs} = \frac{1.928s}{s^2 + \omega_s^2} + \frac{2164.36}{s^2 + \omega_s^2} - \frac{1.928}{s + 1/\tau_1} - \frac{0.00057}{s + 1/\tau_2}$$

The inverse transform is

$$i_{qs} = 1.928 \cos \omega_s t + 5.741 \sin \omega_s t - 1.928 e^{-t/\tau_1} - 0.0057 e^{-t/\tau_2}$$

$$= 6.06 \cos (\omega_s t - 71.44°) - 1.928 e^{-t/\tau_1} - 0.00057 e^{-t/\tau_2}$$

Note that the term with the long time constant τ_2 is negligible. This occurs in that τ_3 and τ_2 are of the same order of magnitude and considerably larger than τ_1. This causes an approximate pole–zero cancellation with respect to the factors containing τ_3 and τ_2. The time constant $\tau_1 = 0.0077$ s is short and the decaying exponential is effectively zero in roughly two cycles.

The current i_{ds} is found similarly.

$$I_{ds} = \frac{5.744s}{s^2 + \omega_s^2} - \frac{727.02}{s^2 + \omega_s^2} - 5.609e^{-t/\tau_1} - 0.1346e^{-t/\tau_2}$$

$$i_{ds} = 5.744 \cos \omega_s t - 1.928 \sin \omega_s t - 5.609e^{-t/\tau_1} - 0.1346e^{-t/\tau_2}$$

$$= 6.06 \cos (\omega_s t + 18.55°) - 5.609e^{-t/\tau_1} - 0.1346e^{-t/\tau_2}$$

The phase currents may be obtained by applying the transformation (6.70b) with $\varphi = 0$. Note that $i_a = i_{qs}$ in this case. In essence, i_a consists of a steady-state component and a short time constant dc offset. The steady-state term could be calculated in a much simpler way using the techniques of Chapter 3 (see Problem 6.11). In particular, the technique used in Example 3.4 is well suited for determining armature currents during motor starting.

Torque transients can be examined by expressing torque as $T = \psi_{ds}i_{qs} - \psi_{qs}i_{ds}$. The flux linkages ψ_{ds} and ψ_{qs} can be determined directly from the armature voltage equations if currents i_{qs} and i_{ds} are known. The armature voltage equations can be expressed as

$$\frac{1}{\omega_s} \frac{d\psi_{qs}}{dt} = v_{qs} - r_s i_{qs}$$

$$\frac{1}{\omega_s} \frac{d\psi_{ds}}{dt} = v_{ds} - r_s i_{ds}$$

The voltages v_{qs} and v_{ds} are specified as inputs and currents have already been determined. Substitution followed by integration term by term yields

$$\psi_{qs} = 0.9682 \sin (\omega_s t + 6.82°) - 0.1122e^{-t/\tau_1} - 0.00268e^{-t/\tau_2}$$

$$\psi_{ds} = 0.9682 \cos (\omega_s t + 6.82°) - 0.3263e^{-t/\tau_1} - 0.6344e^{-t/\tau_2}$$

Solving for torque, we have

$$T = \psi_{ds}i_{qs} - \psi_{qs}i_{ds}$$

$$= 1.193 + 1.193e^{-[(1/\tau_1)+(1/\tau_2)]t}$$

$$- 3.72e^{-t/\tau_1} \cos (\omega_s t + 71.3°)$$

$$- 3.72e^{-t/\tau_2} \cos (\omega_s t - 71.3°)$$

$$+ \text{additional terms of small magnitude}$$

The first term is the steady-state torque at zero speed. It is more easily determined by steady-state analysis techniques, that is, (3.63) (see also, Problem 6.11). The second term is a dc offset which will decay rapidly to zero in that the short time constant τ_1 will dominate. The third term is a damped sinusoid at fundamental frequency. It will decay rapidly to zero with the time constant τ_1. The fourth term is interesting in that it is a damped sinusoid of significant magnitude which will decay relatively slowly with the longer time constant τ_2. It will be apparent in this case for roughly $3\tau_2 \simeq 2.0$ s. The torque waveform is illustrated in Figure 6.33. The rotor would typically accelerate, with speed determined for the most part by the average torque. The inertia of the rotor is sufficiently large to minimize the effect of the torque transients at the fundamental frequency ω_s and the effects of the fast-decaying dc offset term. As the speed changes, the effective accelerating torque will vary with speed approximately in accordance with the machine's steady-state torque–speed curve (see Example 3.4). Although torque transients have little effect on machine speed in many cases, their effects on some mechanical systems may be significant.

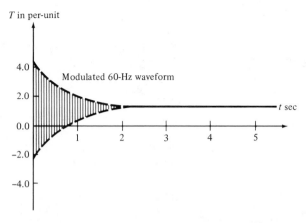

FIGURE 6.33 Torque Waveform for the Induction Machine of Example 6.7 (Rated Armature Voltage Applied at t = 0 with Rotor Speed Assumed to Be Zero)

The analysis described above becomes tedious if a number of different cases are to be examined. It cannot be applied if an assumption of constant speed is inappropriate. As in the case of the synchronous machine, computer-assisted solutions are invaluable and are presented in Chapter 8.

REFERENCES

[1] *Electrical Transmission and Distribution Reference Book.* East Pittsburgh, Pa.: Westinghouse Electric Corporation, 1964.

[2] P. L. DANDENO, P. KUNDUR, and R. P. SHULZ, "Recent Trends and Progress in Synchronous Machine Modeling in the Electric Utility Industry," *Proceedings of the IEEE,* Vol. 62, pp. 941–950, July 1974.

[3] IEEE Joint Working Group, "Supplementary Definitions and Associated Test Methods for Obtaining Parameters for Synchronous Machine Stability Study Simulations," *IEEE Transactions on Power Apparatus and Systems,* Vol. PAS-99, pp. 1625–1633, July/August 1980.

[4] P. L. DANDENO, R. L. HAUTH, and R. P. SCHULZ, "Effects of Synchronous Machine Modeling in Large Scale Stability Studies," *IEEE Transactions on Power Apparatus and Systems,* Vol. PAS-92, pp. 574–582, March/April 1973.

[5] D. R. BROWN AND P. C. KRAUSE, "Modeling of Transient Electrical Torques in Solid Iron Rotor Turbogenerators," *IEEE Transactions on Power Apparatus and Systems,* Vol. PAS-98, pp. 1502–1508, September/October 1979.

[6] P. C. KRAUSE and C. H. THOMAS, "Simulation of Symmetrical Induction Machinery," *IEEE Transactions on Power Apparatus and Systems,* Vol. PAS-84, pp. 1038–1053, November 1965.

[7] P. C. KRAUSE, "Method of Multiple Reference Frames Applied to the Analysis of Symmetrical Induction Machinery," *IEEE Transactions on Power Apparatus and Systems,* Vol. PAS-87, pp. 218–227, January 1968.

[8] R. H. PARK, "Two Reaction Theory of Synchronous Machines: Generalized Method of Analysis—Part I," *Transactions of the AIEE,* Vol. 48, pp. 716–730, July 1929.

[9] W. D. STEVENSON, JR., *Elements of Power System Analysis,* 4th ed. New York: McGraw-Hill Book Company, 1982.

[10] Test Procedures for Synchronous Machines, IEEE Standard 115.

PROBLEMS

6.1 If flux linkages are linearly related to currents, a relationship for the developed torque in terms of currents and machine rotor angle θ_r is [see (6.12)]

$$T = \frac{1}{2} \bar{i}^T \left[\frac{d}{d\theta_r} \bar{L}(\theta_r) \right] \bar{i}$$

Find a similar matrix relationship for the torque if flux linkages and machine rotor angle are chosen as independent variables.

6.2 Show that the product of the transformation matrix $\bar{T}(\varphi)$ given by (6.17) and its inverse $\bar{T}^{-1}(\varphi)$ given by (6.19) yields the identity matrix

$$\bar{u} = \begin{bmatrix} 1 & 0 \\ 0 & 1 \end{bmatrix}$$

for any value of φ.

6.3 For a two-phase machine, the total power at the armature winding terminals is

$$p = v_a i_a + v_b i_b$$

Show that for the transformation given by (6.17) and (6.19),

$$p = v_a i_a + v_b i_b = v_q i_q + v_d i_d$$

6.4 Show that torque as given in general by (6.26),

$$T = \frac{P}{2} (\lambda_d i_q - \lambda_q i_d)$$

is equivalent to the expression given by (6.12),

$$T = M i'_r (\cos \theta_r i_b - \sin \theta_r i_a)$$

for a two-phase two pole round-rotor synchronous machine with a single direct-axis rotor winding.

6.5 The application of a disturbance to a certain two-phase synchronous machine results in the following armature currents for $t \geq 0$:

$$i_a(t) = 5 \sin \omega_s t$$

$$i_b(t) = 5e^{-2t} - 5 \cos \omega_s t$$

Assume that the machine rotor speed remains constant at $\omega_r = \omega_s$ and find $i_d(t)$ and $i_q(t)$. Assume that $\varphi(0) = 0$.

6.6 The flux linkage relationships for a two-phase induction machine were developed in Chapter 3:

$$\begin{bmatrix} \lambda_{as} \\ \lambda_{bs} \\ \lambda'_{ar} \\ \lambda'_{br} \end{bmatrix} = \begin{bmatrix} L_s & 0 & M \cos \theta_r & -M \sin \theta_r \\ & L_s & M \sin \theta_r & M \cos \theta_r \\ & & L'_r & 0 \\ \text{(symmetric matrix)} & & & L'_r \end{bmatrix} \begin{bmatrix} i_{as} \\ i_{bs} \\ i'_{ar} \\ i'_{br} \end{bmatrix}$$

where

$$L_s = L_{\ell s} + M$$

$$L'_r = L'_{\ell r} + M$$

Show that the transformation given by (6.56) and (6.57) results in relationships between flux linkages and currents that are independent of rotor position.

6.7 A three-phase salient-pole synchronous generator delivers its rated kVA at rated voltage at a power factor of 0.707 lagging. The machine has negligible armature resistance and direct- and quadrature-axis reactances of

$$x_d = 1.5 \text{ per unit}$$

$$x_q = 1.0 \text{ per unit}$$

Armature leakage reactance is 0.10 per unit. Find values of ψ_d, ψ_q, i_d, i_q, and i'_r for this operating condition.

6.8 The total instantaneous power at the armature terminals of a three-phase machine is given by

$$p(t) = v_a i_a + v_b i_b + v_c i_c$$

$$= \bar{v}^{abcT} i^{abc}$$

Show that if the transformation given by (6.70) is used, instantaneous power in terms of $qd0$ variables is given by

$$p(t) = \tfrac{3}{2} (v_q i_q + v_d i_d + 2v_0 i_0)$$

6.9 Figure 6.30 relates direct-axis current to direct-axis flux for a synchronous machine with a single direct-axis rotor winding. Show that the operational reactance $X_d(s) = \psi_d(s)/-I_d(s)$ can be expressed as

$$X_d(s) = x_d \frac{1 + \tau'_d s}{1 + \tau'_{do} s}$$

where

$$\tau'_d = \frac{1}{\omega_s r'_r} \left(x'_{\ell r} + \frac{x_{\ell s} x_{md}}{x_{\ell s} + x_{md}} \right)$$

$$\tau'_{do} = \frac{1}{\omega_s r'_r} (x'_{\ell r} + x_{md})$$

6.10 A certain synchronous machine has the following direct-axis data:

$$x_d = 1.5 \text{ per unit}$$

$$x'_d = 0.20 \text{ per unit}$$

$$\tau'_{do} = 4.0 \text{ s}$$

The machine does not have damper windings. Find the short-circuit time constant τ'_d.

6.11 Use steady-state analysis techniques to determine the inrush current and blocked-rotor torque for the induction machine of Example 6.7. Compare the results to the expressions obtained in the example, which include transients.

CHAPTER 7

Special-Purpose Electromechanical and Electromagnetic Devices

INTRODUCTION

In Chapters 1 through 5 we have addressed the fundamentals of steady-state machine and transducer operation, and the topic of ac machine transients was considered in Chapter 6. However, in no way have we exhausted the number of devices that operate based on the principles contained in these chapters. In this chapter we focus our attention on some of these other machines and devices. We consider how they are used and how they can be analyzed, at least in an overall sense, using the principles learned earlier.

This chapter, then, contains a potpourri of devices and systems—some simple, some extremely complex—which illustrate the practical utilization and integration of the basic concepts of machines. Devices are grouped somewhat by complexity and in basically the same order as their fundamental principles are addressed in Chapters 1 through 5. The approach is one of illustration only; each device is really an example of these principles applied to a system other than a classical inductor, transformer, or rotating machine. Topics covered include magnetic amplifiers and transducers, special-purpose motors, and linear machines and levitation. It must be stressed that the student *already* knows enough to understand the devices presented in this chapter. All of these systems have as their basis the simple electromagnetic principles stressed earlier. But the applications know almost no bounds, as the reader will see by investigating the small number of the possibilities presented here for illustration.

THE MAGNETIC AMPLIFIER

The magnetic amplifier, also known as a transductor or saturable reactor, is a device that is primarily of historical interest but which also is a novel extension of the transformer and inductor theory discussed in Chapter 1. Although the theory behind this device has been known since the turn of the century, it only began to achieve practical widespread application in the decade of the 1940s, primarily due to the work of A. U. Lamm of Sweden, and it practically disappeared from general use with the perfection of reliable solid-state technology in the 1960s. Magnetic amplifiers were and in some cases, still can be seen in numerous applications, including theater-light dimmers, strain gauges, thermocouples, inductive pickups, telephone repeaters, autopilots and other servomechanisms, motor controllers, and dc instrumentation. In an earlier age of cumbersome, marginally reliable† vacuum tubes, with their requisite high-voltage power supplies, long warm-up periods, high heat output, and low efficiencies, magnetic amplifiers offered relative simplicity, a lack of moving or mechanically suspended parts subject to vibration damage or mechanical failure (such as plagued the entire contents of a vacuum tube), higher efficiency, relatively low heat loss, no warm-up time, no high-voltage supply requirements, and a unique ability to add or subtract inputs while maintaining complete electrical isolation among all input and output circuits. However, before the reader begins to think that these devices were an absolute panacea, it should be noted that magnetic amplifiers had a significant disadvantage in their extremely long time constants associated with the magnetic circuit, which limited signal frequencies to only a small fraction of the ac supply frequency (another less important disadvantage—an ac supply is required), which in turn was generally restricted to a few kilohertz because of magnetic material limitations. Thus magnetic amplifiers were limited to low-frequency applications (e.g., some servomechanisms) but they saw significant use in these applications. While the discovery of much more sensitive magnetic materials, such as ferrites, allowed the development of significantly improved magnetic amplifiers with much higher frequency response, the concurrent development of solid-state and optoelectronic devices provided significant competition by offering a lower-cost product for similar applications, with equal reliability and much better frequency response and flexibility. Hence solid-state systems have virtually replaced any new uses of magnetic amplifiers, except in a few unique, job-specific applications. However, magnetic amplifiers can be found in older control systems that are still in operation.

†In comparison to their contemporary electromagnetic and electromechanical devices, for example. Some high-reliability industries avoided vacuum tubes for either a real or imagined lack of reliability.

As with vacuum tubes, magnetic amplifiers were developed in a nearly limitless number of configurations. Consequently, this discussion will deal only with the simplest form of magnetic amplifier and is oriented only toward illustrating its characteristics as a true amplifier and its similarity to a transistor or a vacuum tube. In addition, we will discuss the use of a magnetic amplifier as a "dc transformer" furnishing signals to instrumentation, a novel application of transformer theory to accomplish what initially appears to be an impossible task, since transformers simply cannot step direct current or a dc voltage up or down. (Or can they? Read on.)

If we recall from Chapter 1 that a current in a wire gives rise to a magnetic field encircling it, it should be clear that, in Figure 7.1(a), a variable dc current i_{dc} (caused by changing R_{dc} with constant voltage e_{dc}) causes a variable flux φ in the core. Furthermore, for nonlinear core materials, the magnetization curve of Figure 7.1(b) is obtained by varying i_{dc} (this was shown previously in Figure 1.15). The shape of this curve obviously is a function of the core material chosen. Moreover, its instantaneous slope $d\lambda/di$ (the *incremental inductance*) is also a function of the core material but, more important, it is a function of i_{dc}. This slope is referred to in some sources as the *incremental permeability* of the material.

Suppose now that a second coil is added to the core and is excited by a

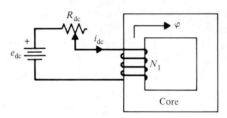

(a) Circuit for excitation of core material

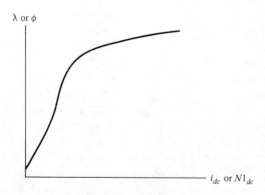

(b) Magnetization curve produced for nonlinear core material by varying i_{dc}

FIGURE 7.1 Operation of Simple Magnetic Circuit

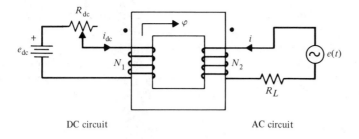

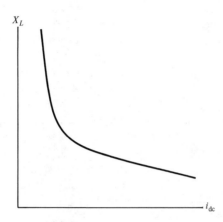

DC circuit   AC circuit

(a) Circuit configuration

(b) Variance of X_L with i_{dc} for constant ac frequency

FIGURE 7.2 Basic Magnetic Amplifier

variable sinusoidal voltage through a load† of resistance R_L causing a sinusoidal current i, as in Figure 7.2(a).

Consider only the ac side of the circuit; the incremental inductance L of the ac coil is

$$L = \frac{d\lambda_{ac}}{di} \tag{7.1}$$

where λ_{ac} is the flux linkage of the ac coil. To determine the form of this derivative, consider the differential flux linkage $d\lambda_{ac}$:

$$d\lambda_{ac} = \frac{\partial \lambda_{ac}}{\partial i_{dc}} di_{dc} + \frac{\partial \lambda_{ac}}{\partial i} di \tag{7.2}$$

Dividing both sides by di yields an expression for L:

$$L = \frac{d\lambda_{ac}}{di} = \frac{\partial \lambda_{ac}}{\partial i_{dc}} \frac{di_{dc}}{di} + \frac{\partial \lambda_{ac}}{\partial i} \frac{di}{di}$$

$$= \frac{\partial \lambda_{ac}}{\partial i} \tag{7.3}$$

†Resistance only is chosen to simplify the discussion. Load impedance may be complex.

since di/di equals 1 and di_{dc}/di equals 0 for constant i_{dc}. However, the value for L in (7.3) is simply the slope of the magnetization curve of Figure 7.1(b) at a point established by i_{dc} if the vertical axis is in terms of λ and the horizontal axis is in terms of $(N_1 i_{dc} + N_2 i)$. This point is the *operating point* of the device. As the operating point is varied by *increasing* i_{dc}, it can be seen from the curve that L *decreases*. Since the inductive reactance of the ac coil is

$$X_L = \omega L = \omega \frac{d\lambda_{ac}}{di} = \omega \frac{\partial \lambda_{ac}}{\partial i} \tag{7.4}$$

where ω is the ac frequency, it follows that, for constant ω, X_L should *also* decrease as i_{dc} is increased, and vice versa. In fact, X_L can be shown to be a function of i_{dc} that exhibits the behavior characterized by the curve of Figure 7.2(b). This curve is asymptotic toward the two axes.

Given the previous simplified description, consider operation of the device in the circuit of Figure 7.2(a). It should be clear that varying the dc current controls X_L. But X_L and R_L together act as a voltage divider and current limiter as well. If the magnitude of the ac voltage is constant, varying X_L can control the voltage across R_L, *and* the current through it. *This is exactly the principle of a power amplifier.* Given an appropriate ratio of turns on the dc and ac coils, a small current (dc, in this case) can cause significant changes in a much larger current (ac), regulating the power dissipated in the load resistor R_L over a wide range. As i_{dc} increases, X_L decreases, causing the power P_L in R_L to increase. As i_{dc} decreases, P_L decreases as well.

The key to amplification with this device is contained in Fig. 7.1(b). If a suitable magnetic material is operated in its linear region (which can be chosen to have almost infinite slope), very small changes in i_{dc} create large changes in X_L, and thus they can control the much higher power delivered to the load. Power gains from 10^3 to more than 10^6 have been reported in the literature (see, e.g., reference 6) for this device.

Magnetic amplifiers have been designed with various coil configurations, primarily to suppress induced ac currents in the dc control windings and to reduce harmonic effects. A common configuration is the series transductor with opposing control windings, shown in Figure 7.3. In this configuration, two coil–core units are wired so that any ac currents induced in the dc control windings are effectively canceled without affecting control of the dc winding over flux in each core. Variations of this design include taking the dc directly from the ac winding, via a diode bridge rectifier, and/or using several dc windings to provide biasing, an important issue in amplifier design.

Taking into account nonlinearities in the core material, curves relating ac current and voltage to dc current can be derived using the magnetization curve. These curves are of the form shown in Figure 7.4. Note that these curves are very similar to those for transistors or tubes used as amplifiers and that the dc current can be considered to act as a bias current. Consequently, the magnetic amplifier can be designed using simple load-line analysis, as is commonly done with conventional transistor and tube amplifiers. The value of the load resistor R_L determines the slope of the load line, and the operating point is determined by the intersection of the load line and the appropriate characteristic curve for

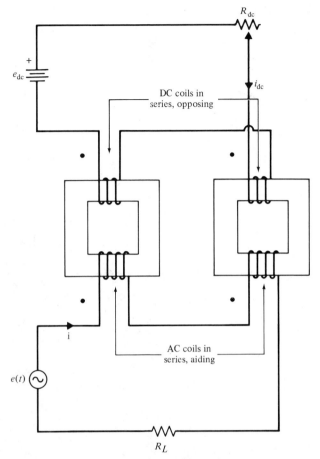

R_{dc}

e_{dc}

DC coils in
series, opposing

i_{dc}

AC coils in
series, aiding

$e(t)$

i

R_L

FIGURE 7.3 Simple Series Magnetic Amplifier with Opposing Control Windings

a given dc control current. This is shown in Figure 7.5. Desirable operation is usually within the linear region, between the dashed lines (corresponding to the active region of a transistor). It can also be seen from this figure, for example, that operation with control currents greater than a certain value (i_5 in this case) result in highly nonlinear ac current–voltage relationships; in fact a lower value of control current (such as i_4) may be a reasonable limit to ensure linear operation. This is also analogous to operation of transistor and tube amplifiers.

Moreover, like other amplifiers, output characteristics can be derived for magnetic amplifiers; Figure 7.6 shows an example of these characteristics. As can be seen from these curves, for a given load, a change in dc current results directly in a change in ac output current, typically "amplified" many times (not shown directly on the plot). These curves change if additional dc bias windings are added (analogous to multigrid tubes, such as pentodes). The only requirement for the device is that $N_1 i_{dc}$ equal $N_2 \hat{i}_{ac}$, where $\hat{i}_{ac}$ is the mean, not the rms, ac current. It can be clearly seen that this device behaves almost exactly like a tube or transistor amplifier (depending on configuration and the number of bias and/or control windings), with the added ability to sum inputs (more than one

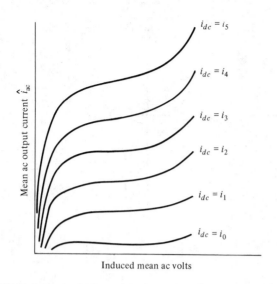

FIGURE 7.4 Typical Magnetic Amplifier Characteristic Curves

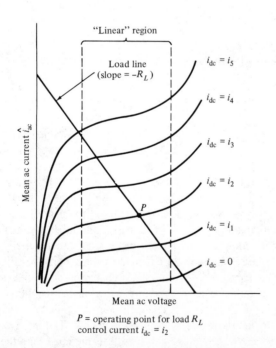

P = operating point for load R_L
control current $i_{dc} = i_2$

FIGURE 7.5 Example of Load Line Analysis to Determine Operating Point of Magnetic Amplifier

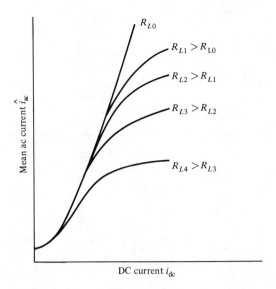

FIGURE 7.6 Typical Output Characteristic for Magnetic Amplifier

dc control winding) and a resultant modulated ac output which can either be used as is or rectified to provide an amplified dc signal which varies proportionally to the control signal. Additional design and operational considerations, such as the use of feedback, are addressed in other works, such as references 1 and 6. It should be clear to the reader at this point that this device could be utilized as an active component in amplifiers, oscillators, and feedback systems. As a final note, the reader should recognize that this device will typically have a long time constant associated with it and thus cannot be used for high-frequency applications. Unlike some conventional amplifiers, however, it can be used to "amplify" direct current or voltage, as will now be discussed.

EXAMPLE 7.1 Using magnetic amplifiers, determine the configuration of a system that will step down dc high voltage and current for measurement by conventional dc meters.

This system, in essence, represents a "dc current transformer" and a "dc potential transformer." Clearly, a transformer per se cannot take a dc voltage or current and step it either up or down. This is evident from Faraday's law, as was shown in Chapter 1. But with the addition of an ac supply, a full-wave rectifier, and a change in core material (magnetic amplifiers generally cannot use conventional transformer steels because they are not nonlinear enough), "transformers" used as magnetic amplifiers can indeed perform this otherwise impossible task, while retaining the required electrical isolation. The resultant circuits, shown in Figure 7.7, actually behave, from a terminal point of view (indicated by the dashed lines) as direct-current transformers!

For the current transformers, the core–coil assemblies are wound on toroidal cores as conventional ac current transformers (see Problem 7.1) and the high voltage dc line passed through the center hole, providing a one-turn dc control winding. The ac winding is wrapped around the core in a conventional manner. The potential transformers utilize conventional magnetic amplifier core-coil assemblies, with the high-voltage dc windings acting as the control circuit.

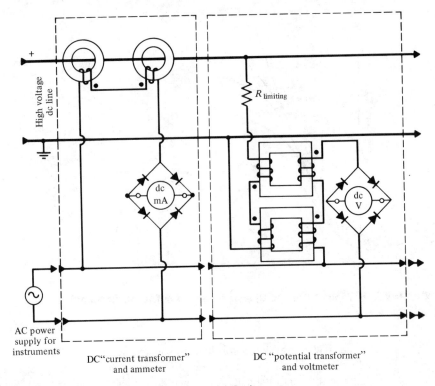

(a) Actual circuit

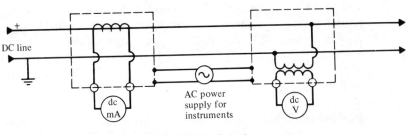

(b) "Effective" circuit

FIGURE 7.7 Use of Magnetic Amplifiers as "DC Transformers" in Instrumentation Applications.

7.3

DYNAMIC MICROPHONES AND PHONOGRAPH PICKUPS

Common "dynamic" microphones and phonograph pickups convert physical motion to electromotive force through the simple interaction of a coil of wire with a static magnetic field. They can be easily analyzed by employing the methods of Chapter 2, provided that reasonable assumptions regarding linearity are made.

A dynamic microphone and phonograph pickup are shown in Figure 7.8. A

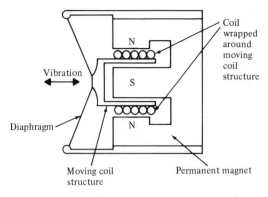

(a) Dynamic microphone

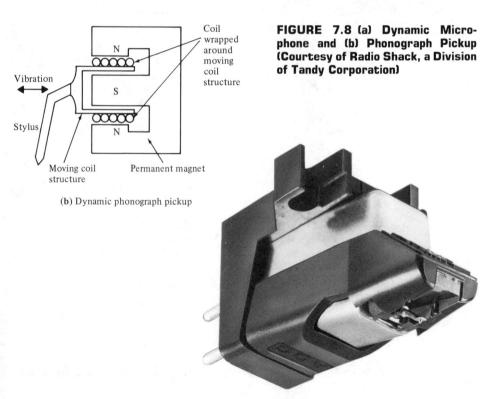

(b) Dynamic phonograph pickup

FIGURE 7.8 (a) Dynamic Microphone and (b) Phonograph Pickup (Courtesy of Radio Shack, a Division of Tandy Corporation)

coil, attached to the mechanical pickup element, is suspended inside a static magnetic field as shown. Motion of the coil will create an emf proportional to its speed and described by

$$e = 2\pi\hat{r}nBv \tag{7.5}$$

where

$\hat{r}$ = mean radius of the coil

n = number of turns

B = magnetic flux density

v = scalar velocity of moving coil (assumed to be in the x direction only for this discussion)

By the right-hand rule, voltage polarity, and thus current direction if loaded, will be dependent on the direction of motion, as shown in Figure 7.9. It is generally more convenient to express (7.5) in terms of position x, where x is defined as the displacement of the coil from its central position x_0. If we recognize that in one dimension velocity v is described by

$$v(t) = \frac{dx(t)}{dt} \tag{7.6}$$

then (7.5) can be restated as

$$e(t) = 2\pi\hat{r}nB\,\frac{dx(t)}{dt} \tag{7.7}$$

Consequently, since all audio signals can be reduced to a series of sinusoids, this equation becomes, in general,

$$e(t) = 2\pi\hat{r}nB\omega x_m \cos(\omega t + \psi) \tag{7.8}$$

for sinusoidal input

$$x(t) = x_m \sin(\omega t + \psi) \tag{7.9}$$

or

$$v(t) = \omega x_m \cos(\omega t + \psi)$$

$$= v_m \cos(\omega t + \psi) \quad \text{(in the x direction only)} \tag{7.10}$$

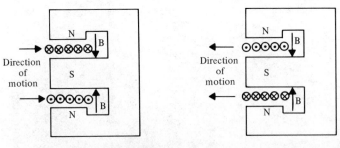

(a) Coil moving into magnet   (b) Coil moving away from magnet

FIGURE 7.9 Current Direction as a Function of Motion (Connected to a Resistive Load).

As can be seen from the equations above, the output of either of these devices is a simple product of the input and a proportionality constant based on its geometry. Thus the devices have essentially linear input/output characteristics, as will now be discussed.

The transfer function of either device, relating electrical output to mechanical input, is easily derived from (7.7), repeated here for continuity:

$$e(t) = 2\pi \hat{r} n B \frac{dx(t)}{dt} \tag{7.7}$$

Upon assuming sinusoidal excitation at constant frequency, this equation corresponds to the following phasor relationship:

$$e(t) \leftrightarrow E = 2\pi \hat{r} n B (j\omega x) \tag{7.11}$$

Thus the transfer function is as follows:

$$\frac{\text{output}}{\text{input}} = \frac{E}{x} = j\omega(2\pi \hat{r} n)B \tag{7.12}$$

where B is a constant (since flux is constant) and ω is the mechanical input frequency in radians per second.

Thus the output voltage for either device, at a given input, is directly proportional to the frequency of excitation and leads the input position by 90°. This means that the device output increases in magnitude with frequency and thus, for a constant output amplitude, the device should be used in a system where either the magnitude of the input signal or its amplification, or both, are increased with decreasing frequency. A microphone should therefore be connected to an amplifier whose gain is inversely proportional to frequency or which amplifies lower tones more than those of higher frequency (called *bass boost*). In the case of a phonograph, some amplification is accomplished using a bass-boost circuit (usually just a tuned amplifier stage); also, records themselves are cut so that the amplitude of the variations in the groove surface increases inversely to frequency, as constrained by groove-spacing considerations. Thus the input signal itself is greater at lower frequencies than at high frequencies, resulting in much improved fidelity, particularly when coupled with bass-boost circuitry in the amplifier.

Analysis of the transfer function for these devices indicates that they have relatively simple equivalent circuits. Since both microphones and phonograph pickups have extremely lightweight coil structures (to increase fidelity, particularly at high frequencies), it follows that inductance will be negligible because very few turns of wire are employed, and therefore coil resistance will be dominant (due to the use of extremely thin wire for light weight). Therefore, the generalized dynamic microphone and phonograph pickups are described by the following terminal equation (in phasor form):

$$V_S = E - RI = j\omega(2\pi \hat{r} n)Bx - RI \tag{7.13}$$

where R is the coil resistance. This equation corresponds to the equivalent circuit shown in Figure 7.10. It can be solved easily if the input impedance Z_i of the amplifier to which the device is connected is known.

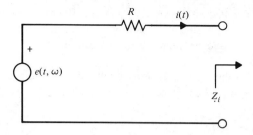

FIGURE 7.10 Microphone or Phonograph Pickup Equivalent Circuit

EXAMPLE 7.2   Suppose that a microphone is excited with a sound pressure which equals

$$p(t) = P_{max} \cos \omega t \qquad N/m^2$$

and that the microphone diaphragm has an effective surface area A and mass m. If the microphone is connected to an amplifier whose input impedance equals twice the microphone resistance, at an angle of $+45°$, find an expression for the output voltage of the microphone as a function of the sound pressure wave and the microphone's geometry.

Since force is

$$f(t) = m \frac{dv(t)}{dt}$$

then

$$f(t) = AP_{max} \cos \omega t = m \frac{dv(t)}{dt}$$

So

$$\frac{dv(t)}{dt} = \frac{AP_{max}}{m} \cos \omega t$$

and therefore the velocity v in the x direction is

$$v(t) = \int_0^t \frac{AP_{max}}{m} \cos \omega t \, dt = \frac{AP_{max}}{\omega m} \sin \omega t + v_0$$

Assuming that the coil is initially at rest ($v_0 = 0$), we find that this reduces to

$$v(t) = \frac{AP_{max}}{\omega m} \sin \omega t$$

So the position of the coil $x(t)$ is

$$x(t) = \int_0^t \frac{AP_{max}}{\omega m} \sin \omega t \, dt = -\frac{AP_{max}}{\omega^2 m} \cos \omega t + x_0$$

and assuming that the coil is initially at $x_0 = 0$ (the midpoint of its range of motion),

$$x(t) = -\frac{AP_{max}}{\omega^2 m} \cos \omega t$$

So, in phasor form,

$$x = -\frac{AP_{max}}{\omega^2 m} \angle 0° = \frac{AP_{max}}{\omega^2 m} \angle 180°$$

SPECIAL-PURPOSE ELECTROMECHANICAL AND ELECTROMAGNETIC DEVICES

Amplifier input impedance Z_i was given as twice R at an angle of $+45°$, so it can be described as

$$Z_i = 2R \underline{/45°}$$

and

$$I = \frac{V_s}{Z_i}$$

So, from (7.13), substituting for I, we have

$$V_s = j\omega(2\pi\hat{r}n)Bx - RI$$

$$V_s = j\omega(2\pi\hat{r}n)Bx - \left(R\underline{/0°} \cdot \frac{V_s}{2R\underline{/45°}}\right)$$

$$= j\omega(2\pi\hat{r}n)Bx - \frac{V_s}{2}\underline{/-45°}$$

Therefore,

$$V_s\left(1\underline{/0°} + \frac{1}{2}\underline{/-45°}\right) = j\omega(2\pi\hat{r}n)Bx$$

Next, inserting all pertinent quantities and dividing gives us

$$V_s = \frac{j\omega(2\pi\hat{r}nB)\ AP_{max}/\omega^2m\ \underline{/180°}}{1.40\ \underline{/-14.6°}}$$

$$= \frac{2\pi\hat{r}nBAP_{max}}{1.40\omega m}\underline{/(90 + 180 + 14.6)°}$$

$$= \frac{4.49\hat{r}nBAP_{max}}{\omega m}\underline{/-76°}$$

So the voltage V_s is described by

$$V_s(t) = \frac{4.49\hat{r}nBAP_{max}}{\omega m}\cos(\omega t - 76°)$$

This value of voltage is proportional to, but shifted in phase $-76°$ from, the input sound pressure, and is inversely proportional to frequency.

7.4

THE DYNAMIC CONE-TYPE LOUDSPEAKER

A conventional, dynamic, cone-type loudspeaker, shown in Figure 7.11, is a dual of the dynamic microphone and phonograph pickup just discussed. Thus the methods in Chapter 2 can also be used to develop equations describing its operation [3, 4, 5, 7].

As can be seen from Figure 7.11, a lightweight voice coil, placed in the field of a permanent magnet, is excited by current from an amplifier. The scalar electromagnetic force on the coil (assumed to be in the x direction only) can be easily shown to be

$$F = 2\pi\hat{r}nBi \qquad (7.14)$$

where

$\hat{r}$ = mean radius of the coil
n = number of turns
B = magnetic flux from the permanent magnet
i = exciting current

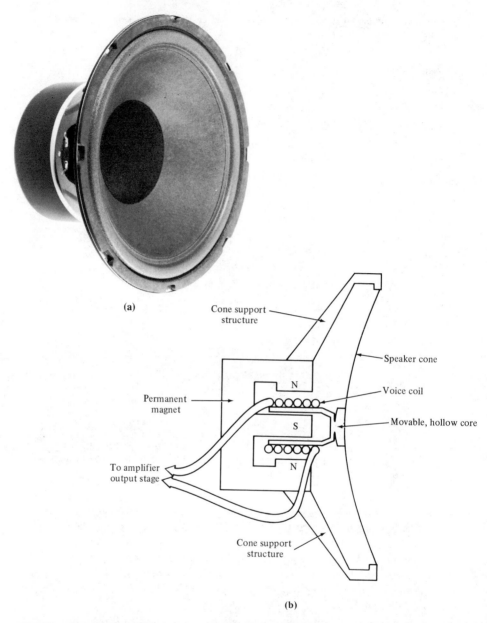

(a)

(b)

FIGURE 7.11 Dynamic Cone-Type Loudspeaker (Courtesy of Radio Shack, a Division of Tandy Corporation)

This is simply a restatement of (2.6) in terms of the coil's measurable parameters. Obviously, the direction of movement will reverse when current reverses, as shown in Figure 7.12. This suggests, then, that the speaker cone will vibrate sinusoidally if the exciting current is sinusoidal and that sound will be produced if the frequency of the current is within the audio-frequency range (generally about 10 Hz to 20 kHz; voice-frequency ranges from about 300 Hz to 3 kHz).

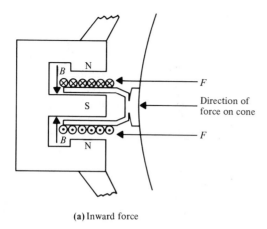

(a) Inward force

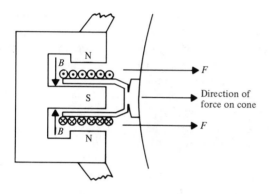

(b) Outward force [current direction reversed
from that in (a)]

FIGURE 7.12 Direction of Force on Speaker Coil for Current Directions Shown

It should be obvious that motion of the voice coil will in turn induce a voltage across it, as in a rotating machine. Analysis of the electromagnetic and mechanical performance of the speaker thus requires solution of the following four simultaneous differential equations, which are similar to those for a machine or the coil and clapper in Example 2.8:

$$F(t) = 2\pi \hat{r} n B i(t) \tag{7.15}$$

$$e(t) = 2\pi \hat{r} n B v(t) \tag{7.16}$$

$$V_s(t) = Ri(t) + L\frac{di(t)}{dt} + e(t) \tag{7.17}$$

$$\dot{F} = \frac{1}{Z_m} \dot{v} \quad \text{(in phasor form)} \tag{7.18}$$

where Z_m is the mechanical impedance of the speaker cone and coil system, m is its mass, v is the speed of the cone–coil in the x direction, n is the number of turns, and V_a, e, r, and L are electrical quantities. In the steady state, these four equations reduce to the following two phasor equations:

$$0 = \frac{v}{Z_m} - 2\pi \hat{r} n B I \qquad (7.19)$$

$$V_a = (R + j\omega L)I + 2\pi \hat{r} n B v \qquad (7.20)$$

The mechanical impedance Z_m of the speaker alone is defined as

$$Z_m = r_m + jX_m \qquad (7.21)$$

which, in terms of measurable quantities, is

$$Z_m = \frac{1}{-\dfrac{j}{\omega C_m} + j\omega m} = \frac{j\omega C_m}{1 - \omega^2 m C_m} \qquad (7.22)$$

The compliance C_m of the speaker in m/N is defined as the reciprocal of the stiffness (or reciprocal of the static force required to cause a linear displacement of one unit, usually 1 cm). In practice, C_m is the reciprocal of the slope of the measured force–displacement curve for the cone–coil structure, usually referred to the linear portion. The speaker has mass m.

Equations (7.19) and (7.20) can be used to find both the speaker input impedance and its transfer function. The input impedance can be found by (1) recognizing that in the steady state, the sum of the forces F is zero (7.19), and (2) then dividing V_a by I while combining (7.19) and (7.20):

$$Z_{in} = \frac{V_a}{I} = R + j\omega L + (2\pi \hat{r} n B)^2 Z_m \qquad (7.23)$$

The unloaded speaker's transfer function or transmittance is defined as

$$\frac{\text{mechanical output}}{\text{electrical input}} = \frac{v}{V_a} = \frac{2\pi \hat{r} n B}{(R + j\omega L)(1/Z_m) + (2\pi \hat{r} n B)^2} \qquad (7.24)$$

These two equations describe the electromechanical equivalent circuit shown in Figure 7.13, where cone speed v is considered to be the analog of applied voltage, since v causes air pressure which is perceived as sound.

Speaker performance can be qualitatively analyzed by observing the structure of the equivalent circuit and (7.23) and (7.24), which describe it. For example, the mechanical design of the speaker clearly influences the input impedance, but the input impedance is not the sum of the electrical and mechanical impedances; electromagnetic interaction must be taken into account, as is done by the transformer in Figure 7.13. Also, the clear LRC nature of the equivalent circuit indicates that the speaker will operate better over some frequency range determined by its electromechanical design and, furthermore, will have a resonant frequency. Modern speakers are designed to reduce significantly the negative effects on fidelity (an increase in audio distortion) caused by this resonance. Moreover, it should be clear that the output of the speaker will be reduced at high frequencies due to the speaker's mass m, but will be nonzero because of

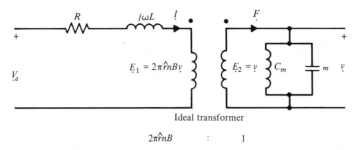

FIGURE 7.13 Electromechanical Equivalent Circuit of a Cone-Type Loudspeaker

the countering effect of stiffness as described by C_m. Consequently, output at high frequencies is enhanced by reducing the mass and increasing the stiffness of the coil–cone system. This enhanced high-frequency response is offset, though, by the conclusion that the mechanical output (audio volume or "power") is enhanced by use of a large, strong magnet and coil structure [Eq. (7.24)] to provide a large motion when excited. This set of conflicting requirements has led to development of various types of conventional loudspeakers, with high-range speakers ("tweeters") characterized by small, light, stiff structures employing expensive high-strength magnets; low-range speakers ("woofers") having large, heavy coil structures and large cones; and midrange speakers sized in between but having limited fidelity at both the upper and lower ends of the audio spectrum.

EXAMPLE 7.3   A cone-type loudspeaker has the following design parameters:

$$\hat{r} = 1.5 \times 10^{-2} \text{ m}$$

$$L = 4.8 \text{ mH (measured)}$$

$$B_{\text{from permanent magnet}} = 2 \text{ W/m}^2$$

$n = 10$ turns AWG No. 18 enameled copper wire

compliance of coil–cone system $C_m = 5.6 \times 10^{-5}$ m/N

mass of core and cone only $m_{\text{core-cone}} = 3.3 \times 10^{-3}$ kg

Compute the input impedance of the speaker in vacuo (i.e., unloaded) and the form of the output if $|I_a| = 10$A and $V_a(t) = 10 \cos 1256.6t$ (200-Hz sinusoid).
    The length of the wire on the coil is

$$\ell = 2\pi\hat{r}n = 2\pi \cdot 1.5 \times 10^{-2} \cdot 10 \approx 1 \text{ m}$$

From wire tables for AWG No. 18 copper wire,

$$R|_{1m} = 1.8 \times 10^{-2} \ \Omega$$

$$m_{\text{wire}} = 6.7 \times 10^{-3} \text{ kg}$$

So

$$m_{\text{wire, coil, and cone}} = 6.7 \times 10^{-3} + 3.3 \times 10^{-3} \text{ kg}$$

$$= 10^{-2} \text{ kg}$$

Therefore, the mechanical impedance of the speaker is

$$Z_m = \frac{j\omega C_m}{1 - \omega^2 m C_m}$$

and since, at 200 Hz, $\omega = 1256.6$, the impedance is as follows:

$$Z_m = \frac{1 \underline{/90°} \cdot 1256.6 \cdot 5.6 \times 10^{-5}}{1 - (1256.6)^2 \cdot 10^{-2} \cdot 5.6 \times 10^{-5}}$$

$$\approx j0.61$$

The input impedance of the unloaded speaker is then found from (7.23):

$$Z = R + j\omega L + (2\pi\hat{r}nB)^2 Z_m$$

$$= 1.8 \times 10^{-2} + j(1256.6 \cdot 4.8 \times 10^{-3}) + j(4 \cdot 0.61)$$

$$= 0.02 + j8.47 \ \Omega \approx j8.5 \ \Omega$$

So the input impedance at 200 Hz is $j8.5 \ \Omega$. The output velocity $v(t)$ of the cone in the x direction would then be found from (7.24):

$$v(t) = \frac{2\pi\hat{r}nBV_a}{(R + j\omega L)(1/Z_m) + (2\pi\hat{r}nB)^2}$$

$$= \frac{2V_a}{\dfrac{1.8 \times 10^{-2} + j6.03}{j0.61} + 4}$$

$$\approx \frac{2V_a}{14} = \frac{1}{7} \cdot 10 \cos 1256.6t$$

So

$$v(t) \approx 1.4 \cos 1256.6t \quad \text{(200-Hz sinusoid)}$$

Since sound pressure is a function of the speed of the cone, the audio output of the speaker would also be a 200-Hz sinusoid, whose amplitude is directly proportional to the input voltage signal $V_a(t)$.

7.5

HOMOPOLAR MACHINES

Homopolar machines, variants of the rotating wire configuration in Problem 2.3, were conceived in their simplest form by Michael Faraday. In the *Faraday disk machine*, the rotating wire of Problem 2.3 is replaced by a conducting metal disk of radius r_o and thickness T, in the configuration of Figure 7.14. In the case of a homopolar dc generator, the disk is rotated in the constant B_z shown in the figure. Brushes are placed in contact with the axle and the outer edge of the disk, and an electrical load is attached.

The induced voltage generated by this device can be found easily from Faraday's law:

$$\oint \overline{E} \cdot d\overline{\ell} = -\int_s \frac{\partial \overline{B}}{\partial t} \cdot d\overline{s} \tag{1.17}$$

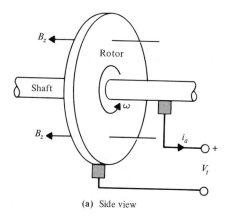

(a) Side view

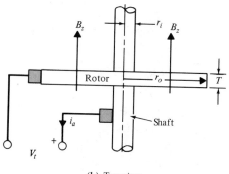

(b) Top view

FIGURE 7.14 Simple Faraday Disk Homopolar Machine

For the device shown, the left-hand side of the equation reduces to

$$\oint \overline{E} \cdot d\overline{\ell} = \int_{r_i}^{r_o} e(r) \, dr = V_t - \int_{r_i}^{r_o} \frac{\rho i_a}{2\pi r T} \, dr \qquad (7.25)$$

where

V_t = terminal voltage

ρ = resistivity of the disk, Ω-m

T = thickness of the disk

r_o = outside radius of the disk

r_i = inside radius of the disk (radius of axle)

i_a = armature current

The right side of (1.17), for the device in the figure, can be shown to equal

$$-\int_s \frac{\partial \overline{B}}{\partial t} \cdot d\overline{s} = -\int_{r_i}^{r_o} -B_z \omega r \, dr \qquad (7.26)$$

where it is assumed that B is constant and oriented only in the z direction (yielding scalar B_z oriented in the z direction into the flat surface of the disk as shown). Equations (7.25) and (7.26) can be combined and the terms regrouped to yield an expression for terminal voltage V_t:

$$V_t = -\int_{r_i}^{r_o} \left(-\omega B_z r + \frac{\rho i_a}{2\pi r T} \right) dr \qquad (7.27)$$

Notice that the first term is the induced voltage e_a resulting from the movement of the conducting disk in the constant field. The second term is the voltage drop due to the resistive nature of the disk material. When (7.27) is solved the classical equation describing the disk machine is derived:

$$V_t = e_a - i_a r_a \qquad (7.28)$$

$$= \frac{\omega B_z}{2} (r_o^2 - r_i^2) - \frac{\rho i_a}{2\pi T} \ln \frac{r_o}{r_i}$$

If the machine is then connected to a load resistance R_L, the armature current i_a can be found to be

$$i_a = \frac{e_a}{R_L + (\rho/2\pi T) \ln (r_o/r_i)} \qquad (7.29)$$

$$= \frac{\omega B_z (r_o^2 - r_i^2)}{2[R_L + (\rho/2\pi T) \ln (r_o/r_i)]} \qquad (7.30)$$

Analysis of these equations indicates that the Faraday disk machine has the equivalent circuit shown in Figure 7.15. Since B_z is constant, (7.29) and (7.30) indicate that this is a dc machine.† The magnetic field of flux density B_z can be developed using either permanent magnets or field coils. If a field coil is used, then, from the configuration shown in Figure 7.16, the flux density can be approximated by (neglecting leakage)

$$B_z = \frac{N_f i_f}{2\pi(r_o - r_i)(\mathcal{R}_{disk} + 2\mathcal{R}_{gap} + \mathcal{R}_{iron})}$$

$$= \frac{N_f i_f}{2\pi(r_o - r_i)\left[\dfrac{T}{\mu_0 \mu_{rd} 2\pi(r_o - r_i)} + \dfrac{2g}{\mu_0 2\pi(r_o - r_i)} + \mathcal{R}_{iron} \right]} \qquad (7.31)$$

$$= \frac{N_f i_f}{T/\mu_0 \mu_{rd} + 2g/\mu_0 + 2\pi(r_o - r_i)\mathcal{R}_{iron}}$$

†Homopolar machines will also operate on ac; this will not be discussed here in the interest of time and space. Homopolar ac machines are sometimes called "asynchronous" machines.

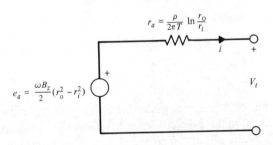

FIGURE 7.15 Equivalent Circuit for Simple Faraday Disk Machine

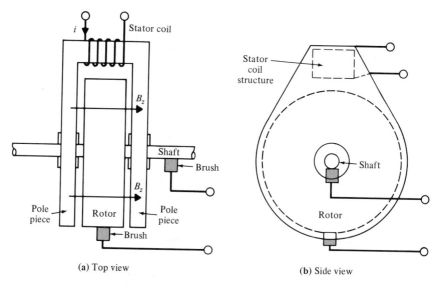

(a) Top view   (b) Side view

FIGURE 7.16 Homopolar Generator with External Stator Coil

where

$$N_f = \text{number of field turns}$$
$$i_f = \text{field current}$$
$$T = \text{thickness of disk}$$
$$g = \text{air gap length}$$
$$r_o = \text{outside radius of disk}$$
$$r_i = \text{inside radius of disk}$$
$$\mu_{rd} = \text{relative permeability of disk}$$
$$\mathcal{R}_{iron} = \text{reluctance of iron flux path (dependent on design)}$$

Inclusion of (7.31) in (7.30) or (7.29) would result in a magnetization curve for the homopolar machine similar to the curve in Figure 5.11 for a separately excited dc machine. Clearly, the homopolar machine can also be used as a motor; this is explored in Problem 7.7.

EXAMPLE 7.4 A Faraday disk machine is constructed by building an aluminum rotor of thickness 10 cm of outside radius $r_o = 0.5$ m on an axle of radius 2 cm. If the rotor is placed in a magnetic field $B_z = 1.0$ W/m^2 and rotated at the extremely high speed of 10,000 rpm, what is the open-circuit voltage generated, the armature resistance r_a, and the short-circuit current i_{sc}?

From material tables

$$\rho_{aluminum} = 2.62 \times 10^{-8} \ \Omega\text{-m}$$

The open-circuit voltage e_a is found from (7.28), with $i_a = 0$:

$$e_a = \frac{\omega B_z}{2} (r_o^2 - r_i^2)$$

where

$$B_z = 1.0 \text{ W/m}^2$$

$$\omega = 2\pi \cdot 10{,}000 \text{ rev/min} \times \frac{1 \text{ min}}{60 \text{ sec}} = 1047 \text{ rad/sec}$$

So, inserting values in the equation, we have

$$e_a = \frac{1047(1.0)}{2}[(0.5)^2 - (2 \times 10^{-2})^2] \quad \text{V}$$

$$= 131 \text{ V}$$

The armature resistance r_a in Figure 7.15 is found to be

$$r_a = \frac{\rho}{2\pi T} \ln \frac{r_o}{r_i} \quad \Omega$$

$$= \frac{2.62 \times 10^{-8}}{2\pi \cdot (0.1)} \ln \frac{0.5}{0.02}$$

$$= 1.34 \times 10^{-7} \, \Omega$$

The short-circuit current will occur if $R_L = 0 \, \Omega$. So i_{sc} is found by dividing e_a by r_a to get

$$i_{sc} = \frac{e_a}{r_a} = \frac{131 \text{ V}}{1.34 \times 10^{-7} \, \Omega} = 9.76 \times 10^8 \text{ A}$$

It can be seen from this somewhat extreme example that a homopolar generator is capable of producing *very* large currents at low voltages. This would imply that such a machine could produce large amounts of power. In fact, the machine described above would have a maximum power output of $e_a i_{sc}$, which is 130,000 MW, or about 30% of the entire installed electric generating capacity in the United States in 1975 [8]! Clearly, a homopolar machine driving such large currents would have enormous $i^2 r_a$ heating losses and could not sustain such a massive power output for very long even if it did stay together at 10,000 rpm. Consequently, homopolar machines of the configuration discussed here are generally useful only in situations where large pulses of power are required.

Because of the limitation on power-output time duration, pulsed homopolar machines find only specific applications. Some proposed applications are as a pulsed power source for nuclear fusion experiments, welding, metal billet heating, pulsed current switch testing, cutting, and high-energy electric arc generation (see, e.g., references 3, 9, 10, and 11). Several variations of the Faraday disk design have been developed for use in these and other applications. Figure 7.17 shows two early disk-type homopolar generators (5 MJ and 10 MJ) similar to the design shown above. Figure 7.18 compares this design to a drum-type unit. In this drum design, longitudinal currents are generated in the drum rather than radially on a disk. The drum-type unit also requires two sets of brushes, since voltages of opposite polarity (and hence currents of opposite direction) are induced between the center and brushes. A third, newer, pulsed homopolar machine design, where the rotor partially surrounds the stator windings, is shown in Figure 7.19. This design is advantageous in some ways because the entire

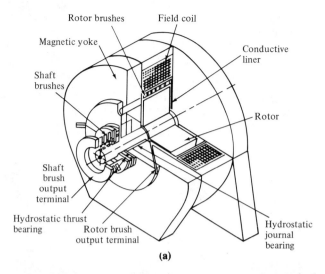

Rotor brushes   Field coil

Magnetic yoke

Conductive
liner

Shaft
brushes

Rotor

Shaft
brush
output
terminal

Hydrostatic thrust
bearing   Rotor brush
output terminal

Hydrostatic
journal
bearing

(a)

(b)

FIGURE 7.17 Two Early Disk-Type Homopolar Generators: (a) 5MJ Machine; (b) 10 MJ Machine (Courtesy of Center for Electromechanics, The University of Texas at Austin)

flux path is enclosed by the rotor; very little stator core material is required. This allows for a reasonably compact, lightweight unit compared to older disk and drum designs, which require considerably more stator core material.

As mentioned earlier, the primary use of this type of dc homopolar machine is to produce large pulses of power at very high currents and low voltage. The rotor is accelerated to running speed either by an outside prime mover or by operating the machine as a motor. It is then switched to the generation mode and rapidly discharged through a load. Consequently, the energy stored (and

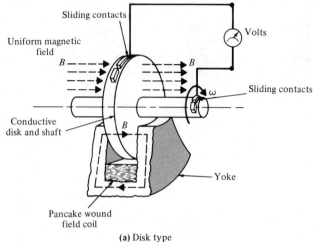

(a) Disk type

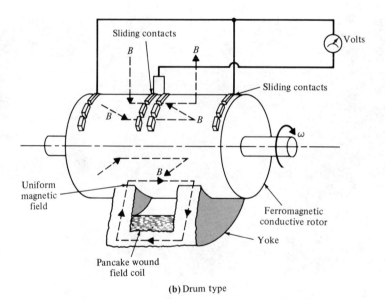

(b) Drum type

FIGURE 7.18 Comparison of Configurations—Disk and Drum-Type Homopolar Generators (Courtesy of Center for Electromechanics, The University of Texas at Austin)

discharged) is simply the inertia of the machine under ideal conditions. In this mode, the device can be modeled as a capacitor of value

$$C = \frac{4\pi^2 J}{\varphi^2} \tag{7.32}$$

where

$$J = \text{polar moment of inertia}$$
$$\varphi = \text{flux } (= B_z \cdot \text{area})$$

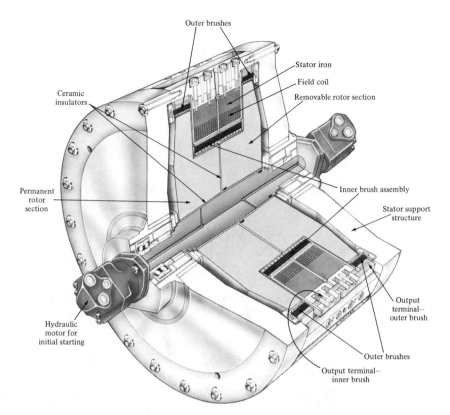

FIGURE 7.19 Compact, Pulsed Homopolar Generator with the Rotor Partially Surrounding the Stator (Courtesy of OIME, Inc.)

This capacitorlike behavior is readily understandable if one considers the actual fundamental characteristics of the machine as an energy storage, rapid discharge device. In a capacitor, the voltage is a linear function of the charge and the stored energy is a function of the square of the voltage. Likewise, in the homopolar machine, the voltage is a linear function of speed and the stored energy is a function of the square of the speed and, thus, the square of the voltage. The relationship of (7.32) can be derived from the stored energy as described above ($C = 2W/V^2$). Moreover, if some practical design assumptions are made (e.g., let everything be constant except quantities that can be directly controlled by an operator), it can be shown that C is approximated by the rotor thickness divided by the square of the flux. This latter relationship is used by many engineers for sizing homopolar machines for various pulsed power applications. The current pulse generated upon machine discharge is similar to Figure 7.20, which is the short-circuit discharge curve for a 6.2-MJ machine, starting at 1360 rpm. The machine produced a voltage of 10.9 V dc and a peak current of 1.02×10^6 A during discharge.

Figures 7.21 and 7.22 show a pulsed homopolar machine used to butt-weld pipe. The high current produced during the machine's discharge pulse greatly

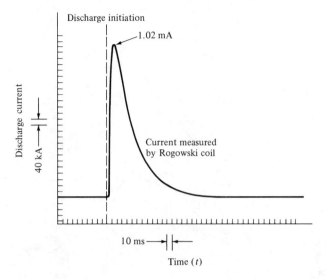

Discharge initiation

1.02 mA

Current measured
by Rogowski coil

40 kA

Discharge current

10 ms

Time (t)

FIGURE 7.20 Discharge Current Pulse from a 6.2-MJ Homopolar Generator (Courtesy of OIME, Inc.)

heats the interface between the two objects to be welded together. During heating, the objects are clamped into solid contact with each other but the joint is lightly loaded mechanically. As the joint is heated, the two pieces are forced together axially and are literally forged into one single piece without melting. This allows for uniform welds of even large, irregular pieces without widespread heating of the metal or other related metallurgical deficiencies. Figure 7.23 shows a longitudinal section of a 4-in. stainless steel pipe welded using a pulsed homopolar machine. Numerous metals have been successfully welded using this process, including titanium, which has been welded successfully in air (see reference 11).

7.6

SYNCHRONOUS RELUCTANCE AND HYSTERESIS MOTORS

Some of the simplest and least expensive synchronous motors available are reluctance and hysteresis motors. These motors generally are of the fractional-horsepower or "subfractional"-horsepower class, and are sometimes seen in clocks and phonograph drives. Although most of these motors are single-phase, some three-phase designs are also in use.

7.6.1 THE POLYPHASE RELUCTANCE MOTOR

A generalized three-phase reluctance motor is shown in Figure 7.24 to illustrate the principle of operation. A reluctance motor is characterized by a salient-pole rotor with no windings on it and it may have a salient-pole stator as well. Salient-pole machines and salient-pole synchronous machines in particular were discussed in Sections 2.4 and 4.4, respectively. It was shown in Example 2.9 that

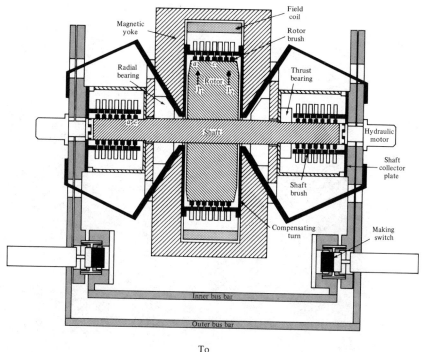

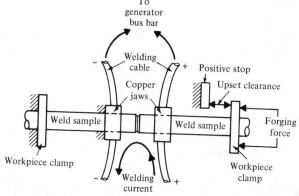

FIGURE 7.21 Schematic Drawing of a 10-MJ Homopolar Generator Welding Circuit (Courtesy of Center for Electromechanics, The University of Texas at Austin)

the torque equation for a salient-pole synchronous machine contains two terms and, further, if the rotor windings were eliminated entirely, a reluctance torque described by the following equation would still be developed:

$$T_e = -i_1^2 \, \Delta L_s \sin 2\theta \tag{7.33}$$

where

i_1 = stator current

ΔL_s = amplitude of sinusoidal component of stator self-inductance

θ = angle between rotor axis and rotating stator field

Pipe sections to be welded

Welding fixture

Homopolar machine

FIGURE 7.22 Commercial Pipe Welding Apparatus Using Homopolar Machine (Courtesy of OIME, Inc.)

FIGURE 7.23 Longitudinal Section Through a 4-in.-Diameter Type 304 Stainless Steel Pipe, Butt-Welded Using a Homopolar Generator (Courtesy of Center for Electromechanics, The University of Texas at Austin)

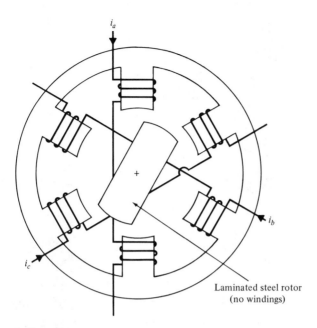

i_a

i_b

i_c

Laminated steel rotor
(no windings)

FIGURE 7.24 Generalized Three-Phase Reluctance Motor

This torque can also be expressed in terms of x_d and x_q as in the second term of (4.81):

$$T_e \approx \frac{V_s^2}{\omega_r} \frac{x_d - x_q}{2 x_d x_q} \sin 2\delta \qquad (7.34)$$

where the d and q axes are defined as in Section 4.4, and δ is the angle between the rotating stator field and the d axis of the rotor.

Clearly, the reluctance motor described above delivers no torque at startup and converts energy only at synchronous speed ($\omega_r = \omega_s$). This poses major operational problems which have been solved in either of two ways. First, some very small reluctance motors are started by hand, although this is somewhat uncommon. Second, the salient-pole rotor can be built inside a set of conventional squirrel-cage bars, as in Figure 7.25, to produce starting torque as an induction machine. The latter starting configuration is no different than for a conventional synchronous motor, which also requires a squirrel-cage winding for startup. This squirrel cage winding has a dragging effect at synchronous speed, which somewhat reduces the delivered torque, if the motor is single-phase. This reduced torque was illustrated for conventional single-phase induction machines in Figure 3.33. A typical torque–speed curve for this machine is also shown in Figure 7.25.

7.6.2 THE SHADED-POLE SINGLE-PHASE RELUCTANCE MOTOR

It should be recalled from Section 3.6 that a single-phase stator winding results in a pulsating, rather than rotating magnetic field, which develops no starting torque regardless of rotor configuration. In Section 3.6, split-phase and capacitor

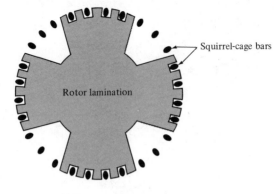

(a) Rotor structure

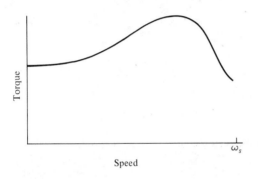

(b) Nominal torque-speed characteristic

FIGURE 7.25 (a) Typical Four-Pole Slotted Reluctance Rotor with Squirrel-Cage Winding and (b) Nominal Torque-Speed Characteristic

motors were discussed as alternative solutions to this problem. In this section we present another solution that is common to fractional- and subfractional-horsepower motors such as the single-phase reluctance motor. It should be noted, however, that pole shading is applicable to other types of single-phase motors as well. It is extremely inexpensive but has the disadvantage (or advantage, depending on application) that a motor so equipped is unidirectional and can be reversed only by mechanical means.

For this discussion we assume a rotor configuration as discussed in the preceding section and deal primarily with the stator. The object is to produce the rotating stator field required to operate the motor by *inducing* an alternating current in a secondary stator winding which is displaced in space from the main winding. To produce a shaded-pole stator, the conventional single-phase salient-pole stator shown in Figure 7.26(a) has slots milled in the pole faces and conducting rings (usually copper or aluminum) applied as shown in Figure 7.26(b). These rings are called *shading rings*. They must be applied to opposite sides of the pole faces as shown.

If the main stator coil has N_1 turns and is excited by a current of frequency ω_s,

$$i_1 = \sqrt{2}\, I_1 \cos \omega_s t$$

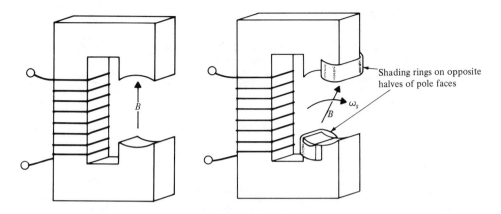

(a) Conventional single-phase stator   (b) Single-phase stator with shading rings

FIGURE 7.26 Example of Shaded-Pole Construction

a sinusoidal field B_1 proportional to i_1 results (assuming for the moment constant core reluctance):

$$B_1 = \frac{N_1 \sqrt{2}\, I_1}{A_{\text{core}} \mathscr{R}_{\text{tot}}} \cos \omega_s t \qquad (7.35)$$

where A_{core} is the effective cross-sectional area of the core. If the slot is milled at the midpoint of the core cross section, the flux passing through the ring is about one-half of the total flux, as in Figure 7.27. Since this flux varies sinusoidally, the shading ring acts as a transformer secondary, and a voltage is induced according to Faraday's law:

$$\oint \overline{E} \cdot d\overline{\ell} = \int_s -\frac{\partial \overline{B}}{\partial t} \cdot d\overline{s} \qquad (1.17)$$

so, for the shading ring, the induced voltage from only I, would be of the form

$$e_{\text{ring}} \propto I_1 \sin \omega_s t$$

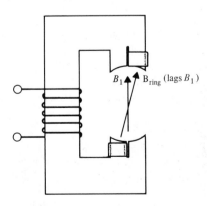

FIGURE 7.27 Flux in a Shaded-Pole Machine

and since the ring is effectively a short circuit, the current due to e_{ring}

$$i_{ring} = \frac{e_{ring}}{|Z_{ring}|} \propto \frac{I_1}{|Z_{ring}|} \sin(\omega_s t + \psi) \qquad (7.36)$$

This current by itself produces a small second flux with flux density B_{ring}, which can be represented by

$$B_{ring} = \frac{N_{ring} i_{ring}}{A' \mathcal{R}'} \propto I_1 \sin(\omega_s t + \psi) \qquad (7.37)$$

where

$\qquad A' =$ effective cross-sectional area enclosed by ring (in this case, $0.5 A_{core}$)

$\qquad \mathcal{R}' =$ reluctance of magnetic circuit through ring and core

$\qquad \psi =$ phase shift due to x/r ratio of conducting ring ($\psi < 90°$)

Comparison of (7.35) and (7.37) yields the following conclusions:

$B_1 \propto I_1 \cos \omega_s t$

$B_{ring} < B_1 \propto I_1 \sin(\omega_s t + \psi) \qquad [= I_1 \cos(\omega_s t - \psi'); \psi' = 90° - \psi]$

So the induced flux represented by B_{ring} lags the main stator flux by some angle ψ' less than 90° and is also angularly displaced from it because the shading ring encloses only half of the pole face. Consequently, this arrangement results in a net flux inside the machine which appears to rotate *toward* the shading ring at synchronous speed ω_s. This rotating field is exactly what is required to start the motor. To address this problem more realistically, it must be solved as a coupled-coil system.

Note that reversal of the shaded-pole motor can occur only if one of the following procedures is executed:

1. The motor is physically turned around and a shaft on the opposite end of the rotor is used.
2. An intermediate idler gear is engaged between the shaft and the load.
3. The shading rings are removed and replaced on the opposite halves of the pole faces from those shown in Figure 7.26(b), providing that reversal, since rotor motion is always toward the rings.

This difficulty in reversing, a general disadvantage, is highly advantageous for certain applications such as clocks and phonograph turntable motors, and shaded-pole reluctance motors are sometimes used in these applications. The overall configuration of such a motor is shown in Figure 7.28. Although other, less obvious configurations also exist, the one shown is quite common.

7.6.3 THE HYSTERESIS MOTOR

A hysteresis motor uses either a polyphase or a single-phase† rotating field to turn a smooth rotor which is made from either a permanent magnet or an easily magnetized (magnetically "hard") iron or steel, as in Figure 7.29(a). Such

---

†Produced by pole shading, or by split-phase or capacitor windings.

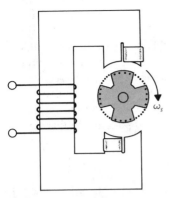

FIGURE 7.28 Typical Configuration for Four-Pole (1800 rpm) Shaded-Pole Reluctance Motor with Squirrel-Cage Winding

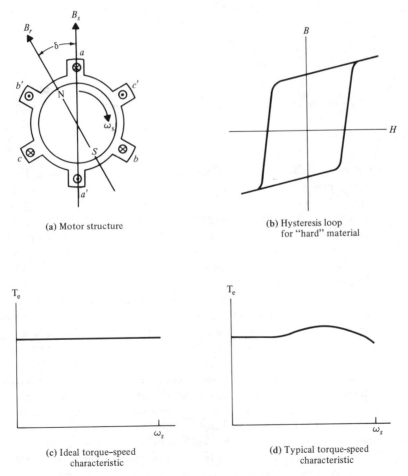

(a) Motor structure

(b) Hysteresis loop for "hard" material

(c) Ideal torque–speed characteristic

(d) Typical torque-speed characteristic

FIGURE 7.29 Hysteresis Motor Configuration and Characteristics

7.6 SYNCHRONOUS RELUCTANCE AND HYSTERESIS MOTORS

material has highly "rectangular" hysteresis loops [Figure 7.29(b) or Figure 1.33] and can retain any magnetization for long periods of time. This magnetic retentivity of the rotor produces a discernible magnetic flux in a single direction, and therefore the rotor will behave like the field coil in a synchronous motor and run at synchronous speed at some torque angle δ. Hysteresis effects occur primarily during startup and acceleration, when the magnetic domains inside the rotor are forced into hysteresis at slip frequency sf_s. As shown in Section 1.3, the size and shape of the hysteresis loop is proportional to this slip frequency, as is the power dissipated due to hysteresis loss:

$$P_h = K_h f (B_{max})^n \qquad (1.68)$$

where f here is the slip frequency sf_s rather than the stator frequency. However, since during acceleration the machine is operating with $s \neq 0$, it can be thought of as being in some sort of "induction" mode, and thus the input power to the rotor P_g can be considered to be proportional to the hysteresis loss P_h in the rotor divided by the slip (as in an induction motor), if we neglect eddy currents and other parasitic effects. Consequently, the power input to the rotor P_g is, from (1.68) and the argument above, simply a constant times stator frequency f_s or stator angular velocity ω_s. The torque for an induction machine operating in this mode can be found from (3.63):

$$T_e = \frac{I_r'^2 r_r'}{s\omega_s} \qquad (3.63)$$

and the power input to the rotor can be found from (3.57):

$$P_g = \frac{I_r'^2 r_r'}{s} \qquad (3.57)$$

By combining these two equations, we can show that the torque T_e is simply the rotor power input P_g divided by stator frequency ω_s. Since, for the hysteresis motor P_g equals a constant times ω_s, the torque T_e must be constant *regardless of slip*. This results in the idealized torque–speed curve shown in Figure 7.29(c). When eddy currents and other parasitic effects are included, though, the non-linear curve of Figure 7.29(d) results. This curve represents relatively constant torque, compared to other machines.

The primary advantage of a hysteresis motor is that, unlike a reluctance motor, which must be started by hand or with an auxiliary squirrel-cage winding on the rotor, it can start a load and accelerate it at essentially constant torque up to synchronous speed, where torque becomes a function of sin δ as in a conventional synchronous motor. Hysteresis motors are quiet and relatively vibration-free due to the smooth rotor design. The average torque output is typically greater than for a reluctance motor of equivalent size. However, because of the materials used in the rotor, flux densities are limited to values several times lower than for conventional induction motors or reluctance motors of the same physical size, whereas magnetizing currents are much higher, resulting in large ratios of losses, reactive power, and size to power output. Consequently, these machines are generally sold only in fractional- and subfractional-horsepower sizes, where they can be manufactured so inexpensively that the reduction in delivered price more than offsets the disadvantages. These motors are usually produced in the shaded-pole configuration (typical application: clocks and small fans) and as capacitor motors (some air-conditioner and fan applications). Capacitor motors typically have higher starting torques due to the more uniformly balanced rotating field, and thus are used in larger applications.

SPEEDOMETERS, AC TACHOMETERS, AND DRAG CUP-TYPE MOTORS

Several unique applications exist for small, very lightweight low-torque subfractional-horsepower induction motors. Such motors are characterized by use of a hollow, lightweight aluminum rotor called a *drag cup,* shown in Figure 7.30. We will look briefly at three applications: automobile speedometers, ac tachometers, and a unique clock motor.

7.7.1. THE AUTOMOBILE SPEEDOMETER

A generalized automotive speedometer system is shown in Figure 7.31. A gear train inside the vehicle transmission's tail shaft rotates the center member of a flexible coaxial shaft (speedometer cable) which connects the transmission to the speedometer inside the dashboard. As in Figure 7.32, the speedometer itself consists only of (1) a permanent magnet connected to the speedometer cable and (2) a drag cup attached to the speedometer needle at the center and to a calibrated spring at the edge.

When the permanent magnet attached to the cable is spun, a rotating field B_r is produced, inducing currents in the stator† drag cup which create a torque in

---

†For ease in analysis, consider this device to be like a generator, where the rotor produces a rotating field. The stator torque is equal and opposite to the "opposing" torque on the rotor, which was discussed in the analysis of synchronous generators in Chapter 4.

FIGURE 7.30 Typical Aluminum Drag Cup

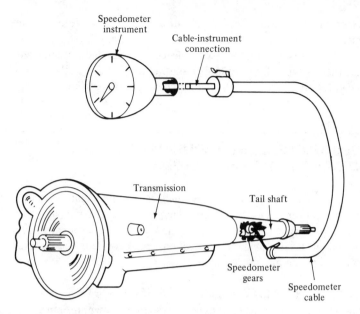

FIGURE 7.31 Automobile Speedometer System

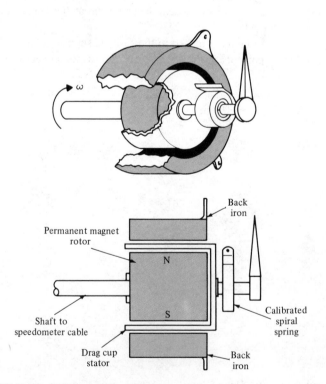

FIGURE 7.32 Speedometer Movement

SPECIAL-PURPOSE ELECTROMECHANICAL AND
ELECTROMAGNETIC DEVICES

the direction of rotation, so the device is basically a simple induction motor that can be analyzed by the methods discussed in Chapter 3.

If the permanent magnet produces a constant maximum flux density B_m and it rotates at an angular frequency ω, the rotating flux B_r created by the rotor is described by

$$B_r = B_m \cos \omega t \tag{7.38}$$

This is analogous to the situation in Figure 3.3 if i_{br} is set equal to zero and B_r is assumed to be equivalent to a flux density resulting from a current $i_{ar} = i_r$ in a coil of $N_r = 1$ turns. This is depicted in Figure 7.33. From this figure it can be seen that the mutual inductances between the stator and rotor are

$$L_{asr} = M \cos \theta_r \tag{7.39}$$

$$L_{bsr} = M \sin \theta_r \tag{7.40}$$

and the L matrix would be

$$\bar{L} = \begin{bmatrix} L_{asas} & 0 & M \cos \theta_r \\ 0 & L_{bsbs} & M \sin \theta_r \\ M \cos \theta_r & M \sin \theta_r & L_{rr} \end{bmatrix} \tag{7.41}$$

The electrical torque exerted on the stator by the moving field B_r can be computed from

$$T_e = \frac{\partial W_c}{\partial \theta_r} = \frac{\partial}{\partial \theta_r} \frac{1}{2} \bar{i}^t \bar{L} \bar{i} \tag{7.42}$$

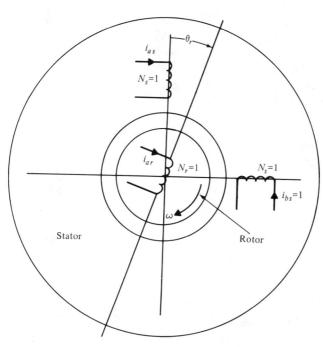

FIGURE 7.33 Coil Diagram for Analysis of Speedometer

assuming linearity of the core material. Performing the matrix operations above and assuming that $|L_{asas}| = |L_{bsbs}| = L$, we get

$$T_e = Mi_r \left[i_{as} \cos \theta_r - i_{bs} \sin \theta_r \right] \qquad (7.43)$$

All that remains is to put T_e in terms of the field B_m. The permanent magnet, represented by a coil of one turn carrying i_r amps rotating at angular frequency ω, is modeled by

$$B_m = \frac{i_r}{A\mathcal{R}} = \frac{i_r}{k_r} \qquad (7.44)$$

where

A = cross-sectional area of flux path

$\mathcal{R}$ = reluctance of flux path

k_r = constant reflecting machine geometry (the two variables above) assuming linear iron and a round, smooth stator and iron backing

Since the two rotor coils are shorted out, we write Kirchhoff's voltage law equations:

$$e_{as} = 0 = R_{as}i_{as} + L \frac{di_{as}}{dt} + L_{asr} \frac{di_r}{dt} + i_r \frac{dL_{asr}}{dt} \qquad (7.45)$$

$$e_{bs} = 0 = R_{bs}i_{bs} + L \frac{di_{bs}}{dt} + L_{bsr} \frac{di_r}{dt} + i_r \frac{dL_{bsr}}{dt} \qquad (7.46)$$

If we assume that the speedometer drag cup is designed so that $|L(di/dt)| << |Ri|$ (generally by using very thin aluminum or other materials with small values of μ relative to iron and a reasonably long current path created by making the cup relatively deep), resistance is uniform and the cup is uniformly cylindrical, (7.45) and (7.46) reduce to the following when combined with (7.39) and (7.40):

$$e_{as} = 0 = Ri_{as} + i_r \frac{dL_{asr}}{dt}$$

$$= Ri_{as} - i_r \omega M \sin \theta_r \qquad (7.47)$$

$$e_{bs} = 0 = Ri_{bs} + i_r \frac{dL_{bsr}}{dt}$$

$$= Ri_{bs} + i_r \omega M \cos \theta_r \qquad (7.48)$$

where

$$\theta_r = \omega t - \theta$$

Combining (7.44), (7.47), and (7.48) with (7.43) yields the electromagnetic torque on the stator:

$$T_e = Mi_r \left[-\frac{\omega M i_r}{R} \cos^2 \theta_r - \frac{\omega M i_r}{R} \sin^2 \theta_r \right]$$

$$= -\frac{\omega M^2 i_r^2}{R}$$

$$= -KB_m^2 \omega \qquad (7.49)$$

where K is a constant based on machine geometry.

So the electromagnetic torque produced is exactly proportional to the rotational frequency of the shaft and therefore is proportional to the speed of the car. If the stator is allowed to rotate but is connected to the instrument panel by a calibrated spring exerting a mechanical torque

$$T_m = -k_{\text{spring}} \theta_r \qquad (7.50)$$

where k_{spring} is the spring constant, then the torque equation of the device (assuming constant rotational frequency ω) is reduced to

$$T_e = -KB_m^2 \omega = T_m = -k_{\text{spring}} \theta_r \qquad (7.51)$$

and therefore, the higher the vehicle speed, the greater the angle of the pointer with respect to the rest position.

7.7.2 THE AC TACHOMETER

A two-phase drag cup induction motor can be used as an ac tachometer by connecting it as shown in Figure 7.34. The phase a stator winding is connected to a constant-frequency ac source and the phase b stator winding feeds an amplifier with extremely high input impedance. This motor can be analyzed in a fashion similar to the single-phase induction machine in Section 3.6, where the main winding is connected to the ac source and the auxiliary winding feeds the amplifier.

In Section 3.6, the equivalent circuit for a two-phase induction motor operating with one winding open was developed. If we assume that the amplifier on the auxiliary winding is of sufficiently high impedance, then $I_a \approx 0$ and circuit development in Section 3.6 is applicable. The equivalent circuit of Figure 7.35 results, where its component parts were defined as follows in Section 3.5:

$$Z_1(s) = R_1 + jX_1 = \frac{jx_m(r_r'/s + jx_{\ell r}')}{r_r'/s + j(x_m + x_{\ell r}')} \qquad (3.116)$$

$$Z_2(s) = R_2 + jX_2 = \frac{jx_m \left(\dfrac{r_r'}{2-s} + jx_{\ell r}' \right)}{\dfrac{r_r'}{2-s} + j(x_m + x_{\ell r}')} \qquad (3.117)$$

From (3.129) it was found that

$$V_m = V_{m1} + V_{m2} \qquad (7.52)$$

$$V_a = -jV_{m1} + jV_{m2} \qquad (7.53)$$

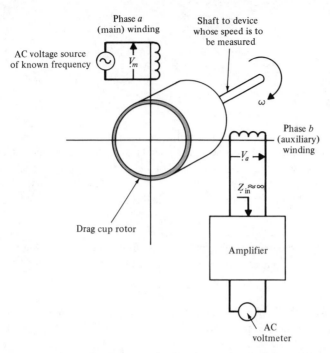

FIGURE 7.34 AC Tachometer Using Two-Phase Drag Cup Induction Motor

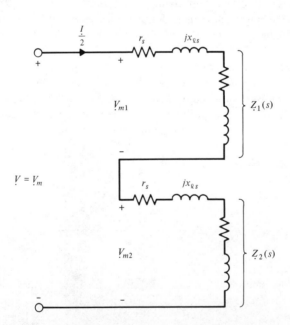

FIGURE 7.35 Equivalent Circuit for AC Tachometer (All Values Referred to Stator Side). Note: $V_a = -jV_{m1} + jV_{m2}$

SPECIAL-PURPOSE ELECTROMECHANICAL AND
ELECTROMAGNETIC DEVICES

and since Figure 7.35 indicates that this device looks like a voltage divider, auxiliary winding output voltage phasor V_a relates to source voltage V_m by the following equation:

$$V_a = jV_m \frac{Z_{tot} - 2Z_{m1}}{Z_{tot}} \qquad (7.54)$$

where

$$Z_{tot} = 2r_s + j2x_{\ell s} + Z_1(s) + Z_2(s)$$

$$Z_{m1} = r_s + jx_{\ell s} + Z_1(s)$$

Equation (7.54) can be rewritten in terms of individual components as

$$V_a = jV_m \frac{Z_2(s) - Z_1(s)}{2(r_s + jx_{\ell s}) + Z_1(s) + Z_2(s)} \qquad (7.55)$$

or

$$V_a = V_m \frac{Z_2(s) - Z_1(s)}{2(r_s + jx_{\ell s}) + Z_1(s) + Z_2(s)} \underline{/90°} \qquad (7.56)$$

Clearly, if (3.116) or (3.117) were inserted in either (7.55) or (7.56), it would be seen that the auxiliary winding output voltage is a function of slip and therefore the speed of the shaft. Determination of the exact form of the voltage is left to the reader (Problem 7.10).

Use of this or similar devices is generally preferred over a speedometer configuration for measuring shaft speeds, as the output is not dependent on use of a spring whose constant k_{spring} can change either with use or with changes in ambient conditions. The two-phase drag cup tachometer offers the ability to control the output and to compensate for any changes in conditions by varying the main winding frequency. Also, since the frequency can be measured and applied to (7.56), it can be used as an input variable by the analog or digital circuit calculating the rotor speed from this equation to provide a higher degree of accuracy than can be obtained otherwise.

7.7.3 A UNIQUE CLOCK MOTOR

Figure 7.36 shows a unique twelve-pole 600-rpm, 60-Hz hybrid hysteresis/drag cup clock motor using a single-phase shaded-pole stator and an aluminum drag cup rotor combined with a steel rotor element. This motor will run at a constant synchronous speed. This motor drives second, minute, and hour hands through a gear train. Rotational speed is reduced approximately 10 times to drive the second hand, implying that torque is magnified approximately the same amount. Torque would thus be magnified 600 times for the minute hand and 36,000 times for the hour hand. This implies that the motor has to deliver only an extremely small torque to operate and, furthermore, the torque amplification of the gear train tends to minimize the effects on the motor of the unequal mechanical torques due to hand position.

The motor is constructed as in Figure 7.37. The aluminum shading disks cause a shift in phase in B on the outer pole pieces relative to the inner ones

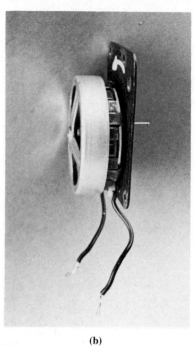

(a)   (b)

FIGURE 7.36 Unique Shaded-Pole Single-Phase Clock Motor. (a) Front View; (b) Side View

(as discussed in Section 7.6.2 for shading rings), causing the rotor to rotate from inner pole piece toward outer pole piece. This rotation is in the direction of the arrow in the figure. The drag cup itself is very lightweight and delivers a small amount of torque during acceleration. A piece of spring steel is force-fit around the inside of the drag cup. This steel develops hysteresis effects that both provide accelerating torque and cause the motor to operate in synchronism with the line. Theoretically, there are no drag cup effects at synchronous speed. This inexpensive wall clock was purchased by the parents of one of the authors and kept accurate time in continuous service for over 20 years until a plastic gear in the mechanism broke, probably from age. The motor itself still runs and shows virtually no signs of wear.

7.8

CONTROL DEVICES: SYNCHROS, DIFFERENTIALS, RESOLVERS, AND STEPPERS

Numerous machine types are found in control loops employing both analog and digital schemes. The most common machines seen in analog control systems are synchros, differentials and control transformers, resolvers, and analog actuators. Digital control systems generally employ steppers, either alone or in conjunction with an optical shaft encoder, and stepper-actuators. Some hybrid

(a)

FIGURE 7.37 Construction of Shaded-Pole Clock Motor

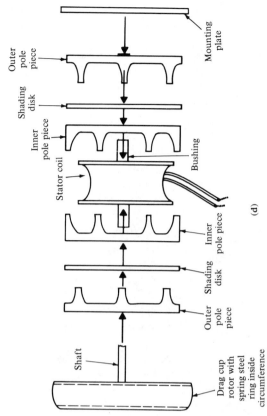

Mounting plate

Outer pole piece

Shading disk

Inner pole piece

Bushing

Stator coil

Inner pole piece

Shading disk

Outer pole piece

Shaft

Drag cup rotor with spring steel ring inside circumference

(d)

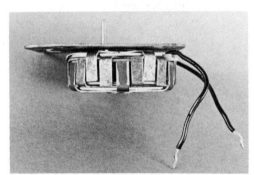

(c)

(b)

7.8 CONTROL DEVICES: SYNCHROS, DIFFERENTIALS, RESOLVERS, AND STEPPERS

digital/analog systems use combinations of all these devices. In this section we investigate how they operate and are used in control applications.

7.8.1 SYNCHROS

Synchros are also called selsyn (or "self-synchronous") machines. The most common types are generators, motors, differential generators, and differential motors. Synchros are designed to determine, transmit, and manipulate shaft-position information. The simplest open-loop synchro system is one in which the shaft position of a motor directly follows the shaft position of a generator as the generator's shaft is moved to various angular positions. More complicated closed-loop and open-loop systems can be developed from this basic open-loop circuit. Operation of these devices and some simple applications will be discussed here; the reader is referred to references 12 and 13 for more detailed information.

Simple selsyn motors and generators are configured as in Figure 7.38. They are salient-pole machines with single-phase rotors and three-phase wye-connected stators. The simplest system possible is shown in Figure 7.39; this is a remote indicator showing the angular position of a shaft. As can be seen from this figure, both the generator and motor rotors are connected to a common single-phase ac supply, and the three-phase stator windings are connected together. The system shown is actually quite complex analytically, requiring a matrix solution for a coupled 8-coil system. Such an analysis is far beyond the scope of this chapter. Instead, we will investigate the general steady-state behavior of this system by making some major simplifying assumptions and treating each device individually.

The Selsyn Generator

Assume that the rotor of the selsyn generator is held at a constant position displaced from the stator a phase winding axis by some angle θ_g, and that the stator b and c phases are displaced from the a phase winding by $-120°$ and $+120°$, respectively, as in Figure 7.40. This arrangement is identical to

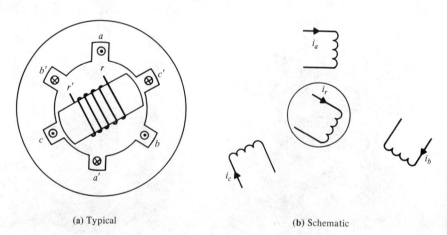

(a) Typical   (b) Schematic

FIGURE 7.38 Simple Selsyn Motor and Generator Configuration

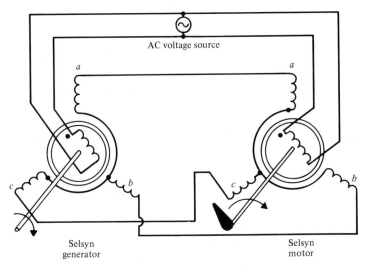

FIGURE 7.39 Basic Selsyn System to Transmit Shaft Angle Information

that of a three-phase, synchronous generator with abc phase rotation, except that now the rotor will be excited with ac and will be attached to a mechanism that may or may not move at variable speed. Assume the generator is unloaded so no stator currents flow. From Faraday's law, the line-to-neutral voltages induced in phases a, b, and c from an ac rotor current

$$i_r = \sqrt{2}\, I \cos \omega t \qquad (7.57)$$

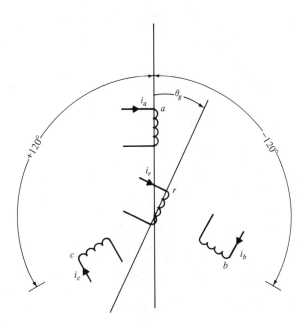

FIGURE 7.40 Selsyn Generator Configuration

7.8 CONTROL DEVICES: SYNCHROS, DIFFERENTIALS, RESOLVERS, AND STEPPERS

will be of the following form:

$$e_{an}(t) = \omega M_g \sqrt{2} I \sin \omega t \cos \theta_g \qquad (7.58)$$

$$e_{bn}(t) = \omega M_g \sqrt{2} I \sin \omega t \cos (\theta_g - 120°) \qquad (7.59)$$

$$e_{cn}(t) = \omega M_g \sqrt{2} I \sin \omega t \cos (\theta_g + 120°) \qquad (7.60)$$

where M_g is the maximum mutual inductance between the generator rotor and the stator coil. Expressed differently, these voltages are

$$e_{an}(t) = \sqrt{2} E_m \sin \omega t \cos \theta_g \qquad (7.61)$$

$$e_{bn}(t) = \sqrt{2} E_m \sin \omega t \cos (\theta_g - 120°) \qquad (7.62)$$

$$e_{cn}(t) = \sqrt{2} E_m \sin \omega t \cos (\theta_g + 120°) \qquad (7.63)$$

For any given angle θ_g, then, the selsyn will generate a set of synchronous, *unbalanced* three-phase voltages whose amplitudes are functions of the angular position of the rotor.

The Selsyn Motor

Assume that the three stator coils of an isolated selsyn motor are connected to an infinite bus that delivers a set of three unbalanced ac voltages of the form:

$$e_{an}(t) = \sqrt{2} E_m \sin \omega t \cos \theta_g \qquad (7.64)$$

$$e_{bn}(t) = \sqrt{2} E_m \sin \omega t \cos (\theta_g - 120°) \qquad (7.65)$$

$$e_{cn}(t) = \sqrt{2} E_m \sin \omega t \cos (\theta_g + 120°) \qquad (7.66)$$

If we ignore the rotor altogether, along with any effects of saliency, it can be shown that the stator currents will *also* be unbalanced and functions of θ_g. Each of these currents contributes to the stator field B_s, and it can be shown that

$$B_s(i, \theta_g) = B_a(i_a, \theta_g) + B_b(i_b, \theta_g) + B_c(i_c, \theta_g) \qquad (7.67)$$

or

$$B_s = f(i, \theta_g) \qquad (7.68)$$

In fact, it can be shown that the three unbalanced currents produce a field B_s that oscillates with frequency ω and *is oriented at an angle θ_g from the a phase*. Furthermore, it can be shown that for the steady-state case, the infinite bus can be replaced by the selsyn generator with virtually identical results: Rotating the generator's rotor to an angle θ_g results in a stator ac field in the motor *oriented at exactly the same angle*.

Consider what would happen if the rotor were now placed in the motor's stator field and excited with an ac current in phase with that of the selsyn generator's rotor. Although this can be analyzed exactly using coupled-coil theory, we will look at it in a more simplistic fashion. Since the torque in a round-rotor machine was found in Chapter 2 to be

$$T_e \propto -i_s i_r \sin \delta \qquad (2.66)$$

where δ is the angle between the stator and rotor fields, it follows that when the rotor is excited by an ac current in phase with the rotor current of the generator, it will experience a torque tending to line it up with the stator field, until the two fields are aligned. Hence under ideal conditions the rotor will also move to an angular position $\theta_m = \theta_g$ with respect to the a phase of the stator, at which point $\delta = 0$, the torque equals zero, the currents equal zero, and it comes to rest. At that point, the rotor will be generating a voltage in opposition to the change in flux just as occurs in a transformer. The position gain of this system is 1.0. Clearly, the actual phenomena occurring in the two machines are more complex than this, involving interactions among all the currents.

Typically, the selsyn motor will include a flywheel on its shaft for damping oscillations caused by position overshoots upon quick changes in angle. This action is like that discussed in Chapter 6 for continuously rotating synchronous machines. A similar torque, opposing motion, is developed in the generator, as in a rotating synchronous generator. Exact formulations describing these effects can be found by solving the eight-coil system mentioned earlier.

The Differential Selsyn Generator and Motor

A differential selsyn (either generator or motor) allows electrical and mechanical angular signals to be added or subtracted. There is essentially no difference in their construction, except that most differential motors are equipped with a flywheel damper to reduce mechanical oscillations. Differential selsyns, unlike standard selsyn generators and motors, have a three-phase rotor as well as a three-phase stator and typically are round-rotor machines with reasonably narrow slots at the surface of the rotor. Differential unit configuration is shown in Figure 7.41.

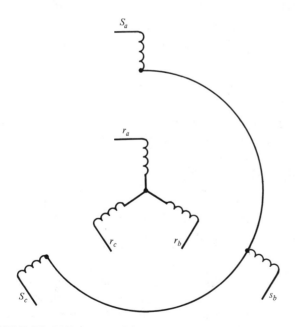

FIGURE 7.41 Differential Selsyn Configuration

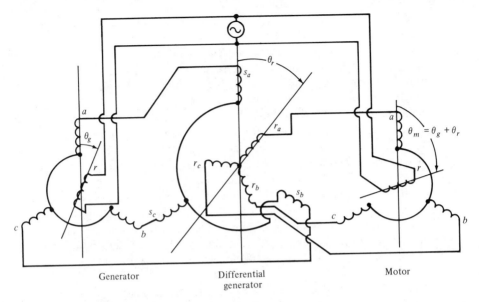

FIGURE 7.42 Simple Open-Loop Addition Circuit

Differential generators add or subtract a mechanical angular signal and an electrical angular signal, and are used in conjunction with conventional selsyn motors and generators; a simple open loop addition circuit is shown in Figure 7.42. In this circuit, angular data from the generator and the differential generator are added and the sum is produced at the motor. We have already concluded that the angular displacement of the generator θ_g results in three-phase voltages as in (7.64) to (7.66) if all other simplifying assumptions are made. These equations are repeated here for reference:

$$e_{an}(t) = \sqrt{2}\, E_m \sin \omega t \cos \theta_g \tag{7.64}$$

$$e_{bn}(t) = \sqrt{2}\, E_m \sin \omega t \cos (\theta_g - 120°) \tag{7.65}$$

$$e_{cn}(t) = \sqrt{2}\, E_m \sin \omega t \cos (\theta_g + 120°) \tag{7.66}$$

These voltages are impressed across the phase windings of the differential selsyn's stator and, because the differential is connected in reverse, they produce a field B_{sd} oriented at an angle $\theta_{sd} = -\theta_g$ with respect to the a phase axis. It should be obvious that, if the rotor of the differential were *also* oriented at the angle $\theta_{rd} = -\theta_g$, then the voltages induced in the three-phase rotor windings would be exactly the same as at the generator and could be represented by (7.58) to (7.60). However, if the rotor is oriented at some angle $\theta_r \neq -\theta_g$, the mutual inductances of the differential selsyn become of the form

$$M_d = M_d \cos \theta_r \tag{7.69}$$

where θ_r is the angle between the a phase of the rotor and the a phase of the stator. Consequently, by making the same simplifying assumptions as before, the induced line-to-neutral voltages in the three-phase rotor of the differential can be shown to be of the form

$$e_{ad}(t) = \sqrt{2}\, E_m \sin \omega t \cos (\theta_g + \theta_r) \qquad (7.70)$$

$$e_{bd}(t) = \sqrt{2}\, E_m \sin \omega t \cos (\theta_g + \theta_r - 120°) \qquad (7.71)$$

$$e_{cd}(t) = \sqrt{2}\, E_m \sin \omega t \cos (\theta_g + \theta_r + 120°) \qquad (7.72)$$

As before, these unbalanced stator voltages can result in currents of the same angular displacement, which are then transmitted to a selsyn motor. Since we have seen that the motor will reproduce its angular input data in the form of shaft motion to a point where θ_m equals the input angle, then for the system in Figure 7.42, it can be shown that

$$\theta_m = \theta_{rsd} = \theta_g + \theta_r \qquad (7.73)$$

where

θ_g = angular displacement of selsyn generator shaft

θ_{rsd} = angular displacement of selsyn differential shaft with respect to

$-\theta_g$, or in other words, $(\theta_g + \theta_r)$

So, the output signal of this device represents the sum of the two input angles.

Since the selsyn differential shaft angle θ_{rsd} is a relative angle with respect to input angle θ_g, it should make sense that selsyn differential generators would be excellent error signal generators. If the differential generator were set up to subtract (Problem 7.12), and the output shaft angle of the selsyn motor were also transmitted to the differential shaft, the differential output would be the error between the input signal θ_g and the output θ_m. Such a system is a simple proportional controller and will be discussed in Example 7.5 after differential motors and transformers are introduced.

A selsyn differential motor is the dual of the generator in that the output is the sum or difference of two input signals. In the case of the motor, both inputs are electrical and the output is the mechanical shaft angle. Consider a differential motor set up to subtract two angular signals from generators g_1 and g_2 connected to the stator and rotor, respectively, as in Figure 7.43. All previous simplifying assumptions hold. Stator and rotor input line-to-neutral voltages will be of the form

$$e_{an_1}(t) = \sqrt{2}\, E_{g1m} \sin \omega t \cos \theta_{g1} \qquad (7.74)$$

$$e_{bn_1}(t) = \sqrt{2}\, E_{g1m} \sin \omega t \cos (\theta_{g1} - 120°) \qquad (7.75)$$

$$e_{cn_1}(t) = \sqrt{2}\, E_{g1m} \sin \omega t \cos (\theta_{g1} + 120°) \qquad (7.76)$$

and

$$e_{an_2}(t) = \sqrt{2}\, E_{g2m} \sin \omega t \cos \theta_{g2} \qquad (7.77)$$

$$e_{bn_2}(t) = \sqrt{2}\, E_{g2m} \sin \omega t \cos (\theta_{g2} - 120°) \qquad (7.78)$$

$$e_{cn_2}(t) = \sqrt{2}\, E_{g2m} \sin \omega t \cos (\theta_{g2} + 120°) \qquad (7.79)$$

By our previous reasoning, the signal from g_1 creates a field in the stator oriented at an angle θ_{g_1} from the a phase. The rotor field will be oriented at an angle θ_{g_2} from the rotor axis, so the axis is oriented $-\theta_{g_2}$ degrees from its own field axis. Since a torque is developed until the two fields line up the rest condition will be when the stator field at θ_{g_1} lines up with the rotor field, leaving the rotor axis at an angular position $(\theta_{g_1} - \theta_{g_2})$, or the difference between the two angular signals. This is seen in Figure 7.43. It should again be evident that such a system could represent an error signal and thus could be used in a servomechanism.

Differential Transformers

A differential transformer or control transformer is similar to a differential generator in operation, and has been described as a transformer with a rotating secondary [13]. The primary difference between a differential transformer and a differential selsyn generator is in construction. As discussed previously, the differential selsyn is constructed like a selsyn generator and therefore develops a torque on the shaft to oppose motion. A differential transformer, on the other hand, is designed to *minimize* any shaft torque. This is done by winding the stator with many turns of fine wire, resulting in a very high input impedance which limits currents to extremely low levels. The rotor is also designed with

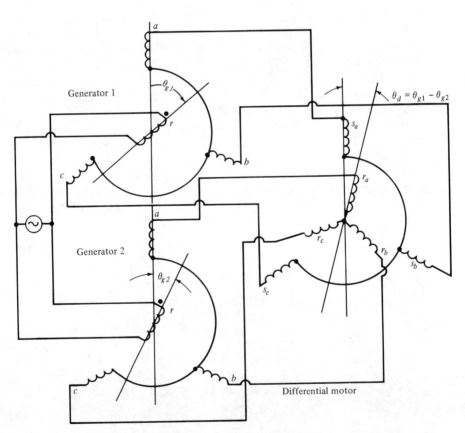

FIGURE 7.43 Differential Motor Set up to Subtract

SPECIAL-PURPOSE ELECTROMECHANICAL AND
ELECTROMAGNETIC DEVICES

the turns buried well below the surface, eliminating any significant surface slots, thus minimizing any potential effects from unavoidable reluctance torques caused by the slots (we have ignored these torques up to this point, but rest assured that they exist in some form in every machine built). A well-built differential transformer, therefore, can be assumed to exert only bearing torque on the shaft. Electromagnetic torque is zero, for all practical purposes, and thus the rotor exerts essentially *no* back pressure on its prime mover *regardless of position*. In addition its stator winding draws virtually no current compared to a differential generator.

The characteristics noted above allow the differential transformer to be used as a highly accurate summer or error signal generator. The output electrical signal from the differential transformer rotor is of the same form as for a differential generator. Angular electrical data on the stator are summed with angular shaft data, and the three-phase rotor produces signals of the same nature as in (7.73) to (7.75). If a selsyn motor's stator windings are connected to the differential transformer's rotor windings (as with a differential generator), the selsyn motor shaft will move to an angular position described by

$$\theta_m = \theta_g \pm \theta_{tr} \qquad (7.80)$$

where θ_{tr} is the angle between differential transformer shaft and the phase a stator axis. The difference is that the error bounds on θ_m are significantly reduced for a differential transformer over a differential generator. Errors in well-designed synchro-transformer systems are usually limited to only a few minutes of arc [1]. This generally means that a control system using differential transformers will reach an equilibrium point faster and with greater accuracy than if a differential selsyn is used. The disadvantage is that a servo amplifier is required to drive the motor.

EXAMPLE 7.5 A very rudimentary analog fly-by-wire closed-loop proportional control system can be designed using two selsyn generators and a differential motor. Suppose that we want to control the elevator of the aircraft with a gain of 1:2 (in other words, each 2° deflection of the control yoke by the pilot results in a 1° deflection in the elevator). The system plan and block diagram for the controller are shown in Figure 7.44. As can be seen, we require a position gain of 0.5 in the forward path and a gain of 2 in the error path to provide the necessary deflection. Since the selsyn units have unity position gain, we determine that the gain in deflection is best accomplished through the use of the linkage shown in Figure 7.45 (see reference 14, for example, with regard to designing such linkages). To provide redundancy, elevator deflection is measured through a gear train attached to the hinge shaft and the shaft of a selsyn generator, and a second generator is attached to the yoke in the cockpit. The complete system is shown in Figure 7.46.

An angular deflection θ_r in the control yoke is added to $-\theta_c$, which is the negative of twice the elevator deflection, by the differential generator. Consequently, the differential motor's stator and rotor fields are displaced by an angle θ_e which is the sum of θ_r and $-\theta_c$ and thus represents the error signal $e(t)$ between the "actual" elevator position and the desired position.† Since $\theta_e \neq 0$, a torque results and the differential motor shaft

---

†The "actual" position is really twice the angular deflection of the elevator, so $e(t)$ is *double* the *actual* error in elevator position relative to what is desired. This is like power steering in some respects and is very forgiving, since large changes in input signal yield small outputs.

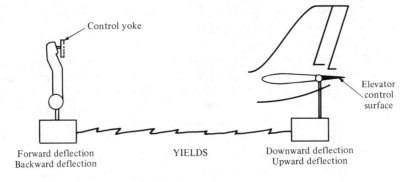

(a) Basic system plan

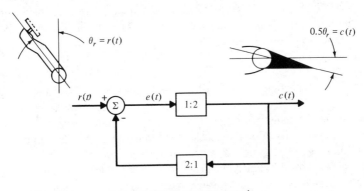

(b) Block diagram showing control
variables and position gains required

FIGURE 7.44 Simple Fly-by-Wire Proportional Control System

rotates until $\theta_e = 0$, at which point the desired position is reached ($\theta_c = \theta_r/2$). This is shown in Figure 7.47.

If the airflow over the elevator then causes its angular position θ_c to change but the yoke angle θ_r remains constant, an error signal is again generated, and a torque is again exerted by the differential motor to correct the position θ_c to the desired angle $\theta_r/2$. This is a simple form of proportional control system. It is more commonly found in ship

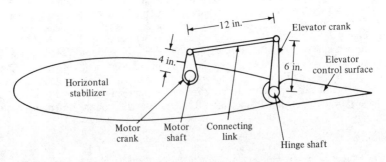

FIGURE 7.45 Linkage for 1:2 Gain in Angular Deflection

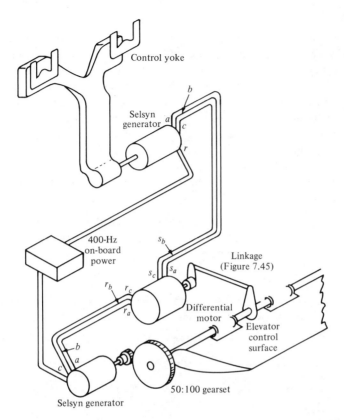

FIGURE 7.46 Proportional Fly-by-Wire Control System for Aircraft Elevator

steering mechanisms rather than aircraft, but is similar to many other aircraft and land-based systems. A dual system using a differential transformer and a selsyn motor is illustrated in Problem 7.13. Dynamic aspects of such systems are treated in depth in control texts such as reference 13.

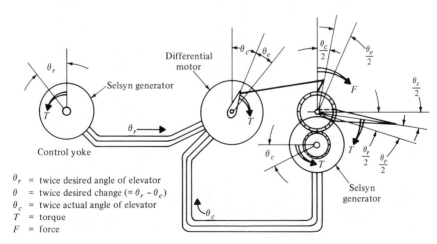

θ_r = twice desired angle of elevator
θ = twice desired change (= $\theta_r - \theta_c$)
θ_c = twice actual angle of elevator
T = torque
F = force

FIGURE 7.47 Operation of Simple Proportional Control System for Aircraft Elevator

7.8.2 THE RESOLVER

The resolver is a synchro device that is used to resolve an input signal into components. Synchros can sometimes be used as resolvers. A simple, two-component resolver, shown in Figure 7.48, takes an ac signal on the rotor and breaks it into two orthogonal components. It can be viewed as a polar-to-rectangular coordinate transform device when used in this application, as will now be shown.

Assume that the rotor of the device in the figure is displaced from stator coil a by some angle θ, and that a voltage $e(t) = E \cos \omega t$ is impressed across the rotor winding. Further, assume symmetry of the coils, neglect any saliency effects, assume that the position of the rotor is fixed and the two stator coils are unloaded and no current flows. The matrix equation for this machine under these assumptions reduces to

$$
\begin{bmatrix} e_a \\ e_b \\ e_r (= E \cos \omega t) \end{bmatrix} = \begin{bmatrix} L_s & 0 & M \cos \theta \\ 0 & L_s & M \sin \theta \\ M \cos \theta & M \sin \theta & L_r \end{bmatrix} \frac{d}{dt} \begin{bmatrix} i_a (= 0) \\ i_b (= 0) \\ i_r (\propto \sin \omega t) \end{bmatrix}
$$

(7.81)

where L_s, L_r, and M are the self- and mutual inductances. Solution of this equation yields expressions for the stator voltages e_a and e_b, which are

$$e_a = k \cos \theta \cos \omega t \tag{7.82}$$

$$e_b = k \sin \theta \cos \omega t \tag{7.83}$$

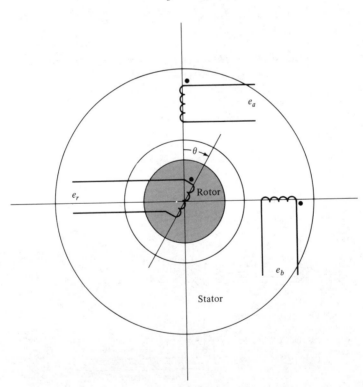

FIGURE 7.48 Simple, Two-Component Resolver

SPECIAL-PURPOSE ELECTROMECHANICAL AND
ELECTROMAGNETIC DEVICES

where k is a constant reflecting inductance, rotor current or voltage, and frequency. Note that these voltages represent orthogonal components of the rotor voltage at angle θ, where the orthogonal axes are aligned with the stator coils (which are at right angles to each other). The same output will be obtained if the device is connected to an amplifier with very high input impedance.

7.8.3 STEPPER MOTORS AND DIRECT DIGITAL CONTROL

Stepper motors are interesting variants of synchronous motor theory. They are excited by dc rather than ac and are designed to rotate in finite steps of a given number of degrees per step each time the stator experiences a dc pulse. Since the angle of rotation per pulse is fixed by design, the motor output has the following unique characteristics:

1. Angular position is predictable with a high degree of accuracy (usually 3 to 5%).
2. Results are therefore repeatable.
3. Error in output position is *not* cumulative over a sequence of motions.
4. A given output position, being the result of a pulse or string of pulses, can be produced directly from a digital input signal into which the correct number of pulses is encoded.

These unique characteristics allow the stepper motor to be used in open-loop (since each position is uniquely encoded and repeatable) or closed-loop, direct digital control (DDC) systems. With proper, but simple interfacing (usually consisting of a pulse encoder and driver, as will be discussed later), the stepper motor can be driven directly from the output data bus of a computer. No analog encoding or decoding circuits or feedback systems, with their incumbent drift and stability problems, are required. Consequently, the stepper motor has seen widespread use in numerous applications, particularly computer peripherals such as disk drives and printers, numerically controlled machine tools, robots, and other DDC applications. Their open-loop capability has eliminated much of the complexity in DDC systems. The computer simply tells the motor what to do, and it does it entirely *without supervision*, with predictable, repeatable results. The computer software is free to assume the final position of the motor after a command, although a simple closed-loop system using an optical shaft encoder and data register, which is routinely polled by the computer, is sometimes utilized to verify shaft position. This verification capability allows the computer to detect motor or shaft failure and although it is a valuable self-diagnosis tool, it is not required to achieve normal operation or positioning of the motor shaft. Consequently, simple stepper circuits under computer control can replace complex analog circuits (such as the synchro circuits in the preceding two sections) with a significant reduction in cost and complexity and a concurrent increase in accuracy, resolution, speed, and flexibility. Unlike the synchro circuits, which must be hard-wired and maintained, stepper controls are generally under direct software supervision and can be reprogrammed and even diagnosed by a minimally trained technician at a keyboard computer terminal.

Stepper motors are manufactured in three configurations: variable reluctance, permanent magnet (PM), and a combination of the two (PM-hybrid). All operate based on the generalized torque equation given in Example 2.9 and repeated here:

$$T_e = -i_1^2 \, \Delta L_s \sin 2\theta - i_1 i_2 L_m \sin \theta \tag{7.84}$$

We will examine very rudimentary forms of steppers and their operation in terms of this equation, and then discuss actual stepper designs and expected performance.

The Simple Variable-Reluctance Stepper

A very crude four-coil variable-reluctance stepper can be constructed as shown in Figure 7.49. A salient-pole rotor made of soft iron is magnetically connected through a shaft to the stator assembly. Suppose that we desire to align the dot on the rotor with coil 1. This desired alignment is called the *detent position*. If the rotor is initially displaced by some angle θ_1, the reluctance $\mathcal{R}(\theta_1)$ in the flux path shown yields an inductance (assuming linearity) described by

$$\Delta L_{\theta_1} = \frac{N_1^2}{\mathcal{R}(\theta_1)} \tag{7.85}$$

Now, if a dc current i_1 is passed through conductor Q_1 to ground through coil 1, a torque will be exerted on the rotor, the magnitude of which can be represented for this discussion by the reluctance torque equation

$$T_{e_1} = -i_1^2 L_{\theta_1} \sin 2\theta_1 \tag{7.86}$$

$$= -i_1^2 \frac{N_1^2}{\mathcal{R}(\theta_1)} \sin 2\theta_1$$

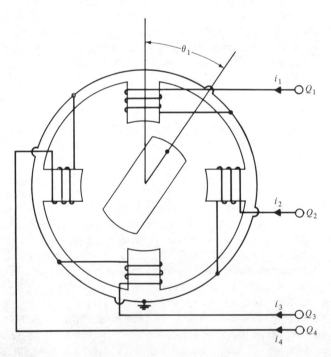

FIGURE 7.49 Simple Variable Reluctance Stepper Motor with Four Stator Coils

and it will be directed toward coil 1. Assuming that the mechanical load torque on the rotor is less than T_{e_1}, the rotor will turn until it aligns itself with the magnetic axis of coil 1 (usually after a small overshoot followed by a damped mechanical oscillation). At that point it has reached a stable equilibrium, and will remain at that position even after current i_1 is switched off, provided that no other forces act on the shaft. Thus a current pulse Q_1 of sufficient magnitude and duration will also cause this motion. Alignment of the rotor at a point midway between coils 1 and 2 would similarly occur if both coils were simultaneously energized with equal currents, creating a magnetic field oriented at an angle of 45°. Similarly, the rotor could be aligned at *any* angle between the two coils by applying unequal currents. The importance of these two considerations will become evident shortly.

From (7.86), it is clear that the variable reluctance stepper has stable equilibrium points at $\theta_1 = 0°$ and $\theta_1 = 180°$, and it has unstable equilibrium points at $\theta_1 = 90°$ and $\theta_1 = 270°$. At any of these four points, a current pulse Q_1 at coil 1 will not cause any motion at all. If it is desired that the rotor come to rest at $\theta_1 = 0°$, some additional actions must occur. Assume now that the dot on the rotor of Figure 7.48 indicates the "top," so that alignment of the dot next to coil 1 represents $\theta_1 = 0°$. The rotor will move to this position upon a current pulse Q_1 and if $\theta_1 < 90°$ or $\theta_1 > 270°$. Outside these bounds, however, and based on the discussion above, it should be evident that, sequentially exciting coil 2, coils 1 and 2 together, then coil 1, will cause the desired rotation, if $90° \leq \theta_1 < 225°$, for example. The necessary pulse train to move the rotor

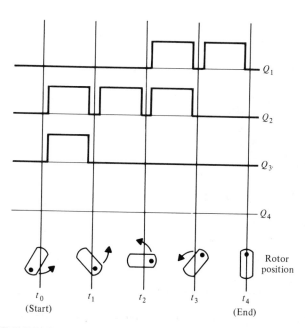

FIGURE 7.50 Example Pulse Sequence to Rotate the Rotor from $90° \leq \theta_1 < 225°$ to Detent Position $\theta_1 = 0°$

from this position to $\theta_1 = 0°$ is shown in Figure 7.50. This process of sequentially stepping the rotor with a pulse train applied to the various stator coils is generally called *slewing,* and usually occurs at high speed with acceleration and deceleration sequences (where the pulse sequence is automatically sped up or slowed down to accelerate or decelerate the rotor) at the beginning of the sequence and immediately prior to the rotor reaching the detent position. Acceleration and deceleration are called *ramping up* and *ramping down,* respectively. Clearly, the motor could be rotated indefinitely by applying a continuous, repeating sequence of pulses to each coil similar to the sequence in Figure 7.50, but also including Q_4. Moreover, the rotor could be made to reverse direction simply by applying the pulse sequence to the coils in reverse order, and the torque can be doubled by pulsing opposite coils or pairs of coils simultaneously.

The Simple Permanent-Magnet Stepper
A rudimentary four-coil PM stepper is shown in Figure 7.51. This machine is structured somewhat like the hysteresis motor seen in Figure 7.29 in that it has a smooth rotor and can have a smooth stator as well. In this case, the second half of the torque equation (7.84) can be considered to describe the electromagnetic torque on the rotor:

$$T_e = -i_1 i_2 L_m \sin \theta \qquad (7.87)$$

If we assume that the permanent magnet can be represented by a coil of

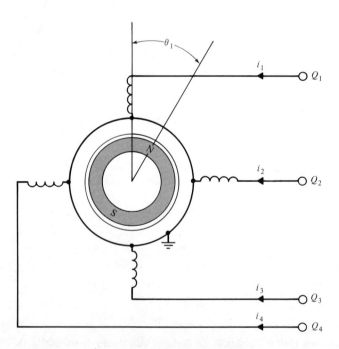

FIGURE 7.51 Simple Permanent-Magnet Stepper

$N_2 = 1$ turns with current i_2, and that the smooth rotor and stator result in a constant reluctance $\mathcal{R}$, the electromagnetic torque T_e would be of the form

$$T_e = -N_s i_s B_r A \sin \theta_1 \tag{7.88}$$

where

$$B_r = \text{flux density from permanent magnet}$$

$$A = \text{cross-sectional area of mean flux path}$$

$$N_s = \text{number of turns in the excited stator winding}$$

$$i_s = \text{dc current in excited stator winding}$$

This torque will be exerted on the rotor for all angles θ_1 except $0°$ and $180°$. In this case, $\theta_1 = 180°$ is unstable and $\theta_1 = 0°$ is stable.

Consider the same situations as discussed earlier, that is, (1) Q_1 excited by i_1 and $\theta_1 < 90°$, followed by (2) a slewing situation. In the first case, Q_1 is pulsed and a torque T_e pulls the rotor from its rest position to a final position of $\theta_1 = 0°$, as before. However, from the argument above, this can now occur if $\theta_1 \leq 180°$, although the torque will be very uneven due to the angular effects. Again, rest positions between coils 1 and 2 can be obtained by concurrently pulsing both and varying the magnitude of i_1 and i_2. In the slewing case, a pulse sequence such as the one in Figure 7.50 could again be used. Reversal could be accomplished as before or by recognizing that reversal of current polarity (say, in coil 1) results in a *negative* torque which would cause the rotor to reach a detent position opposite the excited coil (i.e., in this case, "facing" coil 3 as in Figure 7.52). Consequently, the PM stepper has an additional dimension of control which can be exploited: pulse current direction or polarity.

The Simple PM-Hybrid Stepper

A PM-hybrid stepper is a combination reluctance–permanent magnet machine. Reluctance torque and electromagnetic torque are obtained from the use of a salient-pole, permanent-magnet rotor. A simple four-coil machine is shown in Figure 7.53. The total torque can be represented for our purposes by the combination of (7.86) and (7.88):

$$T_e = -i_s^2 \frac{N_s^2}{\mathcal{R}_1(\theta_1)} \sin 2\theta_1 - N_s i_s B_r A \sin \theta_1 \tag{7.89}$$

where N_s and i_s are the number of turns and current, respectively, in the excited stator coil. This machine obviously produces much higher torque for a given current pulse and equilibrium points other than $\theta_1 = 0$ or $\theta_1 = 180°$ are not likely to exist. Consequently, this machine essentially combines the positive attributes of both the reluctance and PM configurations. Rotor action is as described previously, and slewing is accomplished by sequentially pulsing the stator coils as before. Both positive and negative pulses can be used, with the added benefit that negative pulses result in a reversed torque at significantly lower levels than if the pulse sequence were reversed. This adds another dimension of control that can be exploited.

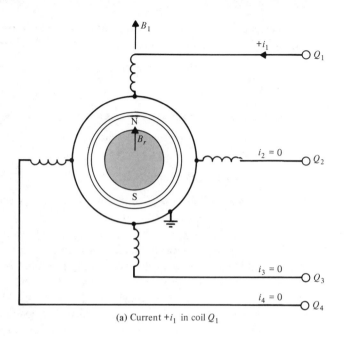

(a) Current $+i_1$ in coil Q_1

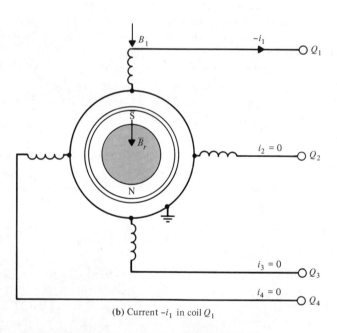

(b) Current $-i_1$ in coil Q_1

FIGURE 7.52 Effect of $(-i_1)$ on Final Rotor Position

Commercial Stepper Designs, Operation, and Specification

All three of the types of stepper motors described above are produced commercially. Table 7.1 provides a general comparison among the three types (see

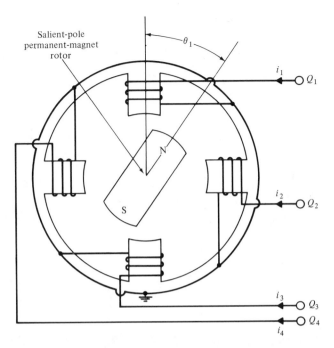

Salient-pole
permanent-magnet
rotor

FIGURE 7.53 Simple PM-Hybrid Stepper Motor Configuration

references 15 to 20, for example, and other manufacturers' data for details). As can be seen from the table, there is a wide range of torques, slew rates, step sizes, and accuracies available. Since most commercial steppers utilize somewhat similar designs, we will concentrate on the very common, high-performance PM-hybrid motor.

Unlike the rudimentary design in Figure 7.53, a typical commercial PM-hybrid motor has a more complicated, differential toothed rotor/stator design shown in Figure 7.54(a). The rotor elements constitute what is called a *stack*.

Table 7.1   General Comparison of Commercial Stepper Motor Types[a]

Stepper Type	Relative Maximum Torque	Typical Step Angle	Relative Maximum Slew Rate	Typical Step Accuracy (%)	Typical Relative Price
Variable reluctance	Lowest	1.8–15°	Slow[b]	±3–10	Lowest
Permanent magnet	Midrange	1.8–90°	Slow[b]	±3–10	Low range to midrange
PM-hybrid	Highest[c]	0.0007°–15°	Fast to very fast[d]	±3–5	Highest

[a]This table was compiled in mid-1983 from a sample of 1978–1982 manufacturers' data listed in references 15 to 20. It is not all-inclusive, nor does it express or imply any endorsement of products or manufacturers listed in these references. Performance and price of specific designs from various manufacturers can vary significantly due to intended end use and design improvements made as the state-of-the-art changes.
[b]Typically, 100 to 350 steps per second have been reported [15].
[c]Torques of in excess of 2000 oz-in. (about 15 N-m) have been reported [15].
[d]Values of 500,000 steps per second have been advertised [19].

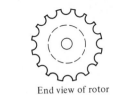

End view of rotor

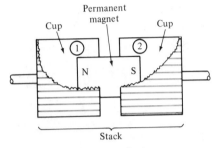

(a) General rotor design

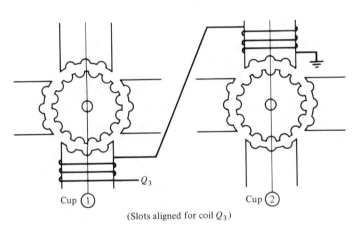

(b) Tooth/slot alignment on typical PM-hybrid rotor

FIGURE 7.54 Typical PM-Hybrid Commercial Design

The rotor teeth are set at a uniform spacing, or pitch, and the teeth on one cup are set to line up with the slots on the other cup, as in Figure 7.54(b). The stator poles are also toothed, but at a slightly different overall pitch so that the rotor teeth line up exactly with the teeth on only one pair of stator poles. The rotor teeth are thus out of line with the teeth in all other poles. Consequently, the basic step angle obtainable is the tooth pitch divided by the number of pole pairs, or

$$\theta_s = \frac{2\theta_t}{P} \tag{7.90}$$

where

$$\theta_s = \text{step angle, deg}$$

$$\theta_t = \text{tooth pitch, deg}$$

$$P = \text{number of poles}$$

and the number of steps S obtainable per revolution is therefore

$$S = \frac{360°}{\theta_s} \tag{7.91}$$

The step angle can be reduced (or the number of steps per revolution increased) in several ways. One way is to increase the number of poles. This can be done by either adding more poles per stack or by adding stacks that are displaced angularly from each other. Another way is to excite two pole pairs simultaneously with precisely controlled unbalanced currents so that the appropriate rotor and stator poles' teeth are pulled slightly out of line by a given amount for each step. Although the latter method is tricky and requires high-precision, low-drift analog electronics to set up the appropriate current pulses, it can be done with modern solid-state devices and is usually the method by which steppers with extremely high numbers of steps per revolution operate.

Torque can be increased either by building a bigger motor with more turns of wire per pole and/or more current capacity, or by adding identical stacks which are not angularly displaced from each other and connecting the stator coils in parallel (or using a long, single set of stator poles). Adding stacks with a single set of stator windings is generally more economical and therefore is most common. The torque typically can be assumed to increase linearly with the number of stacks.

A typical torque–speed curve for a PM-hybrid stepper is shown in Figure 7.55. The motor can operate in either of two modes: error-free start–stop (EFSS) and high-speed slew. Generally, EFSS operation includes pulse rates from zero

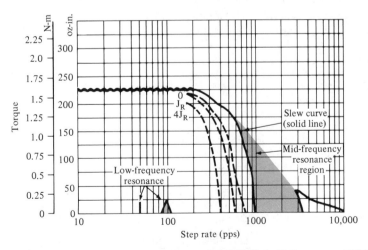

FIGURE 7.55 Typical Torque–Speed Curve for PM-Hybrid Stepper. EFSS and slew curves for a 23-frame single stack stepper. Dashed curves are EFSS curves for zero and two selected values of inertial load (Courtesy of Bodine Electric Company)

up to some fraction (say, about 15%) of rated maximum slew rate. Between the EFSS and high-speed slew modes is a resonance region characterized by noisy operation, loss of control, and even stalling. Such regions can also occur at low slew rates. Care must be taken to specify a motor whose desired output slew rate is not in one of the resonant regions. Resonance is avoided by properly matching the motor to the load and by specifying or designing control electronics and software which are matched to the motor and which will ramp the motor through the resonant regions. A useful guideline for sizing steppers suggested by one manufacturer is to select a motor to drive a load whose inertia J_L is no more than 10 times, and typically 2 to 4 times, the rotor inertia J_R of the motor [19]. The effect of load inertia can be seen on the torque–speed curves of Figure 7.55. As can be seen, EFSS performance is reduced somewhat as load is increased.

Control electronics for open-loop steppers typically convert a parallel-encoded position signal to the number of serial coil pulses necessary to reach the final position. These serial pulses may be variably timed to provide acceleration or deceleration for optimizing motor performance. Since step errors are not cumulative, initial positioning of the stepper rotor to a known position (either manually or electrically) is all that is required. Closed-loop systems typically use two-phase optical shaft encoders to provide digitally encoded shaft position data to the controls, and required movements can thus be computed for each set of input data. Typically, these more complex controls contain separate microprocessors, may handle more than one stepper, and may contain complicated stepper positioning programs and sequences which can be accessed and executed with single-word commands from the host computer.

The controls, either open or closed-loop, feed pulses to a driver or translator that provides the necessary switching sequence to the stepper motor. A typical commercially available driver for a four-coil, eight-pole, two-phase PM stepper operated as a linear actuator,† together with the pulse sequence necessary for a step-out, step-in (forward-reverse) EFSS action, are shown in Figure 7.56. The switching is according to a four-step sequence. This particular driver is contained on a single 16-pin DIP integrated circuit chip.

EXAMPLE 7.6 A certain manufacturer produces two stepper motors, whose torque–speed (slew) curves are given in Figure 7.57. Choose the correct motor to step a load of 600 oz-in. constant torque (regardless of speed) and determine the maximum possible slew rate. If the load and rotor have a combined moment of inertia of 10^4 g-cm², find the instantaneous angular acceleration upon braking in rad/s² if the motor is initially running at the maximum slew rate.

From the figure it is clear that motor 2 cannot deliver a torque of 600 oz-in., so motor 1 is chosen. Again, reading from the graph, if torque is constant, the maximum slew rate occurs where the torque–speed curve crosses the 600-oz-in. line. This is at 6 rps or 360 rpm.

To find the instantaneous acceleration upon braking at the maximum slew rate, first read the acceleration curve for motor 1 at 360 rpm. The maximum braking torque produced is approximately 450 N-cm. The angular acceleration rate in rad/s² is therefore

$$\alpha = -\frac{T_{\text{braking}}}{J} = -\frac{450 \text{ N-cm}}{10^4 \text{ g-cm}^2} = -\frac{450 \times 10^5 \text{ g-cm}^2/\text{s}^2}{10^4 \text{ g-cm}^2} = -4500 \text{ rad/s}^2$$

†By using a threaded shaft that moves in or out depending on the direction of rotor rotation.

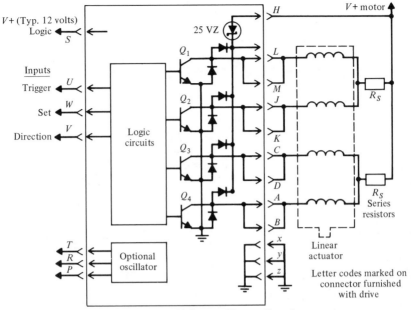

(a) Stepper driver configuration

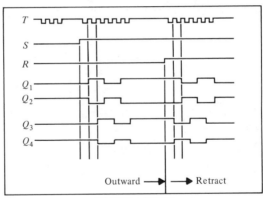

(b) Timing diagram

Four-step switching sequence									
$S = H$ (High level)									
Retract $R = H$ (High level)					Outward $R = L$ (Low level)				
T	Q_1	Q_2	Q_3	Q_4	T	Q_1	Q_2	Q_3	Q_4
0	L	H	L	H	0	L	H	L	H
1	H	L	L	H	1	L	H	H	L
2	H	L	H	L	2	H	L	H	L
3	L	H	H	L	3	H	L	L	H
0	L	H	L	H	0	L	H	L	H

(c) Output switching sequence

FIGURE 7.56 Example of (a) Stepper Driver Configuration, (b) Timing Diagram, and (c) Switching Sequence (Courtesy of Airpax Corporation, a North American Philips Company, Cheshire Division)

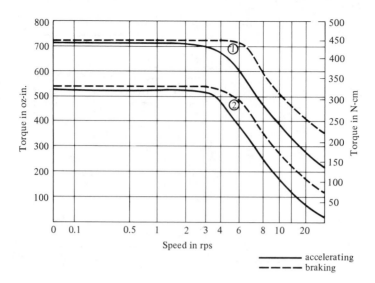

FIGURE 7.57 Torque–Speed Curves for Example 7.6

As a matter of interest, suppose that we assume this deceleration rate is constant (according to the curve, that may be a reasonable assumption) and find the time and angular rotation until the load is fully stopped. From basic rotational kinematics (see reference 2), the equations that must be solved are

$$\theta = \theta_0 + \omega t$$

$$\theta = \theta_0 + \frac{1}{2}(\omega_0 + \omega)t$$

$$\theta = \theta_0 + \omega_0 t + \frac{1}{2}\alpha t^2$$

$$\omega^2 = \omega_0^2 + 2\alpha(\theta - \theta_0)$$

If we assume that $\theta_0 = 0$, final $\omega = 0$, and $\omega_0 = 360$ rpm $= 37.7$ rad/s, these four equations reduce to the following two equations:

$$0 = 37.7 - 4500t$$

$$\theta = 0 + \frac{1}{2}(37.7)t$$

From the first equation, the time t to stop is found to be 8.4 ms and, from the second equation, the angular stopping distance θ is 0.16 rad, or about 9°. This motor can thus stop a light, low-inertia load very rapidly, as is generally characteristic of the performance of stepper motors.

7.9

LINEAR MACHINES AND MAGNETIC LEVITATION

The rotating machines emphasized in this text clearly can be applied to situations where linear motion is required, either through the use of reciprocating mechanisms, helical or worm gear trains, or by rolling a wheel or cog along a surface or guideway. However, in some cases it is advantageous to generate linear motion directly using a machine that is configured to exert linear force rather than angular torque. Generalized analysis of such a machine can be done by utilizing the theory developed in the preceding chapters. Applications of interest include liquid metal pumps, linear translation devices, and high-speed ground transportation.

7.9.1 GENERAL CONSIDERATIONS

It should be clear to the reader that the machines described in Chapters 3 through 5 were defined essentially without regard to specific dimensions. At no point were any requirements given calling for a specific ratio of machine radius to length, for example, nor do any such requirements generally exist. In this section we extend this specific concept to the ultimate degree and analyze, in brief and in very general terms, what happens when a machine's radius is allowed to increase to infinity.

It is evident from geometry that as the radius of a circle or cylinder increases, the curvature of a specific section or arc decreases. In other words, as the radius r is increased, the arc becomes more linear until at $r = \infty$, it is a straight line. There is no reason why a machine's geometry cannot be changed accordingly and, in fact, a linear machine in its most rudimentary form can be viewed (for the moment, at least) as a rotating machine of infinite radius, whose air gap is an arc of infinitesimally small curvature (i.e., a straight line). The linear machine can also be thought of as a rotating machine whose rotor–gap–stator structure is "unrolled" from $-\infty$ to $+\infty$. Both concepts are shown in Figure 7.58.

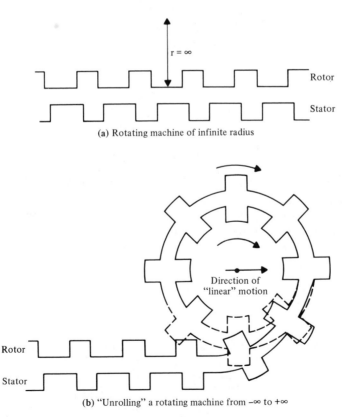

(a) Rotating machine of infinite radius

(b) "Unrolling" a rotating machine from $-\infty$ to $+\infty$

FIGURE 7.58 Analogies for a Linear Machine

The reader should understand that we have not restricted the discussion above to any *specific* type or class of machine. Consequently, *all rotating machines can have linear duals,* regardless of number of phases, poles, or excitation means, and thus one could develop, for example, linear dc, induction and synchronous motors, as well as a host of other configurations, including reluctance, homopolar, and shaded-pole types. Nor can generators be excluded; linear machines will generate electricity from linear motion just as a rotating machine does from rotary motion.

Based on the discussion above and Figure 7.58, it should be evident that, in the most general sense, performance of any linear machine can be described by a relationship that has a dual for a rotating machine. The equations of motion in one dimension are "translated" from rotational to linear by recognizing that (in an ac machine, for example)

$$v = 2pf \qquad (7.92)$$

where

$v =$ velocity or linear speed (considered here to be a scalar)

$f =$ frequency, Hz

$p =$ pole pitch (distance between pole centers)

In a rotating machine, (7.92) was expressed as

$$n = \frac{2 \cdot 60f}{P} = \frac{60\omega}{\pi P} \qquad (7.93)$$

where

$n =$ rotational speed, rpm

$f =$ frequency, Hz

$P =$ number of poles

$\omega =$ rotational frequency, rad/s

Both equations state that the machine must move two pole pitches for each cycle of supply voltage. "Torque" becomes "force" or "thrust" in the linear machine, "rotational frequency" becomes "speed" or "velocity," and everything else essentially remains the same. For the three major types of motors studied, the basic dual equations for torque are as follows:

DC machine:

$$T = K_m \varphi I_a \qquad \text{(rotating)} \qquad (5.8)$$

$$F = K_m \varphi I_a \qquad \text{(linear)} \qquad (7.94)$$

Polyphase induction machine:

$$\left. \begin{array}{l} T = \dfrac{I_r'^2 r_r'}{s\omega_s} \\[2em] \text{where slip } s = \dfrac{\omega_s - \omega_r}{\omega_s} \end{array} \right\} \text{(rotating)} \qquad \begin{array}{l} (3.63) \\[2em] (3.62) \end{array}$$

$$F = \frac{I_r'^2 r_r'}{s v_s} \left.\vphantom{\begin{array}{c} \\ \\ \end{array}}\right\} \text{(linear)}$$ (7.95)

where slip $s = \dfrac{v_s - v_r}{v_s}$ (7.96)

Polyphase synchronous machine (smooth rotor):

$$T = \frac{\sqrt{3}\, V_d I_a}{\omega_s} \sin \delta \qquad \text{(rotating)}$$ (4.5)

$$F = \frac{\sqrt{3}\, V_d I_a}{v_s} \sin \left(x_\delta \cdot \frac{180°}{p} \right) \left.\vphantom{\begin{array}{c} \\ \\ \\ \\ \end{array}}\right\} \text{(linear)}$$

where x_δ = linear dual of
torque angle δ (7.97)

p = pole pitch

By now, however, the alert reader will have recognized that the general motor geometry in Figure 7.58 described by these equations is completely impractical because the equations assume that the moving element (the "rotor" in the figure) is as long as the stationary element. In reality, a linear motor must have one element which is shorter† than the other to produce reasonable motion. Machines with stator longer than rotor (stator fixed, rotor moves) are called *long stator* or *short rotor*. Where the stator moves and the rotor is fixed,‡ the rotor element is longer than the stator and the device is called a *long rotor* or *short stator* machine. Under this system of nomenclature, the elements of the machines generally can be named as in Table 7.2.

Aside from the obvious potential for confusion of the machine elements as illustrated in the discussion above, some very real complications are posed by the significant difference in length between rotor and stator elements in a practical linear machine. In particular, one can easily distribute the total flux in a

---

†Usually, significantly shorter. In transportation applications, the track is one motor element and may be hundreds of miles long, while the vehicle (the other element) is on the order of tens or hundreds of feet.

‡This nomenclature is entirely contradictory to established rotating machine practice, but unfortunately it grew out of linear induction motor development and became accepted. Other nomenclatures are also used.

Table 7.2 Nomenclature for Linear and Conventional Rotating Machines

Type	Nomenclature	
	Rotating Machine	**Linear Machine**
Dc	Armature (usually on rotor)	Rotor
	Field (usually on stator)	Stator
Synchronous	Armature (usually on stator)	Stator
	Field (usually on rotor)	Rotor
Induction	Stator (always wound)	Stator
	Rotor (may be wound)	Rotor

rotating machine so that it is sinusoidal over an entire revolution (and thus from $\theta = -\infty$ to $\theta = +\infty$) by winding it appropriately. In a linear machine this is completely impossible due to the significant difference in length between the rotor and stator. Moreover, the flux about the ends of the vehicle will not be distributed in any sort of uniform fashion, introducing a drag force and other undesirable effects. Although these so-called ''end effects'' have been quantified and general equations developed, analysis usually requires Fourier or equivalent techniques and is far beyond the scope of this chapter. Such devices also have parasitic effects analogous to those from pole shading. These sorts of effects cannot be treated in detail here; the reader is referred to references 21 to 31 for additional insight. An additional effect also covered by these references is the development of normal and lateral forces on the vehicle under linear machine configurations. Although lateral forces can pose operational problems, the existence of a normal force opens avenues whereby the vehicle may be levitated. This will be discussed in general terms later.

Prior to discussing some specific types of linear machines, it is worthwhile to examine another general attribute of these devices. In Figure 7.58(b), a rotating machine was ''unrolled'' or made an infinite radius to create a generalized linear machine. The flux paths are as shown in Figure 7.59. From viewing the latter figure, however, one can also conclude that there is no reason why a second set of stator windings could not be added on the opposite side of the rotor, or vice versa, as long as the flux paths were kept reasonably contiguous; this is shown in Figure 7.60 and is called a *double-sided machine*. When this flexibility is combined with the various stator/rotor configurations possible, a litany of linear machine designs, some of which are shown in Figure 7.61, are possible. Clearly, double-sided machines are advantageous for some applications because the same force or thrust as for a single-sided machine can be obtained using smaller wire and dividing the current between the two sides; a double-sided vehicle (short stator or, if *not* induction, short rotor), is, in fact, a common design. Double-sided, fixed guideways are generally not cost-effective unless they are linear induction squirrel cage or solid aluminum rotor elements. On the other hand, levitation is easily facilitated using a *single-sided* device, as we will see shortly.

With this background, it is useful to look at several examples of linear ma-

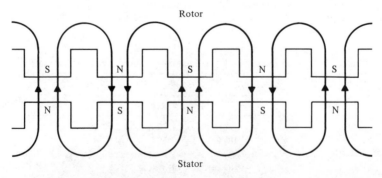

Rotor

Stator

FIGURE 7.59 Flux Paths in Simple Linear Machine with One Set of Rotor Windings and One Set of Stator Windings. Coils are not shown.

SPECIAL-PURPOSE ELECTROMECHANICAL AND
ELECTROMAGNETIC DEVICES

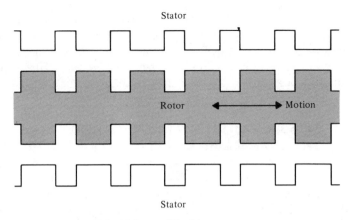

Stator

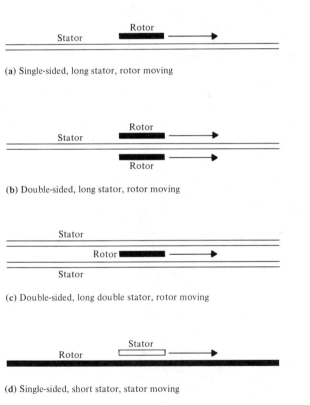

Rotor ←——→ Motion

Stator

FIGURE 7.60 Double-Sided Linear Machine

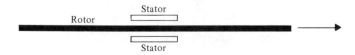

Rotor

Stator ———→

(a) Single-sided, long stator, rotor moving

Rotor

Stator ———→

Rotor ———→

(b) Double-sided, long stator, rotor moving

Stator

Rotor ———→

Stator

(c) Double-sided, long double stator, rotor moving

Stator

Rotor ———→

(d) Single-sided, short stator, stator moving

Stator

Rotor

Stator

(e) Double-sided, short double stator, rotor moving

FIGURE 7.61 Several Linear Machine Designs

chines and how they work. The descriptions given here will not be exhaustive and, in fact, are greatly oversimplified in many respects. The detailed analysis of these devices generally is quite complicated, particularly if end effects are included, and cannot be covered in depth here. Instead, the reader should develop an awareness of the device configurations and their overall operating characteristics based on the following discussions.

7.9.2 LINEAR INDUCTION MOTORS AND THE "IMPELLERLESS" LIQUID METAL PUMP

Linear induction machines are quite common, come in numerous configurations, and have been in existence for many years. Although they have been utilized in or proposed for a number of applications, including transportation, levitation, and industrial mechanisms (e.g., looms), a unique application which will be considered here is the pumping of liquid metals without the use of moving parts. To understand this application, it is useful first to consider a greatly simplified, generalized, linear induction motor (LIM).

Assume that the simple squirrel-cage motor in Figures 3.1 through 3.3 can be "unrolled" to make the linear machine shown in Figure 7.62(a) and, further, that this machine is reconfigured as the short-stator, double-sided type shown in Figure 7.62(b). We will simplify matters at this point by postulating that the double-sided stator produces a single moving field B_s equal to the sum of the two equal fields from each side, that the air gap is completely uniform, that the

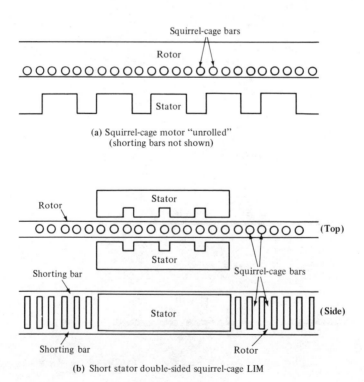

(a) Squirrel-cage motor "unrolled"
(shorting bars not shown)

(b) Short stator double-sided squirrel-cage LIM

FIGURE 7.62 Linear Induction Motor (LIM)

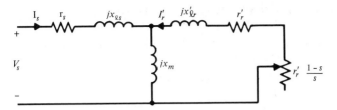

FIGURE 7.63 Equivalent Circuit for LIM Ignoring End and Edge Effects

conductors are infinitesimally thin and fed by a polyphase sinusoidal current source, and that the reluctance of the core material is negligible. The reader should notice that there are essentially the same assumptions made in Chapter 3 for a rotating machine. We will also assume that the rotor is made of aluminum, common for this application. If we also assume for the moment that end effects can be ignored, the linear machine is exactly a dual of the one proposed in Chapter 3 and is thus described by the equivalent circuit in Figure 7.63 and by the following equations (motion in the x direction only):

$$v_s = 2pf \tag{7.98}$$

$$s = \frac{v_s - v_r}{v_s} \tag{7.96}$$

$$F = \frac{I_r'^2 r_r'}{s v_s} \tag{7.95}$$

As in Chapter 3, it can be shown that the machine, under these simplified conditions, has the thrust–speed characteristic of Figure 7.64, which is very similar to the rotating machine's torque–speed curve. Operational characteristics of this machine, such as starting current, maximum thrust, and so on, are computed from the equivalent circuit and the equations above as in Chapter 3.

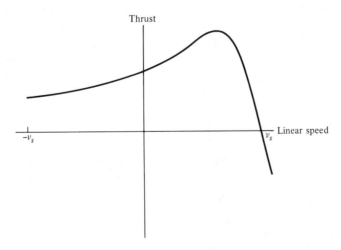

FIGURE 7.64 Generalized Thrust–Speed Characteristic of LIM

Let us now complicate the matter by recognizing the finite length of the stator and including end effects in a very elementary fashion.† Assuming that motion is in the x direction, that the device is fed from a current source such that current density in the windings is sinusoidal, we obtain

$$J_s(x, t) = J_m \cos(\omega t - kx)$$

$$= J_m e^{-jkx} e^{j\omega t} \tag{7.99}$$

and that leakage can be neglected, then from Ampère's law,

$$\oint \bar{H} \cdot d\bar{\ell} = \int_s \bar{J} \cdot d\bar{s} \tag{1.1}$$

it can be shown that the current J_r induced in the rotor satisfies the phasor relationship (if $B = \underset{\sim}{B} e^{j\omega t}$)

$$-\frac{\partial \underset{\sim}{B}}{\partial x} = \frac{\mu_0}{g} [t_r \underset{\sim}{J}_r + \underset{\sim}{J}_s] \tag{7.100}$$

where t_r is the thickness of the rotor and g is the air-gap distance. Combining (7.99) and (7.100), and assuming that Ohm's law and Kirchhoff's laws hold, we can show that

$$\frac{d^3 \underset{\sim}{B}}{dx^3} - v_r \frac{\mu_0 t_r}{\rho g} \frac{d^2 \underset{\sim}{B}}{dx^2} - \left(j\omega \frac{\mu_0 t_r}{\rho g} + v_0^2\right) \frac{d\underset{\sim}{B}}{dx} = \frac{-\mu_0}{g} \left(\frac{d^2 \underset{\sim}{J}_s}{dx^2} - v_0^2 \underset{\sim}{J}_s\right) \tag{7.101}$$

where v_0^2 is a constant associated with machine geometry such that

$$v_0^2 = \frac{4}{\ell_s(\ell_r - \ell_s)} \tag{7.102}$$

relating stator length ℓ_s to rotor length ℓ_r. Note that the physical effect of the stator ends distort B because of the significantly larger air gaps between the rotor and stator, as in Figure 7.65. This can be seen by solving (7.101), assuming sinusoidal currents distributed sinusoidally over x (as in the rotating machine):

$$\frac{d^3 \underset{\sim}{B}_r}{dx^3} - v_r \frac{\mu_0 t_r}{\rho g} \frac{d^2 \underset{\sim}{B}_r}{dx^2} - \left(j\omega \frac{\mu_0 t_r}{\rho g} + v_0^2\right) \frac{d\underset{\sim}{B}_r}{dt} = \frac{\mu_0 t_r}{\rho g} \left(j\omega \frac{d\underset{\sim}{B}_s}{dx} + v_r \frac{d^2 \underset{\sim}{B}_s}{dx^2}\right) \tag{7.103}$$

where B_r is the flux caused by induced currents in the rotor and v_r is the speed of the rotor in the x direction. If $\underset{\sim}{B}_s$ is known, the solution of this equation is of the form

$$\underset{\sim}{B}_r = \underset{\sim}{k}_0 + \underset{\sim}{k}_1 e^{\lambda_1 x} + \underset{\sim}{k}_2 e^{\lambda_2 x} \tag{7.104}$$

where λ_1 and λ_2 are roots of the characteristic equation (or eigenvalues—see Appendix G)

$$\lambda\left(\lambda^2 - v_r \frac{\mu_0 t_r}{\rho g} \lambda\right) - \left(j\omega \frac{\mu_0 t_r}{\rho g} + v_0^2\right) = 0 \tag{7.105}$$

---

†This development is based on the one-dimensional analysis in reference 25. For more complicated and realistic analyses, see this reference and references 21 to 24 and 26 to 31.

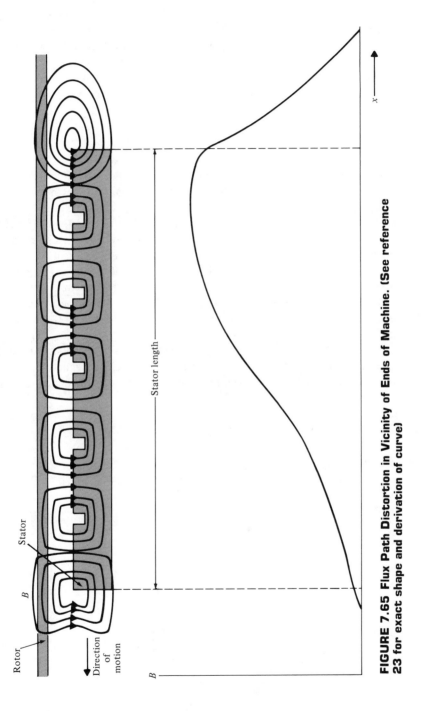

FIGURE 7.65 Flux Path Distortion in Vicinity of Ends of Machine. (See reference 23 for exact shape and derivation of curve)

Notice that this equation is entirely dependent on machine geometry and, in particular, that λ_1 and λ_2 will be functions of the difference in length between the stator and rotor [Eq. (7.102)]. The terms of (7.104) in k_1 and k_2 are said to represent the entrance and exit effects, respectively. It can be shown by solving this equation (see reference 25) that the entrance effect results in a flux wave moving in the $+x$ direction at a speed v_r which is slightly attenuated over distance, and, more important, that the exit effect is a flux wave traveling in the $-x$ direction, *opposite to the direction of motion*, at a speed $-v_r$. Thus the exit flux wave is moving at a slip of

$$ s_{\text{exit}} = \frac{-v_r - v_r}{-v_r} = 200\% \qquad (7.106) $$

which is in the *braking* region. Consequently, *the exit end effect produces a drag or braking force* on the rotor while the remainder of the motor exerts a positive thrust. Fortunately, it can also be shown that the exit effect is highly attenuated on most machines. However, it is still significant, and generalized thrust–speed curves similar to those of Figure 7.66 result when these effects are included. As can be seen from this figure, there is a significant reduction in thrust, both maximum and at rated speed, when end effects are recognized. Losses are also increased. Other effects that significantly influence performance include transverse forces, which act to pull the rotor out of the machine, and other edge effects. Experimental results (see, e.g., reference 32), confirm the negative effects of these phenomena.

Based on this admittedly greatly simplified analysis, it is possible to understand how an ''impellerless'' pump can be built and approximately how it will behave. Linear induction pumps have been in existence for many years and

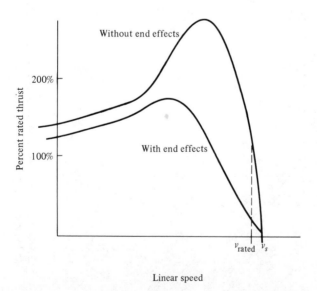

FIGURE 7.66 Comparison of Characteristics of LIM with and Without End Effects Included (Nominal Curves)

became quite popular with the advent of nuclear research and development. Their popularity in certain applications stems from their ability to pump very hot, frequently radioactive, and generally highly corrosive liquid metals without the use of moving parts or any other structures intruding into the fluid stream. Consequently, they are reasonably low-cost, low-maintenance devices which can be serviced easily without disturbing the integrity of the fluid piping system, and thus they prevent any potentially dangerous leaks or exposure of personnel to highly hazardous materials.

Suppose that the rotor in Figure 7.62(b) is replaced by a liquid metal of resistivity approximately equal to that of aluminum and that the air gaps are replaced by an insulating material and an iron or steel alloy pipe as in Figure 7.67. It should be obvious to the reader from a comparison of the two figures that with respect to the magnetic circuit and the electromagnetic phenomena occurring, little or nothing has changed from the classic LIM design. The only difference is that the rotor is a conducting liquid, whose behavior is described by the various laws of hydraulics. Hence, if a polyphase stator produces a traveling flux wave, which passes through the rotor material, then as in the LIM just discussed, currents will be induced and linear motion of the liquid will result.

Consequently, ignoring end and edge effects, the liquid will flow in the x direction at a speed v_r that is related to v_s by the slip equation (7.96) and the thrust imparted to the liquid is as in (7.95). It can be shown that the pressure developed is on the order of

$$\Delta P = \frac{1}{\rho} s v_s p B_{max}^2 \qquad (7.107)$$

where

$$\Delta P = \text{pressure developed}$$
$$\rho = \text{resistivity of liquid metal}$$
$$s = \text{slip}$$
$$v_s = \text{speed of traveling field} \; (= 2pf)$$
$$p = \text{pole pitch}$$
$$B_{max} = \text{maximum value of sinusoidally distributed flux density wave}$$

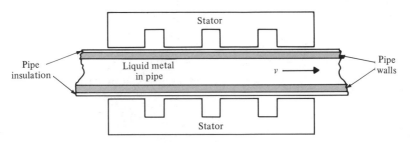

FIGURE 7.67 Flat-Type Linear Induction Pump

Since the flow q equals the speed of the fluid multiplied by the cross-sectional area A of the pipe and since the output power is $\Delta P \cdot q$, then

$$P_{out} = \frac{1}{\rho} sv_s pB_{max}^2 v_r A$$

$$= P_p(1 - s)s$$

(7.108)

where P_p, referred to as the *induction pump parameter*, can be found to be

$$P_p = \frac{1}{\rho} B_{max}^2 Apv_s^2$$

(7.109)

P_p is more commonly expressed by

$$P_p = \frac{1}{2\rho} \mu^2 H_{max}^2 A\Lambda v_s^2$$

(7.110)

where Λ is the "wavelength" of the machine ($= 2p$), μ is the permeability of the fluid ($\mu_0\mu_r$), and H_{max} is the maximum field intensity. It can be shown that the heating loss in the liquid metal (analogous to rotor heating loss) is equal to $P_p s^2$, yielding an ideal pump efficiency of $(1 - s)$, the same as the efficiency for an induction motor. This efficiency will be reduced somewhat by end and edge effects and hydraulic losses.

Liquid metal pumps are generally of the flat or tubular type, as shown in Figure 7.67. Other designs have also been proposed. For example, an annular pump for liquid metal fast breeder reactors has been proposed. This pump, shown in Figure 7.68, is described in detail in reference 33. It is effectively a single-sided LIM, pumping liquid in the outer duct, with the inner return duct separated from the motor by an iron shell which acts as the return path for the stator flux. Such a motor has the advantages that it can expand and contract freely in the longitudinal direction and it can be removed from the pipe loop without opening the pipe.

EXAMPLE 7.7 A small liquid-metal pump is arranged as in Figure 7.67. Liquid sodium at 100°C ($\rho = 9.7 \times 10^{-6}$ Ω-cm) is to be pumped in a pipe whose cross-sectional area is 50 cm^2. The stator of the pump is designed with a pole pitch of 15 cm and is excited by 60 Hz such that the maximum magnetic flux density in the metal is 0.5 Wb. Find the flow rate in m^3/s and the pressure head in N/m^2 if the motor operates at a slip of 28%.

From (7.107),

$$\Delta P = \frac{1}{\rho} sv_s pB_{max}^2$$

and

$$v_s = 2pf = 2 \times 1.5 \times 10^{-2} \times 60 = 1.8 \text{ m/s}$$

$$\rho = 9.7 \times 10^{-8} \text{ } \Omega\text{-m} (= 9.7 \times 10^{-6} \text{ } \Omega\text{-cm})$$

So

$$\Delta P = \frac{0.28 \cdot 1.8 \cdot 1.5 \times 10^{-2} \cdot (0.5)^2}{9.7 \times 10^{-8}} \text{ N/m}^2$$

$$= 1.95 \times 10^4 \text{ N/m}^2 \text{ (about 3 psi)}$$

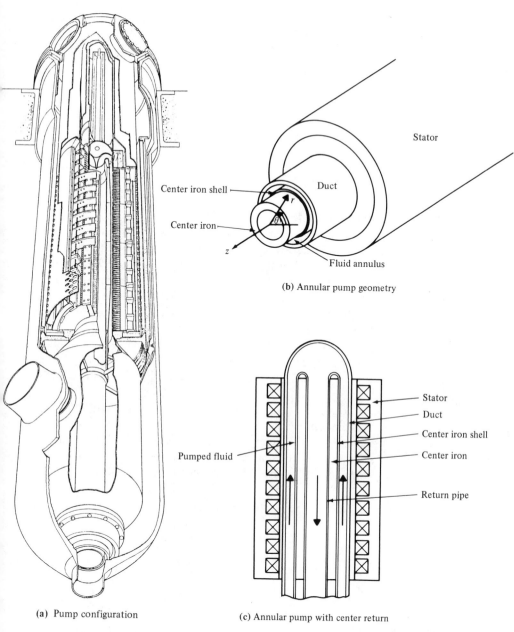

(b) Annular pump geometry

(a) Pump configuration

(c) Annular pump with center return

FIGURE 7.68 Annular-Type Linear Induction Pump (Courtesy of General Electric Company)

and

$$q = v_r \cdot A$$
$$= (1 - s)v_s \cdot A$$
$$= (1 - 0.28)(1.8 \text{ m/s}) \times 50 \times 10^{-4} \text{ m}^2$$
$$= 6.48 \times 10^{-3} \text{ m}^3/\text{s (about 100 gal/min)}$$

It is also useful to find the input and output power expected under ideal conditions. From (7.108),

$$P_{out} = \frac{1}{\rho} s v_s p (B_{max})^2 v_r A \quad \text{W}$$

$$= \frac{0.28 \cdot 1.8 \cdot 1.5 \times 10^{-2}(0.5)^2(1 - 0.28)1.8 \cdot 50 \times 10^{-4}}{9.7 \times 10^{-8}} \quad \text{W}$$

$$= 126 \text{ W}$$

If we assume an efficiency of $1 - s$, then

$$P_{in} = \frac{P_{out}}{1 - s} = \frac{126}{1 - 0.28} = 175 \text{ W}$$

7.9.3 THE LINEAR DC MOTOR

In the preceding section, linear induction machines were discussed, and an example was given of a device whose "stator" windings were fixed in space. Linear dc motors, while reasonably simple to analyze, are somewhat more difficult to envision because the field coils, generally fixed on a rotating machine, are placed on the rotor of a linear machine, and the armature becomes the stator. This is done purely for convenience; rotating dc machines could also be "turned inside out" if desired and, in fact, some electronically commutated, permanent-magnet dc rotating motors are also configured so the field moves and the armature is stationary. The reader should consider this fact carefully prior to continuing to minimize any potential confusion caused by the configuration of this machine.

A permanent-magnet dc machine can be viewed as a separately excited machine where the field flux φ is a constant and is produced by a permanent magnet rather than a coil. In such a configuration, the speed v of the motor in the x direction is a function of the armature voltage, and the torque is a function of armature current only; the motor is therefore described as follows:

$$v = \frac{e_a}{K\varphi} = \frac{e_a}{K_B} \tag{7.111}$$

$$T = K_m i_a \tag{7.112}$$

$$e_a = V_t - i_a r_a \tag{7.113}$$

Since a linear machine is simply the dual of a rotating machine, it makes sense that the same equations would apply, in the general sense, to a permanent magnet linear dc machine.

Such a machine is shown in Figure 7.69. This motor, designed for linear motion servo applications, has the performance characteristics shown in Table 7.3. In this motor, the armature is fixed in the stator, which contains windings, the commutator bars, and power rails (equivalent to slip rings in a rotating machine). The field assembly, consisting of two permanent magnets, is mounted on the slider. Brushes on the slider connect the power rails with the commutator bars such that coils immediately beneath the field poles are energized. This action is identical to the commutator action in Figure 5.4. Figure 5.9, repeated

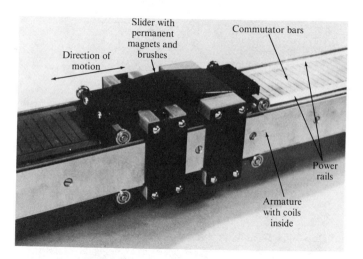

FIGURE 7.69 Linear DC Servomotor (Courtesy of Inland Motor Specialty Products Division, a division of Kollmorgen Corporation)

here as Figure 7.70, describes the flux density for this machine ignoring end effects, and the analysis in Section 5.2 also holds. The power rails on the stator assembly are connected to a power source (in this case, a servo amplifier) and the speed and direction of motion are directly proportional to the magnitude and polarity of the source voltage. The equivalent circuit for this machine, again ignoring end effects, is shown in Figure 7.71. As one might expect, other dc motor connections could also be produced by replacing the permanent magnets with field coils and connecting them in series, shunt, or compound.

Table 7.3 Representative Data for a
Commercial Linear DC Servomotor

Peak force F_p	5.50 lb
Motor constant K_m	2.3 lb/$\sqrt{W}$
Power input P_p at F_p	570 W
Static friction F_F	1.8 lb
Inertia J_s	0.304 lb $\cdot$ s^2/ft
Acceleration α_s	181 ft/s^2
Voltage V_p at F_p	29.3 V
Peak current I_p	19.5 A
Dc resistance R_m	1.50 Ω
Inductance L_m	12 mH
Force sensitivity K_F	2.82 lb/A
Back emf constant K_B	3.82 V per ft/s
Weight of slider	9.8 lb
Weight of armature	52 lb
Length of armature	36 in.

Source: Inland Motor Specialty Products Division,
a Division of Kollmorgen Corporation.

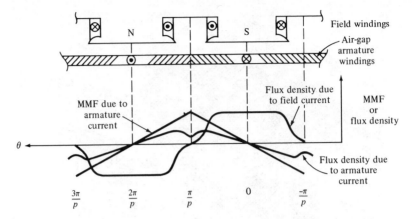

FIGURE 7.70 Air Gap Flux Densities due to Field and Armature Circuits (Figure 5.9)

EXAMPLE 7.8 The linear dc motor described in Table 7.3 is to be operated in a servo application. Draw the equivalent circuit. Compute the no-load speed.
From the table

$$V_p = 29.3 \text{ V}$$

$$\text{dc resistance} = 1.5 \text{ }\Omega$$

$$\text{inductance} = 12 \text{ mH}$$

So the equivalent circuit is as in Figure 7.71, with $R_m = 1.5 \text{ }\Omega$ and $L_m = 12$ mH. By using the back emf constant, K_B, linear speed v can be found from e_a:

$$v = \frac{e_a}{K_B}$$

At no load, assume that the machine is delivering F_p so that $e_a = V_p = 29.3$ V, and therefore

$$v = \frac{29.3}{3.82} = 7.67 \text{ ft/s}$$

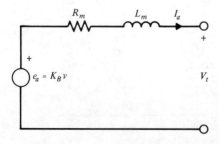

FIGURE 7.71 Equivalent Circuit for Armature of Linear DC Machine

7.9.4 LINEAR SYNCHRONOUS MOTORS AND MAGNETICALLY LEVITATED "SUPER-SPEED" TRAINS

Numerous technologies, both conventional and unconventional, have been proposed for high-speed ground transportation. Developmental work on these various technologies has been under way in many countries, including the United States, Great Britain, Japan, France, West Germany, and Canada, for many years. Conventional rotating machine/wheel-on-rail systems presently operating in revenue service on multiple daily frequencies in Japan (Shin-Kansen, or "bullet train") and France (train à grande vitesse, or TGV) at sustained speeds in excess of 150 mph (about 250 km/h) represent a proven, albeit improving, state of the art. However, it appears that a practical limit on top speed for these conventional high-speed wheeled vehicles is about 225 to 250 mph (about 350 km/h) due to wheel-rail dynamics and electric power pickup considerations. For speeds in excess of 225 mph (the so-called "super-speed" region), research has concentrated on the use of air cushion and magnetically levitated vehicles (called MAGLEV) to provide "wheel-less" propulsion.

Much super-speed research has involved the use of linear motors in one or more forms. Linear motors are clearly well adapted to wheel-less propulsion because they exert a linear force rather than a rotating torque, and their reasonably quiet operating characteristics provide a clear advantage over such competing technologies as propellers and turbojet engines.[†] In particular, LIMs in several vehicle configurations, including air-cushion, levitated, and conventional wheel-on-rail, have been demonstrated successfully. Both single-sided and double-sided LIMs have been operated, generally in the long-rotor configuration. Speed has been controlled by varying the frequency of the stator current, requiring heavy solid-state power conditioning equipment plus a means for generating or picking up the large amounts of power required by the stator on-board the vehicle. These types of vehicles are discussed in detail in many articles and books, including references 21 to 32 and 46.

It was later recognized that this undesirable requirement for on-board equipment and power pickup could be largely eliminated if a single-sided long-stator linear synchronous motor (LSM) as in Figure 7.72 were utilized, since the vehicle would contain (essentially) only the field coils[‡] and the heavy power equipment could be located on the ground, feeding current to the stator windings, which could be located in the "track" structure or guideway. Furthermore, only track sections "contiguous" (usually within a half-mile) with the vehicle itself would have to be energized, resulting in substantial potential economies over other configurations. In addition, LSMs can be made to produce a substantial levitating or "normal" force in addition to their propulsive force. Although it can be shown that this force exists in virtually all linear machines (see references

---

†Although the concept of jet-powered trains may seem somewhat bizarre, the reader should note that the New York Central Railroad, about 20 years ago, actually demonstrated a converted rail passenger car powered by a jet engine from a B-47 bomber. Needless to say, the neighbors were not too impressed with the idea, and there were some technical problems; the concept did not catch on.

‡Plus air conditioning, lights, and other incidental loads grouped into the general category of "house" power.

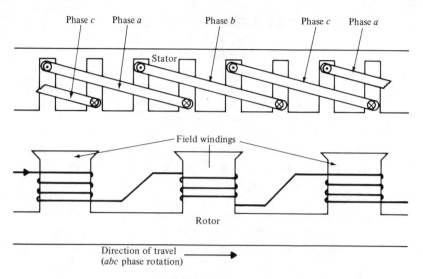

Phase *c*   Phase *a*   Phase *b*   Phase *c*   Phase *a*

Stator

Field windings

Rotor

Direction of travel
(*abc* phase rotation)

FIGURE 7.72 Configuration of Generalized Single-Sided Salient-Pole Linear Synchronous Motor with Long Stator and Moving Rotor

21 to 24), it can be exploited easily and economically in an LSM to provide essentially a "one-machine" noncontact propulsion system which concurrently develops a normal force to lift a commercial-sized vehicle and propel it at speeds of over 250 mph (300 km/h) at an altitude of up to several inches above the track or guideway. These advantages, when combined, have resulted in a major thrust in development work toward a commercial, levitated super-speed train employing LSMs. While the initial research on these vehicles was carried out in several countries, commercial development efforts have been centered in Japan (Japanese National Railways—JNR) and West Germany (Transrapid Consortium, a group of companies headed by Thyssen Henschel, Krauss-Maffei, and Messerschmitt Bölkow Blohm, and sponsored by the West German Federal Ministry of Research and Technology).

Linear synchronous motors can be made to develop a normal force that can be exploited to provide levitation in either of two ways: repulsion or attraction. The repulsion system, presently being developed by JNR in Japan, is highly complex theoretically and is based on the repulsive force generated between a superconducting ac magnet and eddy currents induced in a coil or conducting sheet. The interaction of the magnet and coil can be viewed as being similar to that which occurs in the very common "floating ring" demonstration device in Figure 7.73. The net result is that two opposing fields are set up by the forward motion and these two fields produce a repulsive force (see, as an analogy, Examples 2.1 and 2.2). If this system is used in conjunction with a long stator LSM, the vehicle will move in a linear direction at synchronous speed, providing the velocity necessary for levitation. While this repulsive force can be shown to be *unstable* for a conventional machine, it can be shown that it is *stable* for the linear machine *if the windings on the vehicle are cooled to the superconducting state*. Repulsion levitation systems do not exert a significant normal force until linear speeds in excess of a critical value (about 50 mph or 80 km/h)

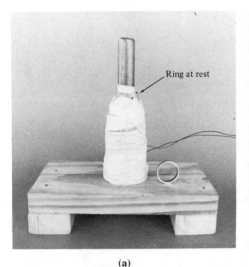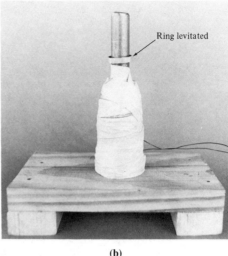

(a)   (b)

FIGURE 7.73 Simple "Floating Ring" Repulsion Levitation Demonstration Device. (a) Coil deenergized; (b) coil energized at 120 VAC

are reached. Consequently, a repulsion levitation vehicle must roll on wheels until this speed is reached after which the levitation force lifts it off the track. The vehicle then "flies" at an altitude of approximately 4 in. (10 cm) above the track. Altitude varies directly with speed. Vehicles of this type are configured in a fashion similar to the one in Figure 7.74. A cryogenically cooled superconducting rotor on-board the vehicle is required in this configuration.

The JNR demonstration MAGLEV system and other proposed repulsion levitation vehicles are discussed at length in references 34 to 39. The actual operation of the system is very complicated and difficult to visualize, particularly the conditions for stability, and is beyond the major thrust of this chapter. With all due respects to the tremendous efforts by JNR in this area, it will not be discussed further here because of the advanced nature of the material, except to state that JNR has a MAGLEV vehicle with a cryogenically cooled rotor, capable of carrying up to eight passengers, which has been successfully tested without passengers at speeds in excess of 300 mph (500 km/h) at their Miyazaki Test Track in Japan, and they are presently seeking government permission to build a larger test track (see reference 49). Similar, impressive results have been obtained with smaller repulsion levitation test vehicles in West Germany [37, 38] at the Erlangen Test Facility.

Attraction levitation, however, can be more readily understood in at least a general way based on the concepts discussed in previous chapters, as it does not involve the special electromagnetic characteristics of superconductors. Attraction levitation systems are presently being developed on a commercial scale in West Germany. We will consider a basic static attraction levitation system and an attraction LSM system in very general terms to understand their basic methods of operation.

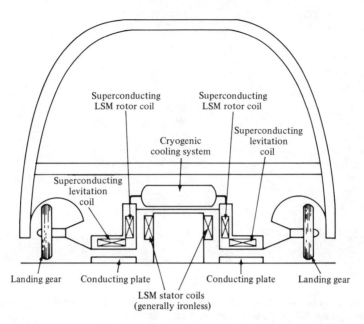

Superconducting
LSM rotor coil

Superconducting
LSM rotor coil

Cryogenic
cooling system

Superconducting
levitation
coil

Superconducting
levitation
coil

Landing gear   Conducting plate   Conducting plate   Landing gear

LSM stator coils
(generally ironless)

FIGURE 7.74 Typical Configuration of Common "Inverted T" Repulsion Levitation Vehicle Drive System Utilizing Long-Stator LSM. Note that this configuration is similar in many respects to the JNR vehicle.

The Simple Static Attraction Levitator

A simple static attraction levitator incapable of propulsion can be demonstrated based on the principles in Chapter 2, particularly Examples 2.7 and 2.8. Suppose that the coil–core system, shown in Figure 2.18, were turned on end as in Figure 7.75. In this figure, the coil structure is movable and will fall to the ground because of the force of gravity unless it is suspended in some fashion. However, if the coil is energized by a current i_f, a steady-state upward force is exerted attracting the coil to the fixed member; the magnitude of that force was found in Chapter 2 to be (reexpressed in the coordinates of Figure 7.75)

$$F_e(i_f, z) = -\frac{(Ni_f)^2}{\mu_0 A}\left(\frac{1}{\mathcal{R}_{core} + 2z/\mu_0 A}\right)^2 \qquad (7.114)$$

where

$$A = \text{cross-sectional area of the core}$$

$$\mathcal{R}_{core} = \text{core reluctance (moving plus fixed element)}$$

$$z = \text{air-gap length}$$

Clearly, this force is opposed by the downward gravitational force F_g on the coil structure:

$$F_g = mg \qquad (7.115)$$

where

$$m = \text{mass of coil structure}$$

$$g = \text{acceleration of gravity } (\approx 9.81 \text{ m/s}^2)$$

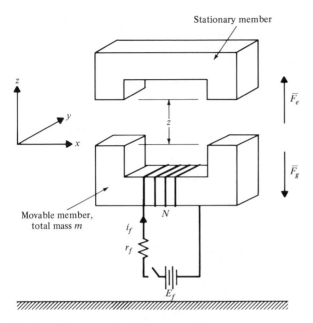

Stationary member

Movable member,
total mass m

i_f

r_f

N

$\overline{F}_e$

$\overline{F}_g$

E_f

FIGURE 7.75 Simple Static Attraction Levitator

For the static case, levitation can occur for any gap length z by adjusting the current i_f so that the sum of the two forces is zero:

$$\overline{F}_e + \overline{F}_g = F_e - F_g = 0 \tag{7.116}$$

Substituting (7.114) and (7.115) into (7.116), we get

$$mg = \frac{(Ni_f)^2}{\mu_0 A} \left(\frac{1}{\mathcal{R}_{core} + 2z/\mu_0 A} \right)^2 \tag{7.117}$$

So the steady-state dc current i_f required to levitate the core for an air-gap distance z under ideal conditions is then found to be

$$i_f = \left(\frac{\mathcal{R}_{core}}{N} + \frac{2z}{\mu_0 AN} \right) \sqrt{\mu_0 Amg} \tag{7.118}$$

For the dynamic case of movement in only the z direction (see Example 2.8), the differential equation that must be solved can be shown to be (ignoring aerodynamic friction and other similar factors):

$$\frac{dv_z}{dt} = \frac{\lambda^2}{mN^2 \mu_0 A} - g \tag{7.119}$$

$$\frac{dz}{dt} = v_z \tag{7.120}$$

$$\frac{d\lambda}{dt} = - \frac{r_f \mathcal{R}_{core}}{N^2} \lambda - \frac{2r_f}{N^2 \mu_0 A} z\lambda + E_f \tag{7.121}$$

This is a nonlinear system, and thus any analysis of a control system to vary E_f to maintain a constant gap length z, must be based on nonlinear control theory. Such an analysis is beyond the scope of this text but is covered in texts such as reference 13.

Suppose now that a coil is also placed on the stationary member in Figure 7.75 and is excited in such a way that additional flux described by B_s is created in the same direction as B_f from the movable member. The attractive force would now be a function of both currents and would be found by differentiating the coenergy W_c of the system with respect to z, or

$$F_e(i_f, i_s, z) = \frac{\partial W_c(i_f, i_s, z)}{\partial z} \tag{7.122}$$

where

$$W_c(i, z) = \frac{1}{2} \bar{i}^t \bar{L}(z) \bar{i} \tag{7.123}$$

and

$$\bar{i} = \begin{bmatrix} i_f \\ i_s \end{bmatrix} \tag{7.124}$$

and assuming linear core material. If the current i_s in the stationary element is assumed to be constant, then again current i_f in the moving element controls the levitating force (see Problem 7.20). The difference is that i_f can now be significantly smaller and can control z more precisely.

The control current i_f in the moving member can be made even smaller (see Problem 7.21) by replacing the moving-core structure (but not the coil) with a permanent magnet aligned in the same direction as the flux from the coil, or by adding a permanent magnet in series with the core, yielding a configuration similar to the one in Figure 7.76. In such a system, if the permanent-magnet flux density B_p were only slightly below the value required to lift the moving member from its resting point, a very small current i_f could increase the force $\bar{F}_e$ enough to levitate what is now a coil–magnet–core structure. Obviously, with such a system one could have very rigid control over the air-gap length z. If this system were duplicated and placed on a mass m which overhangs a stationary supporting T structure, as in Figure 7.77, the attracting force $\bar{F}_e$ would be divided between the two systems and the two currents i_{f_1} and i_{f_2} could be controlled to lift the mass uniformly up to a height $(d - z)$ above the supporting structure. This is the basic principle upon which an attraction-levitated vehicle is lifted above the track. Combined levitation and linear motion in the steady state will now be discussed in very general terms.

LSMs and Attraction Levitation

The levitating force $\bar{F}_e$ in the two-coil or two-coil permanent magnet static attraction levitators was seen to be a function of the interaction of two static magnetic fields and their associated flux linkages. Now, we will go one step further and allow the flux caused by current i_s to travel in the x direction through the use of a polyphase stator winding laid out like the stator of a LIM. This

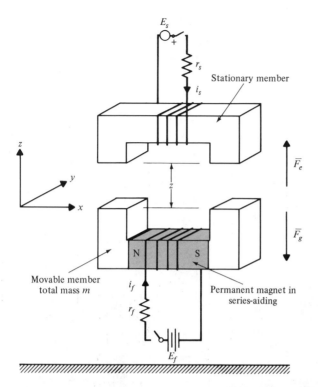

FIGURE 7.76 A Simple Two-Coil Permanent Magnet Static Levitator

geometry should be immediately recognized as the linear dual of a polyphase synchronous motor whose rotor consists of a permanent-magnet fixed field B_p plus a field winding carrying a varying dc current i_f. If this were a rotating machine, we could easily calculate the rotor torque as a function of B_p, i_f, the stator currents, and the torque angle δ. Similarly, one can calculate the forces on the rotor (now in both the x and z directions) for the linear machine, assuming that it is operating in synchronism in the steady state, that the forces can be decomposed into independent, orthogonal elements, and neglecting harmonics, saturation, and end effects. For the track and rotor coil structure in Figure 7.77 it can be shown that magnitude of the propulsion force or thrust F_x in the x direction for a three-phase motor is (see reference 23 for derivation)

$$
F_x = 3 \frac{\partial W_c(i_f, i_s, x, z)}{\partial x}
$$

$$
= 3 \frac{\partial W_c(E_f, i_s, x, z)}{\partial x} \tag{7.125}
$$

$$
= 3 \frac{\omega_s p}{\pi} \frac{V_s(E_f - V_s \cos \delta)R + V_s(E_f \sin \delta)\omega_s L}{R^2 + \omega_s^2 L^2}
$$

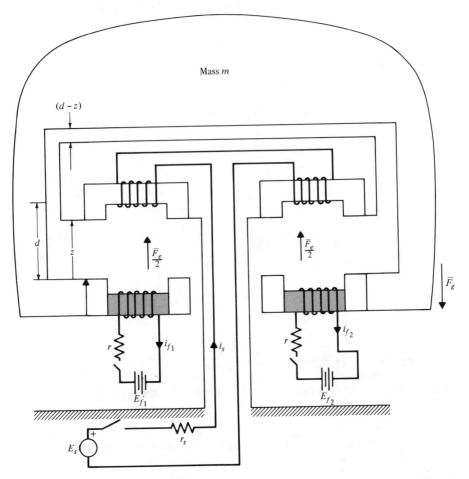

FIGURE 7.77 Simple Two-Coil Permanent-Magnet Static Levitator on T-Structure

where

$$p = \text{pole pitch}$$
$$\omega_s = \text{synchronous frequency}$$
$$V_s = \text{terminal voltage of stator}$$
$$E_f = \text{induced voltage}$$
$$L = \text{self-inductance of stator coils}$$
$$R = \text{resistance of stator coils}$$
$$\delta = \text{torque angle}$$

The levitation force $F_{z_{\text{tot}}}$ at synchronous speed can be approximated on the basis of the static two-coil permanent-magnet model (assuming very small deviations in δ from $0°$ as well as the other conditions discussed previously)

$$F_{z_{\text{tot}}} \approx N_R F_z \cos \delta \qquad (7.126)$$

where

N_R = number of rotor coils

F_z = levitation force for one rotor coil when aligned with the axis of the stator field $(= F_e)$

This is clearly a greatly oversimplified model, but it is useful for our purposes because it shows that the normal force is not a function of speed, so the vehicle can be levitated even while stopped. This is somewhat of an advantage for attraction-type systems, as wheels are not required for routine operation.

In addition to the propulsion and levitation forces above, lateral forces due to mechanical disymmetry and outside causes (such as aerodynamic turbulence) also exist. In an attraction-levitation LSM, the normal force is generally accompanied by lateral forces which are oriented to center the rotor pole faces over the stator poles (called *restoring forces*). These forces are highly complex and will not be discussed here. Attraction-levitation vehicles typically are provided with two sets of lateral stabilization magnets located on the vehicle with poles facing inward toward a steel plate attached to the outside vertical edge of the track and which are excited in tandem by carefully controlled time-varying dc currents in such a way to keep the vehicle centered on the track.

A typical configuration of an attraction-levitation LSM vehicle, based on the discussion above, is shown in Figure 7.78. The vehicle has dc-excited salient

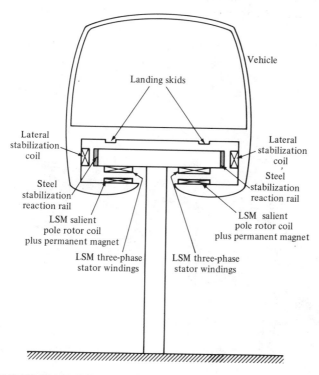

FIGURE 7.78 Typical Configuration of an Attraction Levitation LSM Vehicle. Note that this configuration is similar in many respects to the Transrapid vehicle.

pole permanent-magnet rotor coils and dc stabilization coils. The track structure consists of steel rails on the outside edges to interact with the stabilization coils and a set of three-phase stator windings. The presence of harmonic induced voltages and currents in the salient-pole configuration, common to both linear and rotating machines, is exploited in the vehicle; the small amount of power necessary to provide the control current for the permanent-magnet-control-coil rotor, the stabilization coils, and on-board "house" power can be produced from the harmonic currents induced in the rotor windings by the effects of the salient-pole stator at nonzero speeds. Consequently, no physical power transfer apparatus, such as a set of current collector shoes between the track and the vehicle, are required except when stopped. The permanent magnets are of the very high strength rare-earth type and are usually sized so that the addition of a very small control current will levitate the vehicle. Typically, power requirements of less than 0.5 kW/ton of lift have been reported [40].

The discussion above clearly has been greatly simplified and generalized. Although some aspects of attraction systems are considered proprietary and have not been covered in the literature, the reader is directed to references 38 to 48 for a more detailed discussion of the operation of these and similar systems.

As mentioned earlier, attraction-levitation LSM systems are under development in West Germany by the Transrapid Consortium, and have been operated in a form of revenue service. In 1979, an attraction vehicle of the type described above operated for several months during the International Transportation Exhibition in Hamburg. A track slightly over $\frac{1}{2}$ mile long was constructed between the exhibition area and an amusement park, and a two-section vehicle seating over 70 passengers was successfully operated on the line. Speeds were restricted to 75 kg/h (slightly over 50 mph) by court order [40]. The vehicle was similar to the newer preproduction test vehicle in Figure 7.79. This three-unit vehicle (Transrapid 06) was delivered for testing in mid-1983 and is to be operated on a 19-mile (31 km) test track at Erlangen, West Germany, at sustained speeds

FIGURE 7.79 Super-Speed Magnetic Attraction-Levitated Train (Transrapid 06) (Courtesy of The Budd Company)

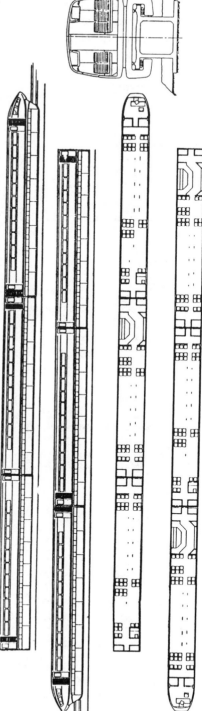

FIGURE 7.80 Proposed Vehicle Designs for Los Angeles-Las Vegas MAGLEV Super-Speed Train (Courtesy of The Budd Company)

of 200 mph (325 km/h) with bursts to 250 mph (400 km/h). This system has been proposed for revenue service in several areas in the United States, primarily Los Angeles to Las Vegas, Milwaukee to Chicago, and Dallas to Houston. A feasibility study for the Los Angeles–Las Vegas line was completed [49] in early 1983. Proposed vehicle design is shown in Figure 7.80. Trip times of 1 h 10 min (average speed 197 mph) are proposed. Twenty-two round trips per day using six-car trains are proposed, yielding a projected annual capacity of 4.8 million passenger round trips. It should be quite interesting to watch the development of both this attraction-type MAGLEV system and the Japanese repulsion-type MAGLEV.

REFERENCES

[1]   H. P. WESTMAN (ED.), *Reference Data for Radio Engineers*. New York: Howard W. Sams & Company, Inc., Publishers, 1972.

[2]   R. RESNICK and D. HALLIDAY, *Physics* (2 vols.). New York: John Wiley & Sons, Inc., 1968.

[3]   H. H. WOODSON and J. R. MELCHER, *Electromechanical Dynamics* (3 vols.). New York: John Wiley & Sons, Inc., 1968.

[4]   H. H. SKILLING, *Electromechanics*. New York: John Wiley & Sons, Inc., 1962.

[5]   W. P. MASON, *Electromechanical Transducers and Wave Filters*. New York: D. Van Nostrand Company, 1948.

[6]   J. H. REYNER, *The Magnetic Amplifier*. London: Stuart and Richards, 1950.

[7]   N. W. MCLACHLAN, *Loud Speakers*. New York: Dover Publications, Inc., 1960.

[8]   .T. E. BROWNE ET AL., *Supply 77—EPRI Annual Energy Supply Forecasts,* EA-634-SR, Electric Power Research Institute, May 1978.

[9]   "Compact Pulsed Homopolar Generator," OIME, Inc., Odessa, Tex.: 1983.

[10] W. F. WELDON, R. E. KEITH, and J. M. WELDON, "Billet Heating with the Homopolar Generator," *Proceedings of the Second Annual Conference on Industrial Energy Conservation Technology,* Houston, Tex., 1980.

[11] T. A. AANSTOOS, R. E. KEITH, and W. F. WELDON, "Application of Homopolar Pulsed Power to Metals Joining," *Proceedings of the Fifth Annual Conference on Industrial Energy Conservation Technology,* Houston, Tex., 1983.

[12] Van Valkenburgh, Nooger and Neville, Inc., *Basic Synchros and Servomechanisms* (2 vols.). New York: John F. Rider Publisher, Inc., 1955.

[13] O. I. ELGERD, *Control Systems Theory*. New York: McGraw-Hill Book Company, 1967.

[14] D. LENT, *Analysis and Design of Mechanisms*. Englewood Cliffs, N.J.: Prentice-Hall, Inc., 1961.

[15] C. BODINE (ED.), *Small Motor, Gearmotor and Control Handbook*. Chicago: Bodine Electric Company, 1978.

[16] "Stepper Motors," *Bodine Motorgram,* Vol. 61, No. 4–6. Chicago: Bodine Electric Company.

[17] Japan Servo Company Ltd., *Stepping Motors,* Bulletin 81EGST-01, 1981.

[18] Compumotor Corporation, *Computer Controlled Motion,* Bulletin CM-153 35K 9/82, 1982.

[19] Compumotor Corporation, *M57 Series Motor/Driver,* Bulletin M57/1-82, 1982.

[20] Airpax Corporation, *Digital Linear Actuators,* Form 117-R2, 1982.

[21] E. R. LAITHWAITE, *Propulsion Without Wheels*. London: English Universities Press Ltd., 1970.

[22] E. R. LAITHWAITE (ED.), *Transport Without Wheels,* Boulder, Colo.: Westview Press, Inc., 1977.

[23] S. A. NASAR and I. BOLDEA, *Linear Motion Electric Machines*. New York: Wiley–Interscience, 1976.

[24] E. R. LAITHWAITE and S. A. NASAR, "Linear-Motion Electrical Machines," *Proceedings of the IEEE*, Vol. 58, pp. 531–542, April 1970.

[25] M. POLOUJADOFF, *The Theory of Linear Induction Machinery*. Oxford: Clarendon Press, 1980.

[26] S. A. NASAR and L. DEL CID, JR., "Certain Approaches to the Analysis of Single-Sided Linear Induction Motors," *Proceedings of the IEE* (Great Britain), Vol. 120, pp. 477–483, April 1973.

[27] S. A. NASAR and L. DEL CID, JR., "Propulsion and Levitation Forces in a Single-Sided Linear Induction Motor for High Speed Ground Transportation," *Proceedings of the IEEE*, Vol. 61, pp. 638–644, May 1973.

[28] H. S. HOLLEY, S. A. NASAR, and L. DEL CID, JR., "Computations of Fields and Forces in a Two-Sided Linear Induction Motor," *IEEE Transactions on Power Apparatus and Systems*, Vol. PAS-92, pp. 1310–1315, July/August 1973.

[29] A. MOR and S. GAVRIL, "Spatial Fourier Techniques and End Effects in the Single-Sided Linear Induction Motor," *Electric Machines and Electromechanics*, Vol. 7, pp. 47–55, January/February 1982.

[30] I. BOLDEA and S. A. NASAR, "Simulation of High-Speed Linear Induction–Motor End Effects in Low-Speed Tests," *Proceedings of the IEE* (Great Britain), Vol. 121, pp. 961–964, September 1974.

[31] I. BOLDEA and S. A. NASAR, "Performance of Ideal-Current Inverter Fed Linear Induction Motors: A Theoretical Study," *Electric Machines and Electromechanics*, Vol. 7, pp. 35–45, January/February 1982.

[32] S. NAKAMURA, Y. TAKEUCHI, and M. TAKAHASHI, "Experimental Results of the Single-Sided Linear Induction Motor," *IEEE Transactions on Magnetics*, Vol. MAG-15, pp. 1434–1436, November 1979.

[33] G. B. KLIMAN, *Large Electromagnetic Pumps*, Report No. 78CRD156. Schenectady, N.Y.: General Electric Company, July 1978.

[34] K. OSHIMA, "Superconducting Magnetic Levitation Train Project in Japan," *IEEE Transactions on Magnetics*, Vol. MAG-17, pp. 2338–2342, September 1981.

[35] T. OHTSUKA and Y. KYOTANI, "Superconducting Maglev Tests," *IEEE Transactions on Magnetics*, Vol. MAG-15, pp. 1416–1423, November 1979.

[36] I. BOLDEA, "Static and Dynamic Performance of Electrodynamic (Repulsion) Levitation Systems (EDS)," *Electric Machines and Electromechanics*, Vol. 6, pp. 45–55, January/February 1981.

[37] G. BOGNER, "Applied Superconductivity Activities at Siemens," *IEEE Transactions on Magnetics*, Vol. MAG-15, pp. 824–827, January 1979.

[38] E. ABEL, J. L. MAHTANI, and R. G. RHODES, "Linear Machine Power Requirements and System Comparisons," *IEEE Transactions on Magnetics*, Vol. MAG-14, pp. 918–920, September 1978.

[39] IEEE Vehicular Technology Group, *IEEE Transactions on Vehicular Technology*, Vol. VT-29, February 1980 (entire issue devoted to magnetic levitation systems).

[40] H. WEH, "Linear Synchronous Motor Development for Urban and Rapid Transit Systems," *IEEE Transactions on Magnetics*, Vol. MAG-15, pp. 1422–1427, November 1979.

[41] R. BORCHERTS, "An Investigation of Two HSGT Magnetic Suspension Systems (Attraction)," *IEEE Transactions on Magnetics*, Vol. MAG-12, pp. 369–372, July 1976.

[42] H. WEH and M. SHALABY, "Magnetic Levitation with Controlled Permanetic Excitation," *IEEE Transactions on Magnetics*, Vol. MAG-13, pp. 1409–1411, September 1977.

[43] G. H. BOHN, "Calculation of Frequency Responses of Electro-magnetic Levitation

Magnets," *IEEE Transactions on Magnetics,* Vol. MAG-13, pp. 1412–1414, September 1977.

[44] I. BOLDEA, "Optimal Design of Attraction Levitation Magnets Including the End Effect," *Electric Machines and Electromechanics,* Vol. 6, pp. 57–66, January/February 1981.

[45] A. EL ZAWAWI, Y. BAUDON, and M. IVANES, "Dynamic Analysis of an Electromagnetically Levitated Vehicle Using Linear Synchronous Motors," *Electric Machines and Electromechanics,* Vol. 6, pp. 129–141, March/April 1981.

[46] S. NAKAMURA, "Development of High Speed Surface Transport System (HSST)," *IEEE Transactions on Magnetics,* Vol. MAG-15, pp. 1428–1433, November 1979.

[47] J. A. WAGNER and W. J. CORNELL, "A Study of Normal Force in a Toothed Linear Reluctance Motor," *Electric Machines and Electromechanics,* Vol. 7, pp. 143–153, March/April 1982.

[48] M. H. NAGRIAL, "Correspondence: Hybrid Linear Reluctance Motor," *Electrical Machines and Electromechanics,* Vol. 6, pp. 281–286, May/June 1981.

[49] The Budd Company, *Las Vegas to Los Angeles High Speed/Super Speed Ground Transportation Feasibility Study,* January 1983 (Executive Summary plus two appendix volumes).

PROBLEMS

7.1 In Problem 1.3, the flux induced in a toroidal core that was excited by a current i_1 in a conductor passed through the center hole was found. If a coil of N turns is wrapped around the toroid as in the accompanying figure and the current i_1 is

$$i_1 = \sqrt{2}\, I \sin \omega t$$

find an expression for the terminal voltage e_{a-b} (neglect coil resistance). If i_1 is set to a value of (at 60 Hz)

$$i_1 = 150 \sin 377t$$

and the coil is connected across a 5-Ω resistor at terminals $a-b$, find the induced current i_2 through the resistor.

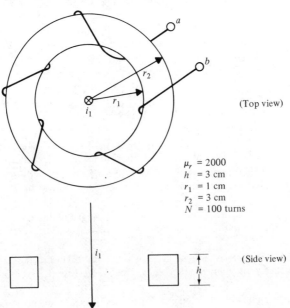

(Top view)

$\mu_r = 2000$
$h = 3$ cm
$r_1 = 1$ cm
$r_2 = 3$ cm
$N = 100$ turns

(Side view)

7.2   Draw a transistor amplifier circuit that is the dual of the magnetic amplifier circuit in the accompanying figure and discuss why it is similar in operation.

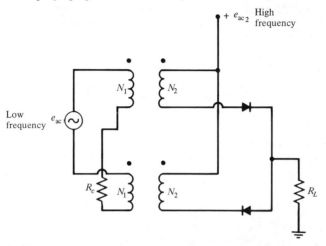

7.3   A "magnetic" phonograph pickup can be configured as in the accompanying figure. This pickup operates on essentially the same principles as a reluctance motor. Show that, for small vibrations of the stylus from side to side, the voltage induced on the coil is of the form

$$e = -N \frac{\mu_0 \mathscr{F} A \omega}{g^2} d_m \cos (\omega t + \psi)$$

for a "linear" sideways displacement d of the stylus which is of the form

$$d = d_m \sin (\omega t + \psi)$$

Assume that the air-gap distance and air gap area A are uniform across the pole face. (HINT: The magnetic circuit looks like a Wheatstone bridge.)

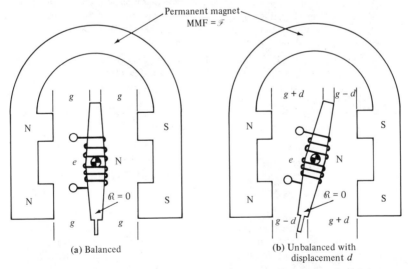

(a) Balanced

(b) Unbalanced with displacement d

7.4   Find an expression for the voltage $V_s(t)$ if the microphone in Example 7.2 is connected to a purely resistive load $Z_i = 2R \underline{/0°}$.

7.5   Explain why a dynamic loudspeaker can also be used as a microphone. Comment on sizing and frequency response to be expected if a dual role is desired.

7.6 Suppose that we replace the permanent magnet in the loudspeaker of Example 7.3 with a coil of 1000 turns of wire excited by a dc current i_{dc}. If the reluctance $\mathcal{R}$ of the coil–core system is assumed to be a constant value of 5×10^4 A-t/Wb and the cross-sectional area of the flux path is 2×10^{-4} m², what value of i_{dc} is required to limit the speaker's input impedance to $j16$ Ω?

7.7 Show that the homopolar generator in Section 7.5 can be used as a motor by computing the torque T_e for a current i_a.

7.8 A 10-MJ homopolar disk machine can fully discharge in 10 ms. If it is initially rotating at 1000 rpm, is made of an aluminum disk of thickness 10 cm on an axle of radius 2 cm, and is excited by a uniform magnetic field $B = 1.0$ W/m², find the approximate diameter of the disk.

7.9 Calculate the maximum torque for a 120-V six-pole single-phase reluctance motor, ignoring squirrel-cage effects, if the reactances x_d and x_q are

$$x_d = 125 \ \Omega$$
$$x_q = 80 \ \Omega$$

7.10 Determine the exact form of the voltage phasor V in (7.55) as a function of the slip s.

7.11 Draw a circuit diagram connecting a selsyn motor and generator in a configuration so that $\theta_m = -\theta_g$.

7.12 Draw a circuit diagram of a selsyn differential generator arranged to subtract its shaft angular position from that of a selsyn generator, and provide the difference signal as input to a selsyn motor.

7.13 Draw a diagram and circuit for the simple fly-by-wire proportional control system of Example 7.5, but with a differential transformer and selsyn motor in place of the differential motor. Assume that an amplifier is used between the motor and the differential transformer.

7.14 For a permanent-magnet stepper motor, prove that (7.88) in the text is correct, assuming the concept of inductance as the ratio of λ/i. (HINT: Go back to Section 2.4.)

7.15 What tooth pitch (in degrees) is required for an eight-pole PM-hybrid stepper motor if it is to have the baseline capability of 1200 steps/rev? Assuming the motor driver has the capability of pulsing two adjacent coils simultaneously, what is the maximum number of steps per revolution possible for a single stack?

7.16 A certain linear induction motor has six poles spaced 24 in. apart in the vehicle and utilizes a solid aluminum track. If it is excited with 60-Hz three-phase power and end effects are ignored, find:
a. The synchronous speed in mph.
b. The speed in mph at 3.5% slip.
c. The approximate efficiency at this value of slip.

7.17 Draw the configuration of a linear induction machine using two three-phase six-pole single-sided LIMs, which will produce oscillatory linear motion over a given distance d, and explain how it operates. (HINT: How can you stop the rotor and make it reverse?)

7.18 Draw the thrust–speed curves for the separately excited, series, and shunt linear dc motors.

7.19 A simple static attraction-levitation system consists of 12 coils of 1000 turns each connected in series. The coils are configured as in the accompanying figure. If these coils are attached to an object so that the total mass is 1000 kg (1 metric ton), how much current is required to lift it to a sustained height z of 2 cm above the surface on which it is resting? Assume that the coils are attracted to a sheet of iron of infinite permeability and that they are wound on cores which are also of infinite permeability. The cross-sectional area of the flux path can be assumed to be a uniform 0.01 m².

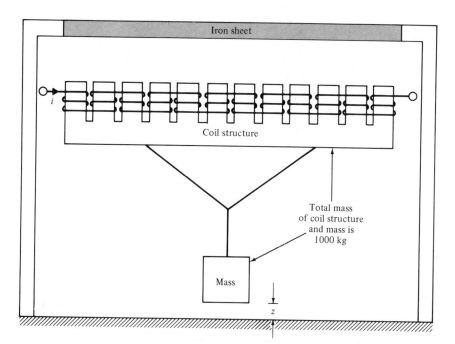

Iron sheet

Coil structure

Total mass of coil structure and mass is 1000 kg

Mass

z

7.20   Find an expression for the levitating force F_e on the static two-coil system described in (7.122) to (7.124) in terms of air-gap length z and cross-sectional area A, core reluctance $\mathcal{R}_{\text{core}}$, and currents i_f and i_m.

7.21   If a permanent magnet of flux density B_p were added in series-aiding with the magnetic circuit in Problem 7.20, find the levitating force F_e and compare it to the previous results (both Problem 7.20 and the text). Assume that the permanent magnet can be represented by one turn of wire carrying current i_p, where

$$i_p = \frac{kB_p}{\mu_0} \quad (k = \text{constant})$$

(HINT: Analyze using coenergy and coupled coil theory.)

CHAPTER 8

Advanced Applications

The fundamental definitions and concepts required to analyze elementary electromechanical energy conversion devices were presented in Chapters 1 and 2. The conventional rotating machines most commonly encountered were introduced in Chapters 3, 4, and 5, where the emphasis was on important steady-state characteristics. Except for the dc machine, rotor dynamics were considered only where a "quasi"-steady-state analysis was deemed appropriate. A more general approach to the analysis of ac machines was presented in Chapter 6. In that chapter, the emphasis was on determination of the complete instantaneous current, voltage, and torque waveforms, including transient effects. In essence, Chapters 1 through 6 provide the necessary background material for generalized analysis of electromechanical energy conversion devices. In Chapter 7 a number of specific devices not generally classified as "conventional" rotating machinery were considered. In this chapter we consider applications generally requiring a more advanced analytical approach. Computer simulation of electromechanical device behavior is an important tool that permits detailed examination of device behavior in a practical manner. Computer simulation techniques will be demonstrated where appropriate. For the most part, the application examples presented are chosen to point out important considerations that must be addressed when devices are utilized in an integrated system.

THE POWER TRANSFORMER

In Chapter 1 we established a suitable model for the power transformer for steady-state operation at a particular frequency. Various aspects of the model must be reexamined in order to address a broader class of problems. In general, the element values in the model are frequency dependent. Distributed capacitance, not previously considered, affects performance at very high frequencies and must be considered in some studies. Magnetic saturation of the core introduces nonlinearities that often cannot be neglected.

Attempts to develop a perfectly general representation for the transformer, suitable for every conceivable analysis, lead to extremely complex models. Such models do not yield results considerably different from the results achieved with simpler models for a particular study of interest if the simpler model is chosen with care. Many practical problems involving switching transients result in currents or voltages whose frequency spectrum is relatively limited. In particular, exponentially damped sinusoids at 60 Hz tend to have significant frequency components predominantly in the vicinity of 60 Hz. Thus a transformer model whose parameters are determined at 60 Hz, and assumed to be invariant with frequency, will yield reasonable results if voltages and currents have frequency components predominantly in the vicinity of 60 Hz. The leakage inductance and winding resistance parameters in the series branches of the transformer equivalent circuit can generally be treated this way.

The shunt impedance branches in the equivalent circuit are somewhat more difficult to model in that they specifically depend on the core material, which is subject to magnetic saturation and in general exhibits the hysteresis phenomenon. Saturation introduces a nonlinearity and the hysteresis phenomenon causes a relationship between B and H (or equivalently λ and i) such that B is a multivalued function of H which depends on past performance of the device. We will examine applications that are very much dependent on the shunt impedance branches of the equivalent circuit and we must, accordingly, examine core behavior in greater detail. To begin, recall that the shunt impedance branches are critical in describing the behavior of the transformer if one of its windings is left an open circuit. If we assume relatively low series branch impedances, the equivalent circuit of Figure 8.1 might be chosen to represent the transformer if one winding is left an open circuit.

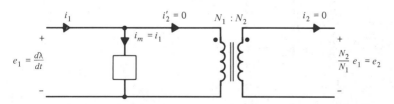

FIGURE 8.1 Equivalent Circuit for a Power transformer with an Open-Circuited Secondary. Impedances of the series branches are assumed to be negligible compared to those of the shunt branches

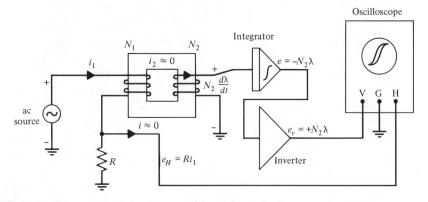

FIGURE 8.2 Method for Experimentally Determining the λ Versus i Curve for a Transformer

A λ versus i curve for the transformer can be determined experimentally with a setup similar to that illustrated in Figure 8.2. In the figure, the input to the integrator is assumed to present a relatively high impedance to the transformer secondary such that $i_2 \approx 0$. Similarly, the horizontal input to the oscilloscope is assumed to present a high impedance so that the voltage e_H is directly proportional to primary current i_1. As shown in Chapter 1, the shape of the λ versus i curve will be similar to the $B\text{--}H$ curve in that (with suitable simplifying assumptions) $\lambda = N\varphi = NA_cB$ and $i = H\ell_c/N$, where A_c is the cross-sectional area of the core and ℓ_c is the mean length of the flux path in the core. It was also shown in Chapter 1 that for sinusoidal excitation the shape of such curves is strongly dependent on the frequency of excitation. Figure 1.33 is repeated here for convenience as Figure 8.3 to illustrate this.

This behavior is consistent with the approximate relationships presented in Chapter 1 for eddy current and hysteresis losses. These relationships are also repeated here for convenience. Losses due to eddy currents per cycle are given approximately by

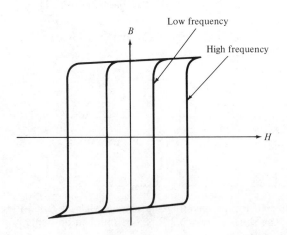

FIGURE 8.3 B versus H at Different Frequencies

$$w_e = K_e f (B_{max} t_\ell)^2 \qquad \text{J/m}^3 \qquad (8.1)$$

where f is frequency and t_ℓ is lamination thickness. The constant K_e depends on conductivity of the core material as shown in (1.64) and (1.65). Losses due to hysteresis per cycle are given approximately by

$$w_h = K_h (B_{max})^n \qquad \text{J/m}^3 \qquad (8.2)$$

Recall that the area inside the B–H loop is proportional to the total energy loss each cycle. Consideration of (8.1) and (8.2) shows that the energy losses per cycle due to eddy currents would increase linearly with frequency, whereas the losses due to hysteresis would be independent of frequency. Thus a loop determined at very low frequency would reflect primarily hysteresis losses. Now consider *power* losses due to eddy currents alone:

$$p_e = f w_e = K_e (f B_{max} t_\ell)^2 \qquad \text{W/m}^3 \qquad (8.3)$$

Thus

$$P_e = K (f B_{max})^2 \qquad \text{W} \qquad (8.4)$$

where the constant K can be chosen to account for the volume of the core. Now if a fixed shunt resistance R is used to model eddy-current effects, losses for sinusoidal excitation would be given by

$$P = \frac{E^2}{R} \qquad (8.5)$$

where E is the rms value of the terminal voltage. Thus

$$E = \frac{\omega N \Phi_{max}}{\sqrt{2}} = \frac{2\pi f N A_c B_{max}}{\sqrt{2}} = 4.44 f N A_c B_{max}$$

[see (1.46)–(1.48)]. Thus (8.5) could be expressed as

$$P = K' (f B_{max})^2 \qquad \text{W} \qquad (8.6)$$

Comparison of (8.4) and (8.6) reveals that a shunt resistance could be chosen to correctly model the eddy-current losses as given by (8.1) and (8.3) assuming sinusoidal excitation at a particular frequency. This tentative conclusion is verified experimentally in an elegant way for a distribution class transformer in reference 1. The shunt branch shown in Figure 8.1 might thus be modeled as shown in Figure 8.4.

In Figure 8.4 a new symbol is introduced in order to represent the nonlinear inductor required to account properly for the effects of saturation and hysteresis. If hysteresis effects cannot be neglected, the model for the nonlinear inductor shown in Figure 8.4 is necessarily complex. References 1 and 2 suggest some possibilities for simulating hysteresis effects. If, over some assumed frequency range, hysteresis power losses are negligible compared to eddy-current losses, it will be sufficient to model the effects of saturation only. This would imply a single-valued relationship between λ and i. The shape shown in Figure 8.5 is an example. A nonlinear inductor with a characteristic similar to that shown in the figure in parallel with a simple linear shunt resistance will yield λ versus i loops which are reasonable provided that hysteresis losses are small compared

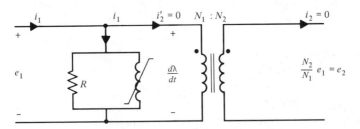

FIGURE 8.4 Equivalent Circuit for a Power Transformer with an Open-Circuited Secondary. The shunt resistance R accounts for eddy-current effects. Saturation and hysteresis effects are accounted for by the nonlinear inductor

to eddy-current losses. It may be helpful here for the reader to review Example 1.4, where the λ versus i loop is developed for the special case of a linear inductor in parallel with a resistance.

The foregoing discussion is based on the simplified expression [(1.65) repeated in this chapter as (8.3)] for eddy-current losses developed in Chapter 1. That development was based on an assumption of uniform flux density over the lamination cross section. The assumption is valid for low frequencies, where skin depth is relatively large compared to lamination thickness. At higher frequencies, the flux penetration in the core material is significantly reduced. It is instructive to consider the ramifications of a solution of the one-dimensional diffusion equation assuming linear material for the core laminations. Figure 8.6 shows a magnetic circuit and a cross-sectional view of its laminated material. If variation in the x and z directions (as defined in the figure) is neglected, a one-dimensional fields problem can be defined. Figure 8.7 shows a single lamination for the magnetic circuit shown in Figure 8.6. Consider the variation of the z component of the magnetic field intensity H as a function of y and time t. It is not difficult to show that $H = H(y, t)$ must satisfy (references 3 and 4)†

$$\frac{\partial^2 H(y, t)}{\partial y^2} = \mu\sigma \frac{\partial H(y, t)}{\partial t} \tag{8.7}$$

†This important relationship is often described as the one-dimensional diffusion equation. It is useful in explaining behavior when an assumption of uniform flux density is inappropriate.

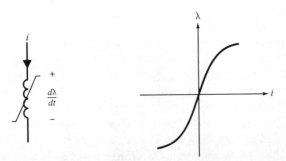

FIGURE 8.5 Nonlinear Inductor and Its Associated λ Versus i Curve. The curve accounts for saturation but neglects hysteresis effects

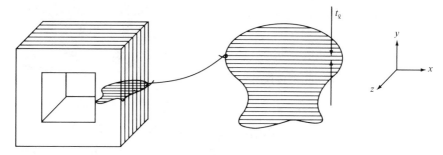

FIGURE 8.6 Magnetic Circuit with Laminated Material

where μ is permeability and σ is conductivity for the assumed linear material. Now if sinusoidal time variation is assumed, the partial differential equation given as (8.7) leads to an ordinary differential equation describing the phasor representation of $H(y, t)$. If the phasor representation of $H(y, t)$ is chosen as

$$H(y, t) \leftrightarrow \dot{H}(y) \tag{8.8}$$

it follows that

$$\frac{d^2 \dot{H}(y)}{dy^2} = j\omega\sigma\mu\dot{H}(y) \tag{8.9}$$

If $\dot{H}(y)$ is assumed uniform in the interlaminar regions at H_0, we have

$$\dot{H}\left(-\frac{t_\ell}{2}\right) = \dot{H}\left(+\frac{t_\ell}{2}\right) = \dot{H}_0 \tag{8.10}$$

Equation (8.9) can be solved subject to the boundary conditions imposed by (8.10). We find that

$$\dot{H}(y) = \dot{H}_0 \frac{\cosh \alpha y}{\cosh (\alpha t_\ell / 2)} \tag{8.11}$$

where

$$\alpha = \sqrt{j\omega\sigma\mu} = \frac{1}{\delta} + j\frac{1}{\delta} \tag{8.12}$$

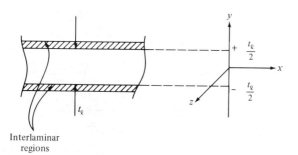

Interlaminar regions

FIGURE 8.7 Single Lamination for the Magnetic Circuit of Figure 8.6

The parameter δ is the "skin depth" and it follows from (8.12) that

$$\delta = \sqrt{\frac{2}{\omega\sigma\mu}} = \sqrt{\frac{1}{\pi f \sigma\mu}} \tag{8.13}$$

Substituting in (8.11) using (8.12) gives us

$$H(y) = H_0 \frac{\cosh\left(\dfrac{y}{\delta} + j\dfrac{y}{\delta}\right)}{\cosh\left(\dfrac{t_\ell}{2\delta} + j\dfrac{t_\ell}{2\delta}\right)} \tag{8.14}$$

Note that for low frequencies, the skin depth is relatively large. If $t_\ell/2 << \delta$ (i.e., if the lamination thickness is small compared to the skin depth), we have $H(y) \approx H_0$ in that $\cosh (0 + j0) = 1$. This would imply a reasonably flat distribution for $H(y)$ or complete penetration. A family of curves showing the magnitude of $H(y)$ as a function of y for various values of the ratio $t_\ell/2\delta$ is shown in Figure 8.8.†

The expressions that give $H(y)$ as a function of y [i.e., (8.11) and (8.14)] can be used to find an appropriate impedance which might be used to represent the shunt branch of a transformer assuming linear laminated material. The following expression for the admittance of the shunt branch is developed completely in reference 5. Notation shown here is changed somewhat from that used in reference 5 in order to maintain consistency with our previous development.

$$Y(j\omega) = \frac{1}{j\omega L_{dc}} + \sum_{n=1}^{\infty} \frac{2/L_{dc}}{j\omega + n^2/\tau} \tag{8.15}$$

†The identity $|\cosh (z)|^2 = \sinh^2 x + \cos^2 y$, where $z = x + jy$ is useful in determining the magnitude of $H(y)$ as given by (8.14).

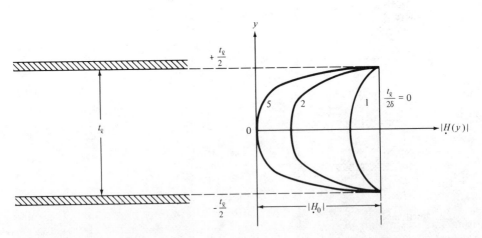

FIGURE 8.8 $|H(y)|$ **Versus y for Various Values of the Ratio $t_\ell/2\delta$. The parameter t_ℓ is lamination thickness and δ is skin depth**

where

$$L_{dc} = \frac{\mu N^2 A_c}{\ell_c} \qquad (8.16)$$

and

$$\tau = \frac{\mu \sigma t_\ell^2}{4\pi^2} \qquad (8.17)$$

Note that the subscript "dc" implies the zero-frequency value. The expression for L_{dc} should be compared to some of our earlier developed simplified expressions for inductance [see (1.27)]. An equivalent circuit is also presented in reference 5 which yields the admittance given by (8.15). In the equivalent circuit of Figure 8.9, the nth resistance is given by

$$R_n = \frac{n^2 L_{dc}}{2\tau} \qquad (8.18)$$

The equivalent circuit of Figure 8.9 shows that an infinite number of parallel lumped element branches are required to represent eddy-current effects if we want a general-purpose representation valid for a wide range of frequencies. As is pointed out in reference 5, the high- and low-frequency values of admittance are easily obtained. For example, at sufficiently low frequencies, the resistance dominates in each RL branch. Neglecting the inductance in such branches, an equivalent resistance is obtained which is the parallel combination of R_1 through R_∞:

$$\frac{1}{R_{eq}} = \sum_n^\infty \frac{1}{R_n} = \frac{2\tau}{L_{dc}} \sum_{n=1}^\infty \frac{1}{n^2} = \frac{\sigma t_\ell^2 \ell_c}{12 N^2 A_c} \qquad (8.19)$$

where

$$\sum_{n=1}^\infty \frac{1}{n^2} = \frac{\pi^2}{6} \qquad (8.20)$$

The value of equivalent resistance given by (8.19) is the value we obtain if we

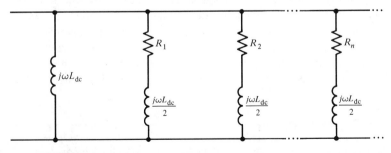

FIGURE 8.9 Equivalent Circuit That Yields the Admittance Given by (8.15) (From J. Avila-Rosales and F. L. Alvarado, "Nonlinear Frequency Dependent Transformer Model for Electromagnetic Transient Studies in Power Systems," *IEEE Transactions on Power Apparatus and Systems*, Vol. PAS-101, pp. 4281–4288, November 1982. © 1982 IEEE

determine a single shunt resistance R based on the approximate expressions for eddy-current losses which assume a uniform flux density distribution (see Problem 8.1). Additional worthwhile insights are presented in reference 5 as well as some methods for incorporating saturation and hysteresis effects in the equivalent circuit of Figure 8.9.

EXAMPLE 8.1 Find a criterion in terms of skin depth δ and lamination thickness t_ℓ which could be used to determine situations where the equivalent circuit of Figure 8.9 could be simplified to a single shunt inductance in parallel with a single shunt resistance.

As discussed above, for sufficiently low frequencies, the resistance dominates in each RL branch of the circuit of Figure 8.9 and the circuit effectively becomes a simplified circuit consisting of an inductance in parallel with a resistance. Examination of Figure 8.9 and the expression for R_n given as (8.18) reveals that if $R_1 \gg \omega L_{dc}/2$ in the first RL branch, we will have $R_n \gg \omega L_{dc}/2$ in the nth RL branch in that $R_n > R_1$. Thus all branches would be predominately resistive if

$$R_1 \geq \frac{10\omega L_{dc}}{2}$$

Equation (8.18) would then require that

$$\frac{L_{dc}}{2\tau} \geq 10\omega \frac{L_{dc}}{2}$$

Substituting using (8.17) gives us

$$\frac{4\pi^2}{\mu\sigma t_\ell^2} \geq 10\omega$$

Now rearrange and substitute using the relationship for skin depth given by (8.13):

$$\delta^2 = \frac{2}{\omega\sigma\mu} \geq \frac{5}{\pi^2} t_\ell^2$$

Thus we should require that

$$\delta \geq 0.7118 t_\ell$$

This criterion should be related to Figure 8.8 in order to see the nature of the distribution of $|H(y)|$ if it is satisfied. Note that the criterion we have developed above would require that $t_\ell/2\delta$ be ≤ 0.702.

Our discussion up to this point suggests that the equivalent circuit shown in Figure 8.10 will be suitable for determination of conditions at the terminals of the transformer in many practical application problems involving transients. It is important to recognize that, in many cases, a simpler model will suffice. For example, in calculating short-circuit currents, the shunt impedance branches have little effect and can generally be neglected. The criterion developed in

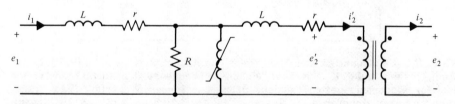

FIGURE 8.10 Equivalent Circuit for a Power Transformer

ADVANCED APPLICATIONS

Example 8.1 illustrates one means of determining when a more detailed representation will be required.

A phenomenon sometimes referred to as transformer inrush current is strongly dependent on the transformer shunt impedance branches. When a transformer is initially energized, a transient inrush current may result depending on the time of circuit breaker closing. The inrush current can be 6 to 10 times higher than the transformer rated current. This initially high current generally decays to normal levels which would be predicted on the basis of a steady-state analysis. A major concern is the possibility that protective devices may remove the transformer from service if an abnormally high current is sensed before a steady-state condition is reached. Increasing the time delay of protective devices is a possible corrective course of action to avoid undesirable "tripping" due to inrush currents.

The inrush phenomenon can be explained by considering energization of an open-circuited transformer. To begin, consider only the nonlinear magnetizing inductance of the transformer assuming negligible eddy-current and hysteresis losses. The equivalent circuit and magnetization curve might be as shown in Figure 8.11. In the figure, variables are assumed to be expressed on a normalized or per unit basis and the change of variables introduced in Chapter 6 ($\psi = \omega_s \lambda$) relating ψ and λ is employed. Note that the magnitude of the voltage source is 1.0, corresponding to energization at rated voltage. The ideal transformer shown in previous equivalent circuits (and by dashed lines in Figure 8.11) is not required if base values are suitably chosen. After the switch is closed we have

$$\frac{1}{\omega_s} \frac{d\psi}{dt} = \sin (\omega_s t + \varphi) \tag{8.21}$$

An expression for $\psi(t)$ is easily obtained by integration:

$$\psi(t) = -\cos (\omega_s t + \varphi) + \psi_0 \tag{8.22}$$

The constant ψ_0 must be chosen to satisfy initial conditions on flux linkages. Assuming that $\psi(0) = 0$, $\psi_0 = \cos \varphi$ and we have

$$\psi(t) = -\cos (\omega_s t + \varphi) + \cos \varphi \tag{8.23}$$

The constant term in (8.23) clearly depends on the angle φ (or, equivalently,

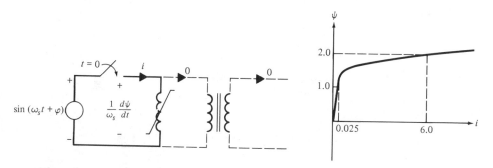

FIGURE 8.11 Simplified Equivalent Circuit for an Open-Circuited Transformer and a Nonlinear Magnetization Curve Relating ψ and i

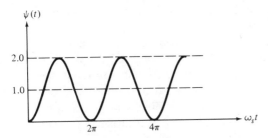

FIGURE 8.12 ψ(t) versus t for the Equivalent Circuit of Figure 8.11 Assuming That φ = 0

the time of circuit breaker closing). This "offset" in $\psi(t)$ would normally decay to zero so that, in the steady state, $\psi(t)$ would exhibit pure sinusoidal behavior. In this case it does not decay because losses have been neglected. Now consider the more-or-less typical magnetization curve of Figure 8.11. Normal steady-state levels of maximum flux would be on the order of 1.0, resulting in perhaps 0.025 or 2.5% magnetizing current. If the angle φ is zero, the offset term is 1.0, resulting in peak values of flux of 2.0 per unit. This is illustrated in Figure 8.12. For the magnetization curve of Figure 8.11, this would result in peak values of current of 6.0, or six times rated transformer current. The nonlinear aspect of the magnetization curve will cause the current waveform to be distorted as well. The actual current waveform could be obtained graphically by following a procedure similar to that illustrated in Figure 1.25.

In order to show the decay of the inrush current, an improved representation of the transformer is required. The equivalent circuit shown in Figure 8.13 would permit determination of a more representative inrush current waveform.

EXAMPLE 8.2 Find a set of equations in state-variable form that could be used to examine the transformer inrush current phenomenon. Choose the equivalent circuit structure shown in Figure 8.13. Assume an empirically determined relationship between ψ and i_m of the form

$$i_m = \alpha\psi + \beta\psi^9$$

Note that the assumed relationship between ψ and i_m is single-valued and thus necessarily neglects hysteresis effects.

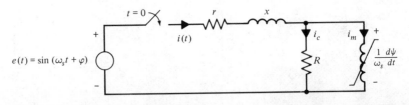

FIGURE 8.13 Improved Equivalent Circuit for Examination of the Inrush Current Phenomenon

If we choose current i and flux linkages ψ as state variables, the desired relationships are easily obtained. We have

$$e(t) = ir + \frac{x}{\omega_s} \frac{di}{dt} + \frac{1}{\omega_s} \frac{d\psi}{dt} \tag{1}$$

$$i_c R = \frac{1}{\omega_s} \frac{d\psi}{dt} \tag{2}$$

The current i_c can be expressed in terms of the chosen state variables:

$$i_c = i - i_m = i - \alpha\psi - \beta\psi^9 \tag{3}$$

Substitute in (2) using (3) to obtain

$$\frac{d\psi}{dt} = \omega_s R(i - \alpha\psi - \beta\psi^9) \tag{4}$$

Now eliminate $d\psi/dt$ in (1) using (4) and rearrange:

$$\frac{di}{dt} = \frac{\omega_s}{x}[R\alpha\psi + R\beta\psi^9 - (r + R)i] + \frac{\omega_s}{x} \sin(\omega_s t + \varphi) \tag{5}$$

Equations (4) and (5) are the desired result. Solution by numerical methods is straightforward, as described in Appendix G.

If winding resistance losses are 5% during normal transformer operation, an appropriate per unit value of r would be $r = 0.05/2 = 0.025$. If core losses are 3% during normal operation, an appropriate value of R would be $1/0.03 = 33.33$. A typical value for leakage reactance x might be 0.05. The parameters α and β in the empirical relationship for the magnetization curve can be determined by identifying two points on the curve. If the curve of Figure 8.11 is chosen,

$$0.025 = \alpha(1) + \beta(1)^9$$
$$6.0 = \alpha(2) + \beta(2)^9$$

and $\alpha = 0.0133$, $\beta = 0.0117$. The waveform for the first few cycles following energization, assuming that $\varphi = 0$, is shown in Figure 8.14.

Ferroresonance in transformers is a phenomenon that is very dependent on the transformer shunt impedance branches. Ferroresonance is a resonance phenomenon involving a saturable iron core inductance. Typically, it occurs when an open-circuited transformer (or a reactor) is capacitively coupled to a source of power. For example, if series capacitors are used to compensate a transmission line, energization of an open-circuited transformer at the end of the line may produce the phenomenon. This particular situation is readily anticipated and generally it is easy to prevent or interrupt a ferroresonant condition. Reference 6 provides additional background in this regard. The more difficult situations to contend with are those where the phenomenon occurs due to inadvertent coupling between an energized circuit and one that is supposedly deenergized. Reference 7 describes a situation where the capacitive coupling between two parallel transmission lines was sufficient to sustain ferroresonance in a transformer connected to one of the lines which, aside from the capacitive coupling, was otherwise disconnected from any source of power. The condition caused overheating in the transformer before being detected and corrected. Prolonged ferroresonance can cause more severe overheating and perhaps degradation of insulating materials.

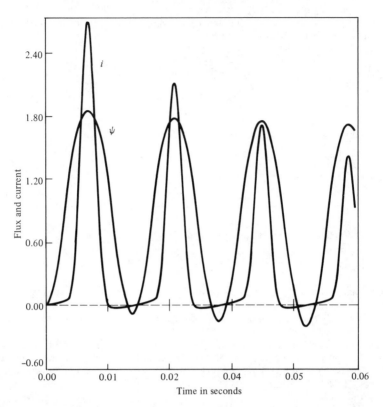

FIGURE 8.14 Transformer Inrush Current Waveform and Flux Waveform as Determined by the Simulation Equations and Data of Example 8.2. Euler's method was used to solve the set of equations numerically

The equivalent circuit of Figure 8.15 illustrates a situation where ferroresonance could occur. Obviously, three-phase systems will present more complicated circuit configurations where line-to-line and line-to-ground capacitance must be considered. Referring to Figure 8.15, the nonlinear inductance would result from an open-circuited transformer as in our previous examples. The situation is somewhat similar to what we encountered when examining transformer inrush currents, in that values of flux linkages ψ above the "knee" of the transformer magnetization curve can cause currents well in excess of rated transformer current. The capacitive coupling may act to sustain an oscillatory mode of operation with excessive peak currents. Consider the magnetization curve of Figure 8.16 and a piecewise linear approximation to the curve. It is

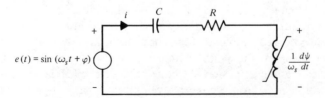

FIGURE 8.15 Simplified Circuit for the Analysis of Ferroresonance

ADVANCED APPLICATIONS

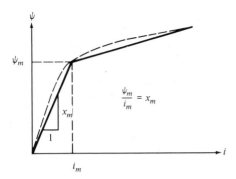

FIGURE 8.16 Magnetization Curve and a Piecewise Linear Approximation to It

possible to identify situations where the transformer will necessarily be driven into saturation with potentially excessive currents. This can be done by assuming that operation will be restricted such that $|\psi| < \psi_m$ or, equivalently, $|i| < i_m$. This would imply that $\psi = x_m i$. The resulting circuit is linear and a steady-state solution is easily obtained. If the solution reveals that maximum values of ψ are in fact greater than ψ_m, operation in the saturated region will necessarily occur. A further effort (*not* based on a linear analysis) is then required to determine the nonsinusoidal waveforms that will result. Unfortunately, it is not possible to rule out a ferroresonant condition if the converse is true, in that a switching transient may result in operation in the saturated region, which is then sustained by capacitive coupling. Because of the many factors that influence the onset of ferroresonance, preventive measures employed often involve "loading" the transformer secondary of an out-of-service transformer in some manner to minimize the effect of the nonlinear transformer magnetization curve. One such approach is described in reference 7. In essence, the "load" impedance is in parallel with and generally much less than the impedance jx_m. The effective impedance is then approximately that of the load impedance. It is noteworthy that ferroresonance rarely occurs for transformers when actually in service.

In order to illustrate some of the foregoing notions we will consider an example.

EXAMPLE 8.3 A transformer connected to a bus that has shunt capacitive compensation is to be removed from service by opening a circuit breaker. The situation is illustrated by the equivalent circuit shown in Figure 8.17. The capacitance C_d shown in

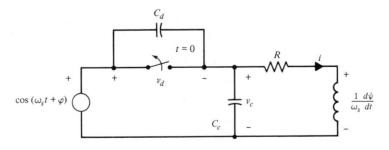

FIGURE 8.17 Equivalent Circuit for Example 8.3

the figure is the stray capacitance across the open contacts of the circuit breaker. The capacitance C_c is the shunt compensation. Assume that the secondary of the transformer has been cleared prior to time $t = 0$. The nonlinear inductor and resistor shown in the figure constitute an equivalent circuit for the unloaded transformer as viewed from the primary. Analyze the circuit for ferroresonance.

In order to demonstrate how a ferroresonant condition might occur, assume a piecewise linear magnetization curve as shown in Figure 8.16. Now assume further that operation will be restricted to $|\psi| < \psi_m$ so that $\psi = x_m i$. For the moment, neglect the resistance R and the stray capacitance C_d. With these assumptions, after the switch is opened we will have

$$\frac{1}{\omega_s}\frac{d\psi}{dt} = v_c$$

and

$$\frac{C_c dv_c}{dt} = -i = -\frac{\psi}{x_m}$$

The two relationships above describe a simple linear oscillator as would be expected for a parallel linear LC circuit without damping. It is not difficult to show that

$$\psi(t) = A \cos \omega_n t + B \sin \omega_n t$$

where

$$\omega_n = \frac{1}{\sqrt{L_m C_c}}$$

and

$$L_m = \frac{x_m}{\omega_s}$$

The constants A and B depend on the angle φ. It is left as an exercise at the end of the chapter to show that $A = \sin \varphi$ and $B = (\omega_s/\omega_n) \cos \varphi$. Now, suppose that $\varphi = 0$. Flux linkages will be given by

$$\psi(t) = \frac{\omega_s}{\omega_n} \sin \omega_n t$$

If we assume that ψ_m (see Figure 8.16) is 1.0, our solution will be valid only if $\omega_s/\omega_n \le 1.0$. If $\omega_s/\omega_n > 1.0$, we necessarily will enter the saturated region of the magnetization curve, possibly with excessive magnetizing currents. The relevant waveforms must be determined by methods that account for the nonlinearity of the magnetization curve if this occurs. In the absence of capacitive coupling, the resistance R would eventually cause oscillations to decay to zero. By assuming a small value of resistance, such as what we might have with a reasonably efficient transformer, little power will have to be transferred via capacitive coupling to sustain a ferroresonant condition. A determination as to whether or not a ferroresonant condition will actually occur can only be made following an analysis that includes the parameters C_d and R.

Given the nonlinear nature of the problem, computer simulation is an attractive option. Simulation relationships can be obtained by finding a set of differential equations in state-variable form. Assume that $i = f(\psi)$, where $f(\psi)$ is a suitable nonlinear relationship for the magnetization curve. From Figure 8.17 it is clear that for $t \ge 0$:

$$\cos (\omega_s t + \varphi) = v_d + v_c \tag{1}$$

$$\frac{1}{\omega_s}\frac{d\psi}{dt} = -iR + v_c \tag{2}$$

$$C_c \frac{dv_c}{dt} = C_d \frac{dv_d}{dt} - i \tag{3}$$

Only two state variables are required to describe the system completely in this case. If we choose ψ and v_c as state variables, dv_d/dt can be eliminated in (3) using (1) and i can be eliminated in (2) and (3) using the nonlinear relationship for the magnetization curve. We find

$$\frac{d\psi}{dt} = -\omega_s R f(\psi) + \omega_s v_c \tag{4}$$

$$\frac{dv_c}{dt} = -\frac{f(\psi)}{C_c + C_d} - \frac{C_d}{C_c + C_d} \omega_s \sin(\omega_s t + \varphi) \tag{5}$$

Equations (4) and (5) can be solved using the techniques described in Appendix G in order to identify parameter values and initial conditions which would produce excessive currents. The existence of such a condition will depend on the angle φ (the time of switch opening) or, equivalently, the initial conditions on v_c and i as well as the parameter values in the circuit. Although solution of (4) and (5) is straightforward using numerical methods, the uncertainties involved will ultimately suggest that we operate in such a way as to avoid the problem if possible; perhaps, in this case, by removing the shunt compensation at light-load conditions.

We have thus far neglected distributed and stray capacitances associated with the transformer itself. Insofar as conditions at the terminals of the transformer are concerned, the effects of capacitance will be minimal except at relatively high frequencies. Thus our previously developed equivalent circuits will be adequate in most power transformer application problems. Small power transformers used in electronic circuits may require more sophisticated models which reflect the effects of capacitance. For example, transformers that couple an audio amplifier to a speaker must exhibit relatively flat frequency-response characteristics over a desired audio-frequency range. In such cases, capacitances can play a significant role at the high-frequency end of the desired range. Because the capacitances are generally "stray" capacitances, say, between portions of a winding and an enclosure, they are difficult to quantify. It may be necessary to determine such parameters by direct measurement. Reference 8 provides some approximate relationships that can be used to estimate various capacitances.

Distributed interwinding capacitance and distributed winding-to-ground capacitance are critical in establishing the initial winding voltage distribution due to a voltage surge at the terminals of a power transformer. Depending on the relative magnitudes of these parameters, most of the initial voltage across the terminals of a winding may be developed across a relatively small number of winding turns, resulting in excessive turn-to-turn voltage stresses. The issue is of concern in the design of winding insulation. This subject is covered in references 9 and 10 and will not be pursued further here.

Our treatment of the power transformer in this chapter has focused primarily on the development of suitable transformer models and their application in a variety of practical problems. It is by no means an exhaustive treatment but rather one that highlights aspects of the transformer which must be considered in explaining or analyzing various aspects of transformer behavior. These same issues must be addressed in the analysis of electromechanical energy conversion devices. Indeed, aside from the relative motion of component parts inherent in the conversion of energy from electrical form to mechanical form, or vice versa, electromechanical energy conversion devices and transformers are conceptually the same, differing only in configuration. Thus, when relatively simple lumped-circuit models for machines do not adequately account for observed behavior, it may be helpful to recall some of the refinements that were required in order to achieve an adequate equivalent circuit for the power transformer.

8.3

SINGLY EXCITED ELECTROMECHANICAL ENERGY CONVERSION DEVICES

Many practical devices fall in the category of singly excited systems. In essence, these devices have a single winding and typically a magnetic circuit whose reluctance varies with translational or rotational displacement of a movable portion of the magnetic circuit. Control relays, electromagnetically actuated switches sometimes referred to as contactors, and solenoid actuated brakes, clutches, and so on, are examples. With canny foresight, we considered a system in Example 2.8 that can be regarded as a prototype for singly excited systems. Figure 2.18 is repeated here as Figure 8.18 for convenience. The system of Figure 8.18 is generally representative in the sense that it requires consideration of:

1. A suitable relationship between flux linkages, current, and displacement.
2. A voltage relationship for the winding or, equivalently, an appropriate circuit equation.
3. A relationship for developed electrical force in terms of circuit variables and displacement.
4. An equation of motion for the movable member in the magnetic circuit.

In Example 2.8, a linear magnetic circuit was assumed. Suitable simplifying assumptions resulted in the following relationship between flux linkages, current, and displacement.

$$\lambda = L(x)i \tag{8.24}$$

$$= \left(\frac{N^2}{R_{\text{core}} + 2x/\mu_0 A}\right)i$$

This relationship implies the set of curves (straight lines) shown in Figure 8.19. With these assumptions, a set of equations in state-variable form was developed in Example 2.8:

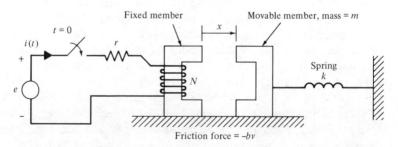

FIGURE 8.18 Singly Excited Electromechanical Energy Conversion Device

ADVANCED APPLICATIONS

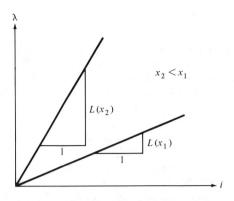

FIGURE 8.19 Graphical Depiction of (8.24). A straight-line relationship between λ and i, for a particular value of x, implies a linear magnetic circuit

$$\frac{dv}{dt} = \frac{\lambda^2}{mN^2\mu_0 A} - \frac{b}{m}v - \frac{k}{m}(x - x_0) \tag{8.25a}$$

$$\frac{dx}{dt} = v \tag{8.25b}$$

$$\frac{d\lambda}{dt} = -\frac{rR_{\text{core}}}{N^2}\lambda - \frac{2r}{N^2\mu_0 A}x\lambda + e \tag{8.25c}$$

It was left as an exercise for the reader (Problem 2.16) to show that if current i rather than flux linkages λ was chosen as a state variable, the following set of equations could be written:

$$\frac{dv}{dt} = -\frac{L^2(x)i^2}{mN^2\mu_0 A} - \frac{b}{m}v - \frac{k}{m}(x - x_0) \tag{8.26a}$$

$$\frac{dx}{dt} = v \tag{8.26b}$$

$$\frac{di}{dt} = -i\frac{r}{L(x)} + \frac{2L(x)}{\mu_0 AN^2}iv + \frac{e}{L(x)} \tag{8.26c}$$

The choice of λ or i as a state variable is arbitrary, but different insights result from the two descriptions. For example, consider (8.26c) if speed v is assumed to be zero or (8.25c) if circuit resistance r is neglected. In each case a familiar relationship results.

With the assumptions above, equation set (8.25) or (8.26) is sufficient to characterize the system. Products of state variables or products of state variables and functions of state variables cause the equations to be characterized as non-linear (even though the magnetic circuit was described as linear; see Figure 8.19) and solution by numerical methods as described in Appendix G is an attractive option.

The magnetic circuit of Figure 8.18 is similar to the magnetic circuit for an unloaded transformer differing only by the variable air gap. It is thus reasonable to consider refinements to the model similar to the refinements developed in

Section 8.2 for the power transformer. Following arguments similar to those made in Section 8.2, we will assume that:

1. Eddy-current losses are sufficiently greater than hysteresis losses, for the frequency range of interest, to justify treating all core losses as eddy-current losses.
2. Eddy-current effects are adequately modeled by a linear circuit element or elements.
3. Skin depth for the frequency range of interest is sufficiently greater than core cross-sectional dimensions (or lamination thickness for a device designed to be operated in a 60-Hz system) to assume uniform flux density distributions.

The assumptions above lead us to an equivalent circuit, for a singly excited system, consisting of a resistance in parallel with an inductor. The first assumption further implies that the λ versus i relationship for the inductor will be a single-valued one in that hysteresis effects are not considered. Saturation is accounted for by using appropriate nonlinear relationships between λ and i. Because of the air gap, we necessarily have flux linkages dependent on both current and displacement. Thus

$$\lambda = \lambda(i, x) \tag{8.27a}$$

or

$$i = i(\lambda, x) \tag{8.27b}$$

Equations (8.27) might lead to a set of curves such as that shown in Figure 8.20.

The curves of Figure 8.20 should be compared to those of Figure 8.19 to emphasize the distinction between a linear and a nonlinear magnetic circuit. An equivalent circuit for a singly excited system is shown in Figure 8.21. It is worthwhile to develop a set of equations in state-variable form to describe the system of Figure 8.18 assuming the equivalent-circuit representation of Figure 8.21. We will do this choosing flux linkages as a state variable. First, assume an empirically determined relationship between flux linkages λ, current i, and displacement x of the form given by (8.27b). Then for the equivalent circuit of Figure 8.21,

$$i_m = i_m(\lambda, x) \tag{8.28}$$

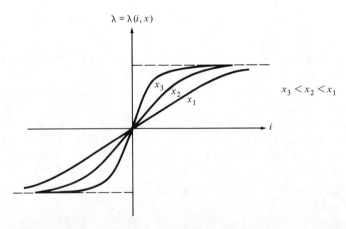

FIGURE 8.20 Graphical Depiction of (8.27a) or (8.27b). A nonlinear relationship between λ and i, for a particular value of x, implies a nonlinear magnetic circuit

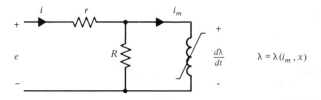

FIGURE 8.21 Equivalent Circuit for a Nonlinear Magnetic Circuit with a Variable Air Gap. The resistance *r* accounts for winding resistance. The resistance *R* accounts for eddy-current effects. Saturation is accounted for by use of a nonlinear inductor. A single-valued relationship between λ and *i* implies that hysteresis effects are neglected

Energy $W = W(\lambda, x)$ is given by

$$W(\lambda, x) = \int_0^\lambda i_m(\tau, x)\, d\tau \qquad (8.29)$$

It may be helpful to consider (8.29) and what it implies in terms of Figure 8.20. The graphical relationship between energy W and coenergy W_c shown in Figure 2.16 should also be reviewed. The electrical force in the direction of increasing x is given by

$$F_e(\lambda, x) = -\frac{\partial W(\lambda, x)}{\partial x} \qquad (8.30)$$

$$= -\frac{\partial}{\partial x} \int_0^\lambda i_m(\tau, x)\, d\tau$$

The equation of motion for the movable member is then

$$m \frac{dv}{dt} = F_e(\lambda, x) - bv - k(x - x_0) \qquad (8.31)$$

From the equivalent circuit of Figure 8.21,

$$e = ir + \frac{d\lambda}{dt} \qquad (8.32)$$

$$= \left(i_m + \frac{1}{R}\frac{d\lambda}{dt} \right) r + \frac{d\lambda}{dt}$$

Since i_m is specifically a function of λ and x, we need only rearrange (8.31) and (8.32) and add the relationship between velocity and displacement to achieve the desired result.

$$\frac{dv}{dt} = \frac{F_e(\lambda, x)}{m} - \frac{b}{m}v - \frac{k}{m}(x - x_0) \qquad (8.33a)$$

$$\frac{dx}{dt} = v \qquad (8.33b)$$

$$\frac{d\lambda}{dt} = -\frac{rR}{r + R}i_m(\lambda, x) + \frac{R}{r + R}e \qquad (8.33c)$$

The empirically determined relationship for $i_m(\lambda, x)$ is required to evaluate equation set (8.33) numerically using the techniques of Appendix G. Problem 2.8 illustrates one such relationship that might be used.

EXAMPLE 8.4 A certain manufacturer of electromagnetically actuated motor starters cites the following data for a NEMA (National Electric Manufacturers' Association) size 5 motor starter:

Coil voltage	120 V, 60 Hz
Coil inrush	3000 VA
Sealed volt-amperes	200 VA
Sealed watts	40 W
Minimum operating voltage	85%
Magnetic circuit air gap	0.25 cm
(in fully open position)	

Determine element values for an equivalent circuit and a suitable relationship between flux linkages, current, and displacement for the inductor of the equivalent circuit. If control wire having a resistance of 2.60 Ω per 1000 ft and negligible reactance is to be used to control the device, determine the maximum permissible length of such wire in

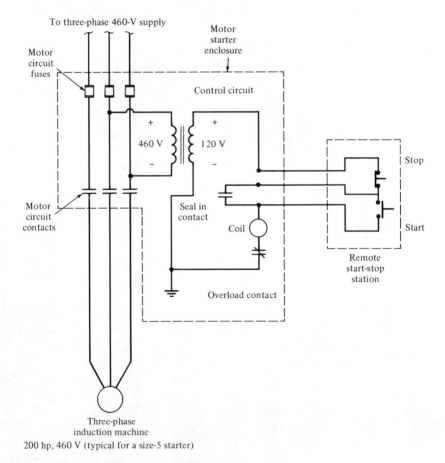

FIGURE 8.22 Elementary Motor Control Circuit

the control circuit so that the device operating voltage is maintained at or above the specified minimum percentage of rated voltage.

A motor starter is an electromagnetically actuated set of contacts (sometimes referred to as a contactor) which is used to start and stop a motor. The device may be thought of as a "switch" whose contacts are designed to switch currents at least as high as the inrush currents that might be anticipated for the motor. A circuit breaker or set of fuses is usually employed in conjunction with the device to provide the capability to interrupt the higher currents that would result if a short circuit were to occur. Figure 8.22 illustrates an elementary motor control circuit.

Referring to Figure 8.22, we see that the motor circuit contacts in the motor supply circuit and the seal-in contact in the control circuit are controlled by the movable member of a magnetic circuit whose coil is shown in the control circuit of the schematic.

Typically, a momentary closure of the start pushbutton contacts will cause the coil to be energized, which causes the magnetic circuit air gap to close. This closes all four contacts and the seal-in contact maintains the control circuit after release of the start pushbutton. The stop pushbutton breaks the control circuit and all contacts are opened by spring action. Vigorous closing and opening of the motor circuit contacts is important to minimize arcing, which reduces contact life. The type of control illustrated is sometimes referred to as "three-wire" control. Note that the motor circuit is a relatively high power circuit with large cross-sectional area conductors and the control circuit is a relatively low power circuit generally utilizing much smaller conductors. The overload contact shown in the control circuit is typically controlled by a temperature-sensitive device which monitors current in one phase of the motor circuit. The device provides overload protection for the motor by interrupting the control circuit in the event that motor current remains above rated motor current for a prolonged period. It is possible to put additional normally closed contacts in series with the stop pushbutton contacts. The additional contacts can be used to interrupt the control circuit in the event that some event occurs which should cause a shutdown. For example, a normally closed contact might be opened by a float switch which senses high water level in a tank being filled by a pump that is driven by the motor.

The data specified are insufficient to develop appropriate nonlinear relationships for the inductor. We will thus assume a linear magnetic circuit. If coil winding resistance is neglected, an appropriate equivalent circuit will be as shown in Figure 8.23. When the magnetic circuit is closed, we have the sealed condition and the inductance $L(x)$ will be at its maximum. The data are similar to those for a no-load transformer test. Assuming 120 V at the terminals of the coil, we find that

$$P = \frac{V^2}{R}$$

Thus

$$R = \frac{V^2}{P} = \frac{120^2}{40} = 360 \ \Omega$$

$$VA = \sqrt{P^2 + Q^2}$$

$$Q = \sqrt{VA^2 - P^2}$$

$$= \sqrt{200^2 - 40^2}$$

$$= 196$$

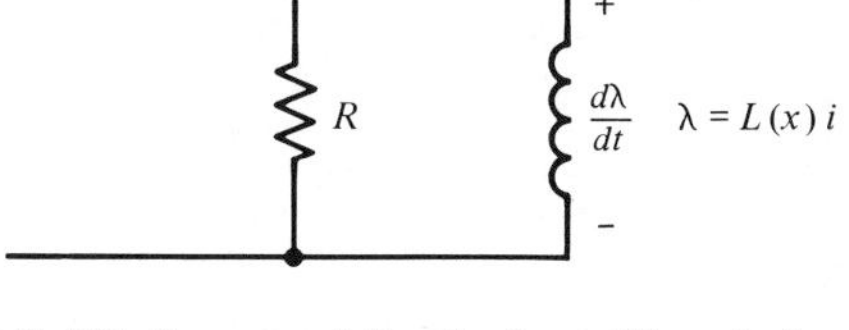

FIGURE 8.23 Assumed Equivalent Circuit for Example 8.4

but

$$Q = \frac{V^2}{X}$$

Thus

$$X = \frac{V^2}{Q} = \frac{120^2}{196} = 73.5 \ \Omega$$

This implies that

$$L_{max} = L(x) \big|_{x=0} = \frac{X}{\omega} = \frac{73.5}{377} = 0.195 \ \text{H}$$

Now assume that when the magnetic circuit is open, the minimum inductance $L_{min} = L(x)\big|_{x=g}$ will be sufficiently small that we can neglect the resistance R.

$$X_{min} = \frac{V}{I} = \frac{V^2}{VA} = \frac{120^2}{3000} = 4.8 \ \Omega$$

This assumption is reasonable in that the current through the resistance R will be $120/360 = 0.333$ A, which is much less than the total inrush current of $3000/120 = 25$ A.

The minimum reactance we have determined implies that

$$L_{min} = L(x)\big|_{x=g} = \frac{X_{min}}{\omega} = \frac{4.8}{377} = 0.0127 \ \text{H}$$

In order to arrive at a suitable relationship between flux linkages, current, and displacement it will be sufficient to find an expression for $L(x)$. Consider (8.24), which applies for the magnetic circuit of Figure 8.18 with suitable simplifying assumptions. This relationship gives

$$L(x) = \frac{N^2}{\mathcal{R}_{core} + 2x/\mu_0 A}$$

Note that $2x/\mu_0 A$ is the reluctance associated with the air gaps in the magnetic circuit and that this reluctance is directly proportional to the gap x. A more general relationship that will apply for an unknown configuration can be written as

$$L(x) = \frac{N^2}{\mathcal{R}_{core} + (x/g) \ \mathcal{R}_{gap}}$$

where in this case $g = 0.25$ cm and $\mathcal{R}_{gap}$ is the air-gap reluctance we have in the fully open position. Now it must be true that

$$\mathcal{R}_{core} = \mathcal{R}_{min}$$

$$\mathcal{R}_{gap} = \mathcal{R}_{max} - \mathcal{R}_{core}$$

$$= \mathcal{R}_{max} - \mathcal{R}_{min}$$

We also have

$$L_{max} = \frac{N^2}{\mathcal{R}_{min}}$$

$$L_{min} = \frac{N^2}{\mathcal{R}_{max}}$$

Thus

$$L(x) = \frac{N^2}{\mathcal{R}_{\text{core}} + (x/g)\,\mathcal{R}_{\text{gap}}}$$

$$= \frac{N^2/\mathcal{R}_{\text{min}}}{1 + \dfrac{x}{g}\left(\dfrac{\mathcal{R}_{\text{max}}}{\mathcal{R}_{\text{min}}} - 1\right)}$$

$$= \frac{L_{\text{max}}}{1 + \dfrac{x}{g}\left(\dfrac{L_{\text{max}}}{L_{\text{min}}} - 1\right)}$$

Substituting for $L_{\text{max}} = 0.195$, $L_{\text{min}} = 0.0127$, and $g = 0.25$ cm, we find that

$$L(x) = \frac{0.195}{1 + 57.42x}$$

where x is assumed to be in centimeters.

In order to determine the maximum permissible length of control wire having a resistance of 2.6 Ω per 1000 ft, recognize that it is the open magnetic circuit condition where $L(x)$ is equal to L_{min} which will produce the greatest voltage drop in the external control wire. The circuit to consider is shown in Figure 8.24. If the minimum operating voltage at the coil V_c must be 85% of the rated voltage, to ensure that the contactor "picks up," we should require that

$$\frac{|V_c|}{|V_s|} = \frac{V_c}{V_s} = \frac{X_{\text{min}}}{\sqrt{r_{\text{line}}^2 + X_{\text{min}}^2}} \geq 0.85$$

If we substitute for X_{min}, the maximum allowable value for r_{line} is found to be

$$r_{\text{line}} = 2.97\ \Omega$$

Thus the maximum length of control wire should be

$$\frac{2.97\ \Omega}{2.60\ \Omega/1000\ \text{ft}} = 1140\ \text{ft}$$

Note that this would imply a maximum distance from the motor starter to the remote start–stop station of $1140/2 = 570$ ft assuming that we have the situation depicted in Figure 8.22. If a greater distance were required, perhaps to accommodate additional normally closed contacts for shutdown (sometimes referred to as "interlocks"), it is generally more economical to use a small relay to control the contactor than to increase the size of the conductors used in the control circuit. In essence, the inrush current of a small relay will be significantly less than that of the contactor coil and will result in much lower voltage drops for the same circuit resistance.

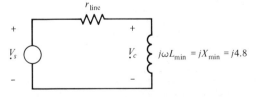

FIGURE 8.24 Equivalent Circuit for the Determination of Maximum Control Wire Length for Example 8.4

THE INDUCTION MACHINE

The induction machine is often the preferred choice in applications requiring a fixed-speed drive. Because of its relatively simple structure, it will have a lower initial cost than a dc or synchronous machine of the same rating. Maintenance requirements are also generally less. The development of efficient solid-state variable-frequency drive systems has significantly increased the range of applications for the induction machine, and we will examine this important area.

The equivalent circuit for the induction machine for steady-state analysis, developed in Chapter 3, is shown in Figure 8.25. In the spirit of the preceding two sections, we start by considering refinements to the equivalent circuit and situations where such refinements might be necessary. Recall that r_s and r_r' are armature winding and rotor winding resistances, respectively. Primed variables or parameters are those referred to the armature winding by an appropriate turns ratio or ratio squared. The reactances $x_{\ell s}$ and $x_{\ell s}'$ are leakage reactances associated with the armature and rotor windings, respectively. The reactance x_m is the magnetizing reactance. It may be helpful to reexamine Figure 3.2, which was the starting point for the development of the equivalent circuit shown in Figure 8.25, and the list of assumptions made at the onset of the development. It is also helpful to recognize the similarity of the equivalent circuit to that of a transformer. For example, if the rotor is blocked, slip s is 1.0 and the circuit suggests a transformer with a short-circuited secondary. If the machine operates unloaded, the slip s will be close to zero, assuming low friction and windage losses, and the equivalent circuit is similar to that of a transformer with an open-circuited secondary. Because of the changing flux in the machine iron, "core" losses are inevitable. These losses can be accounted for by adding a resistance in parallel with the reactance x_m. Alternatively, core losses are often assumed constant and accounted for by including them with friction and windage losses. Blocked-rotor and no-load tests can be used to determine element values for the equivalent circuit of the induction machine. The procedure is quite similar in many ways to the short-circuit and open-circuit tests used to determine transformer parameters (see Problem 3.5).

As discussed in Chapter 3, squirrel-cage motors are generally designed to achieve particular torque–speed characteristics (see Figure 3.14) and current–speed characteristics. The geometry of the rotor bars is critical in this

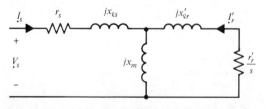

FIGURE 8.25 Steady-State Equivalent Circuit for a Polyphase Induction Machine

regard. In essence, the resistance r'_r and reactance $x'_{\ell r}$ are functions of the frequency of the rotor currents. The rotor resistance at 60 Hz is typically several times higher than that at low frequency. If a blocked-rotor test is used to determine r'_r, the value determined will be suitable for use in determining inrush currents and locked rotor torque, but will be too high for use in determining performance at slip values near rated slip, where the rotor currents are at relatively low frequency. One approach to this dilemma is simply to recognize the requirement to use appropriate values of rotor resistance and reactance in the equivalent circuit of Figure 8.25. The problem is not as complicated, as it might first appear, in that induction machine slip typically varies little from no-load to loads somewhat in excess of rated load (see Table 3.1). This implies that values of $x'_{\ell r}$ and r'_r determined for operation at rated load (relatively low frequency rotor currents) will be suitable for determining performance over a wide range of load conditions. Different values of these parameters should be used in determining blocked rotor performance. Reference 11 describes tests that can be used to determine the two sets of parameter values generally required.

Some machines are designed to have multiple cages each at different radii from the axis of the rotor. An appropriate equivalent circuit is obtained by treating the multiple cages as multiple rotor windings. The equivalent circuit for a triple-squirrel-cage induction machine is shown in Figure 8.26.

If we assume relatively shallow bars, the parameters in the equivalent circuit of Figure 8.26 will not vary significantly with frequency and may be treated as constants. A single-cage configuration with deep bars implies a rotor winding impedance which cannot be matched exactly with a finite number of rotor winding branches as shown in Figure 8.26, but an equivalent circuit with two or three branches can be used to model its performance with reasonable accuracy [11, 12].

Blocked rotor tests are generally used to determine r'_r and $x'_{\ell r}$. In cases where the frequency dependence of r'_r and $x'_{\ell r}$ is important, the tests can be conducted at different source voltage frequencies. Generally, $x_{\ell s}$ is approximately equal to $x'_{\ell r}$ as determined by a blocked rotor test at 60 Hz. Armature resistance r_s can be determined by direct measurements at the armature terminals using a dc source. The magnetizing reactance x_m is generally determined from results of a no-load test.

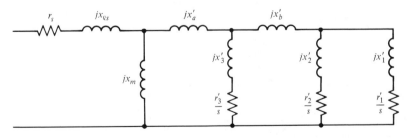

FIGURE 8.26 Equivalent Circuit for an Induction Machine Having a Triple-Cage Configuration

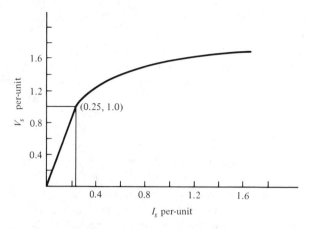

FIGURE 8.27 Armature Voltage Versus Current for an Induction Machine at No-Load

EXAMPLE 8.5 A no-load test conducted on an induction machine yields data as shown in Figure 8.27. Parameter values per unit determined from a blocked rotor test are

$$r_s = 0.03$$

$$r'_r = 0.02 \quad \text{(at 60 Hz)}$$

$$x_{\ell s} = 0.12$$

$$x'_{\ell r} = 0.12 \quad \text{(at 60 Hz)}$$

Determine a value for magnetizing reactance x_m. Assume armature voltage at rated voltage or less. If, for the no-load test, power as measured at the terminals of the machine is 0.050 per unit at 1.0 per unit voltage and 0.039 per unit at 0.8 per unit voltage, estimate core losses at rated voltage and friction and windage losses at speeds near synchronous speed.

The equivalent circuit for the machine operating at no-load will be as shown in Figure 8.28. Upon referring to the figure, it is assumed that friction and windage losses are small enough so that the machine operates with slip $s \approx 0$. If the reactance to be determined is to be used in an equivalent circuit where voltage is assumed to be 1.0 or less, the approximately linear portion of the curve shown in Figure 8.27 applies and a value for x_m can be determined. Typically, the no-load test is conducted by varying voltage magnitude at the terminals of the machine and measuring current. If we assume a source of pure sinusoidal voltage excitation for purposes of running the test, the current measured will be sinusoidal only over the region of the magnetization curve that can be taken as linear. As voltage increases (say, above 1.0 for the curve of Figure 8.27),

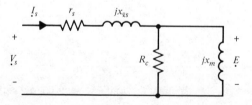

FIGURE 8.28 Equivalent Circuit for an Induction Machine at No-Load

current increases more rapidly due to saturation. The curve then typically reflects the magnitude of the fundamental component of current.

If current through the core loss resistance R_c is neglected, the voltage across the magnetizing reactance x_m will be

$$E = V_s - (r_s + jx_{\ell s})I_s$$

$$\approx V_s - jx_{\ell s}I_s$$

$$= 1 - (j.12)(-j.25)$$

$$= 0.97$$

where data are taken at a voltage of 1.0 per unit. In the calculation above, we have neglected the drop across the armature resistance and accordingly taken current as lagging V_s by 90° for a purely inductive circuit. Our simplifying assumptions can be validated after we have determined x_m. It is easy to see that the voltage E will differ by little from the terminal voltage. The reactance x_m is then given by

$$x_m = \frac{E}{I_s} = \frac{0.97}{0.25} = 3.88 \text{ per unit}$$

Core losses at rated voltage and friction and windage losses at speeds near synchronous speed can be estimated by assuming core losses proportional to voltage magnitude squared and friction and windage losses constant. In essence, speed will be very near synchronous speed for all the data points taken during the no-load test. Note that armature winding losses for the no-load test are relatively small (assuming that voltage ≤ 1.0) in that armature current is small for an unloaded machine. They can thus be neglected. If P_c is core losses at 1.0 per unit voltage and P_{fw} is friction and windage losses, assumed constant, we have for $V_s = 1.0$,

$$P_c + P_{fw} = 0.050$$

and at $V_s = 0.8$,

$$(0.8)^2 P_c + P_{fw} = 0.039$$

Solving the two equations in two unknowns, we find

$$P_c = 0.031$$

$$P_{fw} = 0.019$$

Thus core losses are roughly 3% and friction and windage losses are roughly 2% of the machine's rated power. Note that an appropriate value for R_c would be determined by

$$\frac{V_s^2}{R_c} = \frac{1}{R_c} = 0.031$$

Thus

$$R_c = 32.3 \text{ per unit}$$

It is left as an exercise for the reader to validate the simplifying assumptions we have made in the example.

In most application problems, the nonlinearity of the no-load test voltage–current curve can be neglected. Assuming that a curve is available, such a determination can be made simply by examining the curve for the range of terminal voltages anticipated. As shown in Example 8.5, the curve is determined primarily by the magnetizing reactance over its linear region. It essentially gives the amount of current required to establish the air-gap flux in the machine. This

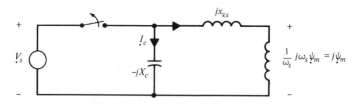

FIGURE 8.29 Approximate Equivalent Circuit for a Lightly Loaded Induction Machine with Capacitive Compensation. Only fundamental frequency components are considered. Harmonics due to nonlinearities are neglected

current is sometimes referred to as *magnetizing current* or *exciting current* and the curve is sometimes referred to accordingly as the *magnetization curve* or *excitation curve* for the machine.

A phenomenon known as *self-excitation* limits the amount of capacitive compensation that can be applied at the terminals of the motor and switched with the motor. Typically, it is not feasible to correct motor power factor to 100% using shunt compensation because of this phenomenon. The nonlinearity of the motor magnetization curve must be considered to appreciate how self-excitation can occur. Figure 8.29 illustrates an approximate equivalent circuit for a lightly loaded induction machine with capacitive compensation. Referring to the figure, as long as the switch is closed, we see that the voltage across the motor terminals and the capacitive compensation is necessarily the same as the source voltage. Currents I_c and I_m are out of phase by 180° but differ in magnitude. If, when the switch is opened, a value of voltage magnitude exists where $|I_m| = |I_c|$, the motor may "self-excite." Figure 8.30, which illustrates a magnetization curve with capacitor characteristics superimposed, shows the manner in which this can occur. Referring to the figure, we note that, at a voltage of 1.6 per unit, the magnetizing current requirement of the motor (fundamental component thereof) is exactly satisfied by the capacitor current corresponding to $X_c = X_{c2}$. As the machine slows down, losses eventually cause the phenomenon to cease. If we

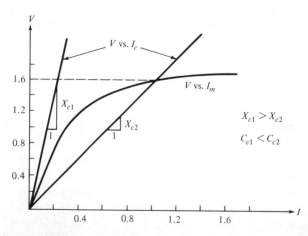

FIGURE 8.30 Induction Machine Magnetization Curve with Reactive Compensation Curves Superimposed

assume that the machine can "coast" at a speed near synchronous speed for an appreciable length of time, the windings will be subjected to abnormally high voltages for prolonged periods of time. If the machine is reconnected to the system when terminal voltage is out of phase with the system voltage, potentially damaging inrush currents may result. The reclosing of the switch shown in Figure 8.29 essentially can produce a situation similar to that which occurs if a synchronous machine undergoes an out-of-phase synchronization. Note that if reactive compensation is decreased, X_c increases and it is possible to preclude self-excitation. The curve of Figure 8.30 corresponding to a capacitive reactance of X_{c1} illustrates this. It is generally possible to correct the power factor of the machine to roughly 95% without risking self-excitation. Reference 13 considers the problem in somewhat greater detail.

The transformation techniques introduced in Chapter 6 can be used to formulate a set of equations in state-variable form which can be used for analysis in general of the induction machine. For a three-phase machine, the appropriate transformations are in terms of the angles defined in Figure 8.31. Referring to the figure, we see that the angle θ_r is the angle that the axis of the phase a winding of the rotor makes with respect to the axis of the phase a winding of the stator. Subscripts s and r refer to stator and rotor variables, respectively. Stator and rotor variables are referred to the common qd reference frame by the transformations

$$\begin{bmatrix} f_{qs} \\ f_{ds} \\ f_{os} \end{bmatrix} = \frac{2}{3} \begin{bmatrix} \cos\varphi_s & \cos(\varphi_s - 2\pi/3) & \cos(\varphi_s + 2\pi/3) \\ \sin\varphi_s & \sin(\varphi_s - 2\pi/3) & \sin(\varphi_s + 2\pi/3) \\ 1/2 & 1/2 & 1/2 \end{bmatrix} \begin{bmatrix} f_{as} \\ f_{bs} \\ f_{cs} \end{bmatrix} \tag{8.34a}$$

$$\begin{bmatrix} f_{qr} \\ f_{dr} \\ f_{or} \end{bmatrix} = \frac{2}{3} \begin{bmatrix} \cos\varphi_r & \cos(\varphi_r - 2\pi/3) & \cos(\varphi_r + 2\pi/3) \\ \sin\varphi_r & \sin(\varphi_r - 2\pi/3) & \sin(\varphi_r + 2\pi/3) \\ 1/2 & 1/2 & 1/2 \end{bmatrix} \begin{bmatrix} f_{ar} \\ f_{br} \\ f_{cr} \end{bmatrix} \tag{8.34b}$$

Equivalent circuits in terms of transformed variables were developed in Chapter 6. For a three-wire machine, zero-sequence variables are zero and, if the change of variables $\psi = \omega_s\lambda$ is employed, the circuits will be as shown in Figure 8.32.

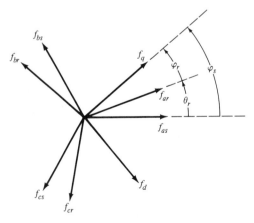

FIGURE 8.31 Angles Associated with the Transformation Relationships for a Three-Phase Induction Machine

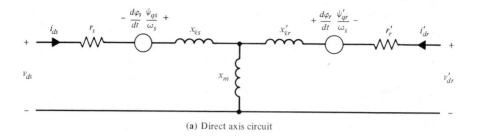

(a) Direct axis circuit

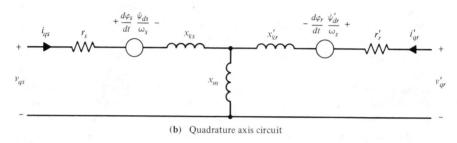

(b) Quadrature axis circuit

FIGURE 8.32 Equivalent Circuits for a Three-Phase Induction Machine. Zero-sequence variables are assumed to be zero

In order to develop a set of equations in state-variable form, several choices must be made. If we align the common reference frame q axis with the axis of the phase a rotor winding, we will have (see Figure 8.31)

$$\varphi_r = 0 \tag{8.35a}$$

$$\varphi_s = \theta_r \tag{8.35b}$$

This implies that in the equivalent circuits of Figure 8.32,

$$\frac{d\varphi_r}{dt} = 0 \tag{8.36a}$$

$$\frac{d\varphi_s}{dt} = \frac{d\theta_r}{dt} = \omega_r \tag{8.36b}$$

Thus the speed–voltages in the rotor circuit equations will be zero with this choice and the speed–voltage terms in the stator circuit equations are related specifically to rotor speed (we will assume that all variables are properly expressed in per unit terms as described in Chapter 6). Now suppose that the machine is connected to a three-phase source so that stator voltages are given by

$$v_{as} = \cos \omega_s t \tag{8.37a}$$

$$v_{bs} = \cos \left(\omega_s t - \frac{2\pi}{3} \right) \tag{8.37b}$$

$$v_{cs} = \cos \left(\omega_s t + \frac{2\pi}{3} \right) \tag{8.37c}$$

This implies a source that might be described alternatively as an "infinite bus." Applying the transformation relationship (8.34a), we find that

$$v_{qs} = \cos (\theta_r - \omega_s t) \tag{8.38a}$$

$$v_{ds} = \sin (\theta_r - \omega_s t) \tag{8.38b}$$

In many cases, the rotor voltages v'_{dr} and v'_{qr} would be taken as zero in that an induction machine typically operates with "short-circuited" rotor windings. If an external secondary controller were used to vary resistance in the rotor windings (see Figures 3.12 and 3.13), we would modify the equivalent circuits of Figure 8.32 to incorporate such resistances. Problem 8.12 addresses this issue. Secondary circuit controllers may employ rotating or static (solid-state-based) devices to effectively provide a variable-frequency supply for the rotor windings. Typically, it would be necessary to model such a device mathematically, then determine its effects if all variables are referred to the chosen common reference frame. Depending on the device employed, the task could be quite involved. To illustrate the basic approach without becoming mired in the process, we will assume that we have an idealized secondary supply that can develop voltages:

$$v'_{ar} = V_f \cos (\omega_f t + \varphi_f) \tag{8.39a}$$

$$v'_{br} = V_f \cos \left(\omega_f t + \varphi_f - \frac{2\pi}{3}\right) \tag{8.39b}$$

$$v'_{cr} = V_f \cos \left(\omega_f t + \varphi_f + \frac{2\pi}{3}\right) \tag{8.39c}$$

If V_f, ω_f, and φ_f can be independently varied, the relationships that result could be used to simulate performance of a machine with a short-circuited secondary simply by setting $V_f = 0$. Applying the transformation relationship (8.34b) with $\varphi_r = 0$ to the voltages given by (8.39), we find that

$$v'_{qr} = V_f \cos (\omega_f t + \varphi_f) \tag{8.40a}$$

$$v'_{dr} = -V_f \sin (\omega_f t + \varphi_f) \tag{8.40b}$$

Either currents or flux linkages can be chosen as state variables. If flux linkages are chosen, it will be necessary to express currents as linear combinations of flux linkages and eliminate them in the equations. The voltage relationships can be written by referring to the equivalent circuits of Figure 8.32 and substituting using (8.38) and (8.40):

$$\frac{1}{\omega_s} \frac{d\psi_{ds}}{dt} = \frac{\omega_r}{\omega_s} \psi_{qs} - i_{ds} r_s + \sin (\theta_r - \omega_s t) \tag{8.41a}$$

$$\frac{1}{\omega_s} \frac{d\psi'_{dr}}{dt} = -i'_{dr} r'_r - V_f \sin (\omega_f t + \varphi_f) \tag{8.41b}$$

$$\frac{1}{\omega_s} \frac{d\psi_{qs}}{dt} = -\frac{\omega_r}{\omega_s} \psi_{ds} - i_{qs} r_s + \cos (\theta_r - \omega_s t) \tag{8.41c}$$

$$\frac{1}{\omega_s} \frac{d\psi'_{qr}}{dt} = -i'_{qr} r'_r + V_f \cos (\omega_f t + \varphi_f) \tag{8.41d}$$

The equation of motion for the rotor assuming negligible damping and no shaft load can be expressed as

$$\frac{H}{\pi f}\frac{d\omega_r}{dt} = T_e = \psi_{ds}i_{qs} - \psi_{qs}i_{ds} \tag{8.41e}$$

Rotor angle is related to rotor speed by

$$\frac{d\theta_r}{dt} = \omega_r \tag{8.41f}$$

In order to eliminate currents in (8.41a) through (8.41e), the flux-linkage relationships must be solved. Referring to Figure 8.32 [also see (6.58)], we have

$$\psi_{qs} = (x_{\ell s} + x_m)i_{qs} + x_m i'_{qr}$$

$$= x_s i_{qs} + x_m i'_{qr} \tag{8.42a}$$

$$\psi_{qr} = x_m i_{qs} + (x'_{\ell r} + x_m)i'_{qr}$$

$$= x_m i_{qs} + x_r i'_{qr} \tag{8.42b}$$

Solving in (8.42) for i_{qs} and i'_{qr}, we obtain

$$i_{qs} = \frac{1}{x_s x_r - x_m^2}(x_r \psi_{qs} - x_m \psi'_{qr}) \tag{8.43a}$$

$$i'_{qr} = \frac{1}{x_s x_r - x_m^2}(-x_m \psi_{qs} + x_s \psi'_{qr}) \tag{8.43b}$$

Similarly,

$$i_{ds} = \frac{1}{x_s x_r - x_m^2}(x_r \psi_{ds} - x_m \psi'_{dr}) \tag{8.43c}$$

$$i'_{dr} = \frac{1}{x_s x_r - x_m^2}(-x_m \psi_{ds} + x_s \psi'_{dr}) \tag{8.43d}$$

Substitution in (8.41) using (8.43) is straightforward and achieves the desired result.

Equations (8.41) with currents eliminated, as described above, can be solved numerically using the techniques described in Appendix G. Although numerical solution of the equations is straightforward, integration step sizes will typically have to be small in order to achieve desired accuracy. If speed can be assumed constant, perhaps in order to examine initial current or torque waveforms, as was done in Chapter 6, the resulting system of equations becomes linear. It is then possible to develop closed-form solutions for flux linkages (and currents which are simple linear combinations of flux linkages) by using eigen-value–eigenvector techniques. Reference 14 describes such procedures in considerable detail. A major benefit of such an approach is the development of expressions for currents, flux linkages, and torque as linear combinations of simple functions such as exponentials and damped sinusoidal time functions.

If speed cannot be assumed constant, (8.41e) and (8.41f) must be retained.

It may be inconvenient to solve using (8.41f) in that the rotor angle θ_r will increase without bound. If the system of equations is augmented with the relationships for a second-order oscillator, (8.41f) can be eliminated. For example, consider (8.41a) and (8.41c), which contain sine and cosine terms involving θ_r. Note further that these are the only equations that require the variable θ_r. By using appropriate trigonometric identities, we have

$$\sin(\theta_r - \omega_s t) = \sin\theta_r \cos\omega_s t - \cos\theta_r \sin\omega_s t \tag{8.44a}$$

$$\cos(\theta_r - \omega_s t) = \cos\theta_r \cos\omega_s t + \sin\theta_r \sin\omega_s t \tag{8.44b}$$

Define auxiliary variables x_1, x_2, x_3, and x_4 as

$$x_1 = \sin\theta_r \tag{8.45a}$$

$$x_2 = \cos\theta_r \tag{8.45b}$$

$$x_3 = \sin\omega_s t \tag{8.45c}$$

$$x_4 = \cos\omega_s t \tag{8.45d}$$

It is easy to see from (8.45) that

$$\frac{dx_1}{dt} = \omega_r x_2 \tag{8.46a}$$

$$\frac{dx_2}{dt} = -\omega_r x_1 \tag{8.46b}$$

$$\frac{dx_3}{dt} = \omega_s x_4 \tag{8.46c}$$

$$\frac{dx_4}{dt} = -\omega_s x_5 \tag{8.46d}$$

Now substitute in (8.41a) and (8.41c) using (8.44) and (8.45), eliminate (8.41f), and augment the system of equations with (8.46). When all variables are properly expressed in per unit, the modification above yields a system of equations that is quite general and suitable for numerical evaluation.

8.5

VARIABLE-SPEED DRIVES FOR INDUCTION MACHINES

Efficient solid-state variable-speed drive systems can make the induction machine a viable option for applications requiring efficient operation over a range of speeds. In essence, such devices constitute a variable-frequency source of power. This is fundamental to the efficient operation of an induction machine over a range of speeds, assuming a capability to adjust speed arbitrarily within the range is desired. To see this, consider (3.61), repeated here for convenience:

$$P_m = (1 - s)P_g \tag{8.47}$$

In (8.47), P_m is mechanical power converted, P_g is power transferred across the air gap, and s is induction machine slip:

$$s = \frac{\omega_s - \omega_r}{\omega_s} \qquad (8.48)$$

Power transferred across the air gap, which is not converted to mechanical power, is converted to rotor winding I^2r losses in the form of heat. That is, rotor circuit losses P_r are given by

$$P_r = P_g - P_m$$
$$= sP_g \qquad (8.49)$$

Equations (8.47), (8.48), and (8.49) suggest that the induction machine must operate at relatively low values of slip for reasonably efficient operation. If, for example, rotor speed $\omega_r = 0.5 \, \omega_s$, slip s is 0.5, and half of the power transferred across the air gap is dissipated as rotor winding losses. At best, efficiency would be 50% assuming that all other losses were negligible. Induction machines typically have rated speeds corresponding to slips ranging from 0.02 to 0.06. If the frequency of the supply is variable, speed changes can be achieved in such a way that slip is kept relatively low. In a limited sense this can be achieved without a variable-frequency supply. For example, if the stator of a machine is wound in such a way that a change of connections at its terminals converts it from a two-pole to a four-pole machine, efficient operation is possible at essentially two "fixed" speeds with a fixed-frequency supply.

From a practical viewpoint a dc machine, with its inherent versatility, should always be considered in applications requiring a variable-speed system. Figure 8.33 contrasts a dc and an ac variable-speed drive system assuming the use of solid-state converters. The dc system typically has a simpler supply in that the ac system requires an inverter in addition to a rectifier. The ac system has the inherent advantages of the squirrel-cage induction machine over a dc machine: lower first cost, simpler machine configuration, and less maintenance. The dc system may not be feasible in some situations. For example, its brush-commutator system with its associated arcing would preclude its use in an explosive environment. It is to be expected that both alternatives will be feasible for the foreseeable future.

A wound-rotor induction machine offers an alternative to the concept illustrated in Figure 8.33, in that the rotor circuits are accessible via slip rings. A variable-frequency device in the machine secondary can be used to control speed. The concept is illustrated in Figure 8.34. If the converter is capable of power flow in either direction, the overall system is referred to as a static Scherbius drive system and the induction machine is said to be "doubly fed." If the power flow can only be from the motor secondary back to the three-phase supply, the system is referred to as a static Kramer drive or slip-power recovery system (Figure 8.35).

EXAMPLE 8.6 A secondary controller for use with a four-pole wound-rotor induction machine has a frequency range of 0 to 20 Hz. The device can accommodate

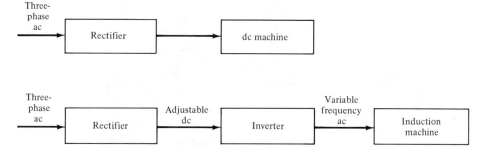

FIGURE 8.33 DC Versus AC Variable-Speed Drive Systems

power flow in either direction. What is the range of possible steady-state speeds at which the motor can operate?

Assume that the frequency range is such that the frequency of the rotor voltages and currents at steady state are bounded by

$$0 \le |f_r| \le 20$$

The field due to stator currents, assuming a 60-Hz supply, will rotate at 1800 rpm for a four-pole machine. If, in the steady state, rotor variables as measured at the slip rings are at $+20$ Hz, the field due to the rotor currents would rotate at $(20/60) \times 1800 = 600$ rpm *with respect to the rotor*. The rotor must then turn at 1200 rpm if the field due to rotor currents is to be in synchronism with the field due to stator currents. Similarly, if the rotor variables are at -20 Hz, in effect requiring a reversal of the phase sequence in the secondary, the rotor must turn at 2400 rpm in order to establish synchronism between stator and rotor fields. Thus the allowable speed range would be

$$1200 \le n \le 2400$$

A number of different drive systems can be used to achieve the required variable-frequency supply for an induction machine. The most commonly ap-

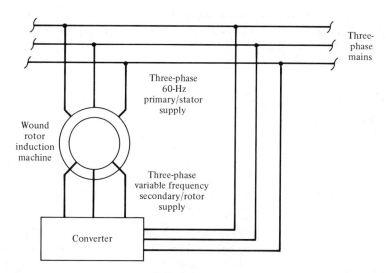

FIGURE 8.34 Wound-Rotor Induction Machine and a Variable-Frequency Secondary Controller

FIGURE 8.35 Solid-State Secondary Controller for a 200-hp Wound-Rotor Induction Machine. The controller is for use in a slip-power recovery system. The allowable speed range is 0 to 99% of rated motor speed. (Courtesy of the Square D Company)

plied systems for supplying the stator windings of a squirrel-cage induction machine are

1. Variable voltage–variable frequency drives supplying square-wave voltage waveforms at the machine terminals.
2. Variable voltage–variable frequency drives supplying pulse-width-modulated (PWM) voltage waveforms at the machine terminals.
3. Variable current–variable frequency drives, sometimes referred to as current source inverters, which supply square-wave current waveforms at the machine terminals.

The waveforms referred to above are not, strictly speaking, square waves but rather stepped waveforms realized by solid-state device switching of currents or voltages. Figure 8.36 illustrates typical waveforms for a three-phase induction machine. A more complete discussion of the characteristics of and the advantages or disadvantages of these drive types as well as some that are less com-

     ADVANCED APPLICATIONS

Line-to-neutral voltage

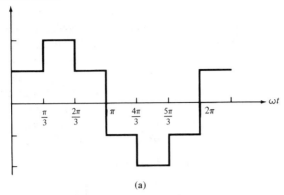

(a)

Line current

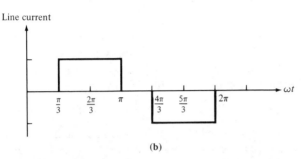

(b)

FIGURE 8.36 Typical Idealized Waveforms at the Terminals of a Three-Phase Induction Machine Supplied by (a) a Variable Voltage–Variable Frequency Variable-Speed Drive, (b) a Variable Current–Variable Frequency Variable-Speed Drive

monly used is contained in the introductory chapter of reference 15. A characteristic, which all such systems have in common, is the presence of harmonics in the voltage and/or current waveforms of the machine they supply. In general, these harmonics detract from the performance that would otherwise be achieved with systems producing only fundamental frequency waveforms. Higher harmonics tend to increase machine losses somewhat, resulting in reduced efficiency and/or higher operating temperatures, which can adversely affect insulation life. A machine for use with a variable-frequency drive system may have to be derated in order to ensure that its expected lifetime is not significantly reduced.

A polyphase induction machine operated in such a way as to have balanced polyphase currents and voltages of fundamental frequency only will have an instantaneous electrical torque that is constant with respect to time. Higher harmonics in current waveforms give rise to unwanted pulsations in the electrical torque waveform. These pulsations can be troublesome at low speeds or may cause vibration problems.

To illustrate one approach to the analysis of an induction machine supplied by a variable-frequency drive, we will consider a two-phase machine supplied by a variable voltage–variable frequency drive system. The same analytical technique is applicable for three-phase machines (recall that a three-phase induction machine can be analyzed as an equivalent two-phase machine as discussed in Section 6.3.8). If nonsinusoidal voltages and currents are periodic waveforms, they can be expressed as sums of sinusoidal waveforms using Fourier analysis. If the machine is assumed to be running at constant speed, a steady-state analysis can be made for each of the various frequencies present in the Fourier series expansion of the input voltage waveform. Currents determined for each of the frequencies can then be summed to obtain total armature currents or total rotor currents using the principle of superposition.

To illustrate, suppose that the voltage for phase a of a two-phase machine can be described as an even function of time with half-wave symmetry, that is,

$$v_a(t) = v_a(-t) \tag{8.50}$$

$$v_a(t) = -v_a\left(t \pm \frac{T}{2}\right) \tag{8.51}$$

It will generally be possible to choose the time origin in such a way as to satisfy (8.50), and most waveforms of interest will satisfy (8.51). From elementary properties of Fourier series, (8.50) implies a series consisting of only cosine functions and (8.51) implies that only odd harmonics will be present. Thus

$$v_a(t) = a_1 \cos \omega t + a_3 \cos 3\omega t + \cdots + a_n \cos n\omega t + \cdots \tag{8.52}$$

The voltage for phase b would be shifted to realize a $\pi/2$-radian phase shift for the fundamental frequency component. Thus if $\omega = 2\pi/T$ is the fundamental radian frequency and T the corresponding time period, we would have

$$v_b(t) = v_a\left(t - \frac{T}{4}\right) \tag{8.53}$$

It is not difficult to show that for $v_b(t)$ we will have [substitute in (8.52) using (8.53)]

$$v_b(t) = a_1 \sin \omega t - a_3 \sin 3\omega t + a_5 \sin 5\omega t + \cdots + a_n \sin \frac{n\pi}{2} \sin n\omega t \cdots \tag{8.54}$$

Upon comparing (8.52) and (8.54), it is evident that the fundamental frequency terms are of positive phase sequence, the third harmonic terms have negative phase sequence, the fifth harmonic terms have positive phase sequence, and so on. It will thus be necessary to consider equivalent circuits for both positive and negative sequence excitation. If the speed of the rotor is assumed constant at ω_r, those components of positive sequence ($n = 1, 5, 9, 13, \ldots$) will have a slip of

$$s_n^+ = \frac{n\omega - \omega_r}{n\omega} = 1 - \frac{\omega_r}{n\omega} \tag{8.55}$$

Those components of negative phase sequence ($n = 3, 7, 11, 15, \ldots$) will have a slip of

$$s_n^- = \frac{-n\omega - \omega_r}{-n\omega} = 1 + \frac{\omega_r}{n\omega} = 2 - s_n^+ \tag{8.56}$$

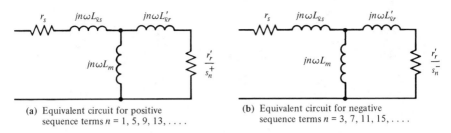

(a) Equivalent circuit for positive
sequence terms n = 1, 5, 9, 13, ...

(b) Equivalent circuit for negative
sequence terms n = 3, 7, 11, 15, ...

FIGURE 8.37 Equivalent Circuits for the nth Harmonic of the Voltage Excitation as Given by (8.52) and (8.54)

The equivalent circuits as developed in Chapter 3 will be as shown in Figure 8.37.

For an assumed voltage waveform, the coefficients $a_1, a_3, a_5, \ldots a_n, \ldots$ can be determined in the usual way using Fourier analysis. For example, if $v_a(t)$ is an even function satisfying (8.50),

$$a_n = \frac{2}{T} \int_0^T v_a(t) \cos n\omega t \, dt \qquad (8.57)$$

where

$$\omega = \frac{2\pi}{T} \qquad (8.58)$$

Armature currents and rotor currents referred to the armature due to each of the harmonics present in the voltage waveforms are determined in a relatively straightforward manner using the equivalent circuits of Figure 8.37. These components may then be added to determine the complete instantaneous current waveforms. The procedure is straightforward conceptually, but computer assistance is helpful in examining the resulting waveforms. Figure 8.39 on page 432 shows armature current waveforms for the special case of square-wave voltage excitation as shown in Figure 8.38. The magnitude of the voltage waveforms shown in Figure 8.38 was chosen so that the fundamental component of the armature terminal voltage would be at rated voltage. The two curves shown in Figure 8.39 compare results if only the fundamental component is considered

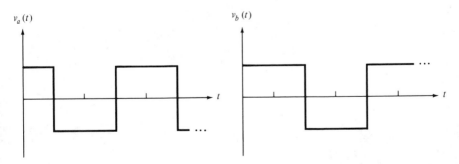

FIGURE 8.38 Assumed Voltages at the Terminals of a Two-Phase Induction Machine

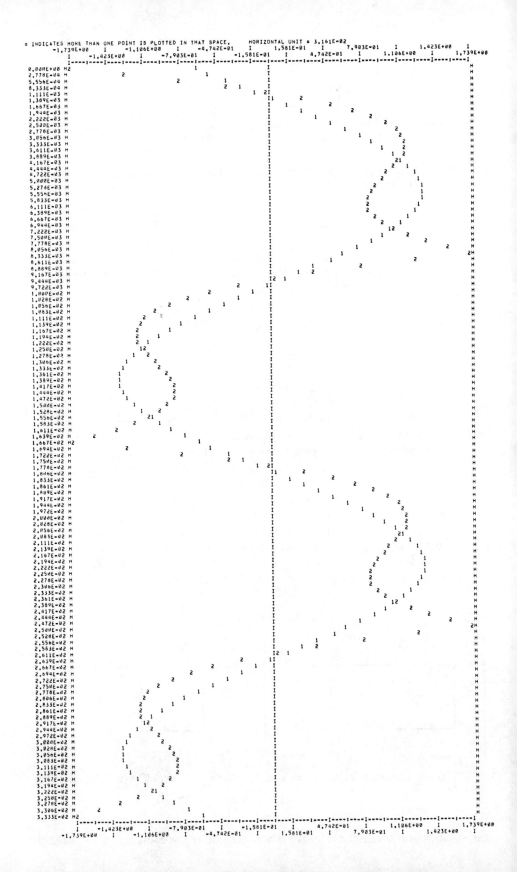

FIGURE 8.39 Armature Current Waveforms Resulting from the Nonsinusoidal Voltage Waveforms Shown in Figure 8.38

versus results if higher harmonics are included. The fundamental current magnitude is thus very close to rated current for the case considered. Time-averaged values for the various resistive power losses, mechanical power converted, and developed torque can be found using the techniques described in Chapter 3. Although, for the particular case considered here, rotor resistance and reactance were assumed constant with frequency, the approach outlined would permit inclusion of the frequency dependence of these parameters.

Other methods of analysis of induction machines assuming nonsinusoidal excitation are presented in references 16 and 17.

REFERENCES

[1] G. W. SWIFT, "Power Transformer Core Behavior Under Transient Conditions," *IEEE Transactions on Power Apparatus and Systems,* Vol. PAS-90, pp. 2206–2210, September/October 1971.

[2] S. N. TALUKDAR and J. R. BAILEY, "Hysteresis Models for System Studies," *IEEE Transactions on Power Apparatus and Systems,* Vol. PAS-95, pp. 1429–1434, July/August 1976.

[3] W. H. HAYT, JR., *Engineering Electromagnetics.* 2nd ed. New York: McGraw-Hill Book Company, 1967.

[4] S. RAMO, J. R. WHINNERY, and T. VAN DUZER, *Fields and Waves in Communications Electronics,* 2nd ed. New York: John Wiley & Sons, Inc., 1965.

[5] J. AVILA-ROSALES and F. L. ALVARADO, "Nonlinear Frequency Dependent Transformer Model for Electromagnetic Transient Studies in Power Systems," *IEEE Transactions on Power Apparatus and Systems,* Vol. PAS-101, pp. 4281–4288, November 1982.

[6] *Distribution Systems Reference Book.* East Pittsburgh, Pa.: Westinghouse Electric Corporation, 1965.

[7] E. J. DOLAN, D. A. GILLIES, and E. W. KIMBARK, "Ferroresonance in a Transformer Switched with an EHV Line," *IEEE Transactions on Power Apparatus and Systems,* Vol. PAS-91, pp. 1273–1280, May/June 1972.

[8] W. T. MCLYMAN, *Transformer and Inductor Design Handbook.* New York: Marcel Dekker, Inc., 1978.

[9] A. GREENWOOD, *Electrical Transients in Power Systems.* New York: Wiley-Interscience, 1971.

[10] L. V. BEWLEY, *Traveling Waves on Transmission Systems,* 2nd ed. New York: John Wiley & Sons, Inc., 1951.

[11] D. G. FINK and H. W. BEATY, *Standard Handbook for Electrical Engineers,* 11th ed. New York: McGraw-Hill Book Company, 1978.

[12] B. ADKINS and R. G. HARLEY, *The General Theory of Alternating Current Machines.* London: Chapman & Hall, Ltd., 1975.

[13] D. F. MILLER, "Application Guide for Shunt Capacitors on Industrial Distribution Systems at Medium Voltage Levels," *IEEE Transactions on Industry Applications,* Vol. IA-12, pp. 449–459, September/October 1976.

[14] S. S. KALSI and T. A. LIPO, "A Modal Approach to the Transient Analysis of Synchronous Machines," *Electric Machines and Electromechanics,* Vol. 1, pp. 337–354, July–Sept. 1977.

[15] B. K. BOSE, *Adjustable Speed AC Drive Systems.* New York: The IEEE Press, 1981.

[16] P. C. KRAUSE, "Method of Multiple Reference Frames Applied to the Analysis of Symmetrical Induction Machinery," *IEEE Transactions on Power Apparatus and Systems,* Vol. PAS-87, pp. 218–227, January 1968.

[17] T. A. LIPO and E. P. CORNELL, "State-Variable Steady-State Analysis of a Controlled Current Induction Motor Drive," *IEEE Transactions on Industry Applications,* Vol. IA-11, pp. 704–712, November/December 1975.

PROBLEMS

8.1 The approximate expression for eddy-current losses developed in Chapter 1 [Eqs. (1.64) and (1.65)] can be used to show that

$$P_e = \frac{\sigma}{6} \pi^2 f^2 B_{max}^2 t_\ell^2 A_c \ell_c$$

where σ is the conductivity, t_ℓ the lamination thickness, A_c the core cross-sectional area, and ℓ_c the mean flux path length. Find an expression for a single shunt resistance R which might be used to model eddy-current losses assuming sinusoidal voltage excitation at frequency f.

8.2 Consider the equivalent circuit shown here for an open-circuited transformer.

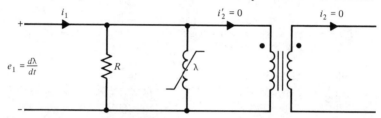

Sketch the general shape of the λ versus i_1 curve that would result if terminal voltage e_1 varies sinusoidally with time and
a. The nonlinear inductor is replaced with a linear inductor having very high inductance.
b. The nonlinear inductor is replaced with a linear inductor having finite inductance.
c. The nonlinear inductor has a characteristic similar to that shown in Figure 8.5.

8.3 Repeat parts (a) and (b) of Problem 8.2 if the voltage e_1 is a square wave.

8.4 The core of a power transformer for operation at 60 Hz is to be constructed of material having a relative permeability of 10,000 and an electrical resistivity of $\rho = 50~\mu\Omega$-cm. What is the skin depth for the material at 60 Hz? (Assume linear material.)

8.5 The skin depth for a particular linear material is known to be 0.50 mm at 60 Hz. What is the skin depth at 400 Hz?

8.6 Starting with (8.11), which describes the manner in which magnetic field intensity varies with position in a linear medium, determine an expression that could be used to describe the variation of current density.

8.7 Consider Figure 8.17, which is the equivalent circuit for the transformer of Example 8.3. If $R = 0$ and $C_d = 0$, the solution for $\psi(t)$ for $t \geq 0$ is given by

$$\psi(t) = A \cos \omega_n t + B \sin \omega_n t$$

Show that

$$A = \sin \varphi$$

$$B = \frac{\omega_s}{\omega_n} \cos \varphi$$

8.8 A singly excited system similar to that shown in Figure 8.18 has the following empirically determined relationship between flux linkages, current, and displacement.

$$\lambda = \lambda_{sat}(1 - e^{-(k/x)i}) \qquad i \geq 0, x \geq 0$$

Find an expression for $F_e = F(i, x)$, that is, where current i and displacement x are regarded as independent variables. What restrictions apply to the relationship?

8.9   Consider the singly excited system of Example 8.4. Determine the time-averaged force developed on the movable member of the magnetic circuit if $x = g = 0.25$ cm. Assume rated voltage at the terminals of the device.

8.10 A blocked rotor test at 60 Hz is conducted on a Y-connected three-phase 75-hp 460-V four-pole induction machine with the following results:

Armature voltage	73.3 V (line to line, reduced)
Armature current	86.5 A (line, rated)
Locked rotor torque	7.93 ft-lb
Power at terminals	4.76 kW (three-phase)

Find $x_{\ell s} = x'_{\ell r}$, and r_s, and r'_r.

8.11 A certain induction machine has core losses of 2000 W at rated voltage and friction and windage losses of 1500 W at or near rated speed. Plot total input power at the machine terminals as a function of voltage magnitude, assuming that the machine is operated at no load.

8.12 A three-phase wound-rotor induction machine is operated with a secondary controller in such a way that

$$v'_{ar} = -i'_{ar}R_r$$

$$v'_{br} = -i'_{br}R_r$$

$$v'_{cr} = -i'_{cr}R_r$$

where R_r is resistance of the secondary controller which may be independently varied. Find appropriate relationships between v'_{dr}, i'_{dr}, and v'_{qr}, and i'_{qr}. Describe how the equivalent circuits of Figure 8.32 might be modified to incorporate the secondary controller.

8.13 The current through a resistance R is given by

$$i(t) = \sqrt{2}\, I_1 \cos \omega t + \sqrt{2}\, I_3 \cos (3\omega t + \varphi_3) + \sqrt{2}\, I_5 \cos (5\omega t + \varphi_5)$$

Find an expression for the *time-averaged* power converted to heat in the resistance.

8.14 A variable-speed drive system of the current source type is used to supply a wound-rotor induction machine. A Fourier analysis of the observed armature current waveforms results in the following expression for armature current for phase a:

$$i_a(t) = \sqrt{2}\, I \left(\cos \omega t - \frac{\cos \omega t}{3} + \frac{\cos \omega t}{5} - \cdots \right)$$

Make reasonable simplifying assumptions and estimate the amount by which a machine should be derated if supplied by this type of drive.

8.15 A six-pole induction machine runs at a steady-state speed of 1140 rpm. The machine is supplied by a variable-speed drive system whose fundamental frequency is at 60 Hz. Determine slip values for

a. s_3^-

b. s_5^+

Selected Units, Constants, and Conversion Factors

Table A.1   Selected Quantities, Symbols, and Units Used in This Book

Quantity	Symbol	SI Units (mks)	Mixed English Units[a]
Voltage	E or V	volts	volts
Current	I	amperes (A)	Amperes (A)
Current density	J	amperes/meter^2	amperes/in.^{2}
Resistance	R	ohms	ohms
Conductance	G	mhos	mhos
Resistivity[b]	ρ	ohm-meters	ohm-circular mils/ft
Inductance	L	henries	henries
Inductive reactance	X_L	ohms	ohms
Impedance	$Z(= R + jX)$	ohms	ohms
Complex power	S or P_a	volt-amperes (VA)	volt-amperes (VA)
Real power	P	watts (W)	horsepower (hp)
Reactive power	Q	vars	vars
Frequency	f	hertz (Hz) or cycles/second	hertz or cycles/second
Angular frequency	ω	radians/second	radians/second
Rotational speed	N	revolutions/minute (rpm)	revolutions/min (rpm)
Force	F	newtons	pounds (lb)
Torque	T	newton-meters	ft-lb
Energy	W	joules (watt-seconds)	hp-hr (also Btu)
Energy density	w	joules/meter^3	hp-hrs/in.^{3}
Moment of inertia	J	kilogram-meters^2 (kg-m^2)	lb-ft^2
Magnetic field intensity	H	ampere-turns/meter	ampere-turns/inch
Magnetic flux	φ	webers	lines or maxwells
Magnetic flux density	B	webers/m^2 or teslas (T)	lines/in.^{2} or maxwells/in.^{2}
Magnetomotive force (mmf)	$\mathscr{F}$	ampere-turns (A · t)	ampere-turns (A-t)
Permeability	μ_0	henries/meter or webers/A · t-m	lines/A-t-inch
Relative permeability	μ_r	[dimensionless]	[dimensionless]
Reluctance	$\mathscr{R}$	A · t/weber	A-t/line
Linkage	λ	weber-turns or henries/ampere	maxwell-turns
Length	ℓ	meters	inches or feet
Area	A	meters^2	inches^2 or feet^2

[a]This book utilizes SI units. Common mixed English units are shown for reference. Ab and stat units for common electrical quantities usually denoted by SI values, such as volts and ohms, have not been in general use among electrical engineers in the United States for many years. A complete treatise on these units with conversion factors is given, for example, in D. Halliday and R. Resnick, *Physics*, Part I, 3rd ed. (New York: John Wiley & Sons, Inc., 1978), Appendix G, or in the *CRC Handbook of Chemistry and Physics*, 62nd ed. (Boca Raton, Fla.: CRC Press, Inc., 1981), Section F.

[b]English units represent common U.S. usage for wires. For bulk materials SI units are commonly used.

Table A.2 Selected Constants

Constant	SI Value	Mixed English Value[a]
Permeability of free space, μ_0	$4\pi \times 10^{-7}$ henries/meter	3.9 lines/A-t-inch
Permittivity of free space, ϵ_0	8.854×10^{-12} farad/meter	2.249×10^{-13} farad/inch
Standard acceleration of gravity, g[b]	9.807 m/s^2	32.1755 ft/sec^2
Velocity of light, c	2.998×10^8 m/s	186,280 mi/sec
Earth's magnetic flux density, B, at Washington, D.C.	5.7×10^{-5} T	3.68 lines/in.^{2}

[a]For explanation of mixed English values, see footnote a of Table A.1.
[b]At 46° latitude. Varies from 9.780 m/s^2 at the equator to 9.832 m/s^2 at the poles. [U.S. Coast and Geodetic Survey data cited in the *CRC Handbook of Chemistry and Physics*, 49th ed. (Boca Raton, Fla.: CRC Press, Inc., 1968).]

Table A.3 Selected Conversion Factors

(1) To Convert from: (To Convert to:)	(2) To: (From:)[a]	Multiply (1) Value by: [Divide (2) Value by:]
meters	inches	39.37
feet	meters	0.305
miles	meters	1609
miles	feet	5280
centimeters	meters	10^{-2}
centimeters^2	meters^2	10^{-4}
inches^2	meters^2	6.452×10^{-4}
feet^2	meters^2	9.290×10^{-2}
centimeters^3	meters^3	10^{-6}
feet^3	meters^3	2.832×10^{-2}
inches^3	meters^3	1.639×10^{-5}
hours	seconds	3600
miles/hr	meters/s	0.447
miles/hr	feet/sec	1.467
feet/sec	meters/s	0.305
hertz (electrical)	radians/sec (electrical)	2π
rpm	radians/sec	0.1047 ($2\pi/60$)
pounds ("mass")	kilograms	0.454
pounds ("mass")	ounces ("mass")	16
metric tons ("mass")	pounds ("mass")	2205
metric tons (mass)	kilograms	1000
tons ("mass" or force)	pounds ("mass" or force)	2000
pounds (force)	newtons (force)	4.448
foot-pound (torque)	newton-meters (torque)	1.355
inch-ounce (torque)	foot-pound (torque)	0.0052
pounds/ft^3 (density)	kg/m^3 (density)	16.02
lb-ft^2 (moment)	kg-m^2 (moment)	0.0422
horsepower	watts	746
Btu/hour	watts	0.2931
horsepower-hours	joules	2.685×10^6
kilowatt-hours	joules	3.6×10^6
Btu	joules	1055
lines or maxwells	webers	10^{-8}
kilolines	lines	1000
lines/in.^{2}	webers/m^2	1.55×10^{-5}

[a]Ounces and pounds "mass" as included in densities are in avoirdupois (common) units. Care should be taken when using conversion factors associated with these quantities, as they are actually weights and thus include the acceleration of gravity. Conversion factors to SI mass or density units may vary slightly relative to latitude (see Table A.2) and are useful only for terrestrial measurement.

AC Steady-State Analysis and Phasors

The steady-state analysis of alternating-current (ac) circuits is considerably simplified through the use of phasors and related steady-state analysis techniques. An ac circuit is said to be in a steady-state condition if all currents and voltages can be described as sinusoidal functions of time at a single fundamental angular frequency ω:

$$e(t) = E_{max} \cos (\omega t + \theta_e) = \sqrt{2}\, E \cos (\omega t + \theta_e)$$

$$i(t) = I_{max} \cos (\omega t + \theta_i) = \sqrt{2}\, I \cos (\omega t + \theta_i)$$

Common U.S. frequency is 60 Hz, yielding an angular frequency ω of approximately 377 rad/s ($\omega = 2\pi f$); 50 Hz corresponds to ω of 314 rad/s.

This description of a steady-state condition implies that many other variables, such as flux linkages, will also be sinusoidal time functions. An exception is instantaneous power, which cannot be described as a fundamental frequency sinusoidal waveform for an ac circuit operating in the steady state. Power relationships are discussed in Appendix C.

Moreover, the definition above assumes that any transient waveforms (resulting from switching actions, major changes in loads, etc.) have decayed to negligible levels. The circuit is therefore assumed to have reached its final steady-state condition. A linear passive circuit driven by sinusoidal sources will exhibit the steady-state behavior described above. Steady-state analysis could be accomplished by evaluating performance in terms of linear combinations of sinusoidal functions as given above (an unwieldy process generally involving use of numerous obscure trigonometric identities), by solving the differential equation, or can be reduced to simple complex arithmetic. By using Euler's formula,

$$e^{j\theta} = \cos\theta + j\sin\theta$$

it is easy to show that the sinusoidal voltage and current waveforms given above can be expressed as

$$e(t) = \sqrt{2}\ \text{Re}\left\{Ee^{j\theta_e}e^{j\omega t}\right\}$$

$$i(t) = \sqrt{2}\ \text{Re}\left\{Ie^{j\theta_i}e^{j\omega t}\right\}$$

The notation Re $\{\cdot\}$ is used to signify that only the real part of the complex expression within the braces is considered. Note that if the frequency ω is known, the only information necessary to describe $e(t)$ and $i(t)$ completely are the rms magnitudes E and I and the phase angles θ_e and θ_i[†]. In fact, any sinusoid of known frequency ω is described by magnitude and phase angle, and hence any shorthand method developed to analyze linear combinations of sinusoids expressed only as magnitudes and angles should reduce to simple arithmetic operations of complex numbers expressed in polar coordinates (magnitude and angle). Such a method is the *method of phasors*.

It is convenient to define a complex number (i.e., $\underset{.}{E}$, $\underset{.}{I}$)[‡] known as a phasor to represent a sinusoid [$e(t)$, $i(t)$] of known frequency ω as

$$e(t) = \sqrt{2}\ E\cos(\omega t + \theta_e) \leftrightarrow \underset{.}{E} = Ee^{j\theta_e} = E\ \underline{/\theta_e}$$

$$i(t) = \sqrt{2}\ I\cos(\omega t + \theta_i) \leftrightarrow \underset{.}{I} = Ie^{j\theta_i} = I\ \underline{/\theta_i}$$

In this text the phasor representations of time functions $e(t)$ and $i(t)$ are $\underset{.}{E}$ and $\underset{.}{I}$. Note that phasors $\underset{.}{E}$ and $\underset{.}{I}$ are simply complex numbers containing only magnitude and angle information; frequency, ω is *understood* in this representation. The symbol "$\leftrightarrow$" means "can be represented by" and implies that the sinusoidal time function and its associated phasor are not mathematically equal to one another. Instead, the mathematical relationship between a sinusoid and a phasor is

$$e(t) = \sqrt{2}\ \text{Re}\left\{Ee^{j\theta_e}e^{j\omega t}\right\} = \sqrt{2}\ \text{Re}\left\{\underset{.}{E}e^{j\omega t}\right\}$$

which is a simple vector space transformation.[§]

[†]The crest values of magnitudes $\sqrt{2}\ E$ and $\sqrt{2}\ I$ can also be used, but the rms (root-mean-square) value is more convenient in the power relationships, as will be shown in Appendix C.

[‡]In this book a dot below a quantity indicates a complex number. Consequently, both phasors and impedances are represented in this fashion.

[§]From this transformation it can be shown that

$$\underset{.}{E}e^{j\omega t} = E\ \underline{/(\omega t + \theta_e)}$$

which is a vector rotating counterclockwise in the complex plane at an angular frequency ω. If this rotating vector is projected, in time, on the real axis, the sinusoid $e(t)$ is produced. The relative position of phasor $\underset{.}{E}$ or rotating vector $\underset{.}{E}e^{j\omega t}$ is constant with respect to any other phasors or rotating vectors in the complex plane (i.e., all of the vectors rotate at the same angular frequency ω). This was depicted in Figure 1.26.

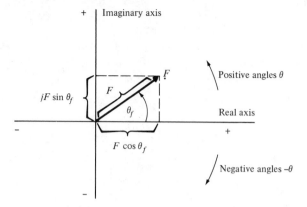

$jF \sin \theta_f$

F

F

Positive angles θ

θ_f

Real axis

$F \cos \theta_f$

Negative angles $-\theta$

FIGURE B.1 Graphical Expression of Any Phasor
$F = F \underline{/\theta_f} \leftrightarrow f(t) = \sqrt{2}F \cos (\omega t + \theta_f)$

Note also that the phasor $\dot{E}$ representing $e(t)$ can be written as

$$\dot{E} = Ee^{j\theta_e} = E \cos \theta_e + jE \sin \theta_e$$

and this is simply the complex number in rectangular coordinates. Four important properties of phasors follow from the two relationships above:

1. Phasors can be expressed in either polar or rectangular form for the purpose of arithmetic manipulations.
2. Addition (or subtraction) of sinusoidal time functions at a common frequency ω corresponds to the addition of their respective phasor representations. Phasor addition and subtraction are simply the addition and subtraction of complex numbers.
3. Differentiation of a sinusoidal time function corresponds to multiplication of its phasor representation by $j\omega$ ($\omega \underline{/90°}$). Integration corresponds to dividing the phasor representation by $j\omega$ [multiplication by $-j(1/\omega)$ or $(1/\omega) \underline{/-90°}$].
4. Phasors can be multiplied or divided by complex numbers, yielding other phasors.

Any phasor can be depicted graphically in the complex plane, as is shown in Figure B.1.

EXAMPLE B.1 Represent the following sinusoid in phasor notion, show it graphically, and determine the real and imaginary parts:

$$\lambda(t) = 170 \cos (\omega t - 36.9°)$$

is represented in phasor notation as

$$\dot{\Lambda} = \Lambda \underline{/-36.9}$$

where

$$\Lambda = \frac{170}{\sqrt{2}} = 120 \quad \text{and} \quad \theta = -36.9°$$

so

$$\dot{\Lambda} = 120 \underline{/-36.9°}$$

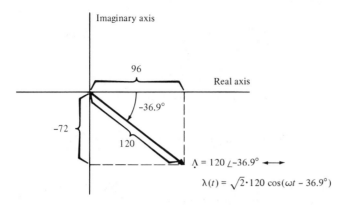

FIGURE B.2 Graphical Expression of Phasor $\underset{\sim}{\Lambda}$ in Example B.1.

The graphical result is shown in Fig. B.2. For the real and imaginary parts,

$$\text{Re} (\underset{\sim}{\Lambda}) = 120 \cos (-36.9) = 120(0.8) = 96$$

$$\text{Im} (\underset{\sim}{\Lambda}) = 120 \sin (-36.9) = 120(-0.6) = -72$$

as shown in Figure B.2, so the phasor represented in rectangular coordinates is

$$\underset{\sim}{\Lambda} = 96 - j72$$

AC STEADY-STATE CIRCUIT ANALYSIS USING PHASORS

The four properties discussed above permit steady-state analysis of ac circuits in a relatively simple manner. Because phasors are defined as complex numbers representing sinusoids, in essence circuits can be solved with algebraic manipulations where either complex trigonometric expressions or the solution of differential equations would otherwise be required. In a practical sense, such an analysis is accomplished in exactly the same fashion as for a purely resistive dc circuit in the steady state, except that complex quantities are used.† Circuit element representations are

$$
\begin{array}{ll}
\text{Voltage} & \underset{\sim}{E} = E \,\underline{/\theta_e} \\[4pt]
\text{Current} & \underset{\sim}{I} = I \,\underline{/\theta_i} \\[4pt]
\text{Resistor} & \underset{\sim}{Z}_R = R + j0 \\[4pt]
\text{Inductor} & \underset{\sim}{Z}_L = 0 + jX_L = 0 + j\omega L \\[4pt]
\text{Capacitor} & \underset{\sim}{Z}_C = 0 - jX_C = 0 - j\dfrac{1}{\omega C} = 0 + j\dfrac{-1}{\omega C}
\end{array}
$$

as discussed in Chapter 1. Ohm's law and Kirchhoff's laws hold. It is assumed that any transients (or constants associated with integration) have decayed to negligible levels.

---

†Voltages and currents being phasors, impedances being complex numbers.

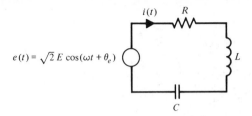

FIGURE B.3 Series *RLC* Circuit Driven by Sinusoidal Voltage Source

For example, the Kirchhoff's voltage law relationship for a series *RLC* circuit driven by a sinusoidal voltage source (Figure B.3) is

$$\sqrt{2}\,E\cos(\omega t + \theta_e) = Ri(t) + L\frac{di(t)}{dt} + \frac{1}{C}\int_0^t i(t)\,dt + v_c(0)$$

The current $i(t)$ could be found by solving this second-order differential equation, provided that initial conditions are known. For the steady state, however, a corresponding phasor relationship can be written immediately using the circuit element representations above, where it is assumed that all constants associated with the integration have decayed to zero:

$$\dot{E} = R\dot{I} + j\omega L\dot{I} - j\frac{1}{\omega C}\dot{I}$$

$$= R\dot{I} + jX_L\dot{I} - jX_C\dot{I}$$

If we assume that $\dot{E}$ is known, the phasor $\dot{I}$ representing current $i(t)$ may be found by dividing as follows:

$$\dot{I} = \frac{\dot{E}}{R + jX_L - jX_C}$$

The concept of complex impedance $\dot{Z}$ at an understood frequency ω also follows from the four properties mentioned earlier, so

$$\dot{Z} = R + jX_L - jX_C = \dot{Z}_R + \dot{Z}_L + \dot{Z}_C$$

and thus

$$\dot{I} = \frac{\dot{E}}{\dot{Z}}$$

As mentioned earlier, the manipulations generally follow the major rules of complex algebra, which are repeated here for reference.
For complex numbers $\dot{A}$ and $\dot{C}$,

$$\dot{A} = \alpha + j\beta = A_m\,\underline{/\theta_a}$$

$$\dot{C} = \gamma + j\delta + C_m\,\underline{/\theta_c}$$

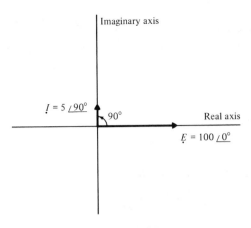

Imaginary axis

$I = 5 \,\underline{/90^\circ}$

90°

Real axis

$E = 100 \,\underline{/0^\circ}$

FIGURE B.4 Phasor Diagram for Example B.2

1. $A = C$ if and only if $\alpha = \gamma$ and $\beta = \delta$.
2. Addition: $A + C = C + A = (\alpha + \gamma) + j(\beta + \delta)$.
3. Subtraction: $A - C = -(C - A) = (\alpha - \gamma) + j(\beta - \delta)$.
[NOTE: Arithmetical addition and subtraction are always done in *rectangular coordinates*.]
4. Multiplication: $AC = A_m \,\underline{/\theta_a} \cdot C_m \,\underline{/\theta_c} = A_m C_m \,\underline{/(\theta_a + \theta_c)}$.
5. Division: $\dfrac{A}{C} = \dfrac{A_m \,\underline{/\theta_a}}{C_m \,\underline{/\theta_c}} = \dfrac{A_m}{C_m} \,\underline{/(\theta_a - \theta_c)}$.
6. Powers: $A^n = (A_m \,\underline{/\theta_a})^n = A_m^n \,\underline{/n\theta_a}$.
7. Roots: $A^{1/n} = (A_m \,\underline{/\theta_a})^{1/n} = A_m^{1/n} \,\underline{/\theta_a/n}$.
[NOTE: Arithmetical multiplication, division, powers, and roots are always done in *polar coordinates*.]

   Solution of steady-state ac circuits with phasors using the rules of complex algebra will be illustrated using the following two examples.†

EXAMPLE B.2   A capacitor with reactance of 20 Ω is excited by a sinusoidal voltage of magnitude 100 V. What is the steady-state current?

$$e(t) = \frac{1}{C} \int_0^t i(t)\, dt + v_c(0)$$

in phasor notation is

$$E = -jX_c I$$

so

$$I = \frac{E}{-jX_C} = \frac{100 \,\underline{/0^\circ}}{-j20} = \frac{100 \,\underline{/0^\circ}}{20 \,\underline{/-90^\circ}} = 5 \,\underline{/90^\circ}$$

so the current $i(t)$ is

$$i(t) = \sqrt{2} \cdot 5 \cos (\omega t + 90^\circ) = -\sqrt{2} \cdot 5 \sin \omega t$$

The voltage and current phasors for this example are shown in Fig. B.4. The current is said to "lead" the voltage by 90° in this case.

   †The concepts of "leading" and "lagging" current shown in these examples are discussed in more detail in Appendix C.

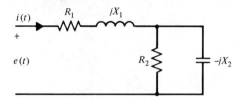

FIGURE B.5 Circuit for Example B.3

EXAMPLE B.3 Solve for the current $i(t)$ in the circuit in Figure B.5 where

$$e(t) = \sqrt{2} \cdot 10 \cos \omega t$$
$$R_1 = R_2 = 2\Omega$$
$$X_1 = 3\Omega$$
$$X_2 = 2\Omega$$

In complex notation,

$$R_2 \| (-jX_2) = \frac{R_2(-jX_2)}{R_2 - jX_2} = \frac{-jR_2X_2}{R_2 - jX_2}$$

and $e(t) \leftrightarrow E = 10 \underline{/0^\circ}$, so

$$I = \frac{10 \underline{/0}}{R_1 + jX_1 + \dfrac{-jR_2X_2}{R_2 - jX_2}} = \frac{10 \underline{/0}}{Z}$$

Solving for Z gives us

$$\frac{-jR_2X_2}{R_2 - jX_2} = \frac{-j2 \cdot 2}{2 - j2} = \frac{4 \underline{/-90}}{2.83 \underline{/-45}}$$

$$= 1.41 \underline{/-45}$$

$$= 1.0 - j1.0$$

$$R_1 + jX_1 + \frac{-jR_2X_2}{R_2 - jX_2} = 2.0 + j3.0 + 1.0 - j1.0$$

$$= 3.0 + j2.0 = 3.61 \underline{/33.7}$$

$$I = \frac{10 \underline{/0}}{3.61 \underline{/33.7}} = 2.77 \underline{/-33.7}$$

This phasor I represents a steady-state current $i(t)$, which is

$$i(t) = \sqrt{2} \cdot 2.77 \cos(\omega t - 33.7^\circ)$$

AC Power Relationships— General

The complex power† S in any circuit is defined as the product of the voltage phasor E (in volts) and the complex conjugate I^* of the current phasor I (in amperes):

$$S = EI^* \text{ (measured in volt-amperes or VA)}$$

where I^* is defined as $I \underline{/-\theta_i}$ if $I = I \underline{/\theta_i}$. Consequently, in shorthand phasor form,

$$S = E \underline{/\theta_e} \cdot I \underline{/-\theta_i}$$
$$= (E \cos \theta_e + jE \sin \theta_e)(I \cos \theta_i - jI \sin \theta_i)$$
$$= EI[(\cos (\theta_e - \theta_i) + j \sin (\theta_e - \theta_i)]$$
$$= EI \underline{/(\theta_e - \theta_i)}$$

The power factor angle $(\theta_e - \theta_i)$ is the angle between voltage and current

---

†The magnitude of complex power is sometimes referred to as *apparent power*.

phasors, which will be called θ for this analysis. Substituting θ for $(\theta_e - \theta_i)$ yields

$$S = EI(\cos \theta + j \sin \theta)$$

$$= EI \cos \theta + jEI \sin \theta$$

The real part of this quantity is the average power P in watts and the imaginary part is the reactive power Q, measured in vars (reactive volt-amperes). Therefore, the steady-state ac power relationship becomes

$$S(\text{VA}) = P(\text{watts}) + jQ(\text{vars})$$

$$= EI \cos \theta + jEI \sin \theta$$

or

$$P = EI \cos \theta$$

$$Q = EI \sin \theta$$

The *power factor* of the system is defined as

$$\text{P.F.} \equiv \cos \theta \qquad (\text{where } \theta = \theta_e - \theta_i \text{ equals the phase angle between voltage and current})$$

and is said to be *leading* if the current leads the voltage $(\theta < 0°)$ as in a capacitive system or *lagging* if the current lags the voltage $(\theta > 0°)$ as in an inductive system. A system with the current in phase with the voltage (either resistive or *RLC* in resonance) is said to have a *unity power factor* $(\theta = 0°)$.

EXAMPLE C.1 Compute S, P, Q, and the power factor at the terminals of the network in Example B.2 (Figure C.1).

In Example B.2, the rms current was computed to be

$$I = 2.77 \underline{/-33.7°}$$

and voltage was given as

$$E = 10 \underline{/0}$$

A comparison of the two phase angles above indicates that terminal current I lags the voltage by an angle of 33.7°. The complex power S is computed to be

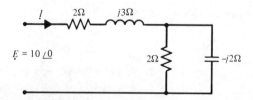

FIGURE C.1 Circuit for Example C.1

$$\underline{S} = \underline{E}\underline{I}^*$$

$$= 10\,\underline{/0°} \cdot 2.77\,\underline{/-(-33.7°)}$$

$$= 10\,\underline{/0°} \cdot 2.77\,\underline{/33.7°}$$

$$= 27.7\,\underline{/33.7°}\ \text{VA}$$

Computing P and Q yields

$$P = EI\cos\theta = 27.7\cos 33.7°$$

$$= 27.7 \times 0.83$$

$$= 23\ \text{W}$$

$$Q = EI\sin\theta = 27.7\sin 33.7°$$

$$= 27.7 \times 0.55 = 15\ \text{vars}$$

The power factor is defined as

$$\text{P.F.} \equiv \cos\theta = \cos 33.7 = 0.83$$

and since $\theta > 0$, it is *lagging*. This was seen earlier by comparing the voltage phase angle to the current phase angle. So

$$\text{P.F.} = 0.83\ \text{lag}$$

for this system.

Power relationships for three-phase circuits are discussed in Appendix D.

Balanced Three-Phase Circuits and the Per Unit System

A balanced three-phase system, in its most rudimentary form, consists of three single-phase circuits in which all voltages or currents are equal in magnitude but displaced in phase by 120° from each other; for example (in phasor notation),

$$E \text{ (phase } a) = E \underline{/0°}$$
$$E \text{ (phase } b) = E \underline{/\pm 120°}$$
$$E \text{ (phase } c) = E \underline{/\mp 120°}$$

This is shown in Figure D.1. The three voltage sources may be separate, or parts of a single device such as a three-phase synchronous generator. In common practice, the return conductors in Figure D.1 can be combined and either grounded or left ungrounded, as in Figure D.2. This return conductor is referred to as the *neutral conductor*. The voltages shown above are measured from the line (phase) conductor to the neutral conductor and are known as *line-to-neutral (L-N) voltages*. Voltages measured from one phase to another are called *line-to-line (L-L) voltages*. Currents, measured in each phase conductor outside the terminals of a load, are known as *line currents* or, in some cases, *phase currents*.† Power

---

†Nomenclature here follows that in the Westinghouse *Electrical Transmission and Distribution Reference Book* (East Pittsburgh, Pa.: Westinghouse Electric Corporation, 1964). The phase current is generally defined as the line current in a wye load and the delta current in a delta load, but will not be used in this appendix to avoid confusion. All issues are discussed in terms of simple line and delta currents only.

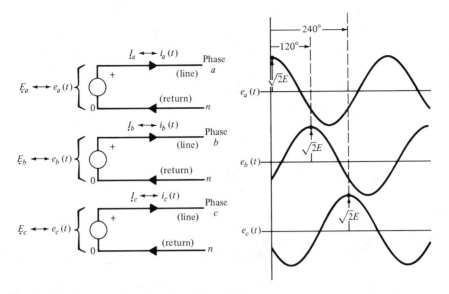

FIGURE D.1 Example of Three-Phase Waveforms

system voltages are generally given in terms of L-L quantities and currents are usually given as line currents. The relationships among these and other relevant quantities will be discussed after phase rotation is explained.

PHASE ROTATION

The electrical system shown in Figures D.1 and D.2 has the voltage phasor diagram in Figure D.3, and L-N voltages can be expressed mathematically as:

$$E_{an} = E \angle 0°$$

$$E_{bn} = E \angle -120°$$

$$E_{cn} = E \angle +120° = E \angle -240°$$

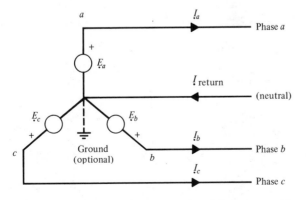

FIGURE D.2 Typical Three-Phase Circuit with Common Neutral

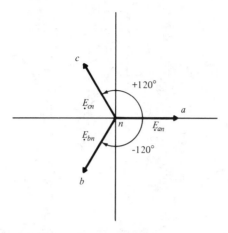

FIGURE D.3 *ABC* **Phase Rotation**

If the phasors are rotated in the counterclockwise direction, it can be seen from Figure D.3 (rotate the figure counterclockwise) that phase *a* passes the reader first, then *b*, then *c*. This is called *abc* phase rotation. It corresponds to the sinusoids of Figure D.1 maximizing, out of phase with each other, in the sequence *a*, *b*, and then *c*. It is the most common phase rotation on U.S. power systems and will be used exclusively in this appendix for simplicity.

If any two of the phases in an *abc*-rotation system are exchanged, the phase sequence reverses and becomes *acb* rotation. For example, if phases *b* and *c* are exchanged, the L-N voltages would be

$$E_{an} = E \underline{/0°}$$

$$E_{bn} = E \underline{/+120°} = E \underline{/-240°}$$

$$E_{cn} = E \underline{/-120°}$$

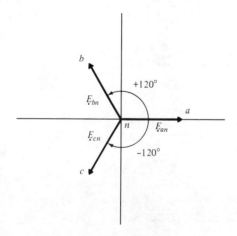

FIGURE D.4 *ACB* **Phase Rotation**

BALANCED THREE-PHASE CIRCUITS AND THE PER UNIT
SYSTEM

and the order of rotation is reversed, as shown in Figure D.4. When dealing with three-phase circuits, it is very important to know the phase sequence before analyzing or adjusting the circuit.

VOLTAGE RELATIONSHIPS IN A BALANCED THREE-PHASE SYSTEM

If we assume *abc* phase rotation, the L-N voltages in a balanced three-phase system are given by

$$E_{an} = E \, \underline{/0°}$$

$$E_{bn} = E \, \underline{/-120°}$$

$$E_{cn} = E \, \underline{/+120°}$$

The L-L voltages are related to the L-N quantities as in Figure D.5:

$$E_{ab} = E_{an} - E_{bn} = E_{an} + E_{nb}$$

$$E_{bc} = E_{bn} - E_{cn} = E_{bn} + E_{nc}$$

$$E_{ca} = E_{cn} - E_{an} = E_{cn} + E_{na}$$

Taking the vector sums, we have

$$E_{ab} = E_{an} + E_{nb} = E + j0 - E(\cos 120° - j \sin 120°)$$

$$= 1.5E + j.87E$$

$$= 1.732E \, \underline{/30°}$$

$$= \sqrt{3} \, E_{an} \, \underline{/30°} \text{ or } \sqrt{3} \, E_{bn} \, \underline{/150°}$$

where $\sqrt{3} \, E_{an} \, \underline{/30°}$ is the complex product

$$E_{an} \times \sqrt{3} \, \underline{/30°}$$

and $\sqrt{3} \, E_{bn} \, \underline{/150°}$ is the complex product

$$E_{bn} \times \sqrt{3} \, \underline{/150°}$$

Similarly, it can be shown that

$$E_{bc} = \sqrt{3} \, E_{bn} \, \underline{/30°} \quad \text{or} \quad \sqrt{3} \, E_{cn} \, \underline{/150°}$$

$$E_{ca} = \sqrt{3} \, E_{cn} \, \underline{/30°} \quad \text{or} \quad \sqrt{3} \, E_{an} \, \underline{/150°}$$

EXAMPLE D.1 What is the magnitude of the line-to-neutral voltage of a 69-kV line?

$$|E_{L\text{-}N}| = \frac{|E_{L\text{-}L}|}{\sqrt{3}} = \frac{69,000}{\sqrt{3}} = 39.8 \text{ kV}$$

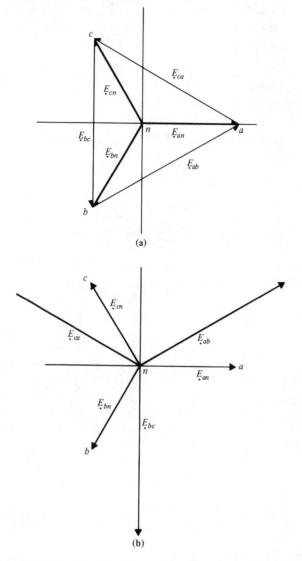

(a)

(b)

FIGURE D.5 Relationship of *L-N* and *L-L* Voltages (Two Equivalent Representations)

This voltage is shifted in phase $-30°$ from its respective line-to-line voltage, so if

$$E_{ab} = 69,000 \underline{/0°}$$

then

$$E_{an} = \frac{E_{ab}}{\sqrt{3}} \underline{/-30°} = 39..8 \text{ kV} \underline{/-30°}$$

DELTA AND WYE CONNECTIONS

It should be evident from the foregoing analysis of voltages on the system in Figures D.1 and D.2 that loads can be connected either between each of the phases (delta or Δ connection) or between each phase and the neutral (wye or Y connection). These connections are shown in Figure D.6. Similarly, voltage sources can also be connected in either delta or wye. Notice that the delta connection has no neutral point.

In the wye connection, the voltage across each complex scalar impedance (e.g., Z_{an}) is the L-N voltage, and the line current flows through the impedance, so

$$I_a = \frac{E_{an}}{Z_{an}}$$

$$I_b = \frac{E_{bn}}{Z_{bn}}$$

$$I_c = \frac{E_{cn}}{Z_{cn}}$$

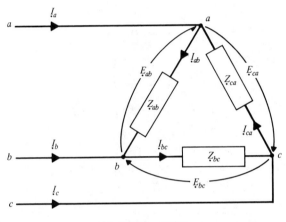

(a) Delta connection.

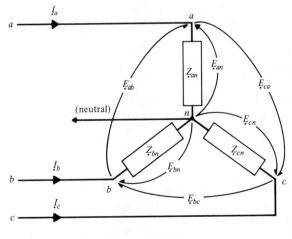

(b) Wye connection.

FIGURE D.6 Example of Delta and Wye Connections

In the delta connection L-L voltages are impressed across each impedance, and the currents flow between pairs of phases, so

$$I_{ab} = \frac{E_{ab}}{Z_{ab}}$$

$$I_{bc} = \frac{E_{bc}}{Z_{bc}}$$

$$I_{ca} = \frac{E_{ca}}{Z_{ca}}$$

These currents are called *delta currents* and are distinct from line currents. Summing currents at each node in Figure D.6(a), for example, yields

$$I_a + I_{ca} - I_{ab} = 0 + j0$$

$$I_b + I_{ab} - I_{bc} = 0 + j0$$

$$I_c + I_{bc} - I_{ca} = 0 + j0$$

Performing the vector additions for the system yields

$$I_a = \sqrt{3}\, I_{ab} \,\underline{/-30°} \text{ or } \sqrt{3}\, I_{ca} \,\underline{/150°}$$

$$I_b = \sqrt{3}\, I_{bc} \,\underline{/-30°} \text{ or } \sqrt{3}\, I_{ab} \,\underline{/150°}$$

$$I_c = \sqrt{3}\, I_{ca} \,\underline{/-30°} \text{ or } \sqrt{3}\, I_{bc} \,\underline{/150°}$$

So for a delta connection, the magnitude of the line current equals $\sqrt{3}$ times the delta current and the angle is shifted $-30°$.

EXAMPLE D.2 A balanced three-phase delta load is served by line currents of 10 A per phase. What is the current in each leg of the delta?

$$|I_\Delta| = \frac{|I_L|}{\sqrt{3}} = \frac{10 \text{ A}}{\sqrt{3}} = 5.8 \text{ A}$$

The delta current is displaced $+30°$ from the respective line current, so if $I_a = I_a \,\underline{/0°}$, then $I_{ca} = (I_a/\sqrt{3}) \,\underline{/+30°} = 5.8 \text{ A} \,\underline{/+30°}$.

It can be shown that for a balanced three-phase system, the sum of either the line currents or the delta currents is zero. For example, let

$$I_a = I \,\underline{/0°} = I + j0$$

$$I_b = I \,\underline{/-120°} = -0.5I - j0.87I$$

$$I_c = I \,\underline{/+120°} = -0.5I + j0.87I$$

$$I_n = I_{\text{neutral}} = I_a + I_b + I_c$$

$$= (I + j0) + (-0.5I - j0.87I) + (-0.5I + j0.87I)$$

$$= 0 + j0 \qquad \text{Q.E.D.}$$

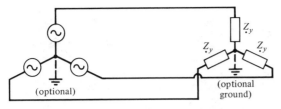

(a) Three-phase, three-wire, Y generator, Y load.

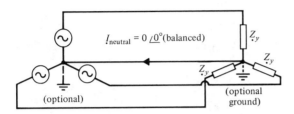

(b) Three-phase, four-wire, Y generator, Y load.

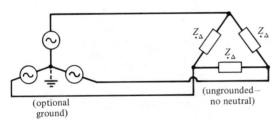

(c) Three-phase, three-wire, Y generator, Δ load.

FIGURE D.7 Examples of Configurations of Simple Balanced Three-Phase Systems

Under balanced conditions no current flows in the neutral, and no net current circulates in a delta connection. This can be seen by summing the currents at any node and showing that i_{loop} equals 0. Consequently, a neutral conductor is not needed in a balanced three-phase system (unless a line-to-neutral voltage is explicitly required), allowing both wye circuits and delta circuits to be connected on the same system, as shown in Figure D.7. Notice also that Figure D.7(b) is a four-wire system, common in many applications because it allows neutral conductors to carry nonzero neutral currents resulting from any imbalances that might occur. Separate additional ground conductors can also be utilized.

DELTA AND WYE IMPEDANCES—THREE-WIRE SYSTEMS

In a three-phase three-wire system, a delta impedance can be replaced by an equivalent wye impedance (complex scalar), and vice versa. From current and voltage relationships (adjusting signs to maintain polarity at the impedance),

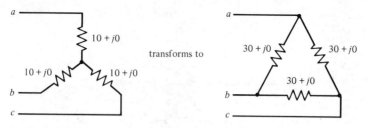

FIGURE D.8 Equivalent Delta and Wye Circuits in Example D.3

$$Z_\Delta = \frac{E_{LL}}{I_\Delta} = \frac{\sqrt{3}\, E_{LN}\, \underline{/30°}}{\frac{1}{\sqrt{3}} I_L\, \underline{/30°}} = 3\frac{E_{LN}}{I_L}\, \underline{/0}$$

$$= 3Z_Y \quad \text{(no additional phase shift introduced)}$$

$$Z_Y = \frac{1}{3} Z_\Delta \quad \text{(no additional phase shift introduced)}$$

EXAMPLE D.3 Three 10-Ω resistors are connected in wye. What is the equivalent delta circuit?

Because $Z_\Delta = 3Z_Y$, the wye circuit transforms as in Figure D.8.

This implies that, for a given balanced set of voltages, the wye load on the left and its equivalent delta load on the right *both draw the same line currents.*

THREE-PHASE POWER

In a balanced three phase (3φ) circuit the total complex power S is equal to three times the per phase complex power. Furthermore, it is independent of time. Therefore,

$$S_{\text{three-phase}} = 3S_{\text{single-phase}} = 3E_{LN}I_L^*$$

$$P_{\text{three-phase}} = 3P_{\text{single-phase}} = 3E_{LN}I_L \cos\theta$$

$$Q_{\text{three-phase}} = 3Q_{\text{single-phase}} = 3E_{LN}I_L \sin\theta$$

where θ is the angle between E_{LN} and I_L or between E_{LL} and I_Δ. So the power factor of the system is the same as that for one phase or one leg of its delta equivalent.

Since most systems specify voltage in terms of L-L quantities and current in terms of line values, it can also be shown that, for any phase,

$$|S_{\text{three-phase}}| = |\sqrt{3}\, E_{LL}I_L^*|$$

This discussion of balanced three-phase circuits leads to the conclusion that if all circuit elements were reduced to wye equivalents and analyzed in terms of line currents and L-N voltages, the only difference between conditions on any two phases at a given time is the angle $\pm 120°$. *Consequently, any balanced three-phase system can be analyzed in terms of any single phase, utilizing L-N and Y quantities,* and values for the other two phases simply derived from the results. This is obviously a very powerful tool, eliminating a very complex multiphase circuit in deltas and wyes and replacing it with one single-phase circuit using wye equivalents.

EXAMPLE D.4   A balanced three-phase circuit is shown in Figure D.9. Solve for I_a, $P_{3\varphi}$, $Q_{3\varphi}$, $S_{3\varphi}$, and P.F.

Assuming *abc* phase rotation, we obtain

$$E_{an} = \frac{1}{\sqrt{3}} E_{ab} \underline{/-30°}$$

$$= \frac{10}{\sqrt{3}} \underline{/-30°} = 5.8 \underline{/-30°}$$

If we convert Z_Δ to Z_Y:

$$Z_Y = \frac{1}{3} Z_\Delta \underline{/0°} = \frac{1}{3} (6 + j3) \underline{/0°}$$

$$= 2 + j1$$

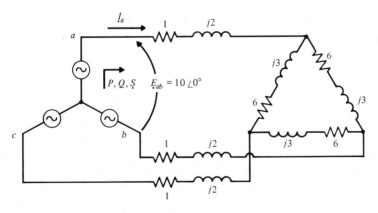

FIGURE D.9 Circuit for Example D.4

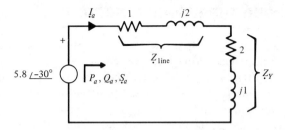

FIGURE D.10 Single-Phase Equivalent Circuit for Example D.4

a single-phase equivalent circuit is in Figure D.10, and

$$Z_{tot} = Z_{line} + Z_Y$$

$$= 1 + j2 + 2 + j1$$

$$= 3 + j3$$

$$I_a = \frac{E_{an}}{Z_{tot}} = \frac{5.8 \,\underline{/-30°}}{1 + j2 + 2 + j1} = \frac{5.8 \,\underline{/-30°}}{3 + j3} = \frac{5.8 \,\underline{/-30°}}{4.2 \,\underline{/45°}}$$

$$= 1.4 \,\underline{/-75°} \quad \text{(note that } I_a \text{ lags } E_{an} \text{ by } 45°)$$

$$S_a = E_{an} I_a^* = 5.8 \,\underline{/-30°} \cdot 1.4 \,\underline{/+75°}$$

$$= 8.1 \,\underline{/45°}$$

$$= 5.8 + j5.8$$

So

$$P_a = 5.8 \text{ W}$$

$$Q_a = 5.8 \text{ vars}$$

$$S = 3S_a = 24.3 \,\underline{/45°}$$

$$P = 3P_a = 17.6 \text{ W}$$

$$Q = 3Q_a = 17.6 \text{ vars}$$

$$\text{power factor} = \cos 45° = 0.71 \text{ lag}$$

[NOTE:

$$P \text{ also equals } 3I_a^2 R = 3 \times 1.4^2 \times 3$$

$$= 17.6 \text{ W}$$

$$Q \text{ also equals } 3I_a^2 X = 3 \times 1.4^2 \times 3$$

$$17.6 \text{ vars}]$$

A system like the one in this example is generally drawn using a *single line diagram*. This diagram is simply a drawing of one phase, ignoring the neutral. As an example the system above would be drawn as in Figure D.11.

Converting a load to an equivalent impedance in ohms from VA, P.F., and voltage is quite simple. Since

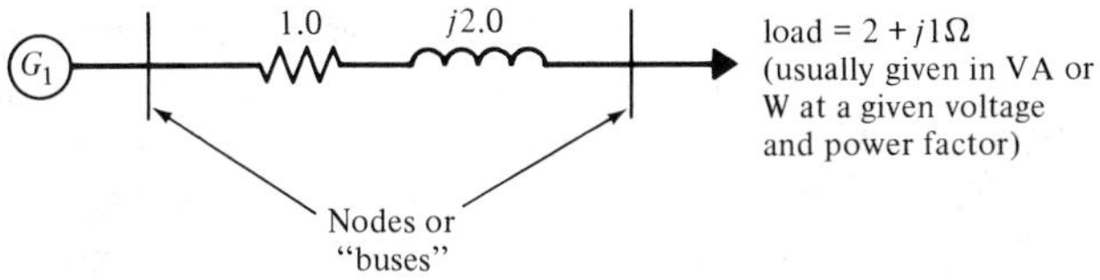

FIGURE D.11 Single-Line Diagram for Example D.4

$$S_{\text{phase}} = \frac{\dot{E}^2}{\dot{Z}}$$

$$P_{\text{phase}} = \frac{|\dot{E}|^2}{|\dot{Z}|} \cos\theta = \frac{|\dot{E}|^2}{|\dot{Z}|} \times (\text{P.F.}) \qquad \text{W}$$

$$Q_{\text{phase}} = \sqrt{|\dot{S}|^2 - P^2} \qquad \text{vars}$$

Then

$$|\dot{Z}_y| = \frac{3|\dot{E}_{\text{LN}}|^2}{|\dot{S}_{\text{three-phase}}|} = \frac{|\dot{E}_{\text{LL}}|^2}{|\dot{S}_{\text{three-phase}}|}$$

$$R_Y = |\dot{Z}_Y| \times \text{P.F.}$$

$$X_Y = \sqrt{|\dot{Z}_Y|^2 - R_Y^2}$$

EXAMPLE D.5   A three-phase load at 240 V_{LL} draws 10 kVA at 0.8 power factor lag. What is its equivalent wye impedance?

$$|\dot{Z}_Y| = \frac{(240)^2}{10,000} = 5.76 \ \Omega$$

$$R_Y = 5.76 \ (0.8)\Omega = 4.61 \ \Omega$$

$$X_Y = \sqrt{5.76^2 - 4.61^2} = +j3.45 \ \Omega \quad \text{(since lagging pf)}$$

$$\therefore \ Z_Y = 4.61 + j3.45 = 5.76 \ \underline{/36.9°}$$

Loads expressed as power at a given power factor and voltage can also be analyzed in a circuit by converting to equivalent line currents at given L-N voltage. The beginner initially may be more comfortable using equivalent impedances but should explore this more powerful method (which eliminates computing the impedance) after practice with impedances.

THE PER UNIT SYSTEM—GENERAL CONSIDERATIONS FOR THREE-PHASE CIRCUIT ANALYSIS

The per unit system (or its equivalent percent system) is generally used to simplify large power system computations when numerous transformers with various turns ratios are in the system. The per unit system measures all system quantities as decimal fractions of a set of base quantities; the percent system does the same except that percentage of base is used. For example, on a 10-Ω base, 9 Ω is 90% or 0.9 per unit (p.u.), and 12 Ω is 120% or 1.2 p.u. Base values are typically chosen based on system quantities for kVA and voltage; base current and impedance are then derived (using three-phase and line-to-line quantities):

$$I_{base} = \frac{kVA_{base}(3\varphi)}{\sqrt{3}\,E_{base}(L - L)}$$

$$Z_{base} = \frac{E_{base}^2(L - L)}{kVA_{base}(3\varphi)}$$

All powers, voltages, currents, and impedances are then computed in per unit or percent by

$$value|_{p.u.} = \frac{value|_{actual}}{value|_{base}}$$

$$value\ (\%) = value|_{p.u.} \times 100\%$$

All impedances in a system must be on the same basis in order to analyze it. In some cases, values may be given in per unit on a basis different from the one used to solve the system. To convert impedance from one basis to another, the following formula is used:

$$Z_{p.u.\ new} = Z_{p.u.\ old} \left(\frac{E_{base\ old}}{E_{base\ new}}\right)^2 \left(\frac{S_{base\ new}}{S_{base\ old}}\right)$$

$$= Z_{p.u.\ old} \times \left(\frac{kV_{base\ old}}{kV_{base\ new}}\right)^2 \left(\frac{kVA_{base\ new}}{kVA_{base\ old}}\right)$$

The concept of selection of base values can be a complex process and is too lengthy to be included here. All major textbooks dealing with power system analysis address this issue in detail.

Three-Phase Transformers—
Balanced Case

This appendix addresses the steady-state operation of transformers in three-phase systems. It is oriented toward a general understanding of terminal characteristics of several very common configurations found on transmission and distribution systems. It is not intended to provide an exhaustive analysis of all configurations under various balanced and unbalanced modes. *The reader is thus cautioned that the concents of this appendix generally are not applicable to systems with significant imbalances in phase voltages and/or currents.*

USE OF THREE-PHASE TRANSFORMERS

Transformers can be included in a three-phase system in either of two ways:

1. Connection of individual single-phase transformers (called a *bank*) in various configurations.
2. Construction of a single three-phase transformer on one core, with the individual coils connected in various configurations.

For the balanced steady-state case it can be shown that the use of three single-phase transformers is essentially equivalent to use of a single three-phase transformer,† except that efficiency is slightly higher in the three-phase unit. In many

---

†This is *not* true for unbalanced systems.

cases the three-phase unit may also be less expensive initially and may require less space than a bank of three single-phase units of equivalent total kVA rating. This advantage may be partially offset, though, by the flexibility inherent in the use of single-phase units and the lower replacement or repair costs if, for example, insulation on a single transformer coil fails. These comparative economics have resulted in the almost universal use of three-phase units on high-voltage transmission systems, whereas on commercial and industrial low-voltage distribution systems a mix of some three-phase and many single-phase banks is used. This use of banks on distribution systems is due primarily to the widespread availability and low price of single-phase distribution transformers, which can be delivered and hooked up in any configuration on very short notice and at a very low price, resulting in significant economies of scale in price for low-voltage distribution units. These economies simply do not exist for large-scale transmission applications of single-phase units.

THREE-PHASE TRANSFORMER CONNECTIONS

For this appendix it is assumed that any three-phase transformer is equivalent to three single-phase transformers of the same rating. This discussion will be in terms of single-phase banks, and is therefore assumed to be applicable, in the balanced steady-state, to three-phase transformers.

As discussed in Appendix D, components in three-phase circuits can be connected in either delta or wye configurations. Single-phase transformers consist of two coils, each of which can be connected either line to line or line to neutral, thus giving rise to the four transformer connections in Figure E.1: wye–wye, wye–delta, delta–wye, and delta–delta. These will now be discussed in terms of ideal transformers.

THE WYE-WYE CONNECTION

In the wye-wye connection, each coil is exposed to the line-to-neutral voltage and line current. Neutral conductors and/or grounded neutrals may or may not be used. Approximate voltage and current relations for phase a, shown in general in Figure E.2, are

$$V_{an2} = \left(\frac{N_2}{N_1}\right) V_{an1}$$

$$V_{ab2} = \left(\frac{N_2}{N_1}\right) V_{ab1}$$

$$I_{a2} = \left(\frac{N_1}{N_2}\right) I_{a1}$$

The wye–wye transformer is sometimes used in high-voltage applications because it must stand only $1/\sqrt{3}$ times the line-to-line voltage. It is usually used on four-wire systems grounded at both ends, as the inclusion of the neutral conductor can be used to stop propagation through the transformer of harmonic frequency components of current caused by core nonlinearities.

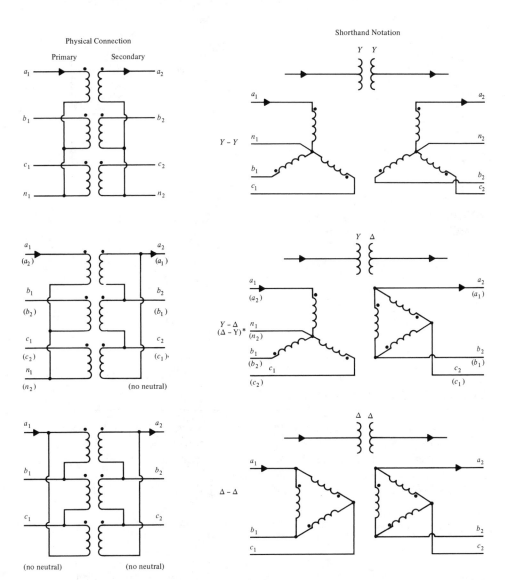

Physical Connection

Primary   Secondary

$Y - Y$

$Y - \Delta$
$(\Delta - Y)^*$

$\Delta - \Delta$

(no neutral)   (no neutral)

The Δ-Y configuration is the same as the Y-Δ configuration except that the predominant current flow is from the Δ-side to the Y-side (right to left in this figure). The phase designation for Δ-Y is shown in parentheses.

FIGURE E.1 Common Three-Phase Two-Winding Transformer Connections. Note: Neutrals may be grounded on the wye sides of the transformers.

EXAMPLE E.1 A 2400/240-V Y–Y transformer bank is connected across a 2400-V line. A current of 1 A flows in each phase on the high-voltage side. Find equivalent L-N and L-L values on the low-voltage side.

$$V_{ab1} = 2400 \underline{/0°} \rightarrow V_{an1} = \frac{2400}{\sqrt{3}} \underline{/-30°} = 1386 \underline{/-30°}$$

$$I_{a1} = 1 \underline{/0°}$$

THE WYE-WYE CONNECTION   **465**

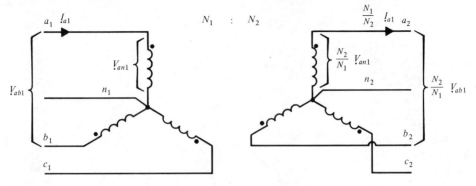

FIGURE E.2 Current and Voltage Relations for Wye-Wye Transformers. (Similar relations for phases *b* and *c*)

so

$$V_{ab2} = \frac{240}{2400} \cdot 2400 \underline{/0°} = 240 \underline{/0°}$$

$$V_{an2} = \frac{240}{2400} \cdot 1386 \underline{/-30°} = 139 \underline{/-30°}$$

$$I_{a2} = \frac{2400}{240} \cdot 1 \underline{/0°} = 10 \underline{/0°}$$

THE WYE–DELTA CONNECTION

Current and voltage relations for one common wye–delta connection are shown in Figure E.3. The primary side is connected from line to neutral and the secondary side is connected line to line, so for the connection shown,

$$V_{ab2} = \left(\frac{N_2}{N_1}\right) V_{an1}$$

$$I_{ab2} = \left(\frac{N_1}{N_2}\right) I_{a1}$$

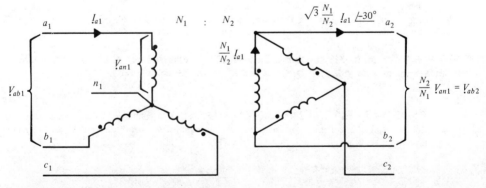

FIGURE E.3 Current and Voltage Relations for the Wye-Delta Transformer. (Similar relations for phases *b* and *c*)

or

$$V_{an2} = \frac{V_{ab2}}{\sqrt{3}} \underline{/-30°} = \frac{1}{\sqrt{3}} \left(\frac{N_2}{N_1}\right) V_{an1} \underline{/-30°}$$

$$I_{a2} = \sqrt{3}\, I_{ab2} \underline{/-30°} = \sqrt{3} \left(\frac{N_1}{N_2}\right) I_{a1} \underline{/-30°}$$

Note that the phase shift can change if the phases on the secondary side are interchanged while keeping the same phase rotation. Wye–delta transformers are generally used to step down transmission voltage (wye side) to low voltage (delta side) because they will not pass ground-fault currents in grounded-neutral systems (such as most high-voltage transmission systems). Thus they offer some degree of protection for low-voltage equipment connected to the Δ side from ground-fault currents occurring from lightning strikes, and so on.

EXAMPLE E.2   For the given data ($V_{L\text{-}L} = 12{,}500 \underline{/0°}$; $I_L = 5 \underline{/-36.9°}$) on the high-voltage side (Y side) of a bank of 2400:12,500-V transformers, compute the corresponding Δ-side quantities:

$$V_{ab1} = 12{,}500 \underline{/0°} \rightarrow V_{an1} = \frac{12{,}500}{\sqrt{3}} \underline{/-30°} = 7217 \underline{/-30°}$$

so

$$V_{an2} = \frac{1}{\sqrt{3}} \left(\frac{N_2}{N_1}\right) V_{an1} \underline{/-30°} = \frac{7217}{\sqrt{3}} \cdot \frac{2400}{12{,}500} \underline{/-30 - 30°} = 800 \underline{/-60°}$$

$$I_{a1} = 5 \underline{/-36.9°} \rightarrow I_{a2} = \sqrt{3}\, \frac{12{,}500}{2400}\, 5 \underline{/(-36.9° - 30°)} = 45 \underline{/-66.9°}$$

THE DELTA–WYE CONNECTION

The delta–wye connection is the opposite of wye–delta, so current and voltage relations, shown in Figure E.4, are

$$V_{an2} = \left(\frac{N_2}{N_1}\right) V_{ab1}$$

$$I_{a2} = \left(\frac{N_1}{N_2}\right) I_{ab1}$$

or

$$V_{ab2} = \left(\frac{N_2}{N_1}\right) \sqrt{3}\, V_{ab1} \underline{/30°}$$

$$I_{a2} = \left(\frac{N_1}{N_2}\right)\left(\frac{1}{\sqrt{3}}\right) I_{a1} \underline{/30°}$$

Delta–wye transformers are generally used as step-up transformers (low-voltage Δ, high voltage Y) to transmission voltages for the same reasons as wye–delta transformers are used to step these voltages down.

Because of the similarity between this configuration and the wye–delta transformer, no example will be given.

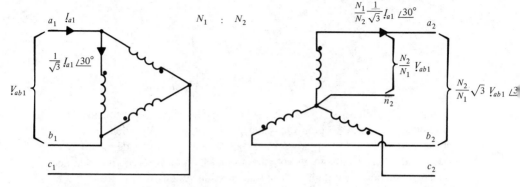

FIGURE E.4 Current and Voltage Relations for the Delta-Wye Transformer. (Similar relations for phases *b* and *c*)

THE DELTA–DELTA CONNECTION

The delta–delta connection has no neutral on either side of the transformer. As is seen by the current and voltage relations in Figure E.5, it is a dual of the wye–wye transformer:

$$V_{ab2} = \left(\frac{N_2}{N_1}\right) V_{ab1}$$

$$V_{an2} = \left(\frac{N_2}{N_1}\right) V_{an1}$$

$$I_{ab2} = \left(\frac{N_1}{N_2}\right) I_{ab1}$$

$$I_{a2} = \left(\frac{N_1}{N_2}\right) I_{a1}$$

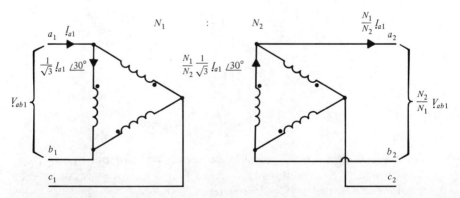

FIGURE E.5 Current and Voltage Relations for the Delta-Delta Connection (Similar relations for phases *b* and *c*)

Neither the delta–delta nor the wye–wye transformer connections cause any phase shift in either voltage or current. The delta–delta connection is commonly used for low- and medium-voltage applications, requiring a higher voltage rating per phase than a wye–wye, since each coil is exposed to full L-L voltage. However, this disadvantage is largely outweighed by the unique capability of a delta–delta connection to be operated at reduced throughput (reduced load) with one of the three transformers physically removed from the circuit, as will now be discussed.

THE OPEN-DELTA CONNECTION—A SPECIAL CASE

The open-delta connection is shown in Figure E.6. One of the three transformers physically is not present or is switched out of the circuit. It can be shown that, under balanced conditions, an open-delta bank can be operated at slightly less than 58% of the throughput of an equivalent delta–delta bank; and the voltages and currents are, for practical purposes, the same as for a delta–delta bank.

Consequently, open-delta banks have two important roles in power distribution. First, they allow operation at reduced initial investment on systems subject to future expansion. A balanced three-phase load can be served with an open-

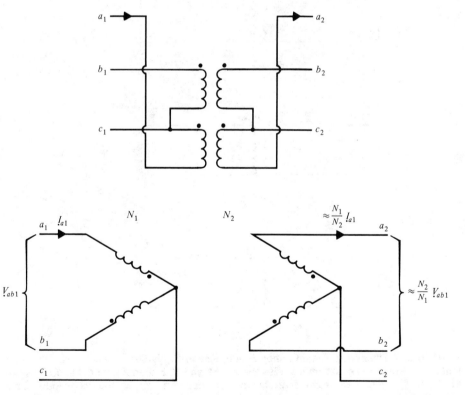

FIGURE E.6 Open-Delta Connection Under Balanced Conditions. (Transformer _A-B_ is eliminated from Δ-Δ to produce an open-Δ)

delta bank, allowing for future expansions in load up to 72% of the open-delta rating, simply by adding the third transformer. Second, a distribution system that is reasonably well balanced can be operated at reduced load while one transformer is being serviced or replaced. This is particularly important in industrial plants that can reduce throughput at low cost but cannot afford periods of shutdown for significant portions of the load served.

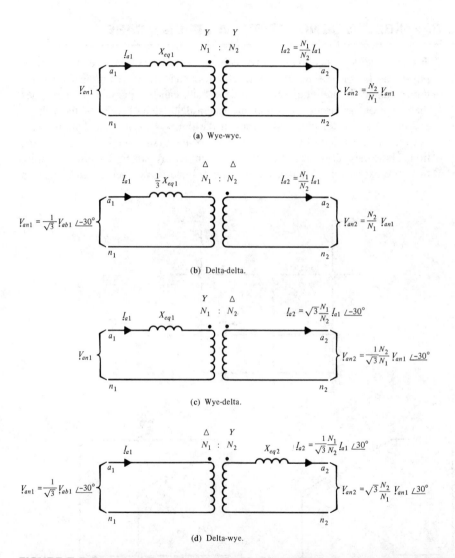

FIGURE E.7 Practical Transformer Single-Phase Equivalent Circuits (L-N equivalents) for Power Transformers. Note: For Δ-Y and Y-Δ transformers, use the value of X_{eq} on the wye side of the transformer. Do not shift X_{eq} to the other side of the ideal transformer once the equivalent circuit is drawn

Power transformers are generally designed with very small core and winding losses, which can be neglected in most analyses. Consequently, the transformer can be represented by a simple series leakage reactance and an ideal transformer, as in Figure E.7(a). Delta–delta transformers have these impedances in the legs of the Δ; as shown in Figure E.7(b), it is generally customary to convert them to Y-equivalent impedances while also converting Δ voltages and current to L-N and line quantities (see Appendix D), thus treating the transformer as an equivalent wye–wye. Delta voltages and currents should also be converted to wye for delta–wye and wye–delta transformers, but the impedance can be referred to the wye side of the transformer and treated directly as in Figure E.7(c) and (d).

Elementary Matrix Operations

This appendix reviews the basic elements of matrix algebra necessary to understand the mathematical analyses of coupled-coil systems in general and machines in particular included in the text. It is not intended to cover all aspects of linear algebra or matrix theory; instead, it approaches the subject from a purely mechanical point of view as it relates to the subjects in the text.

One way to solve many machine and circuits problems is through the use of matrices. Matrices not only allow use of shorthand notation to describe very large systems of linear equations, but also are well suited for solution by computer.

EXPRESSION OF SIMULTANEOUS LINEAR EQUATIONS IN MATRIX FORM

A system of simultaneous linear equations

$$
\begin{aligned}
a_{11}x_1 + a_{12}x_2 + \cdots + a_{1n}x_n &= b_1 \\
a_{21}x_1 + a_{22}x_2 + \cdots + a_{2n}x_n &= b_2 \\
&\vdots \\
a_{n1}x_1 + a_{n2}x_2 + \cdots + a_{nn}x_n &= b_n
\end{aligned}
$$

can be written in matrix form as $\overline{A}\overline{x} = \overline{b}$, where

$$\overline{A} = \begin{bmatrix} a_{11} & a_{12} & \cdots & a_{1n} \\ a_{21} & a_{22} & \cdots & a_{2n} \\ \vdots & & & \vdots \\ a_{n1} & a_{n2} & \cdots & a_{nn} \end{bmatrix}$$

$$\overline{x} = \begin{bmatrix} x_1 \\ x_2 \\ \vdots \\ x_n \end{bmatrix}$$

$$\overline{b} = \begin{bmatrix} b_1 \\ b_2 \\ \vdots \\ b_n \end{bmatrix}$$

For a system of n equations in n unknowns, the matrix $\overline{A}$ is a *square matrix* with n rows and n columns; in this case the number of equations n equals the *order* of the square matrix $\overline{A}$. Matrices $\overline{x}$ and $\overline{b}$ have n rows and one column and are called *vectors*.

EXAMPLE F.1   Express the following system of linear equations in matrix form:

$$2x_1 + 3x_2 + 5x_3 = 17$$
$$x_2 - 4x_3 = 12$$
$$21x_1 - 7x_2 - 2x_3 = 6$$

In matrix form, this system is expressed as $\overline{A}\overline{x} = \overline{b}$, or

$$\begin{bmatrix} 2 & 3 & 5 \\ 0 & 1 & -4 \\ 21 & -7 & -2 \end{bmatrix} \begin{bmatrix} x_1 \\ x_2 \\ x_3 \end{bmatrix} = \begin{bmatrix} 17 \\ 12 \\ 6 \end{bmatrix}$$

where the order or $\overline{A}$ is 3.

MATRIX DIMENSIONS AND ELEMENT NOTATION

The size of a matrix or vector can be described using its dimensions expressed as

$$\text{(number of rows } n_r) \times \text{(number of columns } n_c)$$

or

$$n_r \times n_c$$

For a square matrix, n_r equals n_c. A vector always has either n_r or n_c equal to 1 (i.e., only one row or one column). A *row vector* has $n_r = 1$ and a *column vector* has $n_c = 1$.

EXAMPLE F.2 Find the dimensions of matrices $\overline{A}$, $\overline{b}$, and $\overline{x}$ in Example F.1.

$$\overline{A} = \begin{bmatrix} 2 & 3 & 5 \\ 0 & 1 & -4 \\ 21 & -7 & -2 \end{bmatrix}$$ has three rows and three columns, so it is a 3×3 matrix

$$\overline{b} = \begin{bmatrix} 17 \\ 12 \\ 6 \end{bmatrix}$$ has three rows and one column so it is a 3×1 matrix or a column vector of order 3

$$\overline{x} = \begin{bmatrix} x_1 \\ x_2 \\ x_3 \end{bmatrix}$$ is also a 3×1 matrix or a column vector of order 3

The location of each element, or entry, in the matrix is described using subscripts giving first the row number, then the column number. So, for example, a 2×3 matrix

$$\overline{Q} = \begin{bmatrix} 2 & 6 & 9 \\ 1 & 4 & 7 \end{bmatrix}$$

has elements

$$\begin{bmatrix} q_{1,1} & q_{1,2} & q_{1,3} \\ q_{2,1} & q_{2,2} & q_{2,3} \end{bmatrix}$$

where

$$q_{1,1} = 2 \qquad q_{2,1} = 1$$

$$q_{1,2} = 6 \qquad q_{2,2} = 4$$

$$q_{1,3} = 9 \qquad q_{2,3} = 7$$

For small matrices (less than order 10), the comma between the subscripts is usually omitted.

THE SYMMETRIC MATRIX

In a *symmetric* matrix, each element a_{ij} equals element a_{ji}.

TRANSPOSITION OF A MATRIX

A matrix is *transposed* by interchanging each element in each row n with the corresponding element in column n. Consequently, if $\overline{A}$ is an $i \times j$ matrix (i rows and j columns), then $\overline{A}^t$ is a $j \times i$ matrix (j rows and i columns).

EXAMPLE F.3 Transpose the matrices $\overline{A}$, $\overline{b}$, and $\overline{x}$ in Example F.1. In transposition, each element a_{ij} becomes a_{ji}, b_{ij} becomes b_{ji}, and x_{ij} becomes x_{ji}, so

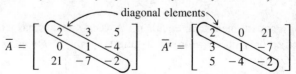

diagonal elements

$$\overline{A} = \begin{bmatrix} 2 & 3 & 5 \\ 0 & 1 & -4 \\ 21 & -7 & -2 \end{bmatrix} \qquad \overline{A}^t = \begin{bmatrix} 2 & 0 & 21 \\ 3 & 1 & -7 \\ 5 & -4 & -2 \end{bmatrix}$$

Note that the main diagonal elements (highlighted) do not change position and that since the matrix is square, its dimensions remain 3×3.

$$\bar{b} = \begin{bmatrix} 17 \\ 12 \\ 6 \end{bmatrix} \qquad \bar{b}' = [17 \quad 12 \quad 6]$$

$$\bar{x} = \begin{bmatrix} x_1 \\ x_2 \\ x_3 \end{bmatrix} \qquad \bar{x}' = [x_1 \quad x_2 \quad x_3]$$

Note here that the column vectors become row vectors when transposed (i.e., $\bar{b}_{3 \times 1}$ becomes $\bar{b}'_{1 \times 3}$). Similarly, a row vector becomes a column vector when transposed:

$$\bar{C} = [\gamma_{11} \quad \gamma_{12} \quad \gamma_{13}] \qquad \bar{C}' = \begin{bmatrix} \gamma_{11} \\ \gamma_{21} \\ \gamma_{31} \end{bmatrix}$$

where $\gamma_{12} = \gamma_{21}$ and $\gamma_{13} = \gamma_{31}$. Other matrices with unequal dimensions transpose similarly:

$$\bar{Q}_{2 \times 3} = \begin{bmatrix} 2 & 6 & 9 \\ 1 & 4 & 7 \end{bmatrix} \qquad \bar{Q}'_{3 \times 2} = \begin{bmatrix} 2 & 1 \\ 6 & 4 \\ 9 & 7 \end{bmatrix}$$

The transpose of a *symmetric* matrix equals the matrix $(\bar{A}' = \bar{A})$.

ADDITION AND SUBTRACTION OF MATRICES

Two or more matrices *of identical dimensions* may be easily added or subtracted by simply adding or subtracting the corresponding elements in each matrix. Addition and subtraction follow the conventional commutative and associative laws as follows:

Addition

$$\bar{A} + \bar{B} = \bar{B} + \bar{A}$$

$$\bar{A} + (\bar{B} + \bar{C}) = (\bar{A} + \bar{B}) + \bar{C}$$

Subtraction

$$\bar{A} - \bar{B} = -(\bar{B} - \bar{A})$$

$$\bar{A} - (\bar{B} - \bar{C}) = (\bar{A} - \bar{B}) - (-\bar{C})$$

EXAMPLE F.4   Add and subtract the two 2×3 matrices $\bar{P}$ and $\bar{Q}$.

$$\bar{P} = \begin{bmatrix} 2 & 1 & 5 \\ 7 & 10 & 12 \end{bmatrix}$$

$$\bar{Q} = \begin{bmatrix} 3 & 8 & 9 \\ 6 & 4 & 11 \end{bmatrix}$$

$$\bar{P} + \bar{Q} = \begin{bmatrix} 2+3 & 1+8 & 5+9 \\ 7+6 & 10+4 & 12+11 \end{bmatrix} = \begin{bmatrix} 5 & 9 & 14 \\ 13 & 14 & 23 \end{bmatrix}$$

$$\bar{Q} + \bar{P} = \begin{bmatrix} 3+2 & 8+1 & 9+5 \\ 6+7 & 4+10 & 11+12 \end{bmatrix} = \begin{bmatrix} 5 & 9 & 14 \\ 13 & 14 & 23 \end{bmatrix} = \bar{P} + \bar{Q}$$

$$\bar{P} - \bar{Q} = \begin{bmatrix} 2-3 & 1-8 & 5-9 \\ 7-6 & 10-4 & 12-11 \end{bmatrix} = \begin{bmatrix} -1 & -7 & -4 \\ 1 & 6 & 1 \end{bmatrix}$$

$$\bar{Q} - \bar{P} = \begin{bmatrix} 3-2 & 8-1 & 9-5 \\ 6-7 & 4-10 & 11-12 \end{bmatrix} = \begin{bmatrix} 1 & 7 & 4 \\ -1 & -6 & -1 \end{bmatrix} = -(\bar{P} - \bar{Q})$$

MULTIPLICATION OR DIVISION OF A MATRIX BY A SCALAR

If a matrix or vector is multiplied or divided by a scalar, then every element is multiplied or divided by the scalar:

$$a\overline{A} \quad \text{where } \overline{A} = \begin{bmatrix} A_{11} & A_{12} \\ A_{21} & A_{22} \end{bmatrix} \quad \text{is} \quad \begin{bmatrix} aA_{11} & aA_{12} \\ aA_{21} & aA_{22} \end{bmatrix}$$

EXAMPLE F.5 Multiply matrix $\overline{P}$ in Example F.4 by 2 and divide matrix $\overline{Q}$ in Example F.4 by 2:

$$2\overline{P} = 2\begin{bmatrix} 2 & 1 & 5 \\ 7 & 10 & 12 \end{bmatrix} = \begin{bmatrix} 2 \cdot 2 & 2 \cdot 1 & 2 \cdot 5 \\ 2 \cdot 7 & 2 \cdot 10 & 2 \cdot 12 \end{bmatrix} = \begin{bmatrix} 4 & 2 & 10 \\ 14 & 20 & 24 \end{bmatrix}$$

$$\frac{\overline{Q}}{2} = \frac{1}{2}\begin{bmatrix} 3 & 8 & 9 \\ 6 & 4 & 3 \end{bmatrix} = \begin{bmatrix} \frac{3}{2} & \frac{8}{2} & \frac{9}{2} \\ \frac{6}{2} & \frac{4}{2} & \frac{11}{2} \end{bmatrix} = \begin{bmatrix} 1.5 & 4 & 4.5 \\ 3 & 2 & 5.5 \end{bmatrix}$$

THE IDENTITY MATRIX

An *identity* matrix is a square matrix where elements $a_{ii} = 1$ (diagonal elements) and $a_{ij} = a_{ji} = 0$, $i \neq j$ (off-diagonal elements). The symbol for an identity matrix is $\overline{I}$. The product of any matrix and the identity matrix is the original matrix.

MULTIPLICATION OF TWO MATRICES

Matrices $\overline{A}$ and $\overline{B}$ can be multiplied to compute $\overline{A} \times \overline{B}$ if the number of columns in $\overline{A}$ equals the number of rows in $\overline{B}$. In general, matrix multiplication is not commutative so (except under very specific circumstances not discussed in this appendix), $\overline{A} \times \overline{B} \neq \overline{B} \times \overline{A}$. Matrix multiplication is both associative and distributive. The matrix $\overline{D}$ resulting from $A_{m \times n} \times B_{n \times p}$ is of dimension m × p. Each element d_{ij} is found to be

$$d_{ij} = \sum_{k=1}^{n} a_{ik}b_{kj}$$

In other words, multiply each element in row i of $\overline{A}$ by each element in column j of $\overline{B}$, adding the products thus obtained, and place the resulting number in matrix $\overline{D}$ as element d_{ij}:

$$\overline{D} = \overline{A} \cdot \overline{B} = \begin{bmatrix} \alpha_{11} & \alpha_{12} & \alpha_{13} \\ \alpha_{21} & \alpha_{22} & \alpha_{23} \end{bmatrix}\begin{bmatrix} \beta_{11} & \beta_{12} \\ \beta_{21} & \beta_{22} \\ \beta_{31} & \beta_{32} \end{bmatrix}$$

$$= \begin{bmatrix} \left(\begin{matrix} \alpha_{11}\beta_{11} + \alpha_{12}\beta_{21} \\ + \alpha_{13}\beta_{31} \end{matrix} \right) & \left(\begin{matrix} \alpha_{11}\beta_{12} + \alpha_{12}\beta_{22} \\ + \alpha_{13}\beta_{32} \end{matrix} \right) \\ \left(\begin{matrix} \alpha_{21}\beta_{11} + \alpha_{22}\beta_{21} \\ + \alpha_{23}\beta_{31} \end{matrix} \right) & \left(\begin{matrix} \alpha_{21}\beta_{12} + \alpha_{22}\beta_{22} \\ + \alpha_{23}\beta_{32} \end{matrix} \right) \end{bmatrix}$$

so

$$d_{11} = \alpha_{11}\beta_{11} + \alpha_{12}\beta_{21} + \alpha_{13}\beta_{31}$$

$$d_{12} = \alpha_{11}\beta_{12} + \alpha_{12}\beta_{22} + \alpha_{13}\beta_{32}$$

$$d_{21} = \alpha_{21}\beta_{11} + \alpha_{22}\beta_{21} + \alpha_{23}\beta_{31}$$

$$d_{22} = \alpha_{22}\beta_{12} + \alpha_{22}\beta_{22} + \alpha_{23}\beta_{32}$$

Note also that since the dimension of $\overline{A}$ is 2×3 and the dimension of $\overline{B}$ is 3×2, the dimension of $\overline{D}$ is 2×2 and each element d_{ij} is made up of three terms. For the product $\overline{A} \times \overline{B}$, $\overline{B}$ is said to be *premultiplied* by $\overline{A}$ and $\overline{A}$ is said to be *postmultiplied* by $\overline{B}$.

Division of two matrices does not exist per se.

THE DETERMINANT OF A SQUARE MATRIX

The determinant of a square matrix is a *scalar* computed as follows: For matrix $\overline{A}$,

$$\det(\overline{A}) \quad \text{or} \quad |\overline{A}| = \sum (-1)^P a_{1i} a_{2j} a_{3k} \cdots a_{mn}$$

where $i, j, k, \ldots, n$ form a permutation of the numbers $1, 2, \ldots, n$, the sum is taken over *all* permutations, and P is the number of digit exchanges necessary to bring the series of digits $i, j, k, \ldots, n$, back into ascending order (e.g., $3124 \rightarrow 1324 \rightarrow 1234$, so $P = 3$). By Laplace's expansion, the determinant of square matrix $\overline{A}$ of order n can be expressed in terms of the sum of determinants of square matrices of order $n - 1$ for any row i or column j:

$$\det(\overline{A}) = \sum_{j=1}^{n} a_{ij} C_{ij} \quad \text{(expansion by rows for any single row } i\text{)}$$

or

$$\det(\overline{A}) = \sum_{i=1}^{n} a_{ij} C_{ij} \quad \text{(expansion by columns for any single column } j\text{)}$$

where the cofactor C_{ij} equals $(-1)^{i+j}$ multiplied by the determinant of the matrix $\overline{A}$ with row i and column j omitted.

EXAMPLE F.6   Compute the determinant of a 2×2 matrix

$$\overline{A} = \begin{bmatrix} \alpha_{11} & \alpha_{12} \\ \alpha_{21} & \alpha_{22} \end{bmatrix}$$

$$\det(\overline{A}) = \alpha_{11}\alpha_{22} - \alpha_{21}\alpha_{12}$$

EXAMPLE F.7   Compute the cofactors for row 1 of the 3×3 matrix $\overline{B}$, then compute $\det(\overline{B})$ using expansion by rows for row 1.

$$\bar{B} = \begin{bmatrix} \beta_{11} & \beta_{12} & \beta_{13} \\ \beta_{21} & \beta_{22} & \beta_{23} \\ \beta_{31} & \beta_{32} & \beta_{33} \end{bmatrix}$$

$$C_{11} = (-1)^{1+1} \det \begin{bmatrix} \beta_{22} & \beta_{23} \\ \beta_{32} & \beta_{33} \end{bmatrix} = \beta_{22}\beta_{33} - \beta_{32}\beta_{23}$$

$$C_{12} = (-1)^{1+2} \det \begin{bmatrix} \beta_{21} & \beta_{23} \\ \beta_{31} & \beta_{33} \end{bmatrix} = (-1)(\beta_{21}\beta_{33} - \beta_{31}\beta_{23})$$

$$C_{13} = (-1)^{1+3} \det \begin{bmatrix} \beta_{21} & \beta_{22} \\ \beta_{31} & \beta_{32} \end{bmatrix} = \beta_{21}\beta_{32} - \beta_{22}\beta_{31}$$

so det $(\bar{B})$, using row 1, is

$$\det (\bar{B}) = \sum_{j=1}^{3} \beta_{1j} C_{1j}$$

$$= \beta_{11}(\beta_{22}\beta_{33} - \beta_{32}\beta_{23}) + \beta_{12}(-1)(\beta_{21}\beta_{33} - \beta_{31}\beta_{23})$$
$$+ \beta_{13}(\beta_{21}\beta_{32} - \beta_{22}\beta_{31})$$

$$= \beta_{11}\beta_{22}\beta_{33} + \beta_{12}\beta_{31}\beta_{23} + \beta_{13}\beta_{21}\beta_{32}$$
$$- \beta_{11}\beta_{32}\beta_{23} - \beta_{12}\beta_{21}\beta_{33} - \beta_{13}\beta_{22}\beta_{31}$$

which is the familiar result.

A square matrix is said to be *singular* if its determinant equals 0 and *nonsingular* otherwise.

THE INVERSE OF A SQUARE MATRIX

Inversion, followed by premultiplication, is the matrix equivalent of division. *The matrix to be inverted must be square and nonsingular (determinant $\neq 0$).* The inverse may be computed by the following formula:

$$\bar{A}^{-1} = \frac{1}{\det (\bar{A})} \cdot \bar{C}_{ij}^{t}$$

where $\bar{C}_{ij}$ is the matrix of cofactors for each element a_{ij} in $\bar{A}$.

EXAMPLE F.8   Invert a nonsingular matrix $\bar{A}$ of order 2.

$$\bar{A} = \begin{bmatrix} \alpha_{11} & \alpha_{12} \\ \alpha_{21} & \alpha_{22} \end{bmatrix}$$

$$\det (\bar{A}) = \alpha_{11}\alpha_{22} - \alpha_{12}\alpha_{21} \ (\neq 0)$$

$$C_{11} = -1^{(1+1)}\alpha_{22} = \alpha_{22}; \qquad C_{21} = -1^{(2+1)}\alpha_{12} = -\alpha_{12}$$

$$C_{12} = -1^{(1+2)}\alpha_{21} = -\alpha_{21}; \qquad C_{22} = -1^{(2+2)}\alpha_{22} = \alpha_{22}$$

so

$$\bar{A}^{-1} = \frac{1}{\alpha_{11}\alpha_{22} - \alpha_{12}\alpha_{21}} \begin{bmatrix} \alpha_{22} & -\alpha_{21} \\ -\alpha_{12} & \alpha_{11} \end{bmatrix}^t$$

$$= \begin{bmatrix} \dfrac{\alpha_{22}}{\alpha_{11}\alpha_{22} - \alpha_{12}\alpha_{21}} & \dfrac{-\alpha_{12}}{\alpha_{11}\alpha_{22} - \alpha_{12}\alpha_{21}} \\ \dfrac{-\alpha_{21}}{\alpha_{11}\alpha_{22} - \alpha_{12}\alpha_{21}} & \dfrac{\alpha_{11}}{\alpha_{11}\alpha_{22} - \alpha_{12}\alpha_{21}} \end{bmatrix}$$

A matrix multiplied by its inverse yields the identity matrix:

$$\bar{A} \cdot \bar{A}^{-1} = \bar{A}^{-1} \cdot \bar{A} = \bar{I}$$

SOLUTION OF A SYSTEM OF SIMULTANEOUS LINEAR EQUATIONS BY MATRIX INVERSION AND MULTIPLICATION

One very basic method for solving a system of simultaneous linear equations using matrix theory is by inversion and multiplication. Consider solving the following matrix equation for $\bar{x}$:

$$\bar{A}\bar{x} = \bar{b}$$

where $\bar{A}$ is a square matrix of coefficients and $\bar{x}$ and $\bar{b}$ are column vectors. To solve for $\bar{x}$, it must be isolated on the left-hand side of the equation by premultiplying both sides by $\bar{A}^{-1}$.

$$\bar{A}^{-1}\bar{A}\bar{x} = \bar{A}^{-1}\bar{b}$$

Notice that the order in which the matrices are multiplied together must be the same on both sides of the equation since multiplication is not commutative. Multiplication results in the desired result:

$$\bar{I}\bar{x} = \bar{A}^{-1}\bar{b}$$

or

$$\bar{x} = \bar{A}^{-1}\bar{b}$$

Remember, in general $\bar{x} \neq \bar{b}\bar{A}^{-1}$ for this problem!

EXAMPLE F.9   Solve the system of equations in Example F.1. From Example F.1, $\bar{A}\bar{x} = \bar{b}$ is

$$\begin{bmatrix} 2 & 3 & 5 \\ 0 & 1 & -4 \\ 21 & -7 & -2 \end{bmatrix} \begin{bmatrix} x_1 \\ x_2 \\ x_3 \end{bmatrix} = \begin{bmatrix} 17 \\ 12 \\ 6 \end{bmatrix}$$

Step 1: Compute det $(\bar{A})$.

$$\det(\bar{A}) = a_{11}a_{22}a_{33} + a_{12}a_{31}a_{23} + a_{13}a_{21}a_{32}$$
$$\qquad - a_{11}a_{32}a_{23} - a_{12}a_{21}a_{33} - a_{13}a_{22}a_{31}$$
$$= 2 \cdot 1 \cdot (-2) + 3 \cdot (-4) \cdot 21 + 0 \cdot (-7)(5)$$
$$\qquad - 5 \cdot 1 \cdot 21 - (-4)(-7)(2) - 0 \cdot 3(-2)$$
$$= -4 + (-252) + 0 - 105 - 56$$
$$= -417 \quad (\text{so } \bar{A} \text{ is nonsingular and thus has an inverse})$$

Step 2: Invert $\bar{A}$.

$$\bar{A}^{-1} = \frac{1}{\det(\bar{A})} \bar{C}^t_{ij}$$

$$= \frac{1}{-417}
\begin{bmatrix}
-30 & -84 & -21 \\
-29 & -109 & 77 \\
-17 & 8 & 2
\end{bmatrix}^t$$

$$= \begin{bmatrix}
0.07 & 0.20 & 0.05 \\
0.07 & 0.26 & -0.18 \\
0.04 & -0.02 & -0.005
\end{bmatrix}^t$$

$$= \begin{bmatrix}
0.07 & 0.07 & 0.04 \\
0.20 & 0.26 & -0.02 \\
0.05 & -0.18 & -0.005
\end{bmatrix}$$

Step 3: Premultiply $\bar{b}$ by $\bar{A}^{-1}$ to find $\bar{x}$.

$$\bar{A}^{-1}\bar{b} = \begin{bmatrix}
0.07 & 0.07 & 0.04 \\
0.20 & 0.26 & -0.02 \\
0.05 & -0.18 & -0.005
\end{bmatrix}
\begin{bmatrix}
17 \\
12 \\
6
\end{bmatrix}$$

$$= \begin{bmatrix}
0.07 \cdot 17 + 0.07 \cdot 12 + 0.04 \cdot 6 \\
0.20 \cdot 17 + 0.26 \cdot 12 + (-0.02) \cdot 6 \\
0.05 \cdot 17 + (-0.18) \cdot 12 + (0.005) \cdot 6
\end{bmatrix} = \bar{x}$$

$$\therefore \bar{x} = \begin{bmatrix}
2.30 \\
6.45 \\
-1.39
\end{bmatrix} \quad \text{or} \quad
\begin{matrix}
x_1 = & 2.30 \\
x_2 = & 6.45 \\
x_3 = & -1.39
\end{matrix}$$

Check:

$$2 \cdot 2.3 + 2 \cdot 6.45 + 5 \cdot (-1.39) = 17$$
$$0 \cdot 2.3 + \quad + 1 \cdot 6.45 + (-4) \cdot (-1.39) = 12$$
$$21 \cdot 2.3 + (-7) \cdot 6.45 + (-2) \cdot (-1.39) = 6$$

The reader should note that numerous other matrix methods exist for solving systems of linear equations. The inverstion/multiplication method was chosen to illustrate the matrix techniques discussed in this appendix. A more detailed discussion of other solution methods is beyond the scope and intent of this book.

Solution of Ordinary Differential Equations in Normal Form

This appendix addresses, in general terms, the solution of ordinary differential equations put in *normal form*. The use of normal form techniques allows the reduction of a high-order differential equation (or series of equations) to a series of first-order equations that can be put in matrix form and solved either analytically (linear differential equations with constant coefficients) or numerically using simulation on a digital computer (linear and nonlinear equations). Examples of both analytical and numerical solutions of linear equations are given in this appendix.

To fully understand the material in this appendix, the student should review the matrix techniques given in Appendix F. As with the latter appendix, the material given here will stress representative techniques rather than theory.

REDUCTION OF A HIGHER-ORDER DIFFERENTIAL EQUATION TO NORMAL FORM

An nth-order differential equation (or series of equations) can be put in *normal form* if it can be decomposed into a set of n first-order differential equations by assigning new variables x_1, x_2, . . . , x_n to system parameters, *and* if these n equations (called *state equations*) completely describe the system. The variables x_1, x_2, . . . , x_n in such a system of state equations are called *state variables*. In general, the state variables must contain enough information about the past history of a system to determine its future behavior. If future inputs are known, they are not necessarily required to represent measurable physical quantities. In some cases it is advantageous to choose state variables to represent sources of stored energy in a system (e.g., voltage across a capacitor, current through an inductor). Methods for making such choices and the resulting mathematical manipulations to obtain the state equations are covered in numerous texts and are beyond the limited scope of this appendix.

In the simplest and most general sense, any differential equation can be reduced to a set of state equations by setting each derivative (including the zero order derivative) up to the $(n - 1)$ order derivative equal to a state variable x_i and then manipulating the resulting n equations into a series of equations involving only first derivatives of the state variables and the variables themselves.

EXAMPLE G.1 Put the following fourth-order linear differential equation with constant coefficients in normal form:

$$u(t) = \alpha \frac{d^4x}{dt^4} + \beta \frac{d^3x}{dt^3} + \gamma \frac{d^2x}{dt^2} + \delta \frac{dx}{dt} + \epsilon x + K$$

Set the state variables as follows:

$$x_1 = x \quad \left(= \frac{d^0x}{dt^0} \right)$$

$$x_2 = \frac{dx}{dt} \quad \left(= \frac{dx_1}{dt} \right)$$

$$x_3 = \frac{d^2x}{dt^2} \quad \left(= \frac{dx_2}{dt} \right)$$

$$x_4 = \frac{d^3x}{dt^3} \quad \left(= \frac{dx_3}{dt} \right)$$

Substituting the state variables yields

$$u(t) = \alpha \frac{dx_4}{dt} + \beta x_4 + \gamma x_3 + \delta x_2 + \epsilon x_1 + K$$

and rearranging to get four first-order differential equations in four unknowns (x_1, x_2, x_3, x_4), the equation is put in normal form:

$$\frac{dx_1}{dt} = x_2$$

$$\frac{dx_2}{dt} = x_3$$

$$\frac{dx_3}{dt} = x_4$$

$$\frac{dx_4}{dt} = \frac{u(t)}{\alpha} - \frac{\beta x_4}{\alpha} - \frac{\gamma x_3}{\alpha} - \frac{\delta x_2}{\alpha} - \frac{\epsilon x_1}{\alpha} - \frac{K}{\alpha}$$

Note that $u(t)$ is the *forcing function* of the equation and can be one or more very complicated functions. The example above considered a *linear* differential equation, that is, one where all terms have only values of x and its derivatives. It should be obvious that even nonlinear equations (equations where functions such as x^2, sin x, cos x, step function of x, products of x and a derivative, etc., are included in the terms) can also be put in normal form.

EXAMPLE G.2 Put the following second-order nonlinear differential equation in normal form:

$$\frac{d^2x}{dt^2} + \sqrt{x} = u$$

Choose the following state variables:

$$x_1 = x$$

$$x_2 = \frac{dx_1}{dt} \left(= \frac{dx}{dt} \right)$$

$$x_3 = \frac{dx_2}{dt} \left(= \frac{d^2x}{dt^2} \right)$$

So, in normal form, the equation becomes

$$\frac{dx_1}{dt} = x_2$$

$$\frac{dx_2}{dt} = x_3$$

$$\frac{dx_3}{dt} = -\sqrt{x_1} + u$$

Notice that the last equation remains nonlinear in this representation.

Analytic solution methods for nonlinear differential equations will not be addressed in this appendix. Numerical solutions are generally applicable. The remainder of this appendix will emphasize linear differential equations.

MATRIX REPRESENTATION OF NORMAL-FORM LINEAR DIFFERENTIAL EQUATIONS

Most analytic and virtually all numerical (computer) solution methods for linear differential equations are based on the matrix representation of the equations in normal form. In general, a linear nth-order differential equation is expressed as

$$\alpha_n \frac{d^n x}{dt^n} + \alpha_{n-1} \frac{d^{n-1}x}{dt^{n-1}} + \cdots + \alpha_2 \frac{d^2x}{dt^2} + \alpha_1 \frac{dx}{dt} + \alpha_0 x$$

$$= u_0(t) + u_1(t) + u_2(t) + \cdots + u_m(t)$$

where for this representation† $x(t)$ is the output variable and there are $m + 1$ input functions $u_0(t)$, $u_1(t)$, . . . , $u_m(t)$.

---

†In the interest of simplicity, this appendix will not address the general output equation $\overline{Y} = \overline{C}x + \overline{D}u$.

Reducing the equation above to normal form yields ($x_1 = x$):

$$\frac{dx_1}{dt} = x_2 \quad \left(= \frac{dx}{dt} \right)$$

$$\frac{dx_2}{dt} = x_3 \quad \left(= \frac{d^2x}{dt^2} \right)$$

$$\vdots$$

$$\frac{dx_{n-1}}{dt} = x_n \quad \left(= \frac{d^{n-1}x}{dt^{n-1}} \right)$$

$$\frac{dx_n}{dt} = -\frac{1}{\alpha_n} (\alpha_{n-1}x_n + \alpha_{n-2}x_{n-1} + \cdots + \alpha_2 x_3 + \alpha_1 x_2$$

$$+ \alpha_0 x_1) + \frac{1}{\alpha_n} (u_0 + u_1 + u_2 + \cdots + u_m)$$

In turn, this system of n equations in n unknowns can be reduced to the matrix equation

$$\frac{d\overline{x}}{dt} = \overline{A}\overline{x} + \overline{B}\overline{u}$$

where

$$\left. \frac{d\overline{x}}{dt} \right|_{n \times 1} = \begin{bmatrix} \dfrac{dx_1}{dt} \\ \dfrac{dx_2}{dt} \\ \vdots \\ \dfrac{dx_n}{dt} \end{bmatrix} \qquad \overline{x}\Big|_{n \times 1} = \begin{bmatrix} x_1 \\ x_2 \\ \vdots \\ x_n \end{bmatrix}$$

$$\overline{u}\Big|_{(m+1) \times 1} = \begin{bmatrix} u_0 \\ u_1 \\ u_2 \\ \vdots \\ u_m \end{bmatrix}$$

$$\overline{A}\Big|_{n \times n} = \begin{bmatrix} 0 & 1 & 0 & \cdots & 0 \\ 0 & 0 & 1 & \cdots & 0 \\ \vdots & \vdots & \vdots & & \vdots \\ -\dfrac{\alpha_0}{\alpha_n} & -\dfrac{\alpha_1}{\alpha_n} & -\dfrac{\alpha_2}{\alpha_n} & & -\dfrac{\alpha_{n-1}}{\alpha_n} \end{bmatrix}$$

$$\overline{B}\Big|_{n \times (m+1)} = \begin{bmatrix} 0 & 0 & 0 & \cdots & 0 \\ 0 & 0 & 0 & \cdots & 0 \\ \vdots & \vdots & \vdots & & \vdots \\ \dfrac{1}{\alpha_n} & \dfrac{1}{\alpha_n} & \dfrac{1}{\alpha_n} & & \dfrac{1}{\alpha_n} \end{bmatrix}$$

Note that since $\overline{B}$ is n rows $\times$ $(m + 1)$ columns and u is $(m + 1)$ rows $\times$ one column, the product $\overline{B}u$ is a column vector with n rows ($n \times 1$ matrix). Thus all terms in the matrix equation have n rows, a requirement for addition. The matrix $\overline{x}$ is the state vector; the matrix $\overline{A}$, the system matrix; $\overline{B}$ is the distribution matrix; and $\overline{u}$ is the input vector.

EXAMPLE G.3 Reduce the following linear differential equation to matrix form:

$$2\frac{d^3x}{dt^3} + 3\frac{d^2x}{dt^2} + 7\frac{dx}{dt} - 5x = 9 \cos t + 11 \sin t$$

First, finding state variables,

$$x_1 = x$$

$$x_2 = \frac{dx_1}{dt}$$

$$x_3 = \frac{dx_2}{dt}\left(= \frac{d^2x}{dt^2}\right)$$

so the equation in normal form is

$$\frac{dx_1}{dt} = x_2$$

$$\frac{dx_2}{dt} = x_3$$

$$\frac{dx_3}{dt} = -\tfrac{3}{2}x_3 - \tfrac{7}{2}x_2 + \tfrac{5}{2}x_1 + \tfrac{9}{2}\cos t + \tfrac{11}{2}\sin t$$

Putting this in the form $d\overline{x}/dt = \overline{A}x + \overline{B}u$ yields

$$
\begin{bmatrix} \dfrac{dx_1}{dt} \\[2mm] \dfrac{dx_2}{dt} \\[2mm] \dfrac{dx_3}{dt} \end{bmatrix}
=
\begin{bmatrix} 0 & 1 & 0 \\ 0 & 0 & 1 \\ \dfrac{5}{2} & -\dfrac{7}{2} & -\dfrac{3}{2} \end{bmatrix}
\begin{bmatrix} x_1 \\ x_2 \\ x_3 \end{bmatrix}
+
\begin{bmatrix} 0 & 0 \\ 0 & 0 \\ \dfrac{9}{2} & \dfrac{11}{2} \end{bmatrix}
\begin{bmatrix} \cos t \\ \sin t \end{bmatrix}
$$

ANALYTIC ANALYSIS OF $d\overline{x}/dt = \overline{A}x + \overline{B}u$

The solution of the normal-form matrix equation $d\overline{x}/dt = \overline{A}x + \overline{B}u$ can be viewed as a sum of solutions to n scalar equations of the form

$$\frac{dx}{dt} = ax + bu$$

Consequently, one would expect a transient solution having terms involving sums and differences of e^{a_it}. In fact, it can be shown that the values a_i are related to another set of values λ_i called *eigenvalues* that satisfy the equation

$$\det(\overline{A} - \lambda\overline{I}) = 0$$

where $\overline{A}$ is from the normal-form equation, λ is a scalar, and $\overline{I}$ is the identity matrix. The determinant equation above is called the *characteristic equation* of the system and it can be proved that the eigenvalues λ_i are the roots.

EXAMPLE G.4 Write as a determinant equation the characteristic equation of the system in Example G.3:

$$\bar{A} = \begin{bmatrix} 0 & 1 & 0 \\ 0 & 0 & 1 \\ \dfrac{5}{2} & -\dfrac{7}{2} & -\dfrac{3}{2} \end{bmatrix}$$

$$\lambda \bar{I} = \lambda \begin{bmatrix} 1 & 0 & 0 \\ 0 & 1 & 0 \\ 0 & 0 & 1 \end{bmatrix} = \begin{bmatrix} \lambda & 0 & 0 \\ 0 & \lambda & 0 \\ 0 & 0 & \lambda \end{bmatrix}$$

So, in determinant equation form, the characteristic equation is

$$\det (\bar{A} - \lambda \bar{I}) = \det \left\{ \begin{bmatrix} 0 & 1 & 0 \\ 0 & 0 & 1 \\ \dfrac{5}{2} & -\dfrac{7}{2} & -\dfrac{3}{2} \end{bmatrix} - \begin{bmatrix} \lambda & 0 & 0 \\ 0 & \lambda & 0 \\ 0 & 0 & \lambda \end{bmatrix} \right\}$$

which reduces to

$$\det \begin{bmatrix} -\lambda & 1 & 0 \\ 0 & -\lambda & 1 \\ \dfrac{5}{2} & -\dfrac{7}{2} & \left(-\dfrac{3}{2} - \lambda \right) \end{bmatrix} = 0$$

The eigenvalues for this system are formed by computing the determinant above and finding the roots λ_1, λ_2, λ_3. These roots may be either real or complex.

EXAMPLE G.5 Determine the eigenvalues of the system in example G.4:

$$\det \begin{bmatrix} -\lambda & 1 & 0 \\ 0 & -\lambda & 1 \\ \dfrac{5}{2} & -\dfrac{7}{2} & \left(-\dfrac{3}{2} - \lambda \right) \end{bmatrix} = 0$$

This determinant equation can be rewritten as

$$-\lambda^3 - \frac{3}{2}\lambda^2 - \frac{7}{2}\lambda + \frac{5}{2} = 0$$

or

$$\lambda^3 + \frac{3}{2}\lambda^2 + \frac{7}{2}\lambda - \frac{5}{2} = 0$$

This is the characteristic equation of the system. The eigenvalues λ_1, λ_2, λ_3 are the three roots of this equation; they can be found by conventional methods:

$$\lambda_1 = 0.54$$

$$\lambda_2 = -1.02 + j1.88$$

$$\lambda_3 = -1.02 - j1.88$$

The eigenvalues determine the behavior of the system, since they represent the poles of the system transfer function. An individual real root indicates a solution term of the form $e^{\lambda t}$. A positive real eigenvalue usually indicates that the system is unstable, that is, the output will grow at an ever-increasing rate in time. All negative real roots tend to indicate that the system is stable. A complex-conjugate pair of roots indicates a product of a sinusoid and an exponential; if the real part is positive, the system is unstable, with the magnitude of the sinusoid increasing in time. A negative real part indicates stability, with decreasing magnitude, and a zero real part is characteristic of an undamped system, with a constant magnitude sinusoidal output term. Multiple complex conjugate eigenvalues yield multiple terms of the form

$$\sum_{i=1}^{k-1} \frac{t^i}{i} e^{\text{Re}(\lambda)t} \sin\left[\text{Im } (\lambda)t + \varphi\right]$$

just as multiple real eigenvalues yield solution terms of the form

$$\sum_{i=1}^{k-1} \frac{t^i}{i} e^{\lambda t}$$

where k is the number of multiple roots. Again, damping can take place, depending on the sign of λ (if real) or Re (λ) if complex.

EXAMPLE G.6   Determine the form of the solution of Example G.5. Eigenvalues are

$$\lambda_1 = 0.54$$

$$\lambda_3 = -1.02 + j1.88$$

$$\lambda_3 = -1.02 - j1.88$$

or one positive real root and one complex-conjugate pair, so the form of the solution is

$$x = k_1 e^{0.54t} + k_2 e^{-1.02t} \sin (1.88t + \varphi)$$

Substitution of initial conditions allows determination of the constants.

NUMERICAL ANALYSIS OF $d\overline{x}/dt = \overline{A}\overline{x} + \overline{B}u$

Numerous computer algorithms employing state-space techniques have been developed to solve systems of linear differential equations in normal form. Typical algorithms utilize such techniques as Euler's method, Runge–Kutta methods, and predictor–corrector methods such as Adams–Moulton. As examples, flowcharts of the Euler and Runge–Kutta fourth-order methods are shown in Figures G.1 and G.2; a listing of a typical Runge–Kutta fourth-order subroutine is shown in Figure G.3. Systems of simultaneous linear differential equations may also be solved using the Runge–Kutta method by successively solving each equation for each time step, as can be seen from the subroutine (for n equations) in Figure G.3. This method is illustrated by the flowchart in Figure G.4.

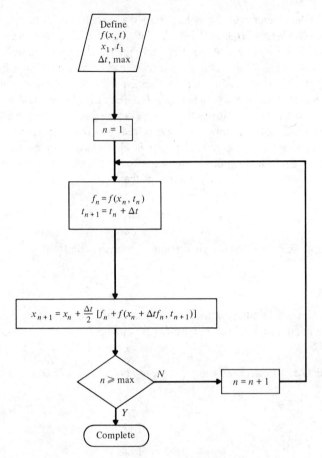

FIGURE G.1 Simplified Flowchart for the Improved Euler Method—One Equation

To demonstrate the use of numerical techniques in the solution of normal-form ordinary differential equations, the simple circuit shown in Figure G.5 will be solved using a Runge–Kutta fourth-order algorithm. The input function is a unit ramp (45° slope) applied at $t = 0.1$ s. The three linear differential equations describing this system are

$$\frac{dx_1}{dt} = \frac{x_3}{3} - \frac{x_2}{3} - \frac{x_1}{3}$$

$$\frac{dx_2}{dt} = \frac{x_1}{2}$$

$$\frac{dx_3}{dt} = -\frac{x_1}{2} + \frac{1}{2}I, \qquad I = \text{unit ramp } R(t - 0.1)$$

Initial conditions are $x_1 = 0$, $x_2 = x_3 = 1.0$. The resulting values of the state variables x_1, x_2, and x_3 are plotted in Figure G.6 for time between 0 and 5 s.

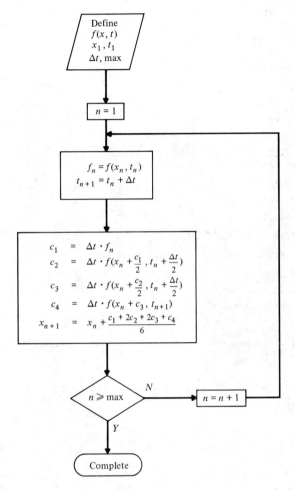

FIGURE G.2 Simplified Flowchart for the Runge–Kutta Fourth-Order Method—One Equation

```fortran
      SUBROUTINE INTRK4(TIME)
C
C
C
C
C     SUBROUTINE INTRK4 SOLVES THE SET OF EQUATIONS FOR EACH VALUE
C     OF TIME USING THE RUNGE-KUTTA 4TH ORDER METHOD. IT MUST BE
C     CALLED AND EXECUTED FOR EACH VALUE OF TIME REQUIRED IN THE
C     ANALYSIS.   VALUES OF THE SOLUTIONS ARE PRINTED UNDER THE
C     VARIABLE NAME RK4.
C
C
      COMMON/DIFEQN/FDOT(5),VAR(5),RK4(5),COEF(5,7),H,TIMAX,ILIM,N,
     1             NOUTPT,XTRANS(9,500),OCOEF(3,7), NPUT,DISP,OMEGA,PHI,
     2             INFLAG
      REAL K
      DIMENSION K(4,5),TEMP(5)
      DO 10 I=1,N
   10 TEMP(I)=RK4(I)
      CALL COMP(TEMP,TIME)
      DO 20 J=1,N
   20 K(1,J)=H*FDOT(J)
      DO 30 J=1,N
   30 TEMP(J)=RK4(J)+K(1,J)/2.
      TIME=TIME+H/2.
      CALL COMP(TEMP,TIME)
      DO 40 J=1,N
   40 K(2,J)=H*FDOT(J)
      DO 50 J=1,N
   50 TEMP(J)=RK4(J)+K(2,J)/2.
      CALL COMP(TEMP,TIME)
      DO 60 J=1,N
   60 K(3,J)=H*FDOT(J)
      DO 70 J=1,N
   70 TEMP(J)=RK4(J)+K(3,J)
      TIME=TIME+H/2.
      CALL COMP(TEMP,TIME)
      DO 80 J=1,N
   80 K(4,J)=H*FDOT(J)
      DO 90 J=1,N
   90 RK4(J)=RK4(J)+K(1,J)/6.+K(2,J)/3.+K(3,J)/3.+K(4,J)/6.
      RETURN
      END



      SUBROUTINE COMP(TEMP,TIME)
C
C
C
C
C     SUBROUTINE COMP GIVES THE DIFFERENTIAL EQUATIONS FOR EACH
C     VARIABLE AND TIME.
C
C
      COMMON/DIFEQN/FDOT(5),VAR(5),RK4(5),COEF(5,7),H,TIMAX,ILIM,N,
     1             NOUTPT,XTRANS(9,500),OCOEF(3,7), NPUT,DISP,OMEGA,PHI,
     2             INFLAG
      DIMENSION TEMP(5)
      DO 50 J=1,N
   50 FDOT(J)=0.
      DO 100 J=1,N
      DO 100 K=1,N
      FDOT(J)=FDOT(J)+COEF(J,K)*TEMP(K)
  100 CONTINUE
      IFLGP1 = INFLAG + 1
      GO TO (160,110,120,130,140,150),IFLGP1
  110 XNPT = STEP(TIME,DISP)
      GO TO 180
  120 XNPT = RAMP(TIME,DISP)
      GO TO 180
  130 XNPT = PULSE(TIME,DISP,OMEGA)
      GO TO 180
  140 XNPT = SINE(TIME,DISP,OMEGA,PHI)
      GO TO 180
  150 XNPT = COSINE(TIME,DISP,OMEGA,PHI)
      GO TO 180
  160 DO 170 J=1,N
  170 FDOT(J)=FDOT(J)+COEF(J,6)
      RETURN
  180 DO 190 J=1,N
  190 FDOT(J)=FDOT(J)+COEF(J,6)+COEF(J,7)*XNPT
      RETURN
      END
```

FIGURE G.3 Listing of a Typical Runge—Kutta Fourth-Order Subroutine for Solving Systems of Linear Differential Equations

SOLUTION OF ORDINARY DIFFERENTIAL EQUATIONS IN
NORMAL FORM

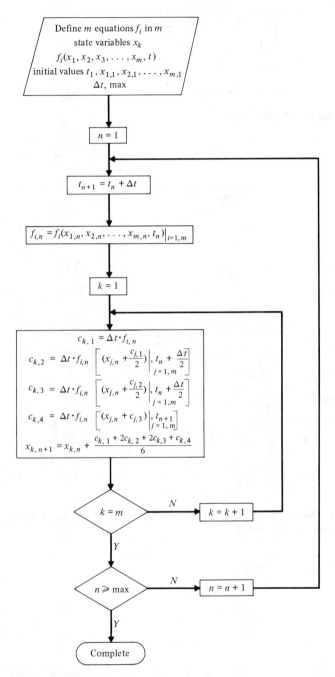

FIGURE G.4 Flowchart for Simultaneous Linear Differential Equations, Runge–Kutta Fourth-Order Method

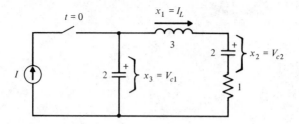

FIGURE G.5 System to Be Modeled

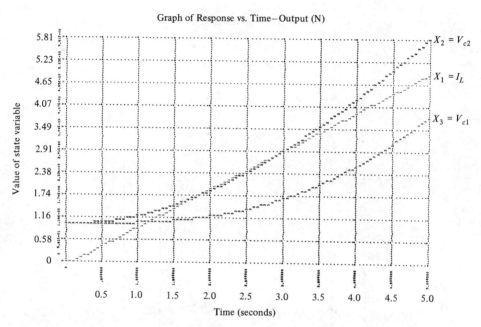

FIGURE G.6 Model Results for an Example System: Runge–Kutta Fourth-Order Simulation (x_1, x_2, x_3 Versus Time)

SOLUTION OF ORDINARY DIFFERENTIAL EQUATIONS IN
NORMAL FORM

INDEX